Recent Developments in Functional Materials for Artificial Photosynthesis

Sustainable Energy

Editor-in-chief:
Zhongwei Chen, *University of Waterloo, Canada*

Series editors:
Glaucia Mendes Souza, *Universidade de São Paulo, Brazil*
Chunyi Zhi, *City University of Hong Kong, Hong Kong*

Titles in the series:
1: Recent Developments in Functional Materials for Artificial Photosynthesis

How to obtain future titles on publication:
A standing order plan is available for this series. A standing order will bring delivery of each new volume immediately on publication.

For further information please contact:
Book Sales Department, Royal Society of Chemistry, Thomas Graham House, Science Park, Milton Road, Cambridge, CB4 0WF, UK
Telephone: +44 (0)1223 420066, Fax: +44 (0)1223 420247
Email: booksales@rsc.org
Visit our website at www.rsc.org/books

Recent Developments in Functional Materials for Artificial Photosynthesis

Edited by

Srabanti Ghosh
Central Glass and Ceramic Research Institute, India
Email: srabanti@cgcri.res.in

and

Qian Wang
Nagoya University, Japan
Email: wang.qian@material.nagoya-u.ac.jp

ROYAL SOCIETY OF CHEMISTRY

Sustainable Energy No. 1

Print ISBN: 978-1-83916-442-2
PDF ISBN: 978-1-83916-776-8
EPUB ISBN: 978-1-83916-777-5
Print ISSN: 2752-6771
Electronic ISSN: 2752-678X

A catalogue record for this book is available from the British Library

The Royal Society of Chemistry is a charity, registered in England and Wales, Number 207890, and a company incorporated in England by Royal Charter (Registered No. RC000524), registered office: Burlington House, Piccadilly, London W1J 0BA, UK, Telephone: +44 (0) 20 7437 8656.

For further information see our website at www.rsc.org

Printed in the United Kingdom by CPI Group (UK) Ltd, Croydon, CR0 4YY, UK

Preface

Generating solar fuels from water and carbon dioxide using photocatalysts has great promise for the development of sustainable energy resources. This book provides a comprehensive description of recent developments in the structural design principles of semiconductor materials and heterostructures, including metal oxide, (oxy)nitride, (oxy)sulfide, (oxy)halide, chalcogenide-based semiconductors, quantum dots, carbon-based nanomaterials and 2D nanomaterials, towards efficient photocatalytic hydrogen evolution, oxygen evolution, overall water splitting and carbon dioxide reduction. The book covers a variety of topics: the photocatalytic mechanism, factors affecting photocatalytic activity, role of co-catalysts, the structure–activity relationship, as well as challenges and strategies. The book also provides theoretical studies that identify the influencing parameters for the photocatalytic performance and provides insight into the different aspects and challenges associated with the realization of sustainable solar hydrogen production systems on a large scale.

Srabanti Ghosh
Qian Wang

Sustainable Energy No. 1
Recent Developments in Functional Materials for Artificial Photosynthesis
Edited by Srabanti Ghosh and Qian Wang

Published by the Royal Society of Chemistry, www.rsc.org

Contents

Sustainable Energy No. 1
Recent Developments in Functional Materials for Artificial Photosynthesis
Edited by Srabanti Ghosh and Qian Wang

Published by the Royal Society of Chemistry, www.rsc.org

CHAPTER 1

Multidisciplinary Approaches to Solar-driven Water Splitting and Carbon Dioxide Conversion

STEFANO PROTTI* AND NICK SERPONE*

Photo Green Laboratory, Department of Chemistry, University of Pavia, Viale Taramelli 12, Pavia 27100, Italy
*E-mail: stefano.protti@unipv.it, nick.serpone@unipv.it

1.1 Introduction

Notwithstanding the declaration of intent of national governments, international organizations, and public opinion (see, for example, the overwhelming success of the ***Fridays for Future*** worldwide protests),[1] the fact remains that the world's energy supply is still mostly dependent on the use of non-renewable resources. It is apparent that scalable and renewable sources of energy, and fuels, are urgently needed to ensure the sustainability of the energy supply for future generations. Indeed, with clean energy systems, the minimization of emissions and waste is a feasible undertaking by taking advantage of energy conservation and recovery.

In his keynote address at the 2011 Sixth International Green Energy Conference, Dincer[2] pointed out that energy sources must exhibit different properties in order to be defined as clean, including, among the others, better effectiveness, high-cost efficiency and safety. At present, the perception that fossil fuels need to be replaced by alternative energy sources appear ingrained

Sustainable Energy No. 1
Recent Developments in Functional Materials for Artificial Photosynthesis
Edited by Srabanti Ghosh and Qian Wang

Published by the Royal Society of Chemistry, www.rsc.org

in the mindset of society. In this context, among the different renewable sources (including hydroelectric, geothermal, and eolic), solar energy is certainly the most promising energy source as the Sun delivers an enormous amount of energy. For instance, the energy that reaches the Earth's surface in one hour (4.3×10^{20} Joules per hour) is comparable to what is consumed globally in one year (5.76×10^{20} Joules per year)[3] so that 1 h of the Sun's energy will satisfy the Earth's energy need for one whole year.

In 1980, Bard[4] pointed to the key importance of suitable forms of interconvertible energy in the hope of developing "*artificial photosynthetic systems for the conversion of abundant materials such as water and carbon dioxide to fuels directly or* (otherwise) *by electrolysis*". Among the available fuels, hydrogen has been considered the elective sustainable energy vector because of its high energy content, because it can be easily stored, and because of the absence of greenhouse gas emissions upon its use. Hay and co-workers[5] reported a growth rate for hydrogen demand as 5.6% in the five years 2011–2016. Hydrogen is produced mainly *via* the steam reforming of fossil resources such as natural gas or liquid hydrocarbons. This process occurs at high temperatures (up to 950 °C) and pressures under nickel catalytic conditions. In contrast, the extent of generation of hydrogen *via* water electrolysis is fairly low at less than 4% of global annual production.[6] Indeed, the water splitting reaction ($H_2O(l) \rightarrow H_2(g) + O_2(g)$) is a process that involves a large positive change of the Gibbs free energy for it to occur ($\Delta G° = 237.13$ kJ mol^{-1}). At the same time, the efficient conversion of CO_2 into chemical fuels such as, among others, methane, and methanol, is limiting as the concentration of such a greenhouse gas in the atmosphere is beyond intriguing. In the most researched technique, however, the efficiency of Carbon Capture and Storage is counterbalanced by environmental and economic limitations, including the risk of leakage from geological storage sites, not least of which is the cost of compression and transportation.[7] In this context, solar-induced photocatalytic and solar–photoelectrochemical processes represent key approaches in harvesting solar energy for energy storage and (solar) fuel production through the exploitation of semiconductors as the photoactive components. The two strategies (photocatalytic and photoelectrochemical) exhibit some similarities (treated together in several articles and reviews)[8] and some differences (described in the next sections) in the requirements needed to optimize performances.

1.2 Photocatalysis

In 2007, Palmisano and co-workers[9] defined photocatalysis as "*a promising route for 21st century organic chemistry*". Initially, such a technology has been applied widely toward environmental remediation that included the photodegradation of organic and photoinduced elimination of inorganic pollutants.[10,11] More rarely, however, it has been used in organic syntheses.[12] Recent decades have witnessed investigations into a plethora of semiconductors that have been used as photocatalytic materials. However, from economic and photostability points of view, titania (TiO_2; anatase polymorph) with a bandgap energy of 3.2 eV has been the semiconductor of choice and the one that has been examined most extensively in terms of its fundamentals and its various applications.

Three fundamental steps are involved in the conversion of sunlight energy to chemical energy: (i) the absorption of solar photons by the semiconductor photocatalysts to cause the electronic transition from the valence band (VB) to the conduction band (CB); (ii) the generation of redox equivalents (an electron in the conduction band, CB, and a positive vacancy or hole in the valence band, VB), which is then followed by; (iii) their migration to the surface to initiate the redox events (reduction and oxidation half reactions).[13,14] Broadly speaking, photocatalytic processes depend both on the wavelength and the irradiance of the light source, which then determine the energy and population of the photogenerated electron–hole pairs. The VB holes (h^+) typically exhibit a strong oxidation potential and can thus react with electron-rich compounds, whereas the electrons located in the CB (e^-) act as reductants. Obviously, the capability of both h^+ and e^- to cause oxidations and reductions depends strictly on the corresponding potentials of the redox couples. However, recombination of the charge carriers[15,16] in a photocatalytic process competes effectively with the redox reactions occurring at the semiconductor particle surface, thereby limiting the efficiency (quantum yield) of the semiconductor photocatalytic process.

1.2.1 Water Splitting

Water splitting (not to be confused with the simple production of hydrogen) is defined as "*the decomposition of water to hydrogen and oxygen in a 2 to 1 ratio*". As predicted, this process is equivalent to the electrolysis of water, whereby water undergoes reduction at the cathode with hydrogen evolution (two reducing equivalents, eqn (1.1)) while oxidation occurs at the anode with release of oxygen (four oxidizing equivalents, eqn (1.2)).[17]

$$2H_2O + 2e^- \rightarrow H_2(g) + 2OH^-(aq) \tag{1.1}$$

$$4OH^-(aq) \rightarrow O_2(g) + 2H_2O(l) + 4e^- \tag{1.2}$$

with the overall reaction being: $2H_2O \rightarrow 2H_2 + O_2$, $\Delta G^\circ = 237.13$ kJ mol^{-1}.

The measured standard free energy for water is 1.23 eV. Therefore, when both the reductive and oxidative events are within the bandgap of the photocatalyst (*e.g.*, titania, Figure 1.1), water splitting becomes a process that is feasible thermodynamically. However, when considering thermodynamic factors, and other losses, it is likely that (electrochemically) the energy required would be greater by 0.3–0.4 eV, namely about 1.5–1.6 eV, and as pointed out by both Gerischer[18] and Nozik[19] adding the need for the overpotential for the

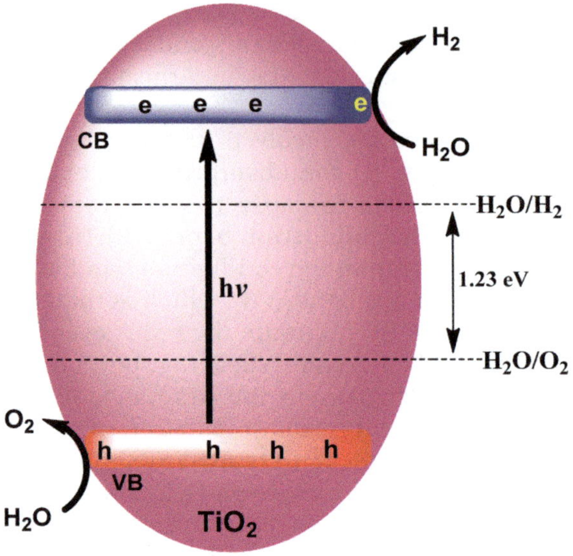

Figure 1.1 An illustration of the thermodynamic feasibility of splitting water into its constituent elements hydrogen and oxygen in the presence of a photoactivated semiconductor such as titania, TiO_2.

oxygen evolution of *ca.* 1 eV would then require approximately 2.5–2.6 eV to achieve water splitting.

As pointed out by Osterloh,[20] only metal oxides with d^0 ion configuration (Ti, Zr, Nb, and Ta) and d^{10} ion (Ga, In, Ge, Sn, and Sb) configuration possess the appropriate activity to allow for the overall photochemical splitting of water. The metal oxides tend to be dominant. However, nitrides and oxynitrides (GaN/ZnO and Ge_3N_4) have also been found to be effective in the process.

Titanium dioxide (TiO_2) was the first semiconductor in photocatalytic water splitting and is still the most widely used. This is due to its efficient photocatalytic activity, its chemical stability, and the relatively low commercial cost. However, titania exhibits a low efficiency in converting solar energy into hydrogen and oxygen *via* the photocatalytic water splitting process and is one the principal limitations for its application. Titania crystallizes in three different polymorphic structures: rutile, anatase, and brookite that display bandgaps in the 3.04–3.23 eV range, depending on the crystal form. As a result, only the UV part of the solar emission spectrum (about 4% of the solar energy at the Earth's surface) can be exploited. Regardless, the list of photocatalysts that can be applied efficiently toward visible light water splitting without additives (*e.g.* electron donors/acceptor) has increased notably in the last decade.[20,21] At present, aside from the historical separate stable self-driven photoelectrochemical cells consisting of the coupled n-TiO_2/p-CdTe, n-TiO_2/p-GaP, n-$SrTiO_3$/p-CdTe and n-$SrTiO_3$/p-GaP systems capable of water photoelectrolysis,[22] novel ones include $\{Zn_{1.44}Ge)(N_{2.08}O_{0.38}\}$ with a bandgap of *ca.* 2.7 eV[23] and {NiO/RuO_2/Ni:$InTaO_4$},[24] together with the tandem systems composed of Pt/WO_3 and Pt/$SrTiO_3$/TaON,[25] the oxysulfide α-$LaOInS_2$,[26] the Milstein's ruthenium complex,[27] the CdS/WO_3/$CdWO_4$ ternary composite,[28] and the carbon dots/cadmium sulfide (CDs/CdS) nanocomposites.[29] Along similar lines, Liu and co-workers[30] proposed a thermal strategy to prepare {reduced graphene oxide/γ-Fe_2O_3–C_3N_4} S-scheme heterojunctions from MIL-101(Fe) and melamine. The rGO/γ-Fe_2O_3–C_3N_4 system presented a relatively high oxygen evolution rate (OER) of 3.85 mmol g^{-1} h^{-1} under visible light irradiation, with an overall H_2O splitting activity that yielded hydrogen evolution (HER) and OER rates of 23.3 and 12 μmol g^{-1} h^{-1}, respectively.

The fast recombination of photogenerated electron–hole pairs is one of the limitations for an efficient water splitting process. Other limitations include the many factors that affect: (1) quantum yields of surface reactions such as (i) the rate constant of either charge transfer to (or from) adsorbed molecules or charge carrier trapping to surface defects to form surface-active centers, (ii) quantum yield of internal photo effects, such as the probability of charge carrier generation caused by absorption of a single photon, (iii) absorption coefficient, (iv) electric field, (v) diffusion length of the charge carriers, (vi) their lifetimes, (vii) depth of surface space charge layer, (viii) charge carrier recombination rate, (ix) size of the semiconductor particulates, (x) concentration of surface-active defects as well as of adsorbed molecules, (xi) diffusion coefficient, (xii) charge carrier mobility and (xiii) the potential surface charge;[31,32] together with (2) back reactions and secondary reactions;[33,34] (3) the photostability of photocatalytic materials;[35] (4) the need for co-catalysts for H_2 and O_2 generation[18,19] and (5) the need to minimize the density of defect states in metal–oxide photocatalysts.[36] To overcome some of these limitations, various strategies have been proposed that include, among others, (a) the tailoring of the surface morphology, (b) engineering the bandgap, and (c) the loading of suitable co-catalysts.[37] Control of the surface morphology plays a key role in both promoting charge separation and suppressing charge recombination at the surface, as demonstrated by Serpone and co-workers[38–41] *via* the inter-particle electron and hole transfer processes between coupled semiconductor photocatalysts (Figure 1.2).

Ternary heterostructures can also be envisaged to separate the photogenerated charge carriers and minimize their recombination as illustrated in Figure 1.3(a) in which, with proper alignment of the VB and CB bands, it is possible to use low-energy visible light to activate wide-bandgap semiconductors utilizing two narrow bandgap semiconductors sensitive to visible light.[40,41] One such example of a ternary heterostructure originates from the work of Jiang and co-workers[42] in which two narrow bandgap systems—graphitic carbon nitride g-C_3N_4 (E_g = 2.66 eV) and indium oxide (E_g = 2.93–3.02 eV)—activate the higher bandgap TiO_2 (E_g = 3.2 eV) for redox reactions (Figure 1.3(b)).

Controlling the morphology can be achieved by the optimization of the synthesis method as well as by adopting doping techniques and surface treatment strategies. A demonstrative example is represented by the single

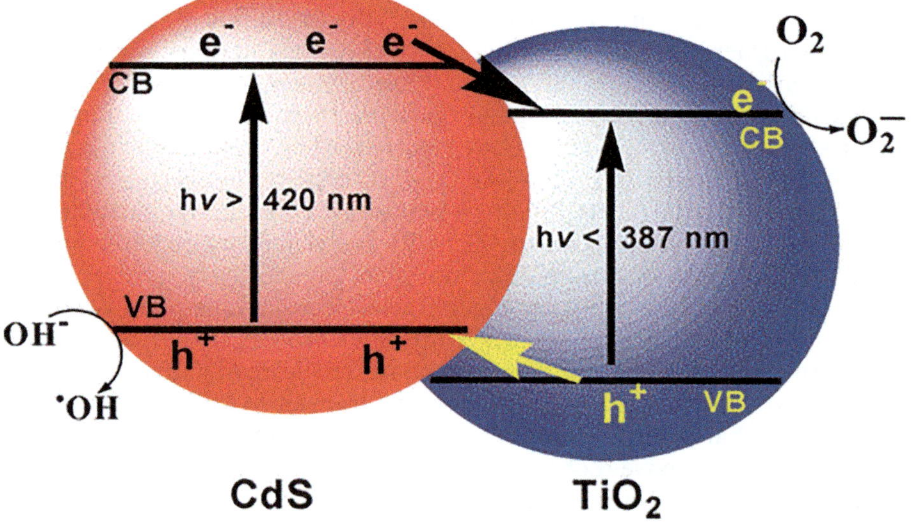

Figure 1.2 Schematic illustration of the coupling of two semiconductors displaying the vectorial inter-particle electron transfer process. Reproduced from ref. 40 with permission from American Chemical Society, Copyright 2012.

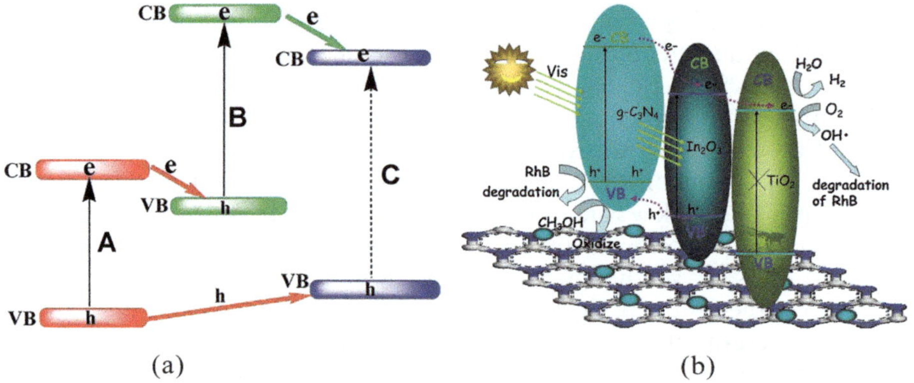

Figure 1.3 Illustration showing ternary heterostructures of coupled semiconductors. (left) Reproduced from ref. 40 with permission from American Chemical Society, Copyright 2012. (right) Reproduced from ref. 42 with permission from Elsevier, Copyright 2015.

crystal Ta_3N_5 nanorods grown on a lattice-matched cubic $KTaO_3$ that, when combined with Rh/Cr_2O_3 as the co-catalyst, resulted in water splitting under both visible light and simulated solar light.[43] Moreover, the use of protectors has often been exploited to stabilize the surface and inhibit the formation of surface defects, thereby preventing or otherwise suppressing any back reaction. As an example, the efficiency of the $Pt/SrTiO_3$ photocatalyst in overall water splitting has been increased dramatically by coating the bare material with a molybdenum-based layer. The so-obtained MoOx/Pt/$SrTiO_3$ is stable in acidic media for more than 15 h and the MoOx layer hinders the gas permeation of O_2 and contact with the active Pt surface.[44]

1.2.2 CO_2 Reduction

The photocatalytic conversion of CO_2 into solar fuels and chemicals using everlasting solar energy is a suitable approach to contemporaneously curb greenhouse gas emissions. However, to develop an efficient protocol, several limitations must be overcome, including: (1) the poor overlap between the absorption spectra of the semiconductor photocatalysts (*e.g.* in the case of titania) and the solar emission spectrum; (2) the inefficient charge carrier separation; (3) the back reactions observed during the CO_2 reduction process; and (4) the low solubility of carbon dioxide in aqueous media (about 33 mmol mL^{-1} in water at 100 kPa and ambient temperature).[45–47] Furthermore, the competition between CO_2 conversion and oxidation of water to hydrogen (eqn (1.9) and (1.10)) must be avoided. The latter point is not a trivial issue, as the reduction of CO_2 into carbon-based fuels requires electrons in the conduction band that exhibit a redox potential more negative than required for water reduction. As highlighted below (see eqn (1.3)–(1.10)), despite the formation of carbon based-fuels from carbon dioxide the process is a multielectron process with the first rate-determining step being the monoelectronic reduction of CO_2 to the corresponding radical anion $CO_2^{\bullet-}$, a monoelectronic reduced species isoelectronic with NO_2 that exhibits a C_{2v} geometry.[48,49]

$$CO_2 + e^- \rightarrow CO_2^{\bullet-} \quad E = -1.90\ \text{V*} \tag{1.3}$$

$$CO_2 + 2H^+ + 2e^- \rightarrow HCOOH \quad E = -0.61\ \text{V} \tag{1.4}$$

$$CO_2 + 2H^+ + 2e^- \rightarrow CO + H_2O \quad E = -0.53\ \text{V} \tag{1.5}$$

$$CO_2 + 4H^+ + 4e^- \rightarrow HCHO + H_2O \quad E = -0.48\ \text{V} \tag{1.6}$$

$$CO_2 + 6H^+ + 6e^- \rightarrow CH_3OH + H_2O \quad E = -0.38\ \text{V} \tag{1.7}$$

$$CO_2 + 8H^+ + 8e^- \rightarrow CH_4 + 2H_2O \quad E = -0.24\ \text{V} \tag{1.8}$$

$$H_2O + 4h^+ \rightarrow O_2 + 4H^+ \quad E = +0.81\ \text{V} \tag{1.9}$$

$$2H^+ + 2e^- \rightarrow H_2 \quad E = -0.42\ \text{V} \tag{1.10}$$

* In aprotic media. $E = -2.21$ V in water.

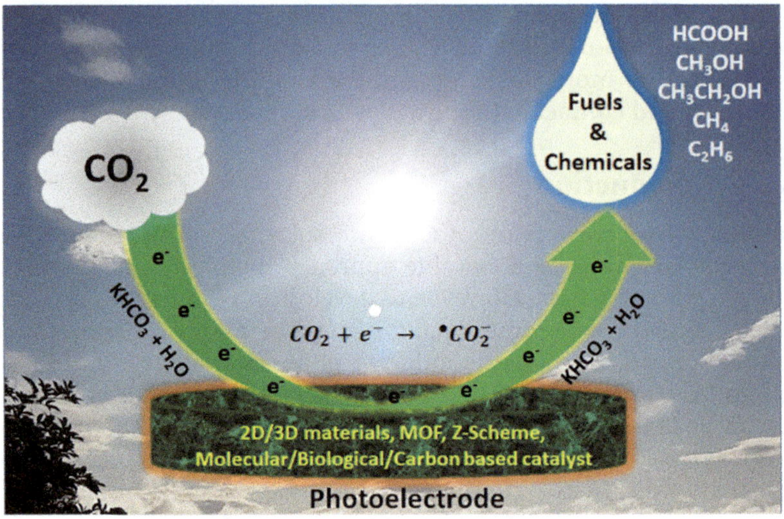

Figure 1.4 Schematic of the PEC CO_2 conversion into fuels and value-added products under solar light irradiation. Reproduced from ref. 51 with permission from American Chemical Society, Copyright 2020.

Once formed, the intermediates can undergo bimolecular decay to form oxalate and formate anions.[50] The pH of the aqueous solution also plays a key role in determining the efficiency of the process, since at pH > 5 carbon dioxide is also present in solution as the carbonate anion (CO_3^{2-}) that acts as an electron donor species toward the photogenerated holes. Recently, different efforts have been carried out to investigate in detail the mechanism of the photochemical process, in order to properly design photocatalysts with improved stability, activity, and high selectivity for CO_2 reduction.

To the extent that the reduction of CO_2 under heterogeneous catalytic conditions occurs at the interface between the catalyst(solid) and the gas, the design of an appropriate photocatalyst plays a fundamental role in the conversion of CO_2 under heterogeneous photocatalytic conditions. In the last decade a plethora of photocatalytic materials has been investigated in this field, including metal oxides (*e.g.*, TiO_2, ZnO, Fe_2O_3, WO_3) and metal sulfides (*e.g.*, CdS), transition metals (Cu, Ni, and Fe), noble metals or alloys thereof (Pt, Au, Rh, Ru, PtCu, AuCu), and several carbon-based materials (*e.g.*, graphene derivatives and the carbon nitride C_3N_4).[51] To date, CO_2 can be photoconverted into a plethora of products (Figure 1.4) that include, among others, CO, alkanes (methane and ethane), olefins (ethylene and propylene), alcohols (methanol and ethanol) as well as organic acids (formic, acetic and malonic acid).[51]

1.3 Photoelectrocatalysis

Research in the field of photoelectrocatalysis is considered to have started with the 1972 seminal paper by Fujishima and Honda.[52] This research reported on the light-driven electrolysis of water to molecular hydrogen and oxygen by means of a photoelectrochemical cell equipped with an n-type rutile-phase titanium dioxide (TiO_2) working electrode under UV irradiation and a platinum (Pt) counter electrode. This ground-breaking study, along with the earlier reports on the electrochemical behavior of metal semiconductors under illumination dating back to the 1950s, have inspired several pioneering studies, starting with the short-circuited photoelectrochemical cell based on TiO_2 and Pt designed by Bard that *de facto* represented the first heterogeneous photocatalytic system.[53,54]

There are significant analogies between photocatalysis (PC) and photoelectrocatalysis (PEC). Indeed, both approaches proceed primarily through three sequential pathways that are fundamental to the solar-to-chemical energy conversion. These are (a) the electron–hole pair generation following the absorption of photons with energies greater than the bandgap energy (E_g) of the photocatalyst employed, (b) the electron–hole separation phenomenon whereby the charges are moved on the surface of the photocatalytic system, and (c) the redox event(s) occurring on the surface of the photocatalyst. Again, the performance of the two systems depends mainly on charge separation and transfer kinetics, which still represent the bottleneck for a large diffusion of the approach for applications such as the solar-to-hydrogen conversion for which the efficiency barely reaches the 10% value.[55–57] In any case, light harvesting and electron transfer occur on the same components of the system for both PC and PEC.

The main difference between the two approaches is a practical one, since the former process is performed in a reaction vessel wherein the (heterogeneous) photocatalytic system is suspended in the reaction solution, while in the second system two separate electrolyte solutions (for reduction and oxidation, respectively) are wired to an external bias to favor the electron transfer (Figure 1.5).

1.3.1 Water Splitting

PEC water splitting is one of the most promising strategies for the generation of hydrogen in a sustainable and economic manner, in view of the small thermodynamic potential required (1.23 eV; however, see prior discussion) and zero CO_2 emissions. As hinted previously in the PEC approach, the photocatalyst is initially prepared on a conductive substrate after which a bias is applied to the resulting photoelectrode. When the cell is irradiated with natural sunlight, the photogenerated charge carriers are generated and separated by the electric field in the space-charge region and the minority charges (holes for an n-type photoanode and electrons for a p-type photocathode) then migrate to the semiconductor electrode–liquid interface for the reaction to occur. One of the main advantages, with respect to the photocatalytic approach, is that no gas separation is needed because the generation of hydrogen and oxygen take place in different electrode compartments (Figure 1.6).[58,59]

The materials used in the preparation of photocathodes (where the reduction of water takes place) should obviously exhibit an edge potential

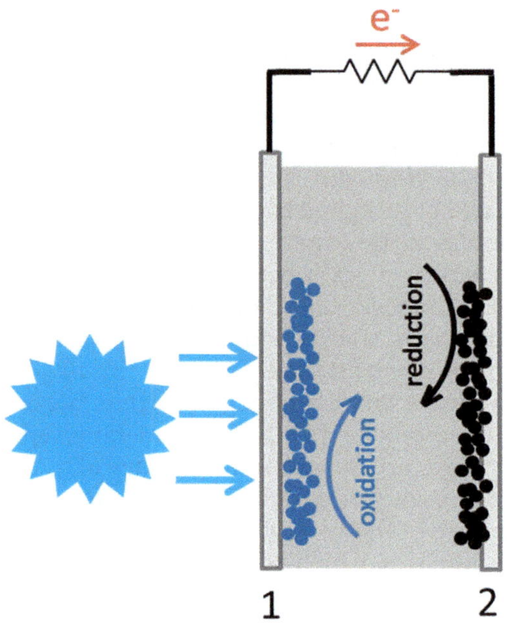

Figure 1.5 Schematic representation of a PEC cell composed of a photoanode electrode equipped with a nanostructured photocatalyst (1) and a cathode (counter) electrocatalyst (2). Reproduced from ref. 57 with permission from Elsevier, Copyright 2017.

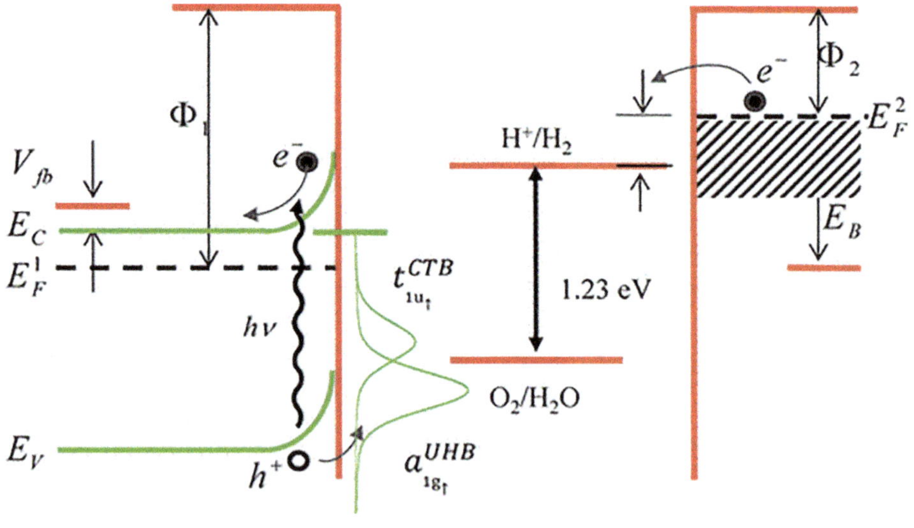

Figure 1.6 Energy diagram for the PEC water splitting at a hematite photoanode with oxygen evolution. Reproduced from ref. 59 with permission from American Chemical Society, Copyright 2012.

for the conduction band more negative than the hydrogen redox potential and an enhanced resistance in aqueous media. In this case, different materials have been tested, including silicon, III–V group semiconductors (InP, GaP, the former treated with a protective layer of TiO_2 to prevent corrosion) and more recently, monometallic oxides (cuprous and cupric oxide, nickel oxide, the latter often sensitized with CdSe quantum dots), bimetallic oxides ($CuFeO_2$, $LaFeO_3$) and chalcogenides ($CuInS_2$, $CuGaS_2$). Photoanodes (where oxygen generation occurs) should be made of a water-stable, photo-responsive n-type semiconductor with a bandgap that allows absorption of a wide range of wavelengths. Different materials were found suitable for this target, including again the monometallic and bimetallic oxides (among others, TiO_2 is the one used mostly, despite its low absorption in the solar emission spectrum), and metal chalcogenides and nitrides (see Table 1.1).[60]

1.3.2 CO_2 Reduction

The PEC reduction of carbon dioxide aims to combine the advantages of both photocatalysis and electrolysis, on the route toward the solar-driven generation of chemical fuels (including, among others, methanol, ethanol, formic acid, formaldehyde, methane, carbon monoxide) from CO_2, thanks

Table 1.1 Photo-responsive materials for water splitting *via* PEC.

Materials	Bandgap (eV)
Photocathode	
Cu_2O	1.9–2.2
CuO	1.2–1.8
NiO	3.6–4.0
$CuFeO_2$	1.5–1.6
$LaFeO_3$	2.3–2.4
$CaFe_2O_4$	1.9
$CuInS_2$	1.5
$CuGaS_2$	2.4
Silicon	1.1
GaP	2.2–2.3
InP	1.35
Photoanode	
TiO_2	3.2
ZnO	3.2
WO_3	2.5–2.8
α-Fe_2O_3	2.0–2.2
$BiVO_4$	2.84
$CuWO_4$	2.3
TaON	2.5
Ta_3N_5	2.1
CdS	2.4
ZnS	3.6

to low cost and easy accessible apparatuses.[61] In addition, the overpotential observed at the electrodes can be reduced smoothly by applying a bias potential that can also be exploited to tune the selectivity of the process and minimize the electron–hole recombination process. A traditional PEC apparatus typically consists of two compartments containing the (photo)electrode separated by a Nafion™ membrane that allows for proton exchange. The carbon dioxide reduction occurs at the cathode, whereas oxidation of the water reaction takes place in the anodic compartment; the protons that are generated in the latter move from the anodic compartment to the cathodic one through the Nafion™ membrane by exploiting the gradient concentration.

As implied previously, the reaction needs the proper combination of reductant equivalents for the carbon dioxide reduction and positive charges (holes) for the water splitting process. For this reason, the choice of (photocatalytic) materials for the photoanode and photocathode is a key issue in the development of a PEC apparatus. At first, the conduction band position of the photoactive semiconductors should be closed to that of the reduction energy level of CO_2 for product selectivity, whereas the valence band position should be more positive than the potential for water oxidation. As in the case of water splitting, the PEC system for CO_2 reduction consists of both working and counter (photo)electrodes. A reference electrode is often employed at the laboratory scale, in order to verify the exact reduction potential of the working electrode in the system. In most cases, n-type semiconductor materials are employed for the photoanode (*e.g.*, TiO_2, $BiVO_4$), whereas photocathodes are constituted by p-type semiconductors. The irradiation strategy determines the overall mechanism. Indeed, when the photoanode is illuminated, the electrons are transferred (after charge separation) toward the cathode (where reduction occurs) while the generated holes are responsible for the oxidation reaction on the surface. However, in a system equipped with a photocathode, the photogenerated electrons will be employed for reducing CO_2 on the surface of the electrode. When light irradiates both photoanode and photocathode, the electrons generated in the photoanode will be transferred to the photocathode *via* the external bias and combine with the electrons produced at the photocathode surface. Recently, Zhou *et al.*[62] replaced the Nafion™ membrane with a bipolar membrane (BPM) composed by an anion-exchange layer and a cation-exchange layer that allows the transfer of protons and hydroxide anions to the cathode and the anode, respectively, so as to maintain the pH in each compartment. In this way, each electrolyte will work under optimal conditions.

1.4 Photovoltaic Electrocatalytically-driven Formation of Solar Fuels

Along with the development of PC and PEC, the conversion of sunlight to electricity based on solar cells has experienced a continuing development. Currently, the power conversion efficiency (PCE) of a single crystal Si solar cell is about 26%.[63] Such impressive success laid the foundation of an alternative approach to solar hydrogen production, namely the photovoltaic (PV) driven electrolysis (PV-EC), and more in general, to the PV-driven electrochemical processes. Contrary to PC and PEC, the PV-EC process is based on separated modules of the PV part (light harvesting and solar-to electricity conversion) and EC part for the redox event(s).

A recent editorial in the journal *ACS Sustainable Chemistry & Engineering*[64] noted that migration from fossil resources to renewable energies is vital if we wish to relieve the impact of energy shortages and environmental costs, particularly as photovoltaic and eolic power have experienced rapid growth and the associated drop in electricity costs might make it possible to replace traditional fossil fuels. This notwithstanding, however, the intermittent nature of many renewable energy sources (*e.g.*, solar and wind power) makes the task of integrating the renewable electricity into existing power grids somewhat challenging. Nonetheless, chemicals and fuels will continue to be needed for some time to come as they are needed for heating, transportation, and industrial production so that conversion and storage of renewable electricity into some suitable form makes sense if the task is to build an energy system based on 100% renewable energy resources (Figure 1.7).[64] Therefore, the introduction of the so-called concept *Power-to-X* (P2X),[65] also dubbed as P2X or PtX, is defined as "a bundle of pathways for the conversion, storage, and reconversion of electric power, especially that generated by renewable energy." Such an umbrella concept, under which X can be taken as heat or such chemicals as hydrogen, syngas, and synthetic fuels, among others, comprises the extent to which most P2X technologies aim at converting renewable electricity into fuels and/or chemicals. Therefore, electrocatalysis plays a key role in the development of P2X strategies.

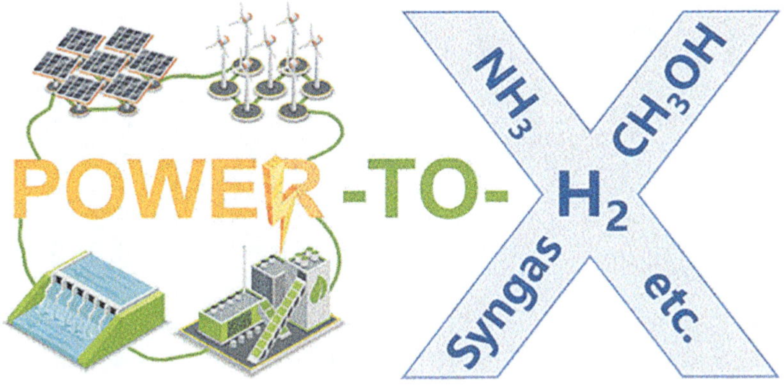

Figure 1.7 Schematic illustration of the Power-to-X umbrella, where X can be hydrogen, syngas, ammonia, methanol, *etc.* Reproduced from ref. 64 with permission from American Chemical Society, Copyright 2021.

1.4.1 Water Splitting

The most mature P2X, namely P2H$_2$, route is water electrolysis, which relies on the use of renewable electricity to drive catalytic water splitting into H_2 and O_2.[66] Generally, three kinds of systems are applied for the electrolysis of water: (1) alkaline electrolysis cell (AEC); (2) solid oxide electrolysis cell (SOEC); and (3) polymer electrolyte membrane electrolysis cell (PEMEC). Liu and co-workers[66] reported, for the first time, the transformation of commercially available stainless steel into useful electrocatalysts for both H_2 and O_2 evolution reactions (HER and OER, respectively) after a facile modification of the surface (Figure 1.8). Specifically, they sulfurized a stainless steel foil (SSFS) and achieved a catalytic current density of 10 mA cm^{-2} at overpotentials of 136 and 262 mV for HER and OER, respectively, in 1.0 M KOH. A water splitting current density of 10 mA cm^{-2} and a robust durability was observed at 1.64 V when the SSFS was employed as the electrocatalytic material for both the cathode and the anode. Figure 1.9 illustrates the stoichiometric quantities of H_2 and O_2 evolved after 167 min in comparison to those values estimated on the basis of charge passed during electrolysis. The nearly perfect overlap between the experimentally measured and the calculated amounts of both gases strongly validated the near unity Faradaic efficiency of their SSFS catalyst couple for overall water splitting.[66]

1.4.2 CO_2 Reduction to Fuels (P2F)

The central P2F process involves the electrocatalytic reduction of CO_2, which has seen significant recent studies and applications. In this regard, power-to-fuels (P2F) processes have recently emerged as a

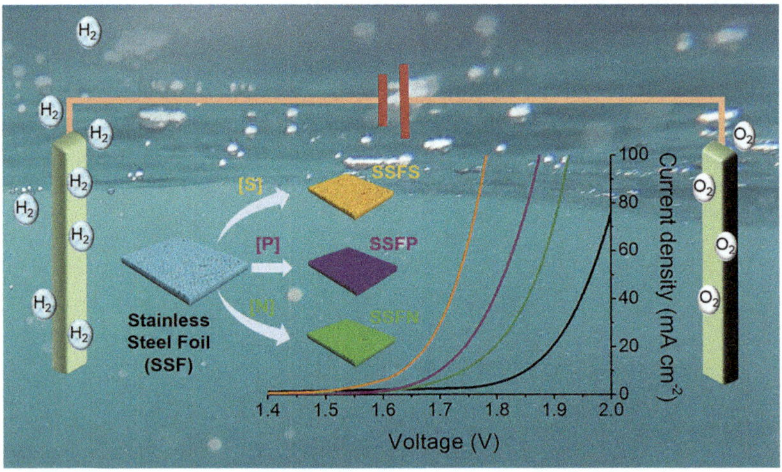

Figure 1.8 System used for the electrocatalytic water splitting for clean H_2 and O_2 production with a modified surface of commercially available stainless steel as the electrocatalyst. Reproduced from ref. 66 with permission from American Chemical Society, Copyright 2017.

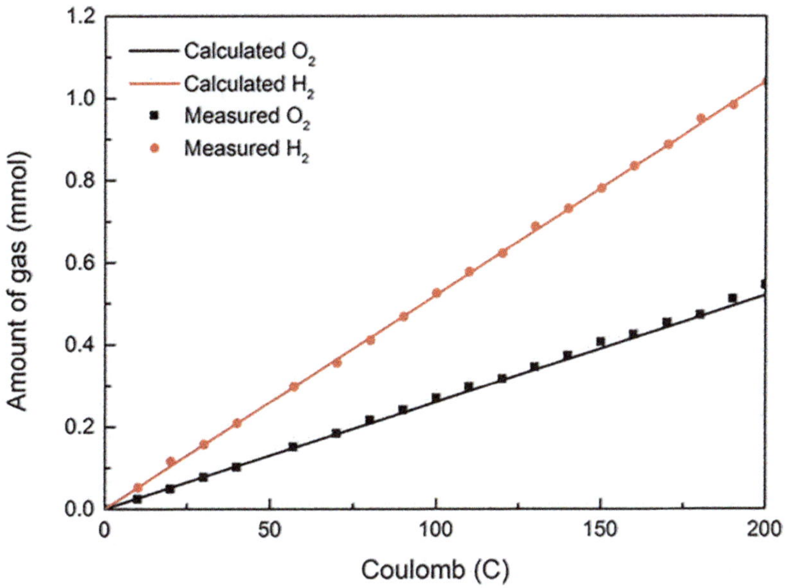

Figure 1.9 Comparison between the amount of chromatography-measured gases and theoretically calculated gases observed in the overall water splitting by the SSFS couple in 1.0 M KOH after 167 min. Reproduced from ref. 66 with permission from American Chemical Society, Copyright 2017.

potential long-term solution for the storage of surplus renewable electricity through the conversion of carbon dioxide into the production of the gaseous fuel methane.[67,68] Such processes are significant for future strategies of energy storage as they address two fundamental issues: (1) electrical grid stability with high share of renewable sources and (2) decarbonization of high-energy density fuels for the transportation sector. A large number of pathways exist to transform the energy from renewable sources into gaseous/liquid fuels through their combination with residual carbon dioxide that have been reviewed in some details by Bailera *et al.* (Figure 1.10).[69]

In addition, one application example that was noted by Bailera and co-workers[69] in their extensive 2017 review of laboratory/pilot/demonstrable plants for storing renewable energy through the transformation of carbon dioxide is the 6-MW Audi e-gas plant, which is currently converting biomass-derived CO_2 to CH_4 at an efficiency of 54% with an expected production of *ca.* 1000 tons per year. The process is based on the catalytic methanation of pure hydrogen and carbon dioxide carried out in a single isothermal fixed-bed reactor; the hydrogen originated from 3×2.0 MW_e alkaline electrolyzers powered by an offshore wind park in the North Sea, and equipped with a 4×3.6 MW_e turbine, while the required CO_2 was separated from the raw biogas of a neighboring biomethane plant.

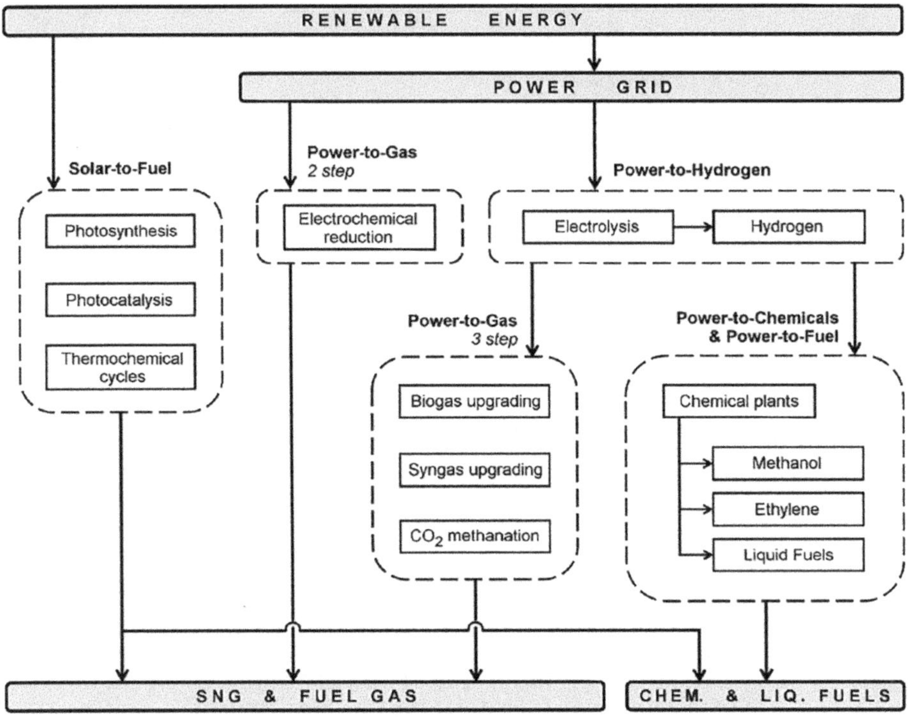

Figure 1.10 Illustration of some renewable energy and CO_2 hybrid storage techniques. Reproduced from ref. 69 with permission from Elsevier, Copyright 2016.

1.4.3 Demonstrated P2F Projects

As pointed out by Thema and co-workers,[70] the coupling of P2F (or in their words, power-to-gas PtG process) and energy storing technology has recently received considerable attention from various areas toward: (1) an integrated future energy systems architecture; (2) focus on the technology; (3) social acceptance; (4) marketing; and (5) policy makers. These authors have described various pilot and demonstration projects because of a significant rise in interest in the PtG technology as it provides an opportunity to convert electricity into chemical energy that can be stored as a combustible gas such as hydrogen and methane. Hydrogen is used either directly or otherwise fed into a downstream methanation process depending on the requirements of the embedding energy system such as hydrogen tolerance of gas networks, gas buffering, mobility and/or heat applications.[70]

Their 2019 research paper[70] gave an exhaustive overview of the active power-to-gas projects for the production of hydrogen or else a renewable substitute natural gas with a particular focus on 153 projects in 22 countries either completed, recent or planned since 1988 in central Europe. The projects were discussed on the basis of plant allocation and size, installed power development, shares, and yields of hydrogen or otherwise substitute natural gas production and product utilization phases. Projects showed various applications in diverse fields, from early research and development level to the adoption of pilot and industrial scale power conversion plants. As well, most projects that were examined in some details were located in Germany, Denmark, the USA, and Canada with the larger number of projects located in Germany and Denmark (see Figure 1.11). Figure 1.12 shows the projects examined on a global scale. Also analyzed were the development costs for electrolysis and carbon dioxide methanation with a projection to 2030 and with an outlook toward the magic year 2050.[70] Results of their analyses showed substantial reduction of the costs for electrolysis and methanation, with a further price decline to less than 500 Euros per KW electric power input for both technologies to 2050, with current projects in 2019 operating at 39 MW.

Along similar lines, Golling and co-workers[71] proposed a roadmap for Germany on the implementation of the PtG technology in which they showed a pathway toward short-term reductions in greenhouse gas emissions of *ca.* 40% until 2020, *ca.* 55% by 2030 and from 80 to 95% in the long term. Such a proposal considered international CO_2-trade, implementation of the hydrogen infrastructure, the intersectoral planning of infrastructure, and the presence of export markets for the PtG technology. The global focus of research and application of the PtG technology rests mostly in Europe, although the USA appears to be catching up.

Table 1.2 summarizes the status of active projects as of three years ago (2019) with regard to conversion efficiencies from PtG in projects that

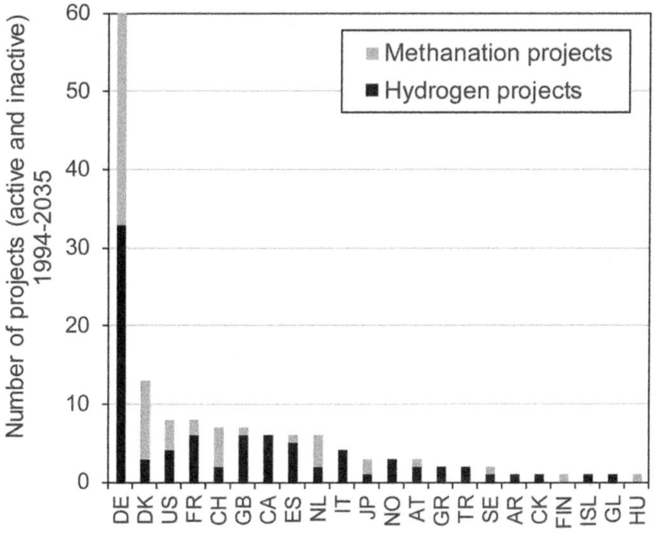

Figure 1.11 PtG project allocation and number of active/inactive (light grey) methane projects and hydrogen projects (black) in various countries. Reproduced from ref. 70, https://doi.org/10.1016/j.rser.2019.06.030, under the terms of the CC BY 4.0 license https://creativecommons.org/licenses/by/4.0/.

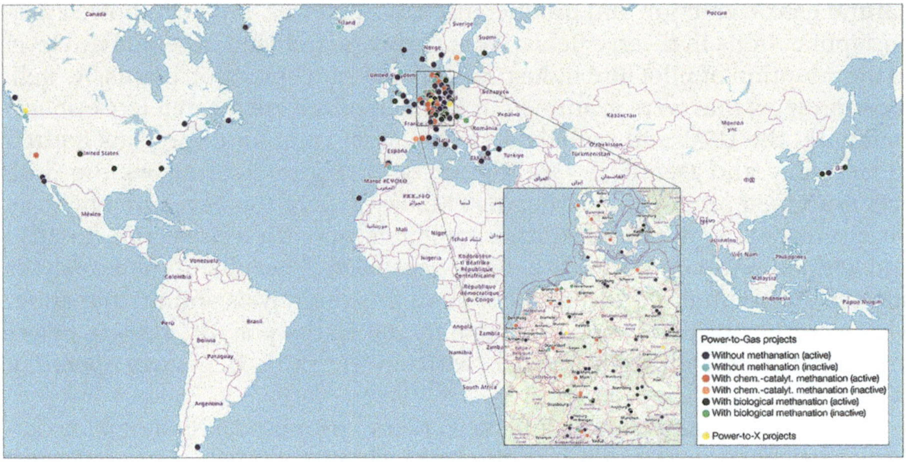

Figure 1.12 PtG project allocation differentiated according to the target products hydrogen and methane as well as activity/inactivity status. Dark green: active PtG with biological CO_2-methanation; light green: inactive PtG with biological CO_2-methanation; red: active PtG with chemical CO_2-methanation; orange: inactive PtG with chemical CO_2-methanation; dark blue: active PtG without methanation; light blue: inactive PtG without methanation. Reproduced from ref. 70, https://doi.org/10.1016/j.rser.2019.06.030, under the terms of the CC BY 4.0 license https://creativecommons.org/licenses/by/4.0/.

Table 1.2 Summary of the status of active projects as of three years ago (2019).[70]

	Hydrogen production	Methane production
Feed-in projects	21	36
Number of active projects (2019)	56	38
Installed production capacity	6205 $m^3\ h^{-1}$ 18.6 $MW_{ch,LHV\text{-}H2}$	590 $m^3\ h^{-1}$ 6 $MW_{ch,LHV\text{-}CH4}$
Installed electrical load	24.1 MW_{el}	14.5 MW_{el}
Efficiency electricity-to-gas	77%	41%

produce hydrogen and methane, as well as the number of plants that feed in their products; also shown are the electrical load of methanation projects related to the electrolysis power needed to feed the methanation unit.[70]

1.5 Concluding Remarks

This chapter has described, albeit not in great detail, the production of solar fuels (hydrogen and methane) *via* the water-splitting process and the methanation of carbon dioxide through the use of three essential technologies: (1) photocatalysis involving photoactivated semiconductor photocatalysts; (2) photoelectrocatalysis also involving semiconductor materials;

and (3) photovoltaic-electrocatalysis to drive the formation of solar fuels with particular attention to the Power-to-X technology. It is evident that much remains to be done in these technologies to diminish the use of non-renewable resources (*e.g.*, fossil fuels) toward sustaining a clean and non-polluting energy supply for years to come.

References

1. See for example, https://www.it.wikipedia.org/wiki/Fridays_for_Future.
2. I. Dincer, Potential options to greenize energy systems, *Sixth International Green Energy Conference*, Keynote Lecture, 5–9 June 2011, Eskisehir, Turkey, 2011.
3. 2019 data from https://ourworldindata.org/energy-production-consumption.
4. A. J. Bard, Photoelectrochemistry, *Science*, 1980, **207**, 139–144.
5. J. X. W. Hay, T. Y. Wu, J. C. Juan and J. Md Jahim, Biohydrogen production through photo fermentation or dark fermentation using waste as a substrate: Overview, economics, and future prospects of hydrogen usage, *Biofuels, Bioprod. Biorefin.*, 2013, **7**, 334–352.
6. G.-L. Chiarello, M. H. Aguire and E. Selli, Hydrogen production by photocatalytic steam reforming of methanol on noble metal-modified TiO_2, *J. Catal.*, 2010, **273**, 182–190.
7. S. Y. Lee, J. U. Lee, I. B. Lee and J. Han, Design under uncertainty of carbon capture and storage infrastructure considering cost, environmental impact, and preference on risk, *Appl. Energy*, 2017, **189**, 725–738.
8. L. M. Peter, Photoelectrochemistry: From Basic Principles to Photocatalysis, in *Photocatalysis: Fundamentals and Perspectives*, 2016, pp. 1–28.
9. G. Palmisano, V. Augugliaro, M. Pagliaro and L. Palmisano, Photocatalysis: a promising route for 21st century organic chemistry, *Chem. Commun.*, 2007, 3425–3437.
10. C. Byrne, G. Subramania and S. C. Pillai, Recent advances in photocatalysis for environmental applications, *J. Environ. Chem. Eng.*, 2018, **6**, 3531–3555.
11. J. Colina-Márquez, F. Machuca-Martínez and G. L. Puma, Radiation Absorption and Optimization of Solar Photocatalytic Reactors for Environmental Applications, *Environ. Sci. Technol.*, 2010, **44**, 5112–5120.
12. N. Hoffmann, Photocatalysis with TiO_2 Applied to Organic Synthesis, *Aust. J. Chem.*, 2015, **68**, 1621–1639.
13. N. Serpone, E. Pelizzetti and H. Hidaka, Heterogeneous Photocatalysis. Issues, Questions, Some Answers, and Some Successes, in *Photochemical and Photoelectrochemical Conversion & Storage of Solar Energy*, ed. Z. W. Tian and Y. Cao, International Academic Publishers, Beijing, China, 1993, pp. 33–73.
14. N. Serpone, A. V. Emeline and S. Horikoshi, Photocatalysis and Solar Energy Conversion (chemical aspects), *SPR - Photochemistry*, RSC Publishing, The Royal Society of Chemistry, London, UK, 2009, vol. 37, pp. 300–361.

15. G. Rothenberger, J. Moser, M. Gratzel, N. Serpone and D. K. Sharma, Charge carrier trapping and recombination dynamics in small semiconductor particles, *J. Am. Chem. Soc.*, 1985, **107**, 8054–8059.
16. N. Serpone, D. Lawless, R. Khairutdinov and E. Pelizzetti, Subnanosecond relaxation dynamics of nanosized TiO_2 colloids (R = 1.0 to 13.4 nm). Relevance to heterogeneous photocatalysis, *J. Phys. Chem.*, 1995, **99**, 16655–16661.
17. S. Protti, A. Albini and N. Serpone, Photocatalytic generation of solar fuels from the reduction of H_2O and CO_2: a look at the patent literature, *Phys. Chem. Chem. Phys.*, 2014, **16**, 19790–19827.
18. H. Gerischer, Semiconductor electrodes and the interaction with light, in *Photoelectrochemistry, Photocatalysis, and Photorereactors – Fundamentals and Developments*, ed. Schiavello, M., NATO ASI Series, Series C: Mathematical and Physical Sciences, Springer Science, Dordrecht, The Netherlands, 1985, vol. 146, pp. 39–106.
19. A. J. Nozik, Photoelectrolysis of water using semiconducting TiO_2 crystals, *Nature*, 1975, **257**, 383–386.
20. F. E. Osterloh, Inorganic Materials as Catalysts for Photochemical Splitting of Water, *Chem. Mater.*, 2008, **20**, 35–54.
21. J. Chen, D. Yang, D. Song, J. Jiang, A. Ma, M. Z. Hu and C. Ni, Recent progress in enhancing solar-to-hydrogen efficiency, *J. Power Sources*, 2015, **280**, 649–666.
22. K. Ohashi, J. McCann and J. O. M. Bockris, Stable photoelectrochemical cells for the splitting of water, *Nature*, 1977, 610–611.
23. Y. Lee, H. Terashima, Y. Shimodaira, K. Teramura, M. Hara, H. Kobayashi, K. Domen and M. Yashima, Zinc Germanium Oxynitride as a Photocatalyst for Overall Water Splitting under Visible Light, *J. Phys. Chem. C*, 2007, **111**, 1042–1948.
24. Y.-C. Chiou, U. Kumar and J. Wu, Photocatalytic splitting of water on $NiO/InTaO_4$ catalysts prepared by an innovative sol–gel method, *Appl. Catal., A*, 2009, **357**, 73–78.
25. K. Sayama, K. Mukasa, R. Abe, Y. Abe and H. Arakawa, A new photocatalytic water splitting system under visible light irradiation mimicking a Z-scheme mechanism in photosynthesis, *J. Photochem. Photobiol., A*, 2002, **148**, 71–77.
26. H. Kabbour, A. Sayede, S. Saitzek, G. Lefèvre, L. Cario, M. Trentesaux and P. Rousse, Structure of the water-splitting photocatalyst oxysulfide α-$LaOInS_2$ and *ab initio* prediction of new polymorphs, *Chem. Commun.*, 2020, **56**, 1645–1648.
27. J. Schneidewind, M. A. Argüello Cordero, H. Junge, S. Lochbrunner and M. Beller, Two-photon, visible light water splitting at a molecular ruthenium complex, *Energy Environ. Sci.*, 2021, **14**, 4427–4436.
28. H. Nagakawa, T. Ochiai, S. Konuma and M. Nagata, Visible-Light Overall Water Splitting by $CdS/WO_3/CdWO_4$ Tricomposite Photocatalyst Suppressing Photocorrosion, *ACS Appl. Energy Mater.*, 2018, **12**, 6730–6735.

29. C. Zhu, C. Liu, Y. Zhou, Y. Fu, S. Guo, H. Li, S. Zhao, H. Huang, Y. Liu and Z. Kang, Carbon dots enhance the stability of CdS for visible-light-driven overall water splitting, *Appl. Catal., B*, 2017, **216**, 114–121.
30. Y. Liu, X. Xu, A. Li, Z. Si, X. Wu, R. Ran and D. Weng, A strategy to construct (reduced graphene oxide, γ-Fe_2O_3)/C_3N_4 step-scheme photocatalyst for visible-light water splitting, *Catal. Commun.*, 2021, **157**, 106327.
31. A. V. Emeline, V. K. Ryabchuk and N. Serpone, Spectral dependencies of the quantum yield of photochemical processes on the surface of nano-/micro-particulates of wide-band-gap metal oxides. 1. Theoretical approach, *J. Phys. Chem. B*, 1999, **103**, 1316–1324.
32. A. V. Emeline, A. V. Frolov, V. K. Ryabchuk and N. Serpone, Spectral dependencies of the quantum yield of photochemical processes on the surface of nano-/micro-particulates of wide-bandgap metal oxides. IV. Theoretical modeling of the activity and selectivity of semiconductor photocatalysts with inclusion of a subsurface electric field in the space charge region, *J. Phys. Chem. B*, 2003, **107**, 7109–7119.
33. Z. Li, B. Tian, W. Zhen, Y. Wu and G. Lu, Inhibition of hydrogen and oxygen recombination using oxygen transfer reagent hemin chloride in Pt/TiO_2 dispersion for photocatalytic hydrogen generation, *Appl. Catal., B*, 2017, **203**, 408–415.
34. A. V. Emeline, A. V. Panasuk, N. Sheremetyeva and N. Serpone, Mechanistic Studies of the Formation of Different States of Oxygen on Irradiated ZrO_2 and the Photocatalytic Nature of Photoprocesses from Determination of Turnover Numbers, *J. Phys. Chem. B*, 2005, **109**, 2785–2792.
35. H. Yu, X. Huang, P. Wang and J. Yu, Enhanced photoinduced stability and photocatalytic activity of CdS by dual amorphous co-catalysts: Synergistic effect of Ti(IV)-hole co-catalyst and Ni(II)-electron co-catalyst, *J. Phys. Chem. C*, 2016, **120**, 3722–3730.
36. T. Hisatomi, T. Minegishi and K. Domen, Kinetic assessment and numerical modeling of photocatalytic water splitting toward efficient solar hydrogen production, *Bull. Chem. Soc. Jpn.*, 2013, **85**, 647–655.
37. Z. Wang, C. Li and K. Domen, Recent developments in heterogeneous photocatalysts for solar-driven overall water splitting, *Chem. Soc. Rev.*, 2019, **48**, 2109–2125.
38. N. Serpone, E. Borgarello and M. Graetzel, Visible Light Induced Generation of Hydrogen from H_2S in Mixed Semiconductor Dispersions: Improved Efficiency through Inter-particle Electron Transfer, *J. Chem. Soc., Chem. Commun.*, 1984, 342–344.
39. N. Serpone, E. Borgarello and E. Pelizzetti, Utilization of the Semiconductor Particle as a Microphotoelectrochemical Cell: Electrochemical evidence for Interparticle Electron Transfer and application to photocatalysis, *J. Electrochem. Soc.*, 1988, **135**, 2760–2766.
40. N. Serpone and A. V. Emeline, Semiconductor photocatalysis – Past, present and future outlook, *J. Phys. Chem. Lett.*, 2012, **3**, 673–677.

41. A. V. Emeline, V. N. Kuznetsov, V. K. Ryabchuk and N. Serpone, On the way to the creation of next generation photoactive materials, *Environ. Sci. Pollut. Res.*, 2012, **19**, 3666–3675.
42. Z. Jiang, D. Jiang, Z. Yan, D. Liu, K. Qian and J. Xie, A new visible light active multifunctional ternary composite based on TiO_2–In_2O_3 nanocrystals heterojunction decorated porous graphitic carbon nitride for photocatalytic treatment of hazardous pollutant and H_2 evolution, *Appl. Catal., B*, 2015, **170–171**, 195–205.
43. Z. Wang, Y. Inoue, T. Hisatomi, R. Ishikawa, Q. Wang, T. Takata, S. Chen, N. Shibata, Y. Ikuhara and K. Domen, Overall water splitting by Ta_3N_5 nanorod single crystals grown on the edges of $KTaO_3$ particles, *Nat. Catal.*, 2018, **1**, 756–763.
44. A. T. Garcia-Esparza, T. Shinagawa, S. Ould-Chikh, M. Qureshi, X. Peng, N. Wei, D. H. Anjum, A. Clo, T.-C. Weng, D. Nordlund, D. Sokaras, J. Kubota, K. Domen and K. Takanabe, An Oxygen-Insensitive Hydrogen Evolution Catalyst Coated by a Molybdenum-Based Layer for Overall Water Splitting, *Angew. Chem., Int. Ed.*, 2017, **56**, 5780–5784.
45. S. Navaln, A. Dhakshinamoorthy, M. lvaro and H. Garcia, Photocatalytic CO_2 Reduction using Non-Titanium Metal Oxides and Sulfides, *ChemSusChem*, 2013, **6**, 562–577.
46. N. Shehzad, M. Tahir, K. Johari, T. Murugesan and M. Hussain, A critical review on TiO_2 based photocatalytic CO_2 reduction system: Strategies to improve efficiency, *J. CO_2 Util.*, 2018, **26**, 98–122.
47. A. A. Khan and M. Tahir, Recent advancements in engineering approach towards design of photoreactors for selective photocatalytic CO_2 reduction to renewable fuels, *J. CO_2 Util.*, 2019, **29**, 205–239.
48. I. Janik and G. N. R. Tripathi, The nature of the CO_2 (-) radical anion in water, *J. Chem. Phys.*, 2016, **144**, 154307.
49. H. He, P. Zapol and L. A. Curtiss, A Theoretical Study of CO_2 Anions on Anatase (101) Surface, *J. Phys. Chem. C*, 2010, **114**, 21474–22148.
50. R. Flyunt, M. N. Schuchmann and C. von Sonntag, A common carbanion intermediate in the recombination and proton-catalysed disproportionation of the carboxyl radical anion CO_2^- in aqueous solution, *Chem. – Eur. J.*, 2001, **7**, 796–799.
51. V. Kumaravel, J. Bartlett and S. C. Pillai, Photoelectrochemical Conversion of Carbon Dioxide (CO_2) into Fuels and Value-Added Products, *ACS Energy Lett.*, 2020, **5**, 486–519 and references therein.
52. A. Fujishima and K. Honda, Electrochemical Photolysis of Water at a Semiconductor Electrode, *Nature*, 1972, **238**, 37–38.
53. J. H. Park and A. J. Bard, Unassisted Water Splitting from Bipolar Pt/ Dye-Sensitized TiO_2 Photoelectrode Arrays, *Electrochem. Solid-State Lett.*, 2005, **8**, G371.
54. J. H. Park and A. J. Bard, Photoelectrochemical Tandem Cell with Bipolar Dye-Sensitized Electrodes for Vectorial Electron Transfer for Water Splitting, *Electrochem. Solid-State Lett.*, 2006, **9**, E5–E8.

55. Y. K. Kim, T. H. Lee, J. Yeop, W. J. Byun, J. H. Kim, J. Y. Kim and J. S. Lee, Hetero-tandem organic solar cells drive water electrolysis with a solar-to-hydrogen conversion efficiency up to 10%, *Appl. Catal., B*, 2022, **309**, 121237.
56. L. Pan, S. Sun, Y. Chen, P. Wang, J. Wang, X. Zhang, J.-J. Zou and Z. L. Wang, Advances in Piezo-Phototronic Effect Enhanced Photocatalysis and Photoelectrocatalysis, *Adv. Energy Mater.*, 2020, **10**, 2000214.
57. P. Lianos, Review of recent trends in photoelectrocatalytic conversion of solar energy to electricity and hydrogen, *Appl. Catal., B*, 2017, **210**, 235–254.
58. K. Sivula, F. Le Formal and M. Grätzel, Solar water splitting: Progress using hematite (α-Fe_2O_3) photoelectrodes, *ChemSusChem*, 2011, **4**, 432–449.
59. A. Braun, K. Sivula, D. K. Bora, J. Zhu, L. Zhang, M. Gratzel, J. Guo and E. C. Constable, Direct Observation of Two Electron Holes in a Hematite Photoanode during Photoelectrochemical Water Splitting, *J. Phys. Chem. C*, 2012, **116**, 16870–16875.
60. B. Y. Alfaifi, H. Ullah, S. Alfaifi, A. A. Tahir and T. K. Mallick, Photoelectrochemical solar water splitting: From basic principles to advanced devices, *Veruscript Funct. Nanomater.*, 2018, **2**, BDJOC3.
61. A. U. Pawar, C. Woo Kim, M.-T. Nguyen-Le and Y. S. Kang, General Review on the Components and Parameters of Photoelectrochemical System for CO_2 Reduction with in Situ Analysis, *ACS Sustainable Chem. Eng.*, 2019, **7**, 7431–7455.
62. X. Zhou, R. Liu, K. Sun, Y. Chen, E. Verlage, S. A. Francis, N. S. Lewis and C. Xiang, Solar-Driven Reduction of 1 atm of CO_2 to Formate at 10% Energy-Conversion Efficiency by Use of a TiO_2-Protected III–V Tandem Photoanode in Conjunction with a Bipolar Membrane and a Pd/C Cathode, *ACS Energy Lett.*, 2016, **1**, 764–770.
63. NRER, National Best research—cell efficiency chart, 2020-04-06, available at website of http://www.nrel.gov.
64. J. Gong, N. J. English, D. Pant, G. R. Patzke, S. Protti and T. Zhang, Power-to-X: Lighting the Path to a Net-Zero-Emission Future, *ACS Sustainable Chem. Eng.*, 2021, **9**, 7179–7181.
65. P. D. Lund, J. Lindgren, J. Mikkola and J. Salpakari, Review of Energy System Flexibility Measures to Enable High Levels of Variable Renewable Electricity, *Renewable Sustainable Energy Rev.*, 2015, **45**, 785–807.
66. X. Liu, B. You and Y. Sun, Facile Surface Modification of Ubiquitous Stainless Steel Led to Competent Electrocatalysts for Overall Water Splitting, *ACS Sustainable Chem. Eng.*, 2017, **5**, 4778–4784.
67. A. S. R. Machado and M. N. da Ponte, CO_2 capture and electrochemical conversion, *Curr. Opin. Green Sustainable Chem.*, 2018, **11**, 86–90.
68. C. Wulf, P. Zapp and A. Schreiber, Review of Power-to-X Demonstration Projects in Europe, *Front. Energy Res.*, 2020, **8**, 191.

69. M. Bailera, P. Lisbona, L. M. Romeo and S. Espatolero, Power to Gas Projects Review: Lab, Pilot and Demo Plants for Storing Renewable Energy and CO_2, *Renewable Sustainable Energy Rev.*, 2017, **69**, 292–312.
70. M. Thema, F. Bauer and M. Sterner, Power-to-Gas: Electrolysis and methanation status review, *Renewable Sustainable Energy Rev.*, 2019, **112**, 775–787.
71. C. Golling, R. Heuke, H. Seidl and J. Uhlig, Roadmap power-to-gas, 19 June 2017, Deutsche Energie-Agentur GmbH (dena), Barbara Schroeder, Chausseestraße 128 a, 10115 Berlin; see https://www.dena.de/newsroom/meldungen/2017/power-to-gas-dena-roadmap-empfiehlt-technologieoffenheit-und-verbesserung-der-rahmenbedingungen/.

CHAPTER 2

$SrTiO_3$-based Perovskite Photocatalysts for Efficient Solar-driven Overall Water Splitting and CO_2 Reduction

QIAN WANG*[a,b] AND ZHENHUA PAN[c]

[a]Graduate School of Engineering, Nagoya University, Furo-cho, Chikusa-ku, Nagoya 464-8603, Japan; [b]Institute for Advanced Research, Nagoya University, Furo-cho, Chikusa-ku, Nagoya 464-8601, Japan; [c]Department of Applied Chemistry, Faculty of Science and Technology, Chuo University, 1-13-27 Kasuga, Bunkyo, Tokyo 112-8551, Japan
*E-mail: wang.qian@material.nagoya-u.ac.jp

2.1 Introduction

Artificial photosynthesis technology that is specifically linked to water (H_2O) splitting and carbon dioxide (CO_2) reduction efficiently converts and stores solar energy in chemicals, such as hydrogen (H_2), syngas, methane (CH_4) and methanol (CH_3OH).[1] For instance, photocatalytic water splitting to produce H_2, which serves as a reducing agent in a variety of chemical transformations as well as a potential zero carbon emission energy carrier that integrates disparate infrastructure systems for increased economic efficiency and flexibility.[2] Artificial photosynthetic processes are thermodynamically uphill reactions with a significant increase in Gibbs free energy (eqn (2.1)–(2.4),

Sustainable Energy No. 1
Recent Developments in Functional Materials for Artificial Photosynthesis
Edited by Srabanti Ghosh and Qian Wang

Published by the Royal Society of Chemistry, www.rsc.org

for example, water splitting: 237 kJ mol^{-1} at 298 K), necessitating the use of semiconductor photocatalysts for driving the reactions.

$$H_2O(l) \rightarrow H_2(g) + \frac{1}{2}O_2(g) \qquad \Delta G = 237 \text{ kJ mol}^{-1} \tag{2.1}$$

$$CO_2(g) \rightarrow CO(g) + \frac{1}{2}O_2(g) \qquad \Delta G = 257 \text{ kJ mol}^{-1} \tag{2.2}$$

$$CO_2(g) + 2H_2O(l) \rightarrow CH_3OH(l) + \frac{3}{2}O_2(g) \quad \Delta G = 703 \text{ kJ mol}^{-1} \tag{2.3}$$

$$CO_2(g) + 2H_2O(l) \rightarrow CH_4(g) + 2O_2(g) \quad \Delta G = 818 \text{ kJ mol}^{-1} \tag{2.4}$$

When the semiconductor photocatalyst is irradiated with energy greater than its bandgap, the electrons (e^-) in the valence band (VB) are excited into the conduction band (CB), while the holes (h^+) remain in the VB (Figure 2.1). The electrons and holes that did not suffer recombination migrate toward the semiconductor surface where they reduce the proton or CO_2 to produce chemical fuels and oxidize the water molecule to form O_2, respectively.[3] Solar fuel generation *via* photocatalytic processes has undergone rapid development and is now being used in both laboratory-scale systems and large-scale demonstrations. Due to the simplicity of this process, which requires far less auxiliary hardware, it is expected to generate solar fuels more cost-effectively on a large scale. Techno-economic analyses estimated that photocatalytic water splitting produces H_2 at a cost of 1.6–3.5 USD kg^{-1}, assuming a solar-to-hydrogen conversion efficiency (STH) of 5–10% and a 5 year system lifetime.[4–6]

Among the numerous photocatalysts discovered and reported for photocatalytic H_2O splitting and CO_2 reduction, a prototypical perovskite oxide in the ABO_3 family, $SrTiO_3$, has recently experienced a tremendous renaissance and demonstrated outstanding photocatalytic ability. While the ideal cubic perovskite structure is rare and the perovskite is commonly slightly

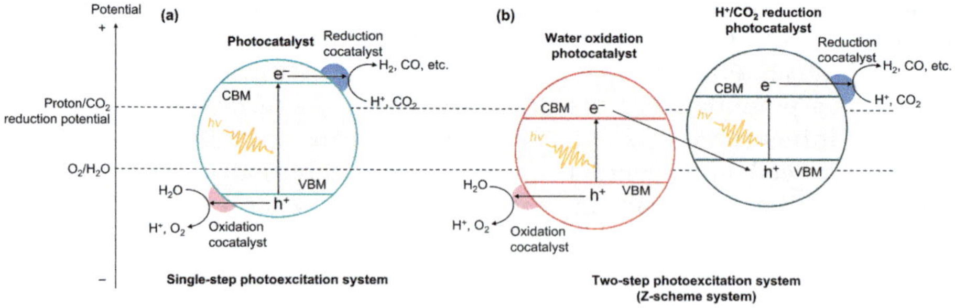

Figure 2.1 Schematic energy diagrams of (a) single-step photoexcitation system and (b) two-step photoexcitation (Z-scheme) system for overall water splitting or carbon dioxide reduction coupled with water oxidation. CBM: conduction band minimum; VBM: valence band maximum.

distorted, $SrTiO_3$ with an idealized cubic structure that consists of a corner-shared network of $[TiO_6]$ octahedral with Sr^{2+} ions in the 12-fold cavities in between the polyhedra is realized at room temperature.[7] $SrTiO_3$ was determined to have an indirect bandgap of 3.25 eV and a direct bandgap of 3.75 eV.[8] The CB bottom in $SrTiO_3$ is composed of mainly Ti 3d, while the upper VB is dominated by O 2p states. The interband transitions from the upper VB to the Ti 3d bands lead to indirect bandgap energy of 3.25 eV and UV light absorption ability.

Domen *et al.* reported the first use of $SrTiO_3$ for photocatalytic overall water splitting in 1980.[9] Under UV light irradiation, simultaneous H_2 and O_2 were evolved from water in a stoichiometric ratio (H_2:O_2 = 2 : 1) using $SrTiO_3$ decorated with NiO as a hydrogen evolution co-catalyst. Recently, many studies have been reported to significantly improve the photocatalytic performance of $SrTiO_3$-based perovskite materials for overall water splitting and CO_2 reduction. In this chapter, we survey these recent developments in $SrTiO_3$-based photocatalysts and systematically illustrate how a typical photocatalytic process can be tuned towards improved light absorption and charge separation efficiency. This chapter aims to inspire effective strategies towards future tailored designs of more efficient photocatalytic solar fuel production systems.

2.2 Single-step Artificial Photosynthetic Reactions

Single-step artificial photosynthetic reactions utilize a single photocatalyst to split water into H_2 and O_2 or reduce CO_2 in conjunction with water oxidation, requiring the semiconductor photocatalyst to have appropriate CB and VB levels for proton/CO_2 reduction and water oxidation, respectively (Figure 2.1(a)). Additionally, the photocatalysts should be photoirradiation stable. While $SrTiO_3$ satisfies these requirements and was demonstrated in 1980 to split water into H_2 and O_2, its photocatalytic activity has been significantly enhanced in very recent years through various material design strategies (Table 2.1), aiming to improve quantum efficiency and extending the usable wavelength region of light.

2.2.1 Defect Engineering

The two distinct cationic sites in $SrTiO_3$ can be replaced by dopants and result in numerous doping combinations. The Goldschmidt tolerance factor, which has been widely used to predict the stability of perovskite structure, is applied to select a dopant with suitable size for site-selective doping.[10] For example, the Sr^{2+} is coordinated by twelve O^{2-} and preferentially occupied by larger cations, whereas the coordination number of Ti^{4+} is 6 and the Ti^{4+} is preferentially occupied by smaller cations. $SrTiO_3$ is thought to contain two types of defects: (1) oxygen vacancies caused by the dissociation of Ti–O bonds, which releases a small amount of lattice oxygen into the gas phase and generates free electrons (eqn (2.5));[11] and (2) Ti^{3+}, which is formed when

Table 2.1 Representative examples of $SrTiO_3$-based photocatalysts for light-driven single-step photocatalytic overall water splitting or CO_2 reduction using water.

Photocatalyst	Bandgap/eV	Co-catalyst	Reactant	Gas evolution rate		Light source	Duration/h	Efficiency	Reference
$SrTiO_3$	3.2	NiO	Water vapor	H_2: 4.4×10^{-3} mL h^{-1}	O_2: 2.2×10^{-3} mL h^{-1}	450 W Hg lamp	~80		9
$SrTiO_3$ (nanoparticle)	3.2	NiO	Water	H_2: ~14 μmol h^{-1}	O_2: ~6 μmol h^{-1}	300 W Xe lamp	24		80
$SrTiO_3$ (mesocrystal)		$Rh_{2-x}Cr_xO_3$	Water			UV-LED source	24	AQY: 6.7% at 360 nm	81
$SrTiO_3$ (3D ordered macroporous)		Pt	Water, CO_2	CH_4: 26.7 μmol h^{-1}; CO: 4.1 μmol h^{-1}; H_2: 5.3 μmol h^{-1}	O_2: 58.1 μmol h^{-1}	300 W Xe lamp	10	AQY: 0.66% 320–780 nm	82
$SrTiO_3$	3.2	Pd/Cr_2O_3,	Water	H_2: ~5.3 μmol h^{-1}	O_2: ~1.7 μmol h^{-1}	300 W Xe lamp			26
$SrTiO_3$	3.2	Rh/Cr_2O_3, $Co_xMn_{1-x}O$	Water	H_2: ~100 μmol h^{-1}	O_2: ~50 μmol h^{-1}	300 W Xe lamp	5		51
$SrTiO_3$:Na	3.2	$Rh_{0.3}Cr_{1.7}O_3$	Water	H_2: 45 mmol h^{-1}	O_2: 23 mmol h^{-1}	450 W Hg lamp	12	AQY: 28% at 365 nm	83
$SrTiO_3$:Li		$Rh_{0.3}Cr_{1.7}O_3$	Water	H_2: ~1650 μmol h^{-1}	O_2: ~850 μmol h^{-1}	450 W Hg lamp			13
$SrTiO_3$:K		$Rh_{0.3}Cr_{1.7}O_3$	Water	H_2: ~1420 μmol h^{-1}	O_2: ~710 μmol h^{-1}	450 W Hg lamp			13
$SrTiO_3$:Rb		$Rh_{0.3}Cr_{1.7}O_3$	Water	H_2: ~800 μmol h^{-1}	O_2: ~410 μmol h^{-1}	450 W Hg lamp			13
$SrTiO_3$:Cs		$Rh_{0.3}Cr_{1.7}O_3$	Water	H_2: ~790 μmol h^{-1}	O_2: ~400 μmol h^{-1}	450 W Hg lamp			13
$SrTiO_3$:In		$Rh_{0.3}Cr_{1.7}O_3$	Water	H_2: ~600 μmol h^{-1}	O_2: ~300 μmol h^{-1}	450 W Hg lamp			13
$SrTiO_3$:In		$Rh_{0.3}Cr_{1.7}O_3$	Water	H_2: ~600 μmol h^{-1}	O_2: ~300 μmol h^{-1}	450 W Hg lamp			13
$SrTiO_3$:Mg		$Rh_{0.3}Cr_{1.7}O_3$	Water	H_2: ~1500 μmol h^{-1}	O_2: ~750 μmol h^{-1}	450 W Hg lamp			13
$SrTiO_3$:Mg	3.2	Ni@NiO	Water	H_2: 39.8 nmol min^{-1}	O_2: 22.8 nmol min^{-1}	Solar simulator (AM 1.5 G)		AQY: 9.1% at 300–400 nm	12
$SrTiO_3$:Mg	3.2	Ni@NiO_x, CrO_y	Water	H_2: 8.4 μmol h^{-1}	O_2: ~4.2 μmol h^{-1}	365 nm LED light	70	AQY: 26% at 365 nm	84
$SrTiO_3$:La	3.2	CoO	Na_2CO_3 aqueous solution	H_2: ~280 μmol h^{-1}	O_2: ~130 μmol h^{-1}	400 W Hg lamp			85

$SrTiO_3$:Ga	3.2	Rh_2O_3, Cr_2O_3	Water	H_2: ~4400 µmol h^{-1}	O_2: ~2100 µmol h^{-1}	450 W Hg lamp	5		10
$SrTiO_3$:B	3.07		Water, CO_2	CO: 21 µmol h^{-1} g^{-1} CH_4: 14 µmol h^{-1} g^{-1}	O_2: 35 µmol h^{-1} g^{-1}	300 W Xe lamp	1	AQY: 1.34% at 365 nm; 0.53% at 420 nm	76
$SrTiO_3$:Al	3.2	Nitrogen-doped graphene quantum dots, CoO_x	Water	H_2: 18.8 µmol h^{-1}	O_2: 9.0 µmol h^{-1}	280 W Xe lamp	4		56
$SrTiO_3$:Al	3.2	CoP_x	Water	H_2: 1.36 mmol h^{-1}	O_2: 0.635 mmol h^{-1}	280 W Xe lamp	4		57
$SrTiO_3$:Al	3.2	CoP, Cr_2O_3	Water	H_2: 3.558 mmol h^{-1} g^{-1}	O_2: 1.722 mmol h^{-1} g^{-1}	300 W Xe lamp	9	AQY: 7.1% at 350 ± 10 nm	86
$SrTiO_3$:Al	3.2	$RhCrO_x$	Water	H_2: ~1.3 mmol h^{-1}	O_2: ~0.65 mmol h^{-1}	300 W Xe lamp	16	AQY: 56% at 365 nm STH: 0.4% at 288 K and 0.6 at 331 K	43
$SrTiO_3$:Al	3.2	$RhCrO_x$, MoO_y	Water	H_2: ~1.9 mmol h^{-1}	O_2: ~0.9 mmol h^{-1}	300 W Xe lamp	16	AQY: 69% at 365 nm	44
$SrTiO_3$:Al	3.2	$RhCrO_x$, CoOOH, TiO_2	Water			Solar simulator (AM 1.5 G)	1300	STH: 0.3%	34
$SrTiO_3$:Al	3.2	Rh/Cr_2O_3, CoOOH	Water	H_2: ~3.54 mmol h^{-1}	O_2: ~1.78 mmol h^{-1}	300 W Xe lamp	12.5	AQY: ~96% at 350–360 nm	52
$SrTiO_3$:Al	3.2	Ag	Water, CO_2	CO: 5.5 µmol h^{-1}	O_2: 2.7 µmol h^{-1}	400 W Hg lamp	5		15
$SrTiO_3$:Al	3.2	Ag, FeOOH	Water, CO_2	CO: 38.2 µmol h^{-1}	O_2: ~19.1 µmol h^{-1}	400 W Hg lamp	5		64
$SrTiO_3$:La, Al	3.2	Rh/Cr_2O_3, CoOOH	Water	H_2: 1.79 mmol h^{-1}	O_2: 0.91 mmol h^{-1}	300 W Xe lamp	18	AQY: ~78.43% at 365 nm	16

Ti^{4+} is reduced by the free electrons (eqn (2.6)). $SrTiO_3$ is demonstrated to be a n-type semiconductor due to the presence of a donor level just below the conduction band bottom caused by Ti^{3+} defects.

$$O_o^{\times} \rightarrow V_O^{\bullet\bullet} + \frac{1}{2}O_2 + 2e' \tag{2.5}$$

$$2e' + 2Ti_{Ti}^{\times} \rightarrow 2Ti_{Ti}' \tag{2.6}$$

As a result of the thermodynamic equilibrium, substituting a lower valence cation for the parent cation will extrinsically introduce oxygen vacancy and inhibit the generation of Ti^{3+} (Figure 2.2(a)). Dopant with a higher valence cation, on the other hand, stabilizes Ti^{3+} without producing oxygen vacancy (Figure 2.2(b)). Therefore, $SrTiO_3$ doped with a variety of dopants was investigated for controlling the defects.[12]

Takata *et al.* observed that doping $SrTiO_3$ with a lower valence cation (such as Ti^{4+} with Ga^{3+} or Sr^{2+} with Na^+) dramatically increased the photocatalytic water splitting activity, whereas doping with a higher valence cation (such as Ti^{4+} with Ta^{5+} or Sr^{2+} with La^{3+}) resulted in an obvious decrease in the gas evolution rate *via* a water splitting reaction.[10] Therefore, they concluded that Ti^{3+} defects deactivated $SrTiO_3$ and that deliberately introducing oxygen vacancies assisted in suppressing Ti^{3+} formation, thereby increasing photocatalytic activity. An impregnation method, in which the $SrTiO_3$ powder was dispersed in an aqueous solution of the metal salts, was used to introduce various metal ions in the $SrTiO_3$ particle homogeneously.[13] The photocatalytic tests for water splitting reaction under UV light irradiation demonstrated that the H_2 and O_2 evolution rates were enhanced by adding Li^+, Na^+,

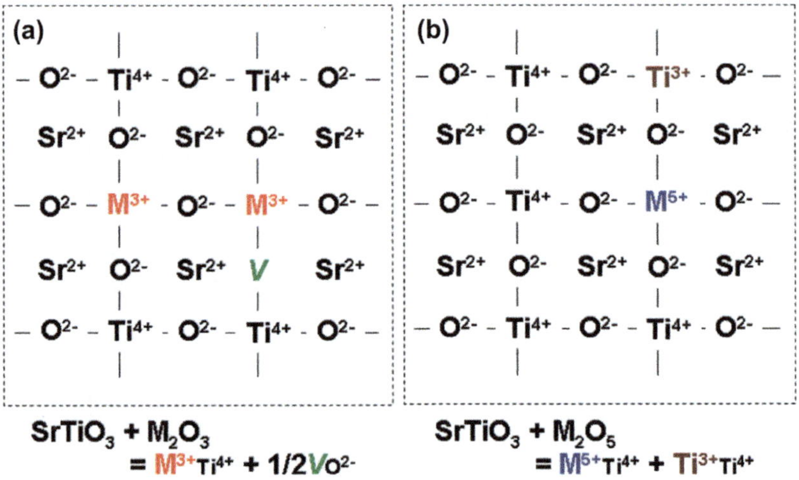

Figure 2.2 Schematic illustration of aliovalent-doped $SrTiO_3$ with a (a) trivalent cation and (b) pentavalent cation. Reproduced from ref. 10 with permission from American Chemical Society, Copyright 2009.

K^+, Rb^+, Cs^+, Mg^{2+}, Al^{3+}, Ga^{3+}, and In^{3+}. In particular, Na^+–$SrTiO_3$ (Na = 2 atm%) decomposed water with an apparent quantum efficiency (AQY) of 16% at 360 nm when $Rh_{0.3}Cr_{1.7}O_3$ was loaded as the co-catalyst.

In recent years, incorporating Al into $SrTiO_3$ ($SrTiO_3$:Al) has been exhibited to efficiently promote the photocatalytic water-splitting reaction. An optimized $Rh_{2-y}Cr_yO_3$-loaded $SrTiO_3$:Al (2[Al]/([Sr]+[Ti]) = 0.31%) photocatalyst achieved an AQY of 30% at 360 nm for overall water splitting.[14] X-Ray photoelectron spectra and density functional theory results confirmed the suppression of Ti^{3+} defect formation by introducing Al^{3+}.[11] The elimination of Ti^{3+} defect gave rise to decreased electron trapping, as observed in electrochemical measurements and surface photovoltage spectra. They also suggested that incorporation of Al^{3+} shifted the Fermi level away from the CB edge of approximately 0.5 eV, diminishing the n-type character of $SrTiO_3$. Therefore, the $SrTiO_3$:Al exhibited a remarkably increased photocatalytic water splitting activity compared to $SrTiO_3$. Apart from the water splitting reaction, the $SrTiO_3$:Al has been applied for photocatalytic CO_2 reduction reaction to produce CO using H_2O as an electron donor under UV light irradiation.[15] Following the loading of Ag nanoparticle as a co-catalyst for CO_2-to-CO conversion, the $SrTiO_3$:Al yielded CO and O_2 in a molar ratio close to the expected stoichiometric value of 2 : 1 under UV light irradiation while attaining 98.5% selectivity for CO formation during the reduction reactions. Very recently, a La and Al co-doped $SrTiO_3$ was reported to produce H_2 and O_2 at a higher rate than $SrTiO_3$:Al during the water splitting reaction, with an AQY of 78.43% at 365 nm.[16] This is due to the La^{3+} doping causing lattice distortion and allowing Sr^{2+} to re-enter the perovskite lattice, which resulted in increased conversion of Ti^{3+} to Ti^{4+} and a decrease in oxygen vacancies.

Normally, a molten salt flux treatment is used to prepare metal ions doped $SrTiO_3$. It has been demonstrated that by using the molten salt flux treatment in the preparation of the photocatalyst, the crystallinity, particle size and grain boundary structure of the resulting material can be tailored, as well as the formation of defects limited.[3,17–19] $SrTiO_3$ prepared *via* flux treatment with molten $SrCl_2$ salt exhibited a factor of approximately 500 increase in photocatalytic water splitting rate.[20] The authors used time-resolved absorption spectroscopy to confirm that the number of deeply trapped electrons was reduced as a result of the decreased defects, thereby leading to an AQY of 30% at 360 nm for H_2 production *via* water splitting.

2.2.2 Co-catalyst and Surface Modification

Co-catalysts not only facilitate charge separation and migration by forming junctions but also act as reaction sites and catalyze the reactions.[21–25] $SrTiO_3$ has been demonstrated to have a low catalytic activity for CO_2/proton reduction reactions.[15,26,27] When $SrTiO_3$ powder was illuminated in an aqueous $AgNO_3$ solution ($AgNO_3$ as electron acceptor), O_2 evolved efficiently, whereas H_2 evolution activity was found to be extremely low in aqueous methanol solution (methanol as electron donor).[27] Co-catalysts for proton or CO_2 reduction are thus required for $SrTiO_3$.

Domen *et al.* loaded NiO on $SrTiO_3$ particles using an impregnation method and obtained a Ni–NiO core–shell structure after reduction–reoxidation treatment (Figure 2.3(a)).[28] After calcination of the resulting $Ni(NO_3)_2$–$SrTiO_3$ prepared by impregnation in air, the predominant Ni species on the $SrTiO_3$ surface was NiO, which exhibited almost no activity for the photodecomposition of water. After treatment with H_2, NiO was reduced primarily to Ni metal, but the catalyst remained ineffective and oxygen did not evolve. When the reduced catalyst was reoxidized with O_2, a NiO/Ni co-catalyst with a core/shell structure formed, capable of decomposing water into H_2 and O_2 under light irradiation. However, when the reduced catalyst was reoxidized excessively at high temperatures, the Ni metal phase dissipated from the $SrTiO_3$ surface, and the remained NiO exhibited very low activity. It can be concluded that the structure and valence state of the Ni species loaded on $SrTiO_3$ surface had a significant effect on its photocatalytic activity in the overall water splitting reaction. NiO has been demonstrated to be a potent co-catalyst capable of significantly increasing the activity of a variety of photocatalysts, including $NaTaO_3$,[29,30] $K_4Nb_6O_{17}$,[31] and $Rb_4Nb_6O_{17}$.[32] One of its advantages is that it is inactive for the typical back reaction of water splitting, which is water formation from produced H_2 and O_2.

Superior electrocatalysts for H_2 evolution reaction from water, in general, also function as highly efficient co-catalysts, exemplified by Pt.[35] Pt is widely used as a co-catalyst for H_2 evolution on photocatalysts in the presence of electron donors.[3] However, due to its catalytic activity for back reactions (oxygen reduction reaction and water formation reaction), Pt is not suitable for use during overall water splitting on single particles. In 2006, Maeda *et al.* developed two-component co-catalysts consisting of Cr and another transition metal, including Ni, Ru, Rh, Ir, and Pt, for efficient photocatalytic water splitting using $(Ga_{1-x}Zn_x)(N_{1-x}O_x)$ photocatalyst.[36] In particular, the

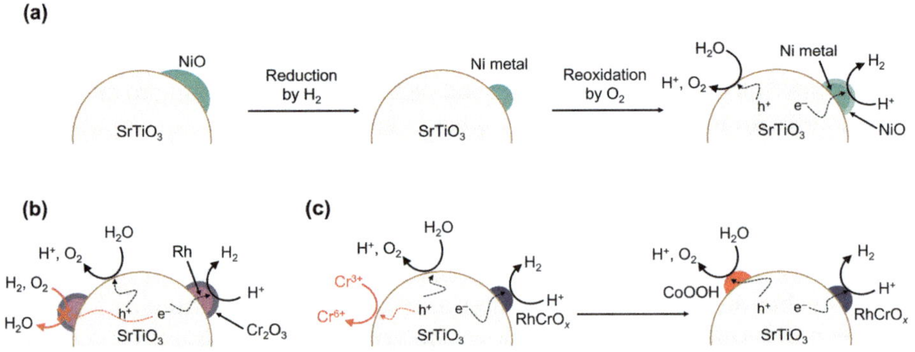

Figure 2.3 Schematic reaction mechanism of photocatalytic overall water splitting on $SrTiO_3$ loaded with (a) NiO (ref. 28), (b) core/shell Cr_2O_3/Rh (ref. 33), and (c) $RhCrO_x$ and CoOOH co-catalysts.[34]

photocatalytic overall water splitting activity was enhanced by two orders of magnitude when the Cr–Rh co-catalyst was used instead of the Rh co-catalyst alone. Co-loading Cr probably promotes the charge transfer from the $(Ga_{1-x}Zn_x)(N_{1-x}O_x)$ bulk to Rh and/or the creation of catalytic sites on the co-catalyst surface. The charge carrier dynamics of $Rh_{2-y}Cr_yO_3$-modified $(Ga_{1-x}Zn_x)(N_{1-x}O_x)$ have been elucidated using time-resolved photoluminescence and transient absorbance spectroscopies, indicating that the electrons efficiently transfer from $(Ga_{1-x}Zn_x)(N_{1-x}O_x)$ to $Rh_{2-y}Cr_yO_3$ on submicrosecond timescales, which is indispensable to decrease the electron concentration within $(Ga_{1-x}Zn_x)(N_{1-x}O_x)$ and facilitate hole accumulation.[37] More importantly, the H_2 and O_2 evolution rates from water using the $Rh_{2-y}Cr_yO_3$-loaded $(Ga_{1-x}Zn_x)(N_{1-x}O_x)$ barely changed with regard to background reaction pressure and the presence of O_2 gas, indicating that $Rh_{2-y}Cr_yO_3$ functions as a co-catalyst for H_2 production from proton while not facilitating the oxygen reduction reaction.[38]

Since then, the $Rh_{2-y}Cr_yO_3$ co-catalyst has been widely loaded onto the photocatalyst surface, including $Ca_2Nb_2O_7$,[39] GaN,[40] $(Zn_{1+x}Ge)(N_2O_x)$,[41] $LaMg_xTa_{1-x}O_{1+3x}N_{2-3x}$,[42] and $SrTiO_3$[14,43] to promote the water cleavage into H_2 and O_2. The addition of $Rh_{2-y}Cr_yO_3$ to the surface of $SrTiO_3$:Al remarkably enhanced water splitting activity, resulting in an AQY of 56% ± 3% in the wavelength range at 365 nm and an STH of 0.4%.[43] The $Rh_{2-y}Cr_yO_3$–$SrTiO_3$:Al particles were drop-cast onto glass plates to form photocatalyst sheets, which were then used to demonstrate photocatalytic water decomposition into a H_2 and O_2 mixture under natural sunlight at a scale of 1 × 1 m.[43] Even at such a large scale and ambient pressure, the intrinsic activity of the $Rh_{2-y}Cr_yO_3$–$SrTiO_3$:Al photocatalyst was retained without deterioration. The AQY was further improved by co-loading MoO_z with $Rh_{2-y}Cr_yO_3$ on $SrTiO_3$:Al, reaching 69% at 365 nm.[44] Mo species were reductively loaded on the $Rh_{2-y}Cr_yO_3$–$SrTiO_3$:Al photocatalyst *via* a photodeposition method, then oxidized to MoO_3 *via* calcination at 573 K, and finally reduced partially during the photocatalytic reaction. Interestingly, the MoO_z loading was found to change the chemical state of the $Rh_{2-y}Cr_yO_3$, rather than catalyze H_2 or O_2 evolution. The function of MoO_z remains a mystery and requires further investigation.

Apart from the incorporation of Cr to form a two-component co-catalyst for inhibiting oxygen reduction on noble metals without interfering with hydrogen evolution, the back reaction can be effectively suppressed *via* a surface modification strategy.[33,45–49] For instance, a Cr_2O_3 layer was developed to cover noble metals such as metallic Ru and Rh nanoparticles and form a core/shell structure *via* photodeposition for efficient photocatalytic water splitting to produce H_2 and O_2 while diminishing back reactions.[33,50] This Cr_2O_3 layer was shown to provide active sites for proton reduction, while allowing protons and hydrogen molecules to pass through but not oxygen molecules, thereby minimizing back reactions involving O_2 molecules on the noble metal (Figure 2.3(b)). Water cleavage into H_2 and O_2 was also accomplished using Cr_2O_3/Rh–$SrTiO_3$:Al.[51,52] The H_2 and O_2 production rates were increased further by coating the $SrTiO_3$:Al surface with a cobalt-doped

Mn_3O_4 nanoparticle that served as an O_2 evolution co-catalyst in addition to Cr_2O_3/Rh for H_2 production, attributed to the improved water oxidation kinetics on the surface reaction.

While the Cr^{3+}-containing transition metal oxide and noble-metal/Cr_2O_3 core/shell nanoparticles demonstrated a strong ability to facilitate photocatalytic overall water splitting, stability issues continue to plague the application of these co-catalysts because the Cr species remained relatively unstable over time due to oxidative dissolution.[53,54] The co-loading of oxygen evolution co-catalyst on the photocatalyst surface was found to prevent the dissolution of Cr species.[34,54] An example is the co-deposition of CoOOH and $Rh_{2-y}Cr_yO_3$ co-catalysts on $SrTiO_3$:Al to facilitate O_2 and H_2 evolution, respectively.[34] After 168 h of UV light illumination, the water splitting rate of $Rh_{2-y}Cr_yO_3$–$SrTiO_3$:Al decreased to 14% of its original value. A dramatic improvement in the stability was observed after depositing CoOOH on $Rh_{2-y}Cr_yO_3$–$SrTiO_3$:Al particles. Without the presence of CoOOH, photoexcited holes oxidize both water molecules and $Rh_{2-y}Cr_yO_3$ (Figure 2.3(c)) and cause the dissolution of Cr. Therefore, back reactions on the Cr-deficient $Rh_{2-y}Cr_yO_3$ co-catalyst may occur, resulting in a decreased water splitting rate. When CoOOH is loaded and functions as the O_2 evolution site, photoexcited holes are collected on it and participate in the water oxidation reaction rather than the oxidative dissolution of Cr, thereby preventing the photocatalyst from deactivating. It has been reported that amorphous TiO_2 forms a selectively permeable layer that completely covers the photocatalyst surface and suppresses the back reactions.[46] For this reason, the durability was further enhanced by the photodeposition of amorphous TiO_2 on CoOOH/$Rh_{2-y}Cr_yO_3$–$SrTiO_3$:Al, which retained 80% of its initial activity after 1300 h of constant simulated sunlight irradiation (1 sun, AM 1.5G) at ambient pressure in a laboratory scale. Impressively, the CoOOH/$Rh_{2-y}Cr_yO_3$–$SrTiO_3$:Al was used as a model material in a 100 m^2 scale prototype photocatalytic solar hydrogen production plant, which consisted of arrayed photocatalytic water splitting panels and an oxyhydrogen gas separation module (Figure 2.4).[55] Solar H_2 production *via* photocatalytic water decomposition and the H_2 separation from the moist gas product mixture was demonstrated to be safe over a period of several months, indicating that this technology has the potential to be a disruptive technology for renewable H_2 production. The solar hydrogen production plant can produce H_2 under natural sunlight with an STH of 0.76% in the summer when sunshine conditions are good.

In recent years, several noble-metal-free materials including nitrogen-doped graphene quantum dots[56] and Co_xP[57] were reported as the co-catalyst on $SrTiO_3$ for promoting the photocatalytic overall water splitting. Nitrogen-doped graphene quantum dots served as a co-catalyst for H_2 evolution, which not only increased the efficiency of photogenerated electron transfer *via* Ti–O–C chemical bonds, but also improved the light-harvesting of the photocatalyst. However, Co_xP dots worked as bifunctional co-catalysts to facilitate both H_2 and O_2 generation reactions.

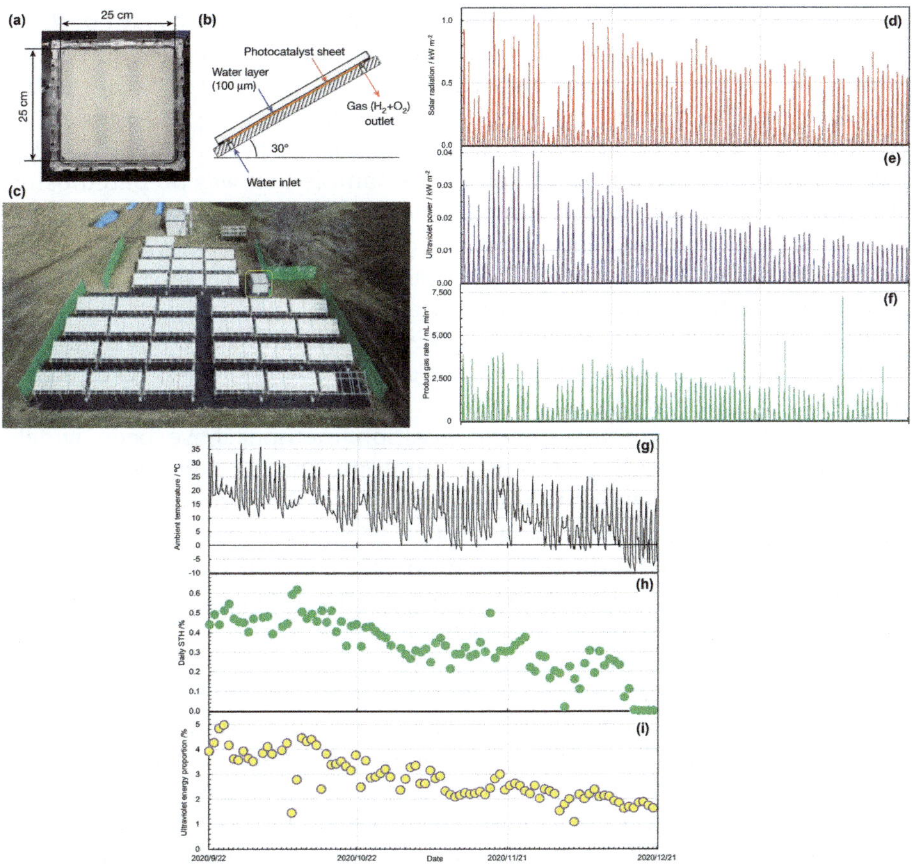

Figure 2.4 (a) A photographic image of a panel reactor unit of 625 cm^2 based on SrTiO3:Al loaded with CoOOH and RhCrO$_x$ co-catalysts. (b) Side-view of the panel reactor unit. (c) A photographic image of the 100 m^2 solar hydrogen production system consisting of 1600-panel reactor units and a gas separation facility indicated by the yellow box. (d)–(i) Long-term records for solar hydrogen production *via* photocatalytic overall water splitting of 100 m^2 system from 22 September to 21 December 2020: solar radiation (d), ultraviolet power (e), oxyhydrogen gas production rate (f), ambient temperature (g), daily STH value (h) and proportion of ultraviolet energy (i). Reproduced from ref. 55 with permission from Springer Nature, Copyright 2021.

2.2.3 Facet Control

The flux treatment is used to synthesize particles with a variety of exposed crystal facets because it involves dissolution and recrystallization mediated by the flux reagents.[58] $SrTiO_3$ crystals, for example, exposed the most stable {100} facets when prepared in molten LiCl, while exposing the second stable {110} facets as well as the {100} facets when prepared in NaCl or $SrCl_2$ salts because $SrTiO_3$ is more soluble in molten LiCl than in NaCl and $SrCl_2$. During the photocatalytic water splitting reaction, the NaCl or $SrCl_2$-treated $SrTiO_3$ showed higher activity than the LiCl-treated sample, indicating that exposing the {110} facets is beneficial for increasing photocatalytic activity. The best performance was exhibited by the sample that was prepared using KCl as the molten salt, which exposed the characteristic high-index facets, {120} and {121}, due to a capping effect of K^+ ions. Therefore, the authors concluded that synthesizing $SrTiO_3$ crystals with high-index facets is critical for achieving a high photocatalytic activity. Crystals with distinct facets are supposed to exhibit a change in activity because electrons and holes migrate to distinct crystal faces due to the difference in energy levels, resulting in effective electron and hole separation.[59]

Impregnation and *in situ* photodeposition methods have been widely used for co-catalyst loading onto photocatalysts. The impregnation method results in random dispersion of the co-catalysts on the photocatalyst surface and the co-catalysts require activation, such as reduction or oxidation, to gain high activity. *In situ* photodeposition, however, enables the co-catalysts to be loaded selectively at reaction sites without the need for activation. Since the reactions occurring at the surfaces or interfaces of the materials are incredibly susceptible to the exposed surface atomic structures and corresponding physicochemical properties, their crystal planes also dictate their photocatalytic performance. Various single crystals have been found to exhibit spatial charge separation between their exposed facets. For example, metallic Pt particles acting as H_2 evolution photocatalyst preferentially photodeposits on $BiVO_4$ {101} facets, whereas MnO_x particles acting as O_2 evolution photocatalyst preferentially photodeposits on {110} facets, implying that these facets provide reduction and oxidation sites, respectively.[60] This Pt/MnO_x/$BiVO_4$ with site-selective co-catalyst deposition performed better for oxygen production in the presence of $NaIO_3$ as the sacrificial electron acceptor during photocatalytic water oxidation reaction than the sample with random co-catalyst loading *via* impregnation. This observation, combined with the simulation results from Takanabe *et al.*,[61] indicated that the photocatalytic performance can be maximized only when reduction/oxidation co-catalysts are selectively deposited on their preferred reaction facets.

Anisotropic photoinduced charge transfer is strongly dependent on the built-in electric fields in the space charge regions of different facets, which has been confirmed by the spatially resolved surface photovoltage spectroscopy.[62] The separate distribution of reduction and oxidation catalytic sites on the anisotropic facets was observed after cubic $SrTiO_3$ was facet engineered to expose both {110} and {001} facets.[63] By selectively photodepositing metallic Pt at {001} facets to facilitate H_2 generation and Co_3O_4 at the {110} facets to facilitate O_2 production on the 18-facet $SrTiO_3$, the AQY for photocatalytic water splitting was increased more than five-fold over that obtained using six-facet $SrTiO_3$ with isotropic {001} facets. The underlying principle leading to this high-level performance of 18-facet $SrTiO_3$ is that the asymmetric assembly of co-catalysts enhances the symmetry of the surface built-in electric fields, thereby increasing the driving force for charge separation and photocatalytic performance.

The Cr_2O_3/Rh–$SrTiO_3$:Al with spatial separation of reduction and oxidation sites has further validated this theory.[52] A superior photocatalytic activity of Cr_2O_3/Rh–$SrTiO_3$:Al prepared by photodeposition with respect to $Rh_{2-y}Cr_yO_3$–$SrTiO_3$:Al prepared by impregnation has been reported. The photocatalyst loaded with Rh and subsequently with Cr_2O_3 *via* two-step photodeposition. The $SrTiO_3$:Al particles synthesized using flux treatment exposed various non-equivalent facets. Rh/Cr_2O_3 core/shell composites were preferentially deposited on the {100} crystal facets. Further additional photodeposition of CoOOH onto Cr_2O_3/Rh–$SrTiO_3$:Al, which promotes oxygen generation and selectively photodeposited on the {110} direction resulted in a dramatic improvement for cleaving water into H_2 and O_2 under UV light irradiation (Figure 2.5). The resulting photocatalyst achieved an external quantum efficiency of up to 96% in the wavelength range of 350–360 nm and internal quantum efficiency of almost unity.

Similarly, photocatalytic conversion of CO_2 by H_2O was investigated using an anisotropic facet-exposed $SrTiO_3$:Al photocatalyst modified with Ag-metal (Fe, Co, Ni, and Pt) dual co-catalysts.[64] Among the various combinations, Ag and FeOOH coloaded $SrTiO_3$:Al exhibited the highest selectivity and activity for CO_2-to-CO conversion in water without a sacrificial reagent under UV light irradiation. Ag and FeOOH were selectively photodeposited on the {100} and {110} facets and acted as active sites for CO_2RR and water oxidation, respectively, resulting in the simultaneous and stoichiometric generation of CO and O_2, with approximately 100% selectivity toward CO evolution in the reaction reactions.

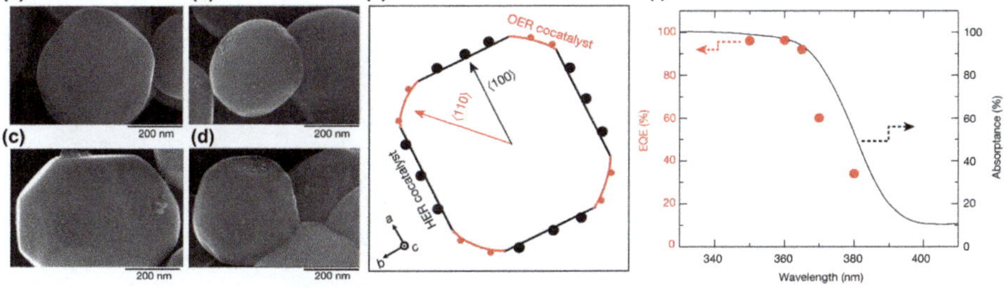

Figure 2.5 SEM images of (a) $SrTiO_3$:Al and $SrTiO_3$:Al loaded with (b) Rh, (c) Rh/Cr_2O_3 and (d) Rh/Cr_2O_3/CoOOH. (e) Selected-area electron diffraction pattern obtained from $SrTiO_3$:Al loaded with Rh/Cr_2O_3/CoOOH. (f) Ultraviolet-visible diffuse reflectance spectrum of bare $SrTiO_3$:Al (black line) and wavelength dependence of external quantum efficiency (EQE) during water splitting on $SrTiO_3$:Al loaded with Rh/Cr_2O_3/CoOOH (red dots). Reproduced from ref. 52 with permission from Springer Nature, Copyright 2020.

2.2.4 Spectral Absorption Expansion

Due to the large bandgap (3.25 eV), $SrTiO_3$ can absorb only UV light, which accounts for only 5% of the sunlight radiation reaching the Earth's surface, yet it efficiently cleaves water into H_2 and O_2 after appropriate modifications. As aforementioned, an STH of 5–10% is indispensable to realize economically viable solar hydrogen production. A theoretical 5% STH can only be obtained with single absorbers of a bandgap smaller than 2.7 eV.[65] Doping by introducing foreign elements to host materials is a feasible and effective approach to expand spectral absorption. A bandgap can be narrowed by the introduction of metal cations with a similar radius to the host ion due to the interaction of cation states with VB or CB or by the formation of intra-bandgap levels.[3] When this occurs, donor dopants donate electrons to the CB to generate a hole, whereas acceptor dopants accept electrons from the VB to create a hole.

An example is the study carried out by Konta *et al.* in which various metal ions, such as Ir, Ru, Rh and Mn have been introduced into $SrTiO_3$ to produce an optical response shift in an attempt to promote H_2 or O_2 production in the presence of sacrificial reagents under visible light.[66] The evident shift occurs as a result of excitation from discontinuous levels created by the dopants to the CB of the $SrTiO_3$ host. As compared with other doped specimens, Rh-doped $SrTiO_3$ ($SrTiO_3$:Rh) showed the highest hydrogen evolution rate using methanol as the electron donor. The authors used diffuse reflectance spectroscopy to confirm the presence of at least two distinct species of doped Rh in the $SrTiO_3$:Rh, designated as Rh^{3+} and Rh^{4+} (Figure 2.6).[67,68] The absorption band in the visible light region up to 520 nm is attributed to the transitions from an electron donor level created by Rh^{3+} ions to the CB of $SrTiO_3$, which contributed to photocatalytic H_2 production. By contrast, the absorption at 580 nm is due to an electronic transition from the VB of $SrTiO_3$

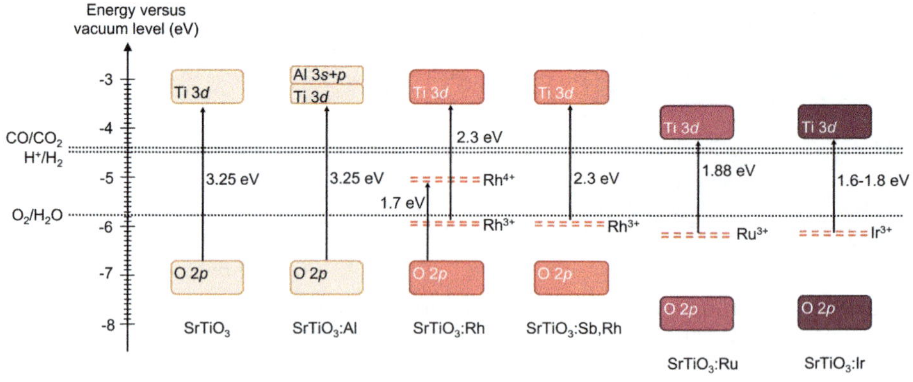

Figure 2.6 Bandgaps and band positions of metal ion doped $SrTiO_3$. Data were respectively obtained from ref. 11 and 72 ($SrTiO_3$), ref. 11 ($SrTiO_3$:Al), ref. 66 and 67 ($SrTiO_3$:Rh), ref. 73 ($SrTiO_3$:Sb,Rh), ref. 74 ($SrTiO_3$:Ru) and ref. 75 ($SrTiO_3$:Ir).

to the acceptor level formed by the Rh^{4+}, which is inactive for H_2 production when exposed to visible light but provides recombination sites for photoexcited electrons and holes, as confirmed by time-resolved infrared absorption study.[68,69] Co-doping with Sb^{5+} at Ti^{4+} sites was shown to ensure that Rh remained in the trivalent state as predicted by the cation charge compensation scheme (Figure 2.6), which was critical for guaranteeing the O_2 evolution *via* water oxidation.[69–71] The resulting $SrTiO_3$:Sb,Rh split water into H_2 and O_2 with an AQY of 0.1% at 420 nm when IrO_2 was loaded as co-catalyst to promote both H_2 and O_2 production.[70]

It has been reported that a co-catalyst-free B-doped $SrTiO_3$ absorbed visible light and reduced CO_2 to form CO and CH_4, while also producing O_2 from water.[76] The B-doping resulted in a dramatically improved photocatalytic activity for CO_2 reduction, up to three times that obtained using bare $SrTiO_3$, in conjunction with AQYs of 1.34% at 365 nm and 0.53% at 420 nm. The authors attributed the improved performance to promoted charge separation and the enhanced CO_2 surface adsorption.

Despite that numerous dopants have been examined for effectively extending the optical absorption of $SrTiO_3$ (Figure 2.6), including Cr (bandgap: 1.9 eV),[77] Cr and Sb (bandgap: 2.4 eV),[73] Cr and B (bandgap: 2.07 eV),[78] Ir (bandgap: ~1.6 eV)[79] and Ru (bandgap: 1.88 eV),[74] the examples of splitting water into H_2 and O_2 using a single light absorber have remained scarce due to the stringent thermodynamic requirements.

2.3 Z-scheme Artificial Photosynthetic Systems

As aforementioned, doping an oxide with foreign elements to form an energy level in the forbidden band is one of the most well-recognized approaches to narrow its bandgap and make it visible-light-responsive. However, the dopants usually generate a discreet energy level rather than a continuous energy band and so the dopants will hinder the migration of photogenerated charge carriers.[87] Moreover, the introduced dopants may disturb the charge balance and disorder the crystal structure. These effects usually generate defects that act as charge-recombination centres. Since visible-light-responsive $SrTiO_3$ developed by doping usually suffer from poor charge separation, they cannot be used for one-step photoexcitation water splitting except in one rare case.[70] Yet, most of them are active for H_2 evolution in the presence of an electron donor and so they can be possibly applied to two-step photoexcitation (Z-scheme) water splitting as tabulated in Table 2.2.[73,88,89]

2.3.1 Redox Mediator

In contrast to a one-step photoexcitation system where proton/CO_2 reduction and water oxidation occur on a single type of photocatalyst, a Z-scheme system separates the proton/CO_2 reduction and water oxidation on two different types of photocatalyst, and the remaining charge carriers on each type

Table 2.2 Visible-light-responsive $SrTiO_3$-based Z-scheme systems for photocatalytic water splitting or CO_2 reduction.

Photocatalyst			Quantum efficiency at	
H_2 evolution	O_2 evolution	Mediator	420 nm/%	Reference
$Pt/SrTiO_3$:Cr,Ta	PtO_x/WO_3	IO_3^-/I^-	1	94, 117 and 118
$Pt/SrTiO_3$:Cr,Ta	PtO_x/H-Cs-WO_3	IO_3^-, I_3^-/I^-	1.5	119
$Pt/SrTiO_3$:Cr,Ta	Pt/WO_3	IO_3^-/I^-	1.52	120
$Pt/SrTiO_3$:Rh	WO_3	Fe^{3+}/Fe^{2+}	0.5	89
$Ru/SrTiO_3$:Rh	PtO_x/WO_3	$[SiW_{11}O_{39}Mn^{III}(H_2O)]^{5-}/[SiW_{11}O_{39}Mn^{II}(H_2O)]^{6-}$		121
$Ru/SrTiO_3$:Rh	Fe-H-Cs-WO_3	VO_2^+/VO^{2+}		122
$Pt/SrTiO_3$:Rh	$Ru/SrTiO_3$:In,V	IO_3^-/I^-	0.056	123
$Ru/SrTiO_3$:Rh	$IrO_x/SrTiO_3$:Rh,Sb	Fe^{3+}/Fe^{2+}		71
$Ru/SrTiO_3$:Rh	$Pt/SrTiO_3$:Ir	Fe^{3+}/Fe^{2+} or $[Co(byp)_3]^{3+/2+}$		75
$Pt/SrTiO_3$:Rh	$BiVO_4$	Fe^{3+}/Fe^{2+}	0.4	89
$Ru/SrTiO_3$:Rh	$BiVO_4$	Fe^{3+}/Fe^{2+}	0.3	97 and 124
$Ru/SrTiO_3$:Rh	$BiVO_4$	$[Co(byp)_3]^{3+/2+}$ or $[Co(phen)_3]^{3+/2+}$	2.1	99 and 124
$Ru/SrTiO_3$:Rh	TiO_2:Rh,Sb	Fe^{3+}/Fe^{2+} or $[Co(byp)_3]^{3+/2+}$		75
$Ru/SrTiO_3$:Rh	RuO_2/TiO_2:Ta,N	Fe^{3+}/Fe^{2+}	0.5	125
$Ru/SrTiO_3$:Rh	$NaNbO_3$:Rh,Ba	Fe^{3+}/Fe^{2+} or $[Co(byp)_3]^{3+/2+}$		75
$Ru/SrTiO_3$:Rh	$La_{0\cdot5}Sr_{0\cdot5}Ta_{0\cdot5}Ti_{0\cdot5}O_2N$	$[Co(byp)_3]^{3+/2+}$	< 0.1	100
$Ru/SrTiO_3$:Rh	$BiVO_4$	None	1.7	68 and 124
$Ru/SrTiO_3$:Rh	$Ir/CoO_x/Ta_3N_5$	None	< 0.3	102
$Ru/SrTiO_3$:La,Rh	$Ir/CoO_x/Ta_3N_5$	None	1.1	103
$Ru/SrTiO_3$:La,Rh	$NaNbO_3$:Rh,Ba	None		75
$Ru/SrTiO_3$:Rh	TiO_2:Rh/Sb	None		75
$Ru/SrTiO_3$:Rh	$BiVO_4$	Composite	1.6	105
$Pt/SrTiO_3$:Rh	$BiVO_4$	Composite	~8	126
$Ru/SrTiO_3$:Rh	$BiVO_4$	RGO	1.03	107
Ru complex/$SrTiO_3$:Rh	$BiVO_4$	RGO		127
$Ru/SrTiO_3$:Rh	$CoO_x/BiVO_4$:Mo	RGO		128
$Pt/SrTiO_3$:La,Rh	$BiVO_4$	RGO	~1.4	129
$Ru/SrTiO_3$:La,Rh	$BiVO_4$:Mo	Au nanoparticle	~3	108
$Ru/SrTiO_3$:Rh	$BiVO_4$:Mo	Au nanoparticle		130
$Cr_2O_3/Ru/SrTiO_3$:La,Rh	$BiVO_4$:Mo	ITO nanoparticle	10.2 at 331 K and 91 kPa	53
$Ru/SrTiO_3$:La,Rh	$BiVO_4$	Au layer	5.9	112

Table 2.2 (*continued*)

Photocatalyst			Quantum efficiency at 420 nm/%	Reference
H_2 evolution	O_2 evolution	Mediator		
Cr_2O_3/Ru/ $SrTiO_3$:La,Rh	$BiVO_4$:Mo	Au layer	33 at 331 K and 10 kPa	108
Cr_2O_3/Ru/ $SrTiO_3$:La,Rh	$BiVO_4$:Mo	Carbon layer	26 at 331 K and 10 kPa	115
Cr_2O_3/Ru/ $SrTiO_3$:Rh	$CoO_x/BiVO_4$:Mo	In/InO_x layer		131

Photocatalyst			Quantum efficiency at 420 nm/%	Reference
CO_2 reduction	O_2 evolution	Mediator		
Au/$SrTiO_3$:Rh	$BiVO_4$	None		132
CotpyP/ $SrTiO_3$:La,Rh	$RuO_2/BiVO_4$	Au layer	2.6	116

of photocatalyst are recombined with the aid of a mediator (Figure 2.1(b)).[90] Typically, the mediator is a redox couple like IO_3^-/I^- in an aqueous solution.[91,92] Though the operation mechanism of a Z-scheme system for overall water splitting is similar to a tandem photoelectrochemical cell, the former does not require a p-type and a n-type photocatalyst for H_2 and O_2 evolution, respectively.[93] As long as the photocatalyst is active for H_2 or O_2 in a redox couple, it can be applied to a Z-scheme overall water system with the corresponding redox couple.

In 2001, the first visible-light-responsive Z-scheme system which consists of Cr and Ta co-doped $SrTiO_3$ for H_2 evolution, WO_3 for O_2 evolution, and IO_3^-/I^- as a mediator was developed for water splitting with an AQY of ~0.1% at 420 nm (Figure 2.7(a)).[94] Inspired by such progress, the Kudo group carefully investigated metal ion-doped $SrTiO_3$ with intense absorption in visible light regions.[66,95] Among them, $SrTiO_3$:Rh was the most active for photocatalytic H_2 evolution in methanol aqueous solution, giving an AQY of 5.2% at 420 nm. It was revealed that the energy levels generated by Rh dopant were basically discrete, but they partly overlapped with the VB composed of O 2p orbital.[66,96] For this reason, the mobility of photogenerated charge carriers was maintained to some degree.

The $SrTiO_3$:Rh was further applied to a Z-scheme system with $BiVO_4$ for O_2 evolution and Fe^{3+}/Fe^{2+} as a mediator, giving an AQY of 0.3% at 420 nm and an STH of 0.02%.[89,97] While a solid-state-reaction method is usually applied to synthesize $SrTiO_3$:Rh, soft chemical synthesis like a hydrothermal method is advantageous for distributing Rh dopant homogenously. With $SrTiO_3$:Rh

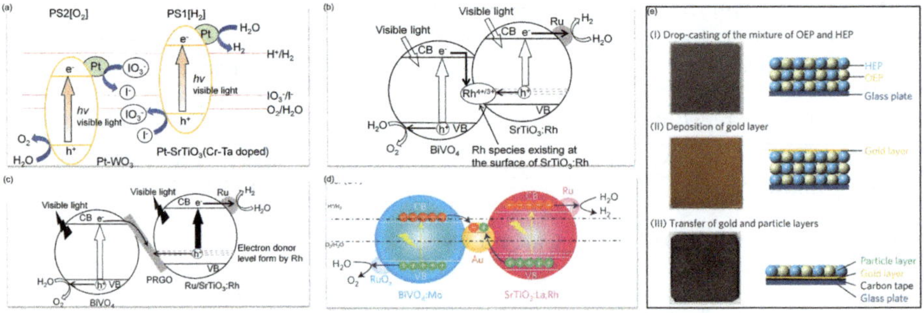

Figure 2.7 (a) A redox-containing Z-scheme system consisting of Cr and Ta co-doped $SrTiO_3$ for H_2 evolution, WO_3 for O_2 evolution and IO_3^-/I^- as a mediator. Adapted from ref. 94 with permission from the Royal Society of Chemistry. (b) A redox-free Z-scheme system consisting of Rh-doped $SrTiO_3$ for H_2 evolution, $BiVO_4$ for O_2 evolution. Reproduced from ref. 68 with permission from American Chemical Society, Copyright 2009. (c) An all-solid-state Z-scheme system consisting of Rh-doped $SrTiO_3$ for H_2 evolution, $BiVO_4$ for O_2 evolution and reduced graphene oxide (PRGO) as a solid mediator. Reproduced from ref. 107 with permission from American Chemical Society, Copyright 2011. (d) A photocatalyst sheet consisting of La- and Rh-co-doped $SrTiO_3$ for H_2 evolution, Mo-doped $BiVO_4$ for O_2 evolution and a gold layer as a solid mediator. (e) Preparation of the photocatalyst sheet in (d) by particle transfer. Adapted from ref. 108 with permission from Springer Nature, Copyright 2016.

prepared by a hydrothermal method, the AQY and STH were improved to be 3.9–4.2% at 420 nm and 0.1%, respectively.[98] Other redox couples like IO_3^-/I^- and $Co(bpy)^{3+/2+}$ were also studied on $SrTiO_3$:Rh, confirming that $SrTiO_3$:Rh can be combined with other O_2 evolution photocatalysts exemplified by WO_3 for Z-scheme water splitting.[99,100]

2.3.2 Redox-free Z-scheme System

While much progress has been made on redox-containing Z-scheme systems, their performance is inevitably limited by the redox-involving backward reactions, *e.g.*, the oxidation of reduced redox ions by the O_2 evolution photocatalyst. To avoid such backward reactions, $SrTiO_3$:Rh and $BiVO_4$ were used to construct a redox-free Z-scheme system where interparticle charge transfer was achieved by simple physical contact (Figure 2.7(b)).[68,101] The critical points for constructing this Z-scheme system were: (1) adjusting the pH of reactant solution to aggregate the particles; and (2) providing the Rh dopant with reversible valence states to enable interparticle charge transfer. Following these studies, another redox-free Z-scheme system based on $SrTiO_3$:Rh and Ta_3N_5 was also developed.[102]

Considering that the Rh dopant with reversible valence states is critical for interparticle charge transfer and such charge transfer proceeds with surface contact, core/shell structured $SrTiO_3$:Rh was prepared by a heterogeneous doping process based on thermal diffusion of Rh_2O_3.[103] Different from the typical homogenous doping process, this heterogenous process generated Rh dopant-rich surfaces that allow photogenerated charge carriers to migrate to the reaction sites efficiently. Furthermore, La was co-doped with Rh to $SrTiO_3$ in an attempt to maintain the charge balance. Transient absorption spectroscopy further revealed that La co-doping induced a Rh oxidation state-dependent reorganization of the electronic structure, involving the elimination of vacant Rh 4d mid-gap states and fixation of Rh at 3+ state.[72] As a result, the lifetime of photogenerated electrons was significantly enlarged.[72,104] With the positive effects of heterogeneous doping and La co-doping, the redox-free Z-scheme based on La and Rh codoped $SrTiO_3$ ($SrTiO_3$:La,Rh) and Ta_3N_5 showed 3.8 times higher water splitting activity than that used $SrTiO_3$:Rh.[103]

Although these non-redox Z-scheme systems can be easily constructed by aggregating the photocatalysts with an acid solution, the contact between photocatalysts was still not sufficient for interparticle charge transfer and so their performance fell behind that of the redox-containing Z-scheme systems. Intuitively, the interparticle charge transfer can be improved if photocatalysts were combined as composites by growing one type of photocatalyst on another one or annealing their mixture.[105,106] However, these methods cannot preserve the photocatalytic activity of each type of photocatalysts effectively. Therefore, the Z-scheme system only showed moderate performance.

2.3.3 All-solid-state Z-scheme System

Besides being active for photocatalytic H_2 evolution, $SrTiO_3$:Rh is confirmed to be a p-type semiconductor. For this unique property as a visible-light-responsive oxide photocatalyst, it was used as a photocathode for H_2 evolution with a conductive substrate.[109,110] Furthermore, it was used to construct a novel all-solid-state Z-scheme system using $BiVO_4$ as the O_2 evolution photocatalyst and reduced graphene oxide (RGO) as a solid mediator (Figure 2.7(c)).[107] A similar system using indium tin oxide as an alternative solid mediator was also developed. Since the indium tin oxide is transparent and electrochemically inert, its mixture with photocatalysts was printed on a substrate as a proof-of-concept model for cost-effective and large-scale water splitting.[53,111] This system showed an STH of 0.4%, one of the highest values among Z-scheme overall water splitting systems. Generally, this kind of Z-scheme system with a solid mediator can be regarded as micro-sized tandem p/n photoelectrodes. To this end, the H_2- and O_2-evolution photocatalysts need to be p- and n-type semiconductors, respectively. Moreover, their cathodic and anodic photocurrents need to overlap a certain potential for non-biased water splitting.[75]

Similar to the redox-free Z-scheme systems, the all-solid-state Z-scheme systems are still limited by the interparticle charge transfer since photocatalysts and the solid mediator only have a weak physical contact by adsorption or compression. To improve the contact between photocatalysts and a solid mediator, a photocatalyst sheet based on a particle-transfer method was developed (Figure 2.7(e)).[108,112,113] During the preparation of a photocatalyst sheet, two types of photocatalysts were firstly mixed in suspension and drop cast on a glass substrate. Then, a conductive layer (for example, Au) was loaded on the mixed photocatalyst by physical vapour deposition. At last, the conductive layer with photocatalysts immobilized on it was lift-off to expose the photocatalysts. Benefiting from the physical vapour deposition, an intimate mechanical and electrical contact between photocatalysts and the conductive layer was formed, which greatly facilitated charge transfer in the Z-scheme system. Furthermore, since the photocatalysts were located with spacing on the order of micrometres, the solution resistance and the concentration overpotentials were significantly minimized.[114] Taking advantage of such merits, the optimized photocatalyst sheet composed of $SrTiO_3$:La,Rh for H_2 and Mo doped $BiVO_4$ ($BiVO_4$:Mo) for O_2 evolution on a gold or carbon layer was allowed to splitting water with an AQY over 30% at 420 nm and an STH over 1%, representing the state-of-the-art Z-scheme systems until now (Figure 2.7(d)).[108,115] This photocatalyst sheet was also combined with a molecular catalyst (phosphonated Co(II) bis(terpyridine)) for photoconversion of CO_2 into formate coupled with the water oxidation, providing a solar energy conversion efficiency of 0.08% with selectivity for formate of 97%.[116]

Doped $SrTiO_3$ is not only applicable to a Z-scheme system for H_2 evolution, but also for O_2 evolution. Particularly, $SrTiO_3$:Sb,Rh and $SrTiO_3$:Ir have been carefully investigated for O_2 evolution in the Z-scheme system containing redox mediators, combined with $SrTiO_3$:Rh for H_2 evolution.[71,75] However, the O_2-evolution activities of these doped $SrTiO_3$ were still low mostly because the photogenerated holes had low mobilities on the discreet energy levels generated by dopants.

2.4 Summary and Perspective

Considerable efforts have been devoted to enhance the photocatalytic activity of $SrTiO_3$-based materials for solar hydrogen production *via* water splitting and understanding the underlying principle leading to this high level of performance. $SrTiO_3$ material has been modified by: (1) Al doping to reduce the defects that acted as the recombination sites for photogenerated electron–hole pairs; (2) successful co-catalyst and surface modification to provide effective active sites and diminish the back reactions; and (3) facet engineering to spatially separate reduction and oxidation sites, allowing exploitation of the built-in electric fields of nanoscale and microscale junctions to facilitate the charge separation, thereby attaining a quantum efficiency of nearly 100% at 350–360 nm and an STH of 0.76% on a 100 m^2 scale and setting the benchmarks in this field.

Considering the effective $SrTiO_3$ photocatalyst evolutionary paths (Figure 2.8), the conclusion is that fundamental improvements are only possible if the photons are capable of harvesting light without jeopardizing the

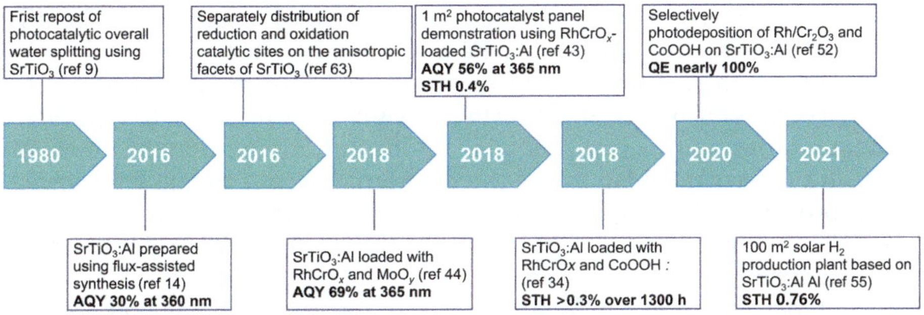

Figure 2.8 Timeline showing the key developments in $SrTiO_3$-based photocatalyst for overall water splitting.

probability of charge separation. As such, certain material strategies, such as flux-assisted synthesis and defect engineering for crystallinity improvement and defect reduction, are required to produce long-lived charge separation states. Additionally, co-catalysts are typically required for the majority of photocatalysts to drive the redox reactions efficiently because the reactions on semiconductors are otherwise slow, yet only a few co-catalysts have been found to be active for water splitting reaction, including RuO_x, NiO_x, and $RhCrO_x$ because the catalyzation of back reactions is largely avoided. The surface modification strategy, which involves covering the noble metallic co-catalyst with an amorphous (oxy)-hydroxide layer, successfully inhibits back reactions on the co-catalyst without interfering with hydrogen evolution. This strategy, in conjunction with spatial separation of co-catalysts *via* facet engineering, could enable the development of more co-catalyst materials for overall water splitting.

$SrTiO_3$:Al has been most successfully demonstrated for photocatalytic overall water splitting with a record 100% quantum efficiency measured under UV light. Its STH is limited by the large intrinsic bandgap and still far below the requirements for worthwhile efficiency gains because only 5% energy distribution of UV in the solar spectrum can be harvested. Improvement in sunlight absorption efficiency is primarily needed to enable photocatalytic materials to be used in real-world applications. The bandgap can be modulated by doping, however, precise control is required to avoid deep trapping and recombination sites. Attempts to incorporate these doped $SrTiO_3$ materials that exhibit broad spectral absorption into Z-scheme systems have been particularly successful. Apart from the quality of the photocatalysts themselves (fewer defects and improved light absorption), the interparticle electron transfer is also critical for high performance in Z-scheme systems. Kinetically, forward electron transfers must be competitive or dominant.

In comparison to photocatalytic water splitting to generate hydrogen, reports of selective and efficient photocatalytic CO_2 reduction coupled with water oxidation remain scarce. The strategies for increasing overall water splitting activity, as illustrated in this review, would be also applicable to CO_2 reduction. Although the photocatalytic technology for solar fuels production has demonstrated the possibility for practical application, consideration of engineering design requirements to optimize implementation of this technology is important. A requirement for clean H_2O and CO_2 streams also raises the economic barrier to entry for artificial photosynthesis. Engineering designs can be implemented strategically by considering the design requirements and available power from the solar spectrum.[133]

Recent developments include increased interest in achieving overall water splitting using materials other than oxides. Such materials include (oxy)nitrides,[134] (oxy)sulfides,[135–137] oxyhalides,[138,139] and conjugated polymers.[140] All of these materials have narrow bandgaps and appear to be the standout candidates in the revolutionary category. We hope this work has provided insight into the rational design of these materials for highly efficient photocatalytic activity for solar fuel production from CO_2 and H_2O in the future.

Acknowledgements

This work was supported by The JSPS Leading Initiative for Excellent Young Researchers program (to Q.W.), the JST Fusion Oriented REsearch for disruptive Science and Technology program (to Q.W.), the JSPS Grant-in-Aid for Young Scientists (Start-up) (no. 21K20485; to Q.W.; no. 20K22556; to Z.P.).

References

1. S. Zhu and D. Wang, *Adv. Energy Mater.*, 2017, **7**, 1700841.
2. O. J. Guerra, J. Eichman, J. Kurtz and B.-M. J. J. Hodge, *Joule*, 2019, **3**, 2425.
3. Q. Wang and K. Domen, *Chem. Rev.*, 2020, **120**, 919.
4. T. Setoyama, T. Takewaki, K. Domen and T. Tatsumi, *Faraday Discuss.*, 2017, **198**, 509.
5. T. Hisatomi and K. Domen, *Nat. Catal.*, 2019, **2**, 387.
6. B. A. Pinaud, J. D. Benck, L. C. Seitz, A. J. Forman, Z. Chen, T. G. Deutsch, B. D. James, K. N. Baum, G. N. Baum and S. Ardo, *Energy Environ. Sci.*, 2013, **6**, 1983.
7. M. Johnsson and P. Lemmens, in *Handbook of Magnetism and Advanced Magnetic Materials*, ed. H. Kronmüller and S. Parkin, John Wiley & Sons Ltd, England, 1st edn, 2007, vol. 4, ch. 3, p. 1.
8. K. v. Benthem, C. Elsässer and R. H. French, *J. Appl. Phys.*, 2001, **90**, 6156.
9. K. Domen, S. Naito, M. Soma, T. Onishi and K. Tamaru, *J. Chem. Soc., Chem. Commun.*, 1980, 543.

10. T. Takata and K. Domen, *J. Phys. Chem. C*, 2009, **113**, 19386.
11. Z. Zhao, R. V. Goncalves, S. K. Barman, E. J. Willard, E. Byle, R. Perry, Z. Wu, M. N. Huda, A. J. Moulé and F. E. Osterloh, *Energy Environ. Sci.*, 2019, **12**, 1385.
12. K. Han, Y.-C. Lin, C.-M. Yang, R. Jong, G. Mul and B. Mei, *ChemSusChem*, 2017, **10**, 4510.
13. Y. Sakata, Y. Miyoshi, T. Maeda, K. Ishikiriyama, Y. Yamazaki, H. Imamura, Y. Ham, T. Hisatomi, J. Kubota, A. Yamakata and K. Domen, *Appl. Catal., A*, 2016, **521**, 227.
14. Y. Ham, T. Hisatomi, Y. Goto, Y. Moriya, Y. Sakata, A. Yamakata, J. Kubota and K. Domen, *J. Mater. Chem. A*, 2016, **4**, 3027.
15. S. Wang, K. Teramura, T. Hisatomi, K. Domen, H. Asakura, S. Hosokawa and T. Tanaka, *ChemistrySelect*, 2020, **5**, 8779.
16. Y. Qin, F. Fang, Z. Xie, H. Lin, K. Zhang, X. Yu and K. Chang, *ACS Catal.*, 2021, **11**, 11429.
17. D. G. Porob and P. A. Maggard, *J. Solid State Chem.*, 2006, **179**, 1727.
18. M. Hojamberdiev and K. Kawashima, *Energy Rep.*, 2020, **6**, 13.
19. E. K. Akdogan, R. E. Brennan, M. Allahverdi and A. Safari, *J. Electroceram.*, 2006, **16**, 159.
20. A. Yamakata, H. Yeilin, M. Kawaguchi, T. Hisatomi, J. Kubota, Y. Sakata and K. Domen, *J. Photochem. Photobiol., A*, 2015, **313**, 168.
21. E. Pastor, F. Le Formal, M. T. Mayer, S. D. Tilley, L. Francàs, C. A. Mesa, M. Grätzel and J. R. Durrant, *Nat. Commun.*, 2017, **8**, 14280.
22. J. Zhu, S. Pang, T. Dittrich, Y. Gao, W. Nie, J. Cui, R. Chen, H. An, F. Fan and C. Li, *Nano Lett.*, 2017, **17**, 6735.
23. B. H. Meekins and P. V. Kamat, *J. Phys. Chem. Lett.*, 2011, **2**, 2304.
24. Z. Pan, J. Rohr, Z. Ye, Z. Fishman, Q. Zhu, X. Shen and S. Hu, *Sustainable Energy Fuels*, 2019, **3**, 850.
25. Z. Pan, R. Yanagi, Q. Wang, X. Shen, Q. Zhu, Y. Xue, J. A. Röhr, T. Hisatomi, K. Domen and S. Hu, *Energy Environ. Sci.*, 2020, **13**, 162.
26. T. Kanazawa, S. Nozawa, D. Lu and K. Maeda, *Catalysts*, 2019, **9**, 59.
27. K. Domen, A. Kudo and T. Onishi, *J. Catal.*, 1986, **102**, 92.
28. K. Domen, A. Kudo, T. Onishi, N. Kosugi and H. Kuroda, *J. Phys. Chem.*, 1986, **90**, 292.
29. H. Kato, K. Asakura and A. Kudo, *J. Am. Chem. Soc.*, 2003, **125**, 3082.
30. Q. Zhang, Z. Li, S. Wang, R. Li, X. Zhang, Z. Liang, H. Han, S. Liao and C. Li, *ACS Catal.*, 2016, **6**, 2182.
31. A. Kudo, A. Tanaka, K. Domen, K.-i. Maruya, K.-i. Aika and T. Onishi, *J. Catal.*, 1988, **111**, 67.
32. K. Sayama, A. Tanaka, K. Domen, K. Maruya and T. Onishi, *J. Catal.*, 1990, **124**, 541.
33. K. Maeda, K. Teramura, D. Lu, N. Saito, Y. Inoue and K. Domen, *Angew. Chem., Int. Ed.*, 2006, **45**, 7806.
34. H. Lyu, T. Hisatomi, Y. Goto, M. Yoshida, T. Higashi, M. Katayama, T. Takata, T. Minegishi, H. Nishiyama, T. Yamada, Y. Sakata, K. Asakura and K. Domen, *Chem. Sci.*, 2019, **10**, 3196.

35. K. Takanabe, *ACS Catal.*, 2017, **7**, 8006.
36. K. Maeda, K. Teramura, N. Saito, Y. Inoue and K. Domen, *J. Catal.*, 2006, **243**, 303.
37. R. Godin, T. Hisatomi, K. Domen and J. R. Durrant, *Chem. Sci.*, 2018, **9**, 7546.
38. K. Maeda, K. Teramura, H. Masuda, T. Takata, N. Saito, Y. Inoue and K. Domen, *J. Phys. Chem. B*, 2006, **110**, 13107.
39. A. Nakamura, O. Tomita, M. Higashi, S. Hosokawa, T. Tanaka and R. Abe, *Chem. Lett.*, 2015, **44**, 1001.
40. M. Kazuhiko, T. Kentaro, S. Nobuo, I. Yasunobu and D. Kazunari, *Bull. Chem. Soc. Jpn.*, 2007, **80**, 1004.
41. X. Wang, K. Maeda, Y. Lee and K. Domen, *Chem. Phys. Lett.*, 2008, **457**, 134.
42. C. Pan, T. Takata, M. Nakabayashi, T. Matsumoto, N. Shibata, Y. Ikuhara and K. Domen, *Angew. Chem.*, 2015, **127**, 2998.
43. Y. Goto, T. Hisatomi, Q. Wang, T. Higashi, K. Ishikiriyama, T. Maeda, Y. Sakata, S. Okunaka, H. Tokudome, M. Katayama, S. Akiyama, H. Nishiyama, Y. Inoue, T. Takewaki, T. Setoyama, T. Minegishi, T. Takata, T. Yamada and K. Domen, *Joule*, 2018, **2**, 509.
44. T. H. Chiang, H. Lyu, T. Hisatomi, Y. Goto, T. Takata, M. Katayama, T. Minegishi and K. Domen, *ACS Catal.*, 2018, **8**, 2782.
45. M. Yoshida, K. Maeda, D. Lu, J. Kubota and K. Domen, *J. Phys. Chem. C*, 2013, **117**, 14000.
46. T. Takata, C. Pan, M. Nakabayashi, N. Shibata and K. Domen, *J. Am. Chem. Soc.*, 2015, **137**, 9627.
47. J. A. Bau and K. Takanabe, *ACS Catal.*, 2017, **7**, 7931.
48. A. T. Garcia-Esparza, T. Shinagawa, S. Ould-Chikh, M. Qureshi, X. Peng, N. Wei, D. H. Anjum, A. Clo, T. C. Weng and D. Nordlund, *Angew. Chem.*, 2017, **129**, 5874.
49. M. Qureshi, T. Shinagawa, N. Tsiapis and K. Takanabe, *ACS Sustainable Chem. Eng.*, 2017, **5**, 8079.
50. K. Maeda, K. Teramura, D. Lu, N. Saito, Y. Inoue and K. Domen, *J. Phys. Chem. C*, 2007, **111**, 7554.
51. T. Yoshinaga, M. Saruyama, A. Xiong, Y. Ham, Y. Kuang, R. Niishiro, S. Akiyama, M. Sakamoto, T. Hisatomi, K. Domen and T. Teranishi, *Nanoscale*, 2018, **10**, 10420.
52. T. Takata, J. Jiang, Y. Sakata, M. Nakabayashi, N. Shibata, V. Nandal, K. Seki, T. Hisatomi and K. Domen, *Nature*, 2020, **581**, 411.
53. Q. Wang, S. Okunaka, H. Tokudome, T. Hisatomi, M. Nakabayashi, N. Shibata, T. Yamada and K. Domen, *Joule*, 2018, **2**, 2667.
54. T. Ohno, L. Bai, T. Hisatomi, K. Maeda and K. Domen, *J. Am. Chem. Soc.*, 2012, **134**, 8254.
55. H. Nishiyama, T. Yamada, M. Nakabayashi, Y. Maehara, M. Yamaguchi, Y. Kuromiya, Y. Nagatsuma, H. Tokudome, S. Akiyama, T. Watanabe, R. Narushima, S. Okunaka, N. Shibata, T. Takata, T. Hisatomi and K. Domen, *Nature*, 2021, **598**, 304.

56. Y. Liu, X. Xu, S. Lv, H. Li, Z. Si, X. Wu, R. Ran and D. Weng, *Catal. Sci. Technol.*, 2021, **11**, 3039.
57. S. Lv, M. Pei, Y. Liu, Z. Si, X. Wu, R. Ran, D. Weng and F. Kang, *Nanoscale*, 2022, **14**, 2427.
58. H. Kato, M. Kobayashi, M. Hara and M. Kakihana, *Catal. Sci. Technol.*, 2013, **3**, 1733.
59. T. Ohno, K. Sarukawa and M. Matsumura, *New J. Chem.*, 2002, **26**, 1167.
60. R. Li, F. Zhang, D. Wang, J. Yang, M. Li, J. Zhu, X. Zhou, H. Han and C. Li, *Nat. Commun.*, 2013, **4**, 1432.
61. A. T. Garcia-Esparza and K. Takanabe, *J. Mater. Chem. A*, 2016, **4**, 2894.
62. R. Chen, F. Fan, T. Dittrich and C. Li, *Chem. Soc. Rev.*, 2018, **47**, 8238.
63. L. Mu, Y. Zhao, A. Li, S. Wang, Z. Wang, J. Yang, Y. Wang, T. Liu, R. Chen, J. Zhu, F. Fan, R. Li and C. Li, *Energy Environ. Sci.*, 2016, **9**, 2463.
64. S. Wang, K. Teramura, T. Hisatomi, K. Domen, H. Asakura, S. Hosokawa and T. Tanaka, *ACS Sustainable Chem. Eng.*, 2021, **9**, 9327.
65. K. Maeda and K. Domen, *J. Phys. Chem. Lett.*, 2010, **1**, 2655.
66. R. Konta, T. Ishii, H. Kato and A. Kudo, *J. Phys. Chem. B*, 2004, **108**, 8992.
67. T. Shinoda, Y. Yamaguchi, A. Kudo and N. Murakami, *Chem. Commun.*, 2020, **56**, 14255.
68. Y. Sasaki, H. Nemoto, K. Saito and A. Kudo, *J. Phys. Chem. C*, 2009, **113**, 17536.
69. K. Furuhashi, Q. Jia, A. Kudo and H. Onishi, *J. Phys. Chem. C*, 2013, **117**, 19101.
70. R. Asai, H. Nemoto, Q. Jia, K. Saito, A. Iwase and A. Kudo, *Chem. Commun.*, 2014, **50**, 2543.
71. R. Niishiro, S. Tanaka and A. Kudo, *Appl. Catal., B*, 2014, **150–151**, 187.
72. B. Moss, Q. Wang, K. T. Butler, R. Grau-Crespo, S. Selim, A. Regoutz, T. Hisatomi, R. Godin, D. J. Payne, A. Kafizas, K. Domen, L. Steier and J. R. Durrant, *Nat. Mater.*, 2021, **20**, 511.
73. H. Kato and A. Kudo, *J. Phys. Chem. B*, 2002, **106**, 5029.
74. S. Suzuki, A. Iwase and A. Kudo, *Catal. Sci. Technol.*, 2020, **10**, 4912.
75. A. Kudo, S. Yoshino, T. Tsuchiya, Y. Udagawa, Y. Takahashi, M. Yamaguchi, I. Ogasawara, H. Matsumoto and A. Iwase, *Faraday Discuss.*, 2019, **215**, 313.
76. J. Shan, F. Raziq, M. Humayun, W. Zhou, Y. Qu, G. Wang and Y. Li, *Appl. Catal., B*, 2017, **219**, 10.
77. D. Wang, J. Ye, T. Kako and T. Kimura, *J. Phys. Chem. B*, 2006, **110**, 15824.
78. H. Yu, J. Wang, S. Yan, T. Yu and Z. Zou, *J. Photochem. Photobiol., A*, 2014, **275**, 65.
79. S. Suzuki, H. Matsumoto, A. Iwase and A. Kudo, *Chem. Commun.*, 2018, **54**, 10606.
80. T. K. Townsend, N. D. Browning and F. E. Osterloh, *ACS Nano*, 2012, **6**, 7420.
81. P. Zhang, T. Ochi, M. Fujitsuka, Y. Kobori, T. Majima and T. Tachikawa, *Angew. Chem., Int. Ed.*, 2017, **56**, 5299.

82. X. Wu, C. Wang, Y. Wei, J. Xiong, Y. Zhao, Z. Zhao, J. Liu and J. Li, *J. Catal.*, 2019, **377**, 309.
83. J. Jiang, K. Kato, H. Fujimori, A. Yamakata and Y. Sakata, *J. Catal.*, 2020, **390**, 81.
84. K. Han, D. M. Haiber, J. Knöppel, C. Lievens, S. Cherevko, P. Crozier, G. Mul and B. Mei, *ACS Catal.*, 2021, **11**, 11049.
85. Y. Qin, G. Wang and Y. Wang, *Catal. Commun.*, 2007, **8**, 926.
86. S. Zong, L. Tian, X. Guan, C. Cheng, J. Shi and L. Guo, *J. Colloid Interface Sci.*, 2022, **606**, 491.
87. G. Liu, H. G. Yang, J. Pan, Y. Q. Yang, G. Q. Lu and H.-M. Cheng, *Chem. Rev.*, 2014, **114**, 9559.
88. T. Ishii, H. Kato and A. Kudo, *J. Photochem. Photobiol., A*, 2004, **163**, 181.
89. H. Kato, M. Hori, R. Konta, Y. Shimodaira and A. Kudo, *Chem. Lett.*, 2004, **33**, 1348.
90. A. J. Bard, *J. Photochem.*, 1979, **10**, 59.
91. K. Maeda, M. Higashi, D. Lu, R. Abe and K. Domen, *J. Am. Chem. Soc.*, 2010, **132**, 5858.
92. R. Abe, M. Higashi and K. Domen, *ChemSusChem*, 2011, **4**, 228.
93. T. Hisatomi, J. Kubota and K. Domen, *Chem. Soc. Rev.*, 2014, **43**, 7520.
94. K. Sayama, K. Mukasa, R. Abe, Y. Abe and H. Arakawa, *Chem. Commun.*, 2001, 2416.
95. R. Niishiro, H. Kato and A. Kudo, *Phys. Chem. Chem. Phys.*, 2005, **7**, 2241.
96. S. Kawasaki, K. Akagi, K. Nakatsuji, S. Yamamoto, I. Matsuda, Y. Harada, J. Yoshinobu, F. Komori, R. Takahashi, M. Lippmaa, C. Sakai, H. Niwa, M. Oshima, K. Iwashina and A. Kudo, *J. Phys. Chem. C*, 2012, **116**, 24445.
97. Y. Sasaki, A. Iwase, H. Kato and A. Kudo, *J. Catal.*, 2008, **259**, 133.
98. H. Kato, Y. Sasaki, N. Shirakura and A. Kudo, *J. Mater. Chem. A*, 2013, **1**, 12327.
99. Y. Sasaki, H. Kato and A. Kudo, *J. Am. Chem. Soc.*, 2013, **135**, 5441.
100. H. Kumagai, R. Aoyagi, K. Kato, A. Yamakata, M. Kakihana and H. Kato, *ACS Appl. Energy Mater.*, 2021, **4**, 2056.
101. R. Han, M. A. Melo, Z. Zhao, Z. Wu and F. E. Osterloh, *J. Phys. Chem. C*, 2020, **124**, 9724.
102. S. S. K. Ma, K. Maeda, T. Hisatomi, M. Tabata, A. Kudo and K. Domen, *Chem. – Eur. J.*, 2013, **19**, 7480.
103. Q. Wang, T. Hisatomi, S. S. K. Ma, Y. Li and K. Domen, *Chem. Mater.*, 2014, **26**, 4144.
104. D. H. K. Murthy, H. Matsuzaki, Q. Wang, Y. Suzuki, K. Seki, T. Hisatomi, T. Yamada, A. Kudo, K. Domen and A. Furube, *Sustainable Energy Fuels*, 2019, **3**, 208.
105. Q. Jia, A. Iwase and A. Kudo, *Chem. Sci.*, 2014, **5**, 1513.
106. M. A. Melo, Z. Wu, B. A. Nail, A. T. De Denko, A. F. Nogueira and F. E. Osterloh, *Nano Lett.*, 2018, **18**, 805.

107. A. Iwase, Y. H. Ng, Y. Ishiguro, A. Kudo and R. Amal, *J. Am. Chem. Soc.*, 2011, **133**, 11054.
108. Q. Wang, T. Hisatomi, Q. Jia, H. Tokudome, M. Zhong, C. Wang, Z. Pan, T. Takata, M. Nakabayashi, N. Shibata, Y. Li, I. D. Sharp, A. Kudo, T. Yamada and K. Domen, *Nat. Mater.*, 2016, **15**, 611.
109. K. Iwashina and A. Kudo, *J. Am. Chem. Soc.*, 2011, **133**, 13272.
110. S. Kawasaki, K. Nakatsuji, J. Yoshinobu, F. Komori, R. Takahashi, M. Lippmaa, K. Mase and A. Kudo, *Appl. Phys. Lett.*, 2012, **101**, 033910.
111. M. Ebihara, T. Ikeda, S. Okunaka, H. Tokudome, K. Domen and K. Katayama, *Nat. Commun.*, 2021, **12**, 3716.
112. Q. Wang, Y. Li, T. Hisatomi, M. Nakabayashi, N. Shibata, J. Kubota and K. Domen, *J. Catal.*, 2015, **328**, 308.
113. T. Minegishi, N. Nishimura, J. Kubota and K. Domen, *Chem. Sci.*, 2013, **4**, 1120.
114. Q. Wang, T. Hisatomi, M. Katayama, T. Takata, T. Minegishi, A. Kudo, T. Yamada and K. Domen, *Faraday Discuss.*, 2017, **197**, 491.
115. Q. Wang, T. Hisatomi, Y. Suzuki, Z. Pan, J. Seo, M. Katayama, T. Minegishi, H. Nishiyama, T. Takata, K. Seki, A. Kudo, T. Yamada and K. Domen, *J. Am. Chem. Soc.*, 2017, **139**, 1675.
116. Q. Wang, J. Warnan, S. Rodríguez-Jiménez, J. J. Leung, S. Kalathil, V. Andrei, K. Domen and E. Reisner, *Nat. Energy*, 2020, **5**, 703.
117. K. Sayama, K. Mukasa, R. Abe, Y. Abe and H. Arakawa, *J. Photochem. Photobiol., A*, 2002, **148**, 71.
118. R. Abe, K. Sayama and H. Sugihara, *J. Phys. Chem. B*, 2005, **109**, 16052.
119. Y. Miseki, S. Fujiyoshi, T. Gunji and K. Sayama, *Catal. Sci. Technol.*, 2013, **3**, 1750.
120. W. Chen, H. Liu, X. Li, S. Liu, L. Gao, L. Mao, Z. Fan, W. Shangguan, W. Fang and Y. Liu, *Appl. Catal., B*, 2016, **192**, 145.
121. K. Tsuji, O. Tomita, M. Higashi and R. Abe, *ChemSusChem*, 2016, **9**, 2201.
122. Y. Miseki, S. Fujiyoshi, T. Gunji and K. Sayama, *J. Phys. Chem. C*, 2017, **121**, 9691.
123. S. Hara, M. Yoshimizu, S. Tanigawa, L. Ni, B. Ohtani and H. Irie, *J. Phys. Chem. C*, 2012, **116**, 17458.
124. H. P. Duong, T. Mashiyama, M. Kobayashi, A. Iwase, A. Kudo, Y. Asakura, S. Yin, M. Kakihana and H. Kato, *Appl. Catal., B*, 2019, **252**, 222.
125. A. Nakada, S. Nishioka, J. J. M. Vequizo, K. Muraoka, T. Kanazawa, A. Yamakata, S. Nozawa, H. Kumagai, S.-i. Adachi, O. Ishitani and K. Maeda, *J. Mater. Chem. A*, 2017, **5**, 11710.
126. J. Wang, M. Kuo, P. Zeng, L. Xu, S. Chen and T. Peng, *Appl. Catal., B*, 2020, **279**, 119377.
127. T. M. Suzuki, A. Iwase, H. Tanaka, S. Sato, A. Kudo and T. Morikawa, *J. Mater. Chem. A*, 2015, **3**, 13283.
128. Y. Wei, J. Wan, J. Wang, X. Zhang, R. Yu, N. Yang and D. Wang, *Small*, 2021, **17**, 2005345.
129. A. Iwase, Y. Udagawa, S. Yoshino, Y. H. Ng, R. Amal and A. Kudo, *Energy Technol.*, 2019, **7**, 1900358.

130. S. Okunaka, H. Kameshige, T. Ikeda, H. Tokudome, T. Hisatomi, T. Yamada and K. Domen, *ChemSusChem*, 2020, **13**, 4906.
131. B. Zhang, K. Liu, Y. Xiang, J. Wang, W. Lin, M. Guo and G. Ma, *ACS Catal.*, 2022, **12**, 2415.
132. S. Yoshino, K. Sato, Y. Yamaguchi, A. Iwase and A. Kudo, *ACS Appl. Energy Mater.*, 2020, **3**, 10001.
133. Q. Wang, C. Pornrungroj, S. Linley and E. Reisner, *Nat. Energy*, 2022, **7**, 13.
134. Z. Wang, Y. Inoue, T. Hisatomi, R. Ishikawa, Q. Wang, T. Takata, S. Chen, N. Shibata, Y. Ikuhara and K. Domen, *Nat. Catal.*, 2018, **1**, 756.
135. Q. Wang, M. Nakabayashi, T. Hisatomi, S. Sun, S. Akiyama, Z. Wang, Z. Pan, X. Xiao, T. Watanabe, T. Yamada, N. Shibata, T. Takata and K. Domen, *Nat. Mater.*, 2019, **18**, 827.
136. Z. Pan, Q. Xiao, S. Chen, Z. Wang, L. Lin, M. Nakabayashi, N. Shibata, T. Takata, T. Hisatomi and K. Domen, *J. Catal.*, 2021, **399**, 230.
137. Z. Pan, H. Yoshida, L. Lin, Q. Xiao, M. Nakabayashi, N. Shibata, T. Takata, T. Hisatomi and K. Domen, *Res. Chem. Intermed.*, 2021, **47**, 225.
138. G.-J. Lee, Y.-C. Zheng and J. J. Wu, *Catal. Today*, 2018, **307**, 197.
139. H. Fujito, H. Kunioku, D. Kato, H. Suzuki, M. Higashi, H. Kageyama and R. Abe, *J. Am. Chem. Soc.*, 2016, **138**, 2082.
140. G. Zhang, Z.-A. Lan, L. Lin, S. Lin and X. Wang, *Chem. Sci.*, 2016, **7**, 3062.

CHAPTER 3

Recent Developments in (Oxy)nitride Photocatalysts With Narrow Bandgaps for Solar-driven Water Splitting

YING LUO[a], ZHENG WANG*[a] AND HONG HE[a]

[a]Laboratory of Atmospheric Environment and Pollution Control, Research Center for Eco-Environmental Sciences, Chinese Academy of Sciences, Beijing 100085, China
*E-mail: zhengwang@rcees.ac.cn

3.1 Introduction

3.1.1 Thermodynamics and Kinetics of Water Splitting

Environmental protection and sustainable energy supply are two major issues for future world development. The current energy consumption of our society mainly relies on non-renewable fossil fuels, which not only produce a large amount of greenhouse gases (*e.g.*, CO_2) but also some harmful gases such as SO_x and NO_x.[1–3] Owing to their massive consumption, non-renewable fossil fuel reverses in nature will not meet the demands of the next century. To maintain the sustainability of society, the development and utilization of renewable and clean energy to replace fossil fuels has become extremely important. As the most abundant and clean energy resource, solar energy

Sustainable Energy No. 1
Recent Developments in Functional Materials for Artificial Photosynthesis
Edited by Srabanti Ghosh and Qian Wang

Published by the Royal Society of Chemistry, www.rsc.org

reaching the surface of earth is roughly 1.3 × 10^5 TW larger than current global energy consumption.[4] If it can be successfully harvested and utilized, both environmental and energy issues could be alleviated. Therefore, numerous methods have been explored to utilize solar energy effectively. Among them, solar-driven water splitting based on particulate photocatalysts as a promising approach to produce renewable and clean hydrogen energy has attracted much attention from researchers, due to the obvious advantages of low capital cost, simple operation, and easy scale-up.[5–8]

Thermodynamically, the decomposition of water into H_2 and O_2 is an uphill reaction that involves a positive change of Gibbs free energy (ΔG° = 237 kJ mol^{-1}) as shown in eqn (3.1), meaning that this reaction requires a lot of energy and therefore cannot occur spontaneously.[9–12]

$$H_2O \rightarrow H_2 + \frac{1}{2}O_2 \qquad \Delta G^{\circ} = 237\ \text{kJ mol}^{-1} \tag{3.1}$$

To drive this reaction, semiconductor photocatalysts should be excited by photons with energies higher than or equal to the bandgaps. The electrons of the valance band will be excited and moved into the conduction band, leaving holes in the valance band. As a result, photoexcited electrons and holes are generated in the conduction band and the valance band, respectively, which have strong reduction or oxidation potentials. If these electrons and holes are thermodynamically favourable, they can be used for H^+ reduction and H_2O oxidation, respectively, as shown in eqn (3.2) and eqn (3.3). Accordingly, the process of water splitting can be divided into three steps: (1) the generation of electrons and holes pairs by photoexcitation; (2) charge transfer; and (3) surface catalytic reaction, as shown in Figure 3.1. For each step, there are some requirements and attention as discussed below.

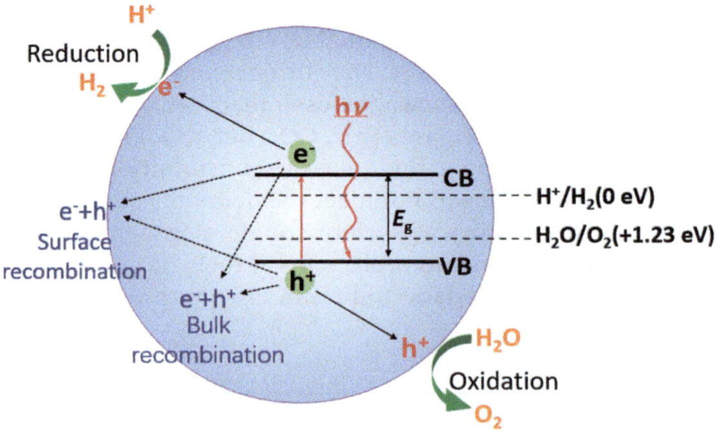

Figure 3.1 Schematic diagram of the photocatalytic overall water splitting reaction process on a photocatalyst. VB = valance band; CB = conduction band; E_g = bandgap of semiconductor.

$$2H^+ + 2e^- \rightarrow H_2 \qquad E^\circ = 0 \text{ V} \tag{3.2}$$

$$2H_2O + 4h^+ \rightarrow O_2 + 4H^+ \quad E^\circ = 1.23 \text{ V} \tag{3.3}$$

In the first step, the bandgap of the photocatalyst is required to be smaller than the incident photon energy. Therefore, the bandgap is the important parameter for the selection of photocatalyst and strongly determines the sunlight absorption range. The requirement for band position of the photocatalyst is that the top of the valance band should be lower than the potential of H_2O oxidation (1.23 V *vs.* NHE at pH 0), while the bottom of the conduction band should be higher than the potential of H^+ reduction (0 V *vs.* NHE at pH 0). Therefore, the thermodynamic requirement of a minimum bandgap for a photocatalyst is 1.23 eV. For example, the widely studied TiO_2 and $SrTiO_3$ have wide bandgaps of 3.2–3.4 eV and can only respond to wavelengths less than 365–390 nm in the ultraviolet (UV) light region.[13–15] To utilize the visible range of sunlight, the bandgap of the semiconductor should be as narrow as possible. Doping foreign elements to introduce additional electronic state(s) is widely used for adjusting the bandgap of the original semiconductor, such as Cr/Ta co-doped $SrTiO_3$ can absorb light with wavelengths more than 550 nm. Ru-, Rh-, and Ir-doped $SrTiO_3$ also exhibit visible light absorption.[16,17] With a decrease of bandgap, however, the redox driving force of the photocatalyst weakens. Therefore, the suitable bandgap energy and band position are the keys for designing visible-light-responsive photocatalysts. In the second step, the separation and migration of photoexcited charges play a crucial role in the performance of water splitting. Although there are many materials with suitable bandgap energies and band positions, a few materials accomplish overall water splitting due to other factors, such as crystallinity, crystal size, crystal structure, surface morphology, *etc.* Generally, high crystallinity is beneficial to the photocatalytic performance because of a low density of defects, which would act as the recombination center for electrons and holes. In addition, small crystal size is expected for higher activity as the migration distance of electrons and holes would be shortened, and some special morphologies (platy, tunnel, pillared, *etc.*) with a large surface area usually have a positive effect on the separation and migration of electrons and holes.[18,19] In the third step, electrons and holes that survived and successfully migrated to the surface would undergo the reduction and oxidation reactions at the active sites of the photocatalyst. The bare photocatalyst always suffers from lacking active sites for photocatalytic reaction. Some typical noble metals (*e.g.*, Pt, Au, Ag, Rh, Co) or transition metal oxides (*e.g.*, RuO_2, CoO_x, MnO_x, IrO_x, NiO_x) are generally located on the surface of photocatalyst as co-catalysts with nanoparticle sizes ranging from 1–50 nm, which can provide active sites to reduce the activation energy of chemical reactions.[20–22]

At present, there are two kinds of successful photocatalytic water splitting systems: a one-step photoexcitation water splitting system using a single photocatalyst and a two-step photoexcitation water splitting system using two

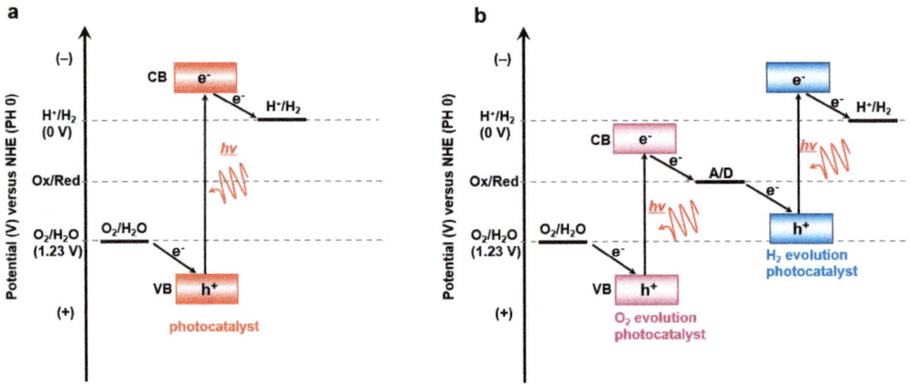

Figure 3.2 Photocatalytic overall water splitting for (a) one-step photoexcitation system and (b) two-step photoexcitation system. A = electron acceptor; D = electron donor.

different photocatalysts, as shown in Figure 3.2.[23,24] For the direct one-step overall water splitting, the photocatalysts should have a bandgap to absorb sufficient photons, suitable band positions for water reduction and oxidation and stability to resist photocorrosion (Figure 3.2(a)). A large number of metal oxides with d^0 or d^{10} electronic configuration have been explored for one-step overall water splitting to date, and some have exhibited extremely high quantum efficiency. However, the bottom of the conduction band and the top of the valence band of these metal oxides mainly consist of metal d^0 or d^{10} orbitals and O 2p orbitals located at ~ +3V or higher (*vs.* NHE), respectively. If the bottom of the conduction band is located at a potential equal to or higher than the water reduction potential, the bandgap of the oxide material would be larger than 3 eV, resulting in only UV light response. Since UV light accounts for less than 5% of the incident light on the Earth's surface, and visible light accounts for about 50%, the development of a visible-light-responsive photocatalyst is essential to improve the conversion efficiency of solar energy. However, narrowing the bandgap of a photocatalyst to harvest visible light inevitably decreases the driving force of band electrons and holes for water reduction and oxidation. Most visible-light-driven photocatalysts such as chalcogenides suffer from the self-oxidation by valence band holes during the water splitting reaction, due to the difficulty of water oxidation.[25,26] In addition, the backward reaction between H_2 and O_2 to form water easily takes place in the one-step excitation water splitting, this is because H_2 and O_2 are simultaneously generated on the surface of one single photocatalyst particle. Although these requirements are stringent, one-step excitation overall water splitting on a single particulate photocatalyst is regarded as the simplest way to produce H_2 from solar energy and water, and the improvement of the photocatalytic efficiency of the one-step excitation system is intensively investigated.

Two-step photoexcitation water splitting system, the so called Z-scheme water splitting, consists of two different photocatalysts for H_2 and O_2 evolution with a suitable redox shuttle or mediator or electron-transfer conductor (Figure 3.2(b)).[27,28] On the H_2-evolution photocatalyst (HEP) side of this system, the protons are reduced by the photoexcited electrons on the conduction band, and the electron donor (D) is oxidized by photoexcited holes on the valence band converting to electron acceptor (A), as shown in eqn (3.4) and (3.5). On the O_2-evolution photocatalyst (OEP) side, the water is oxidized by the valence band holes and the A is reduced by the conduction band electrons to form D, as shown in eqn (3.6) and (3.7). Therefore, a cycle of redox mediators (A and D) occurs, and the overall water splitting based on two different photocatalysts is achieved. This system has the advantage of expanding the selection of photocatalysts that are not necessary to meet both water reduction and oxidation potentials, and it is possible to directly separate H_2 and O_2. Many semiconductors with either a suitable water reduction or oxidation potential have been considered as one side in the Z-scheme water splitting. WO_3 and $BiVO_4$ are good photocatalysts for O_2 production, even though they are unable to reduce water.[29–32]

$$2H^+ + 2e^- \rightarrow H_2 \text{ (photoreduction of } H^+ \text{ to } H_2) \tag{3.4}$$

$$D + nh^+ \rightarrow A \text{ (photooxidation of D to A)} \tag{3.5}$$

$$A + ne^- \rightarrow D \text{ (photoreduction of A to D)} \tag{3.6}$$

$$2H_2O + 4h^+ \rightarrow O_2 + 4H^+ \text{ (photooxidation of } H_2O \text{ to } O_2) \tag{3.7}$$

Because of the different photocatalytic reaction conditions among different research groups, including the amounts of photocatalysts, irradiation conditions, reactor size and shape, it is not helpful to directly compare the photocatalytic activity (*e.g.*, the total amounts and evolution rates of gases). The apparent quantum yield (AQY) and solar-to-hydrogen (STH) energy conversion efficiency are the standard parameters to evaluate the photocatalytic performance of a particulate photocatalyst during overall water splitting, which are defined in eqn (3.8) and (3.9), respectively.

$$AQY = \frac{nR}{I} \tag{3.8}$$

Where n, R and I denote the number of electrons or holes involved in the production of one H_2 or O_2 molecule, the formation rate of gas molecules, and the incident photon flux, respectively. The constant of n for H_2 and O_2 evolution are two and four in one-step photoexcitation overall water splitting, and four and eight in two-step Z-scheme overall water splitting, respectively. It should be noted that the AQY is strongly dependent on the incident wavelengths because of the different absorption coefficients and activities for the photocatalyst at different irradiation wavelengths. Therefore, it is helpful to confirm the AQY as a function of irradiation wavelength.

$$\mathrm{STH} = \frac{R \times \Delta G}{P_{\mathrm{sun}} \times \Delta S} \tag{3.9}$$

Where, R, ΔG, P_{sun} and S represent the evolution rate of H_2 gas, the Gibbs energy for water splitting, the energy flux of sunlight at the AM1.5 global tilt (1.0×10^3 W m^{-2}), and the area of irradiated photocatalyst, respectively. This STH energy conversion efficiency is only applicable when H_2 and O_2 gases are evolved in a stoichiometric ratio of water decomposition under sunlight irradiation.

According to the technique and economy analyses, H_2 production from water splitting on particulate photocatalysts can be applied in industry when the STH of water splitting reaches 5–10%. The STH is strongly related to the light absorption of photocatalysts because photons with energy greater than the band gap of photocatalyst can be converted into chemical energy *via* water splitting.[33,34] As shown in Figure 3.3, the STH conversion efficiency goes up as the increase of light wavelength available for the one-step excitation overall water splitting reaction. The theoretical maximum STH is calculated to be 2, 8 and 16%, respectively, when photons with wavelength up to 400, 500 and 600 nm are perfectly utilized in the water splitting reaction (quantum efficiency reaches 100%).

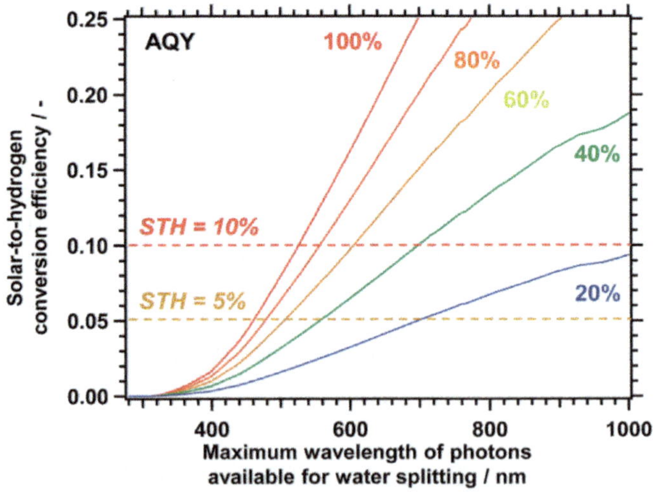

Figure 3.3 Relationships between the STH values and wavelengths of photons available for water splitting at different AQYs. It is assumed that two water molecules are split into one oxygen molecule and two hydrogen molecules with a four-photon process. Reproduced from ref. 34 with permission from Springer Nature, Copyright 2014.

Therefore, the development of efficient photocatalysts with narrow bandgaps is the most important research target for applicable photocatalytic water splitting.

3.2 Development of Photocatalysts

In 1972, Fujishima and Hondo achieved water splitting using TiO_2 under UV irradiation.[35] Since then, many oxide materials have been developed and designed as photocatalysts for overall water splitting under UV light irradiation. In early research, primarily TiO_2 and various titanates were studied.[36–38] From 1980–2010, many Ta-, Zr- and Nb-based oxides were reported to be active for water splitting.[39,40] $ATaO_3$ (A = Li, Na, K), $MgTa_2O_6$ and $BaTa_2O_6$ exhibited photocatalytic activity for water splitting even without co-catalysts under UV irradiation.[41] After modification with nickel oxide or nanosized gold particles, the photocatalytic performance was further enhanced.[42–44] Particularly, a high AQY of 56% at 270 nm for overall water splitting was obtained on the La-doped $NaTaO_3$ photocatalyst.[45] Bare ZrO_2 also exhibited high photocatalytic activity and stability for overall water splitting in aqueous Na_2CO_3 or $NaHCO_3$ solution under UV irradiation.[46,47] A ZrO_2 with a high surface area prepared by precipitation method has also shown enhanced photocatalytic activity.[48] NiO-loaded ZrO_2 with a clean metal-support interface *via* cold plasma treatment gave higher photocatalytic performance for water splitting.[49] Subsequently, lanthanide zirconium oxides $Ln_2Zr_2O_7$ (Ln = La, Ce, Nd, and Sm) were studied for water splitting.[50] In addition, bulk Nb_2O_5 cannot split water into H_2 and O_2.[51] However, mesoporous Nb_2O_5 showed H_2 evolution from the half reaction 20 times higher than that of bulk Nb_2O_5.[52] Besides, many niobates, such as K_4NbO_{17}, Rb_4NbO_{17}, $ANbO_3$ (A = Li, Na, K), Cs_2NbO_{11}, $SrNb_2O_6$, $Sr_2Nb_2O_7$, $Sr_5Nb_4O_{15}$, $Ca_2Nb_2O_7$, $Ba_5Nb_4O_{15}$, $A(M_{n-1}Nb_nO_{3n+1})$ (A = H, Na, K, Rb, Cs; M = La, Ca *etc.*) and so on were active in water splitting to H_2 and/or O_2 upon UV irradiation.[53–62]

These photocatalysts consisting of transition metal cations (Ti^{4+}, Zr^{4+}, Ta^{5+}, Nb^{5+}, and W^{6+}) with empty d orbitals are defined as having d^0 electronic configurations, and another class of photocatalysts made up of typical metal cations (Ga^{3+}, In^{3+}, Ge^{4+}, Sn^{4+}, and Sb^{5+}) with filled d orbitals are defined as having d^{10} electronic configurations. NiO-loaded Ga_2O_3 was one of the promising photocatalysts for the simultaneous decomposition of water into H_2 and O_2 in stoichiometry.[63] After the incorporation of Ca, Cr, Zn, Sr, Ba, and Ta ions, the photocatalytic activity of Ga_2O_3 was effectively enhanced.[64] Particularly, the doping of Zn ions remarkably enhanced the photocatalytic activity with an AQY of 20%. Moreover, $ZnGa_2O_4$, Zn_2GeO_4, $M_2Sb_2O_7$ (M = Ca, Sr), $LiInGeO_4$, $CaIn_2O_4$, $SrIn_2O_4$, and $LnInO_3$ (Ln = La, Nd) were also examined with a RuO_2 co-catalyst for water splitting under UV irradiation.[65–71]

To improve the conversion efficiency of solar energy, it is important to develop visible-light-responsive photocatalysts. Since 2000, (oxy)sulfides and (oxy)nitrides have been used as new non-oxide photocatalysts for visible-light-driven water splitting. Due to the potential of S 3p and N 2p orbitals higher than that of O 2p orbital, the (oxy)sulfides and (oxy)nitrides have sufficiently small bandgaps and can be activated by visible light. Domen and co-workers have extensively investigated such materials for overall water splitting.[72] For example, the Ti-based oxysulfide $Sm_2Ti_2S_2O_5$ with layered perovskite structure functioned as a stable photocatalyst for H_2 or O_2 production in the presence of the corresponding sacrificial electron donor (Na_2S–Na_2SO_3 or methanol) or acceptor (Ag^+) under visible light irradiation (440 nm ≤ λ ≤ 650 nm).[73] When compared with sulfides, oxysulfides are more stable in the reaction likely owing to the hybridization of O 2p and S 2p orbitals in the valence band. Other oxysulfides, $Ln_2Ti_2S_2O$ (Ln = Pr, Tb, Nd, Gd, Ho, Dy, and Er) with similar layered perovskite structures, also exhibited photocatalytic activity in the aqueous solution containing the corresponding sacrificial reagents.[74]

The (oxy)nitrides containing metal cations with d^0 electronic configurations (*e.g.*, Ti^{4+}, Ta^{5+}, Nb^{5+}) and d^{10} electronic configurations (*e.g.*, Ga^{3+}, Ge^{4+}) mostly have suitable band potentials for water splitting and appropriate bandgaps for wide-wavelength sunlight absorption, which can be obtained by nitriding the corresponding oxides in an NH_3 atmosphere.[75–77] The nitridation process usually involves the replacement of oxygen atoms in the oxide precursor with nitrogen atoms from NH_3 gas or relative NH and NH_2 radicals at high temperatures. For example, TaON and Ta_3N_5 as the simplest (oxy) nitrides, were converted from the Ta_2O_5 precursor under an NH_3 flow.[78] The transformation process is shown in eqn (3.10). The final products are determined by the nitridation conditions. TaON is formed at a relatively mild condition such as an NH_3 flow rate of 20 mL min^{-1}, holding temperature of 1123 K and time of 15 h. Ta_3N_5 is produced under more severe conditions, an NH_3 flow rate up to 500 mL min^{-1} and a nitridation time of 20 h. Beside NH_3 as the nitrogen source, other nitrogen sources such as amines and urea have also been employed in hydrothermal, sol–gel, and other synthetic methods for (oxy)nitrides preparation. However, the decomposition of amines or urea unavoidably introduces carbon impurity species into the final products. Recently, many (oxy)nitrides have been reported as efficient photocatalysts for water splitting, which will be discussed in detail in the following section.

$$Ta_2O_5 \rightarrow TaON \rightarrow Ta_3N_5 \tag{3.10}$$

3.3 (Oxy)nitride Photocatalysts with Visible Light Absorption

3.3.1 Tantalum-based (Oxy)nitrides

(Oxy)nitrides are promising photocatalyst materials for solar water splitting systems, because of their stable chemical structure, light-harvesting properties and band positions suitable for both water reduction and oxidation.[79–81]

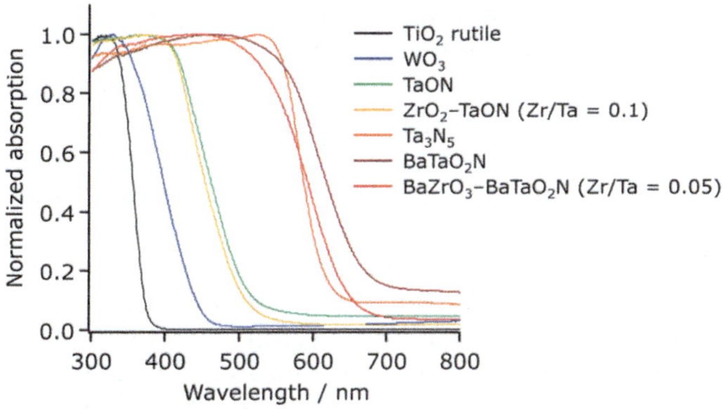

Figure 3.4 UV-vis diffuse reflectance spectra for some (oxy)nitrides. Reproduced from ref. 81 with permission from American Chemical Society, Copyright 2013.

Figure 3.4 shows the UV-vis diffused reflectance spectra (UV-vis DRS) of some intensively-studied tantalum-based (oxy)nitrides, which possess excellent wide-visible-light absorption ability.[81] The potential energy of N 2p orbitals is higher than that of O 2p orbitals, the partial or full substitution of N atoms for O atoms makes the valence band maximum of the (oxy)nitrides shift up to higher energy without affecting the conduction band minimum. Consequently, the application of the wide-visible-light responsive (oxy)nitrides in the solar water splitting is highly expected to achieve practical energy conversion efficiency.

TaON with a bandgap of 2.5 eV (absorption edge at 520 nm) is a popular oxynitride photocatalyst having the baddeleyite-type structure. It has been demonstrated that proton reduction or water oxidation was able to proceed on TaON in the presence of suitable electron donors and acceptors, respectively. The AQY for H_2 evolution on Pt-modified TaON from methanol solution was 0.2% and the value for O_2 evolution on bare TaON from $AgNO_3$ solution was 34%.[82] It has been also demonstrated that the two-step excitation water splitting under visible-light irradiation was achieved when the Pt-loaded TaON was combined with PtO_x-loaded WO_3 in the aqueous IO_3^-/I^- shuttle redox solution.[83] Pt-loaded TaON could efficiently promote photocatalytic H_2 evolution which was accompanied by the oxidation of I^- to IO_3^-. The formation of IO_3^- obviously suppressed the H_2 evolution activity. This is because of the competitive consumption of photoexcited electrons by the reduction of IO_3^- on the Pt-loaded TaON. When RuO_2 or IrO_2 was modified on TaON as an O_2-evolving co-catalyst, the photocatalytic activity for O_2 evolution from the aqueous IO_3^- solution was enhanced. The Z-scheme water splitting was successfully constructed on RuO_2-modified TaON as O_2-evolving photocatalyst and Pt-loaded TaON as H_2-evolving photocatalyst.[84] These results indicated that appropriate co-catalyst modification is an effective way to control the reaction selectivity of photocatalysts with redox mediators in the Z-scheme water splitting systems. Moreover, when compared to being loaded with the conventional RuO_2 co-catalyst, a TaON photocatalyst co-loaded with $Ru(OH)_xCl_y$ as the reduction sites and Co species as the oxidation sites exhibited more than a two-fold increase in O_2 evolution activity from aqueous IO_3^- solution.[85] Such dual co-catalyst modification enables the potential use of TaON as an O_2 evolution photocatalyst towards efficient Z-scheme water splitting.

The low photocatalytic efficiency of TaON is also attributed to the inherently high defect density.[79,81,86] The TaON material is generally synthesized through high temperature nitridation of an Ta_2O_5 precursor in an NH_3 atmosphere. Thus, the Ta^{5+} cations in Ta_2O_5 are easily reduced into other Ta species such as Ta^{3+} and Ta^{4+} by NH_3. Simultaneously, anionic vacancies are formed to maintain the lattice charge balance (shown in Figure 3.5). Such defect states existing in bulk or on the surface of TaON usually act as recombination centers of photoexcited electron–hole pairs that are detrimental to the photocatalytic performance. It is difficult to eliminate defects in TaON by refining the crystallinity of TaON, because TaON is an intermediate phase between Ta_2O_5 and Ta_3N_5 and its thermal stability is quite low. Therefore, surface modification is important to reduce the surface defect states of TaON. One effective strategy was developed by decorating a monoclinic ZrO_2 protector on TaON.[86] Because the nitridation of ZrO_2 is more difficult at high temperatures, when compared with Ta_2O_5, and the interaction between Ta cations at the surface of Ta_2O_5 or TaON and ZrO_2 species maintains Ta more cationic, the pre-loaded well-dispersed ZrO_2 nanoparticles on the surface of the Ta_2O_5 precursor functions as a protector to suppress the reduction of Ta^{5+} cations at the TaON surface layer during high-temperature ammonolysis

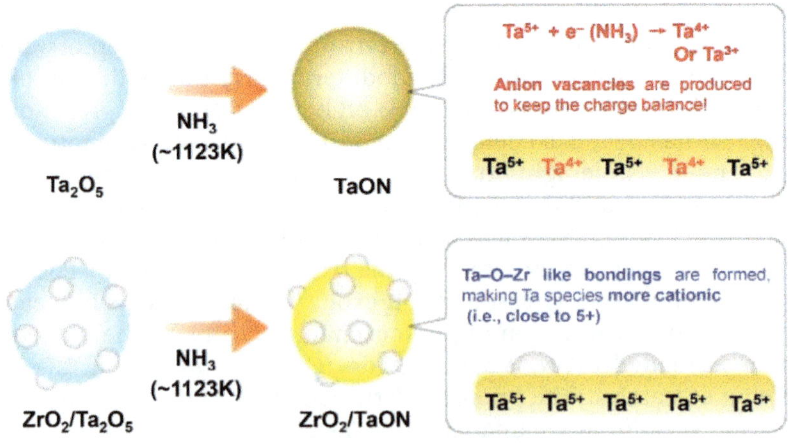

Figure 3.5 Schematic illustration showing the nitridation of Ta_2O_5 and ZrO_2/Ta_2O_5 composite to suppress the production of reduced Ta species (defect sites) near the surface. Reproduced from ref. 86 with permission from Chemical Society of Japan, Copyright 2008.

(shown in Figure 3.5). Upon modification with ZrO_2, the photocatalytic performance for H_2 evolution or O_2 evolution in the presence of the respective sacrificial reagents was improved. Furthermore, an AQY of Z-scheme water splitting reached 6.3% at 420 nm irradiation, when using Pt-loaded ZrO_2/TaON as the H_2-evolving photocatalyst, PtO_x/WO_3 as the O_2-evolving photocatalyst and IO_3^-/I^- redox couple.[87] This was attributed to the enhancement of H_2 evolution ability for the ZrO_2/TaON photocatalyst with the decreased defect states that functioned as undesirable electron–hole recombination centers. The decoration of $Rh_yCr_{2-y}O_3$ co-catalyst instead of Pt co-catalyst enabled the efficient extraction of photogenerated electrons from ZrO_2/TaON and effectively restrained the water formation reaction between H_2 and O_2. Therefore, highly efficient Z-scheme water splitting with an optimal AQY of 10.3% at 420 nm and 1.9% at 500 nm was successfully assembled with $Rh_yCr_{2-y}O_3$-modified ZrO_2/TaON, Au/CoO_x co-loaded facet-controlled $BiVO_4$ and $[Fe(CN)_6]^{3-}/[Fe(CN)_6]^{4-}$ as the H_2-evolution photocatalyst, O_2-evolution photocatalyst and redox mediator, respectively.[88] Interestingly, one-step excitation overall water splitting under visible light irradiation was first accomplished on ZrO_2/TaON as a non-oxide photocatalyst with d^0 electronic configuration, when modifying with Cr_2O_3/RuO_x as a H_2-evolving co-catalyst.[89] There are several key factors contributing to the one-step overall water splitting reaction. The separation and migration of photogenerated charge carriers were promoted in the less-defective ZrO_2/TaON. RuO_x as co-catalyst effectively captured photogenerated electrons from ZrO_2/TaON and Cr_2O_3 shell coated on RuO_x restrained the backward reaction of the evolved H_2 and O_2. Although the water splitting rate by $Cr_2O_3/RuO_x/ZrO_2$/TaON decreased gradually, further modification of $Cr_2O_3/RuO_x/ZrO_2$/TaON with colloidal IrO_2 as the water oxidation site improved the stability.

A $MgTa_2O_{6-x}N_y$/TaON heterostructure was obtained in a one-pot nitridation of a $MgTa_2O_6$-modified Ta_2O_5 precursor.[90] According to the flat-band potential and UV-vis DRS, the typical type II band alignment between $MgTa_2O_{6-x}N_y$ and TaON was formed and the valence band maximum of TaON was approximately 0.32 eV more positive than that of $MgTa_2O_{6-x}N_y$ as shown in Figure 3.6. Thus, the heterostructure of suitable band alignment enhanced the photogenerated electrons transfer from $MgTa_2O_{6-x}N_y$ to TaON. The spatial separation of charge carriers was confirmed by the location of Pt nanoparticles that were selectively photodeposited on TaON. Moreover, the modification of Ta_2O_5 with $MgTa_2O_6$ prevented the formation of defects in TaON during nitridation. Both the heterojunction and the reduced structural defect density effectively suppressed the photogenerated charge recombination in $MgTa_2O_{6-x}N_y$/TaON photocatalyst. By using Pt-loaded $MgTa_2O_{6-x}N_y$/TaON as a H_2-evolution photocatalyst, Z-scheme overall water splitting system with an AQY of 6.8% at 420 nm was constructed with PtO_x-loaded WO_3 as an O_2-evolution photocatalyst and IO_3^-/I^- as a redox mediator, the improved photocatalytic performance was around 7 or 360 times better than using Pt-loaded TaON or Pt-loaded $MgTa_2O_{6-x}N_y$ as an H_2-evolving photocatalyst, respectively.

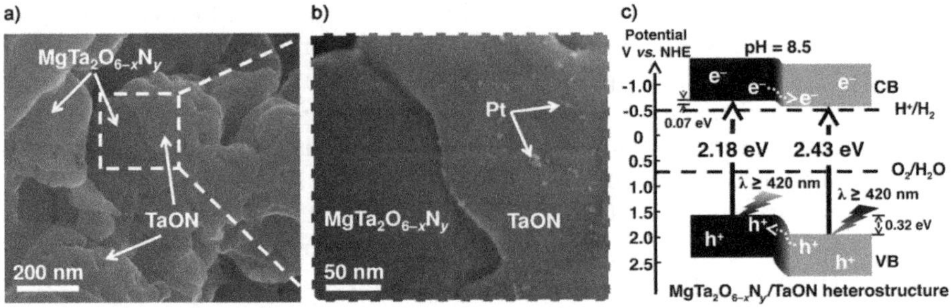

Figure 3.6 (a) and (b) Field emission scanning electron microscopy (FESEM) images of a photoreduced Pt-loaded $MgTa_2O_{6-x}N_y$/TaON photocatalyst. (c) The estimated relative band positions of the $MgTa_2O_{6-x}N_y$/TaON heterostructure. Reproduced from ref. 90 with permission from John Wiley & Sons, Copyright © 2015 WILEY-VCH Verlag GmbH & Co. KgaA, Weinheim.

Ta_3N_5 has a simple chemical composition with an anosovite structure and a narrow bandgap energy of 2.1 eV (absorption edge at 600 nm) with conduction and valence band edges at approximately −0.4 and +1.7 V *vs.* NHE at pH 0, respectively. It has been intensively studied as a promising photocatalyst candidate for solar hydrogen production, such as photocatalytic H_2 or O_2 evolution from half reactions,[91,92] one-step excitation overall watersplitting,[93] Z-scheme water splitting[94] and photoelectrochemical water oxidation.[95] This material was first reported to have photocatalytic activity of H^+ reduction or H_2O oxidation for H_2 or O_2 evolution individually from aqueous solutions containing sacrificial reagents in 2002.[91] The AQY of H_2O oxidation was estimated to be approximately 10% (420 nm < λ < 600 nm), whereas the value for H^+ reduction reaction between 420 nm < λ < 600 nm was 0.1%. Since then, tremendous efforts have been made to improve the photocatalytic H_2 evolution and O_2 evolution abilities of Ta_3N_5. A facile modification of the Ta_2O_5 precursor with alkali metal salts was valid to tailor the morphology and enhance the photocatalytic performance of Ta_3N_5.[96] It is suggested that the surface-modified alkali metal species such as Na_2CO_3 facilitated the partial dissolution of Ta_2O_5 and the nucleation of $NaTaO_3$ in the early stages of nitridation. This gave rise to the characteristic particle morphologies, improved the crystallinity and decreased the surface defects of Ta_3N_5. The Na_2CO_3-modified Ta_3N_5 combined with a CoO_x co-catalyst showed improved photocatalytic O_2 evolution with a high AQY of 5.2% at 500–600 nm. Moreover, the surface modification of Ta_3N_5 with MgO species and a CoO_x co-catalyst largely promoted the AQY for photocatalytic O_2 evolution to 11.3% under 500–600 nm illumination.[97] The modified MgO species on Ta_3N_5 acted as a passivation layer to decrease surface defect sites of Ta_3N_5. This inhibited the recombination of photoinduced carriers. Moreover, the hydrophilic surface of MgO-modified Ta_3N_5 facilitated a uniform deposition of the CoO_x co-catalysts with an intimate contact interface and large contact area. This led to a better charge transfer from Ta_3N_5 to CoO_x. However, the photocatalytic H_2 evolution ability of Ta_3N_5 is far behind the O_2 evolution ability in most cases. Conventional Ta_3N_5 photocatalysts are typically prepared *via* the nitridation of powdered Ta_2O_5 precursors under an NH_3 flow at high temperatures for a prolonged time. As a result, Ta_3N_5 tends to consist of aggregated polycrystalline particles with crystal defects and impurity energy levels, both of which are detrimental to the photocatalytic H_2 evolution activity of the material. Therefore, it is desirable to obtain good-quality Ta_3N_5 single crystals. Metallic Ta particles can be regarded as a promising material for the growth of well-crystallized Ta_3N_5 photocatalysts. This is because the oxidation reaction of Ta metal with NH_3 by high-temperature nitridation does not involve O-to-N substitution unlike conventional Ta_2O_5 precursors, which largely shortens the nitridation process. A Mg–Zr-co-doped, monodispersed, single-crystalline Ta_3N_5 nanoparticles were synthesized using a NaCl-mixed Ta nanopowder by brief nitridation (4 h) under a flow of gaseous NH_3 at 1173 K.[98] As shown in Figure 3.7, the incorporation of Mg and Zr into Ta_3N_5 eliminated the Ta^{3+} defect species and increased the number of O in the Ta_3N_5 structure, resulting in populous photoexcited electrons with extended lifetimes. Upon the highly dispersed Pt deposition by two steps (impregnation-reduction of Pt, followed by *in situ* photodeposition of Pt), the doped Ta_3N_5 nanoparticulate single crystals exhibited photocatalytic water reduction activity 45 times

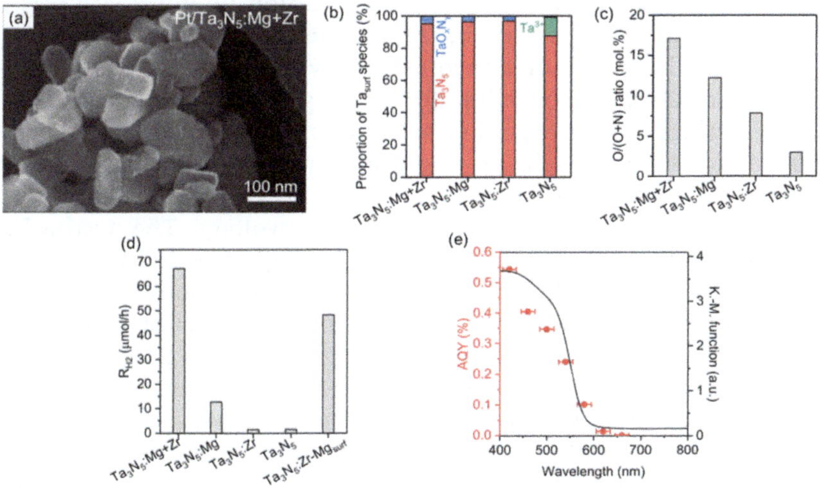

Figure 3.7 (a) FESEM images of Pt-loaded MG-Zr-co-doped Ta_3N_5 photocatalyst. (b) and (c) Proportions of different surface Ta species and O-to-anion molar ratios. (d) Initial photocatalytic H_2 evolution rates over different Ta_3N_5 specimens loaded with Pt. (e) Apparent quantum yield values for photocatalytic H_2 evolution over the Pt-loaded Mg-Zr-codoped Ta_3N_5 at various wavelengths. Reproduced from ref. 98 with permission from American Chemical Society, Copyright 2021.

higher than that of undoped Ta_3N_5, and a significant AQY of 0.54% (at 420 nm) for photocatalytic H_2 evolution among previously reported Ta_3N_5 in the literature.

Owing to the good photocatalytic performance for water oxidation, Ta_3N_5 has been applied as an O_2-evolution building block in Z-scheme water splitting. When Ta_3N_5 was modified with MgO, the Z-scheme water splitting was successfully achieved using Ir/MgO/Ta_3N_5, Pt/ZrO_2/TaON and IO_3^-/I^- as O_2-evolution photocatalyst, H_2-evolution photocatalyst and redox couple, respectively.[99] Moreover, Ta_3N_5-based Z-scheme water splitting without a redox shuttle mediator was constructed in combination with Ru/$SrTiO_3$: Rh.[100] In this case, the Ta_3N_5 photocatalyst was modified with CoO_x as water oxidation sites and metallic Ir as an electron sink, so that the efficient interparticle electron transfer was accomplished through the physical contact between Ir/CoO_x/Ta_3N_5 and Ru/$SrTiO_3$: Rh. Although Z-scheme overall water splitting using Ta_3N_5 as a H_2-evolution photocatalyst is still challenging because of its low proton reduction ability, the construction of a heterostructure to build up an interfacial electron field facilitating the spatial charge separation is an effective way to enhance the photocatalytic performance of Ta_3N_5 for H_2 evolution in Z-scheme systems. The Ta_3N_5/$BaTaO_2N$ heterostructure was created through the *in situ* formation of $BaTaO_2N$ on the surface of Ta_3N_5 by a one-pot nitridation of a Ba-impregnated Ta_2O_5 precursor.[94] As a result, the H_2 evolution rate of Ta_3N_5/$BaTaO_2N$ in the presence of methanol under visible light was enhanced and a photocatalytic Z-scheme overall water splitting system using the Ta_3N_5/$BaTaO_2N$ as a H_2-evolution photocatalyst was successfully constructed. In addition, the heterostructure $BaMg_{1/3}Ta_{2/3}O_{3-x}N_y$/$Ta_3N_5$ photocatalyst also functioned as a H_2-evolution photocatalyst in visible-light-driven Z-scheme overall water splitting.[101]

One-step excitation overall water splitting on a particulate photocatalyst is more difficult than Z-scheme water splitting, because of the stringent requirement to trigger the H^+ reduction and H_2O oxidation reactions without the assistance of sacrificial reagent and bias voltage. The narrow bandgap photocatalyst has the decreased driven energy supplied to the H_2 and O_2 evolution reactions, therefore the restriction of photoexcited charges recombination at defect sites and the control of redox reaction on the surface become more essential to realize H_2 and O_2 evolution on a particulate photocatalyst. Conventional Ta_3N_5 suffering long-term nitridation at a high temperature usually contains many anionic defects and aggregated particles that are detrimental to the photocatalytic activity. An innovative crystal growth procedure by brief nitridation of $KtaO_3$ to produce highly-crystallized Ta_3N_5 nanorod photocatalyst active for the overall water splitting without any sacrificial reagent is reported for the first time.[93] $KtaO_3$ has a cubic structure with unit cell parameters $a = b = c = 3.98$ Å, and $\alpha = \beta = \gamma = 90°$, which is naturally matchable to Ta_3N_5 structure with symmetric space group and lattice parameters $a = 3.89$ Å, $b = 10.21$ Å, $c = 10.26$ Å and $\alpha = \beta = \gamma = 90°$. It is highly possible to create a Ta_3N_5 nanostructure based on a $KtaO_3$ structure. Moreover, the volatilization of the K compound from the $KtaO_3$ crystal lattice enables the

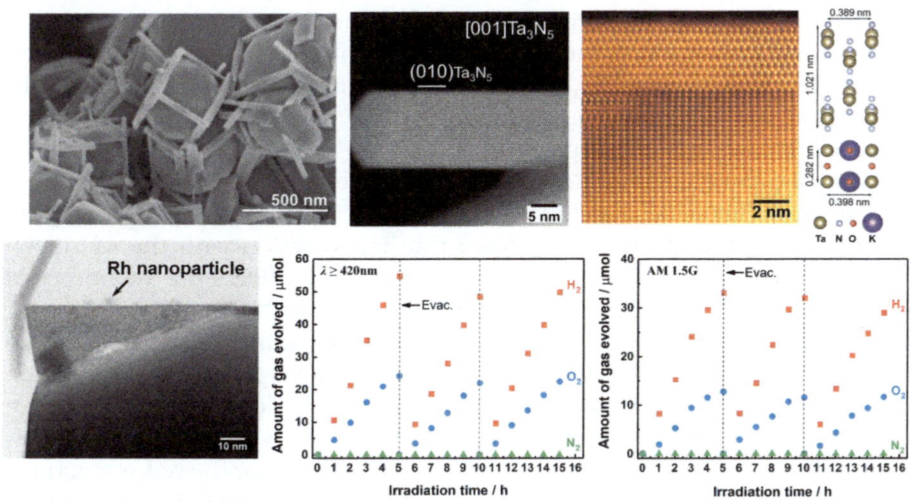

Figure 3.8 Growth of Ta_3N_5 single crystal nanorods on $KtaO_3$ cubic particles and one-step excitation overall water splitting over Rh/Cr_2O_3-modified Ta_3N_5 nanorods under visible light and simulated sunlight. Reproduced from ref. 93 with permission from Springer Nature, Copyright 2018.

fast removal of O atoms and the rapid insertion of N anions into the Ta atom matrix. This promotes the formation of high-quality Ta_3N_5 during the high temperature nitridation. As shown in Figure 3.8, single-crystalline Ta_3N_5 nanorods were preferentially grown on the edge (110) faces of $KtaO_3$ cubes without any atomic distortion and grain boundary from the surface to the interior in a very short-term (15 min) nitridation. After selective deposition of Rh/Cr_2O_3 core–shell structure nanoparticles as H_2-evolution co-catalysts, the high-crystalline and low-defective Ta_3N_5 nanorods grown on $KtaO_3$ showed photocatalytic activity for one-step excitation overall water splitting under visible light or simulated sunlight irradiation.

$AtaO_2N$ (A = Ca, Sr, Ba) are a kind of typical perovskite photocatalysts with suitable conduction and valence bands for water reduction and oxidation. The bandgap energies for $CaTaO_2N$, $SrTaO_2N$ and $BaTaO_2N$ are 2.4, 2.2, 1.9 eV, respectively. These bandgap energies reduce with the increase of ionic radius of A^{2+}, *e.g.*, Ca^{2+} (1.34 Å), Sr^{2+} (1.44 Å) and Ba^{2+} (1.61 Å). $SrTaO_2N$ is unstable during the photocatalytic reaction when compared with $CaTaO_2N$ and $BaTaO_2N$. This is due to the easy decomposition of N anions in the $SrTaO_2N$ structure in the presence of sacrificial reagents. Thus, $CaTaO_2N$ and $BaTaO_2N$ are intensively applied in photocatalytic water splitting reactions. It is reported that $CaTaO_2N$ was prepared by the nitridation of a $Ca_2Ta_2O_7$ precursor from a polymerized complex method and applied as a H_2-evolution photocatalyst in Z-scheme water splitting with Pt-modified WO_3.[102] Moreover, a heterostructure of $CaTaO_2N$ nanosheets and Ta_3N_5 nanorods by directly nitriding a unique $Kca_2Ta_3O_{10}$ precursor under the assistance of K_2CO_3 flux acted as an effective role for H_2 evolution in steady Z-scheme water splitting.[103] Importantly, through the surface modification of amorphous Ti oxyhydroxide to suppress the N_2 evolution from self-decomposition and the reverse reaction between H_2 and O_2, one-step excitation overall water splitting was achieved on $RhCrO_y$-loaded $CaTaO_2N$.[104]

$BaTaO_2N$ with wide adsorption of *ca.* 660 nm is a promising material that shows a theoretical STH of 20.6%.[102,105] When compared with other similar perovskites, $CaTaO_2N$ and $SrTaO_2N$, $BaTaO_2N$ exhibited the highest photocatalytic H_2 evolution activity from the half reaction with the sacrificial reagent. This is because the large Ba^{2+} cation size affects the symmetry of the perovskite crystal lattice.[102] When combining with Pt-modified WO_3 for O_2 evolution, $BaTaO_2N$ exhibited Z-scheme water splitting activity with an AQY of 0.1% at 420–440 nm in the presence of IO_3^-/I^- as a redox mediator. The modification of $BaTaO_x$ with $BaZrO_x$ has been demonstrated to be an effective approach to improve the photocatalytic activity of $BaTaO_2N$ for H_2 evolution from water. This is because the driving force for the redox reactions increased with the incorporation of $BaZrO_3$ into the $BaTaO_2N$ crystal lattice. Also, the formation of a large number of defects in the $BaZrO_3$–$BaTaO_2N$ solid solution was suppressed which minimized undesirable photogenerated charge recombination.[105,106] The solar-driven Z-scheme water splitting systems using Pt/$BaZrO_3$–$BaTaO_2N$ as the H_2-evolution photocatalyst were successfully constructed with Pt/WO_3 or rutile TiO_2 as the O_2-evolution photocatalyst, in which STHs were 0.0067% and 0.014%, respectively.[106]

Conventional $BaTaO_2N$ particles are produced by high-temperature nitridation of Ta_2O_5 and $BaCO_3$ mixture or Ba–Ta oxides from solid-state reaction method or polymerized complex method. This generally involves structural defects and mid-gap states working as recombination and trapping centers for photogenerated electron–hole pairs. The flux-assisted nitridation method has been demonstrated to be effective for the synthesis of high-quality $BaTaO_2N$ crystals during the thermal procedure, in which a high-temperature molten salt is used as the solvent for crystallization. It is reported that a $BaTaO_2N$ photocatalyst with reduced defect density and increased surface area was synthesized by a KCl flux-assisted nitridation of a Ba-rich $LiBa_4Ta_3O_{12}$ precursor.[107] When compared with $BaTaO_2N$ from the nitridation of a $Ba_2Ta_2O_7$ precursor prepared by a conventional solid-state reaction method, the $BaTaO_2N$ made using the flux method exhibited much better photocatalytic H_2 evolution performance in the sacrificial half reaction or Z-scheme water splitting. $BaTaO_2N$ synthesized by direct nitridation of Ta_2O_5 and $BaCO_3$ with KCl as a flux showed cubic morphology with both {100} and {110} facets.[108] Benefiting from efficient spatial charge separation between the {100} and {110} facets of the cubic $BaTaO_2N$ particles, the photocatalytic activity for H_2 production on $BaTaO_2N$ with co-exposed {100} and {110} facets was nearly 10 times higher than that of $BaTaO_2N$ with only {100} facets and that of the conventional irregularly shaped sample. Moreover, a highly-crystalline $BaTaO_2N$ crystal with a plate-like structure from a K_2CO_3–KCl binary flux also exhibited an excellent photocatalytic H_2 evolution activity under visible light irradiation.[109] Upon the optimization of the flux species,

RbCl-assisted nitridation provided well-crystalized $BaTaO_2N$ single crystals with a low density of structural defects and mid-gap states. This resulted in the suppression of charge recombination and the promotion of photoexcited electron transfer.[110,111] When decorating the uniform Pt nanoparticulate co-catalysts highly dispersed on RbCl-assisted $BaTaO_2N$ with intimate contact interface through stepwise deposition procedure, such co-catalyst/photocatalyst reciprocal structure facilitated very efficient electron transfer from $BaTaO_2N$ to Pt co-catalyst.[111] Therefore, the Pt-loaded $BaTaO_2N$ photocatalyst showed H_2 evolution over 100 times more efficient than before, with an AQY of 6.8% at 420 nm from a methanol aqueous solution, and an STH of 0.24% in Z-scheme water splitting (Figure 3.9). This is by far the most efficient bias-free Z-scheme water splitting system involving particulate photocatalysts harvesting visible light up to 650 nm.

$LaTaON_2$ has a perovskite structure like $ATaO_2N$ (A = Ca, Sr, Ba), and the O–N ratio in the material changes to 1 : 2 to maintain the charge balance. $LaTaON_2$ harvesting intense visible light up to 640 nm can photocatalytically produce H_2 and O_2 from aqueous solutions containing appropriate sacrificial reagents.[112,113] However, conventional $LaTaON_2$ photocatalysts obtained from solid-state reactions exhibit low photocatalytic activity owing to high densities of defects associated with anion vacancies and reduced Ta species generated during the harsh nitridation process. It is reported that high-quality $LaTaON_2$ shells with a low density of reduced Ta species were successfully

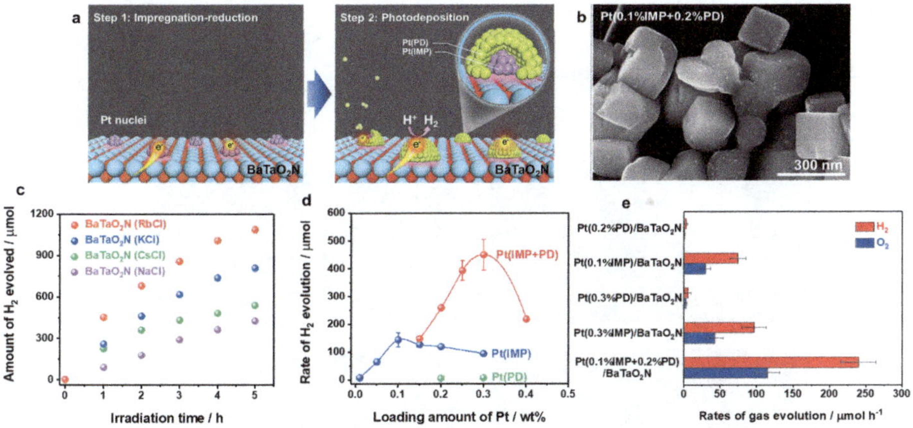

Figure 3.9 (a) Schematic of stepwise Pt co-catalyst deposition on $BaTaO_2N$. (b) SEM image of Pt-modified $BaTaO_2N$ produced by stepwise deposition method. (c) Visible-light-driven H_2 evolution on different flux-assisted $BaTaO_2N$ photocatalysts from an aqueous methanol solution. (d) H_2 evolution rates as a function of the total Pt loading during visible-light-driven H_2 production reactions from an aqueous methanol solution. (e) H_2 and O_2 evolution rates during Z-scheme water splitting reaction using Pt-modified $BaTaO_2N$ (RbCl). Reproduced from ref. 111 with permission from Springer Nature, Copyright 2021.

grown on layered $LaKNaTaO_5$ that has lattice parameters that match those of $LaTaON_2$ and contains volatile components by mild nitridation.[112] Thus, the photocatalytic H_2 evolution activity of $LaKNaTaO_5/LaTaON_2$ is four times greater than that obtained from conventional irregular-shaped $LaTaON_2$ powders when modified with Rh nanoparticles as the H_2-evolving co-catalyst. The formation of solid solution is a functional pathway to operate band structure and control defect states of the original oxynitride. $LaMg_{2/3}Ta_{1/3}O_3$ is isostructural with $LaTaON_2$, with two-thirds of the Ta sites in $LaTaON_2$ replaced by Mg^{2+}. Using a homogeneously mixed La–Mg–Ta oxide precursor prepared from a polymerized complex method, $LaMg_xTa_{1-x}O_{1+3x}N_{2-3x}$ can be produced by thermal ammonolysis of the corresponding oxides.[114] As shown in Figure 3.10, with the increase in Mg–Ta ratio, the wavelength of the absorption edge of $LaMg_xTa_{1-x}O_{1+3x}N_{2-3x}$ exhibited a monotonic blue shift from 640 to 525 nm. Simultaneously, the increased O–N ratio in the solid solution structure led to the downward shift of the valence band maximum, which provided sufficient driving force for the water oxidation reaction. When coating the entire surface with amorphous Ti and Si oxyhydroxides

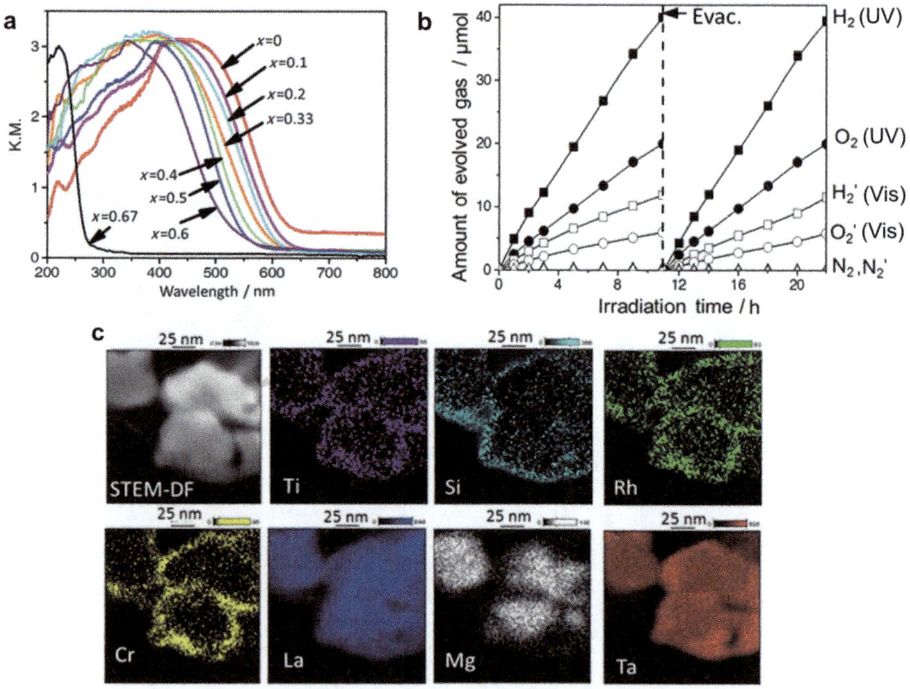

Figure 3.10 (a) UV-vis DRS of the $LaMg_xTa_{1-x}O_{1+3x}N_{2-3x}$ series. (b) Gas evolution during water splitting on $TiOXH/SiOXH/RhCrO_y/LaMg_xTa_{1-x}O_{1+3x}N_{2-3x}$ under UV-vis irradiation. (c) Scanning transmission electron microscopy with energy dispersive X-ray spectroscopy (STEM-EDS) elemental maps of $TiOXH/SiOXH/RhCrO_y/LaMg_xTa_{1-x}O_{1+3x}N_{2-3x}$. Reproduced from ref. 114 with permission from John Wiley & Sons, Copyright © 2015 WILEY-VCH Verlag GmbH & Co. KGaA, Weinheim.

and loading $RhCrO_y$ as the H_2-evolving co-catalyst, $LaMg_xTa_{1-x}O_{1+3x}N_{2-3x}$ solid solution steadily decomposed water into H_2 and O_2 under visible light irradiation (Figure 3.10). This is because the surface modification can efficiently suppress N_2 evolution from the self-oxidation of the photocatalyst and O_2 reduction in competition with water splitting. Based on the bandgap engineering and surface modification, one-step excitation overall water splitting was also achieved on the $LaSc_xTa_{1-x}O_{1+2x}N_{2-2x}$ ($x \geq 0.5$) solid solution photocatalyst.[115]

3.3.2 Niobium- and Titanium-based Oxynitrides

Nb-based perovskite oxynitrides, such as $BaNbO_2N$[116] and $LaNbON_2$,[117] and Ti-based perovskite oxynitrides like $LaTiO_2N$,[118] are also attractive materials for solar energy conversion, because of their narrow bandgap energies. However, photocatalytic activities for H_2 or O_2 evolution using Nb- and Ti-based perovskite oxynitrides are inefficient when compared with Ta-based oxynitride compounds. This is because Nb^{5+} or Ti^{4+} species are reduced more readily to Nb^{4+}, Nb^{3+} or Ti^{3+} during high temperature nitridation in an NH_3 atmosphere. These lead to the formation of surface anion defects to maintain charge balance and/or surface impurity phases.[116,119] Therefore, a novel synthesis method and surface treatment are required to obtain high-quality Nb- and Ti-based perovskite oxynitrides. It is reported that well-crystallized $BaNbO_2N$ was prepared by the nitridation of $Ba_5Nb_4O_{15}$ with additional $BaCO_3$, which suppressed the possible generation of impurity niobium oxynitride species.[120] When loading CoO_x as the O_2-evolving co-catalyst, $BaNbO_2N$ generated oxygen from an aqueous $AgNO_3$ solution under visible light irradiation up to 740 nm, the longest wavelength ever reported for (oxy) nitride photocatalysts. Moreover, highly crystalline $BaNbO_2N$ cubic particles were synthesized using NaCl or CsCl as the flux, while truncated octahedral shapes were obtained using KCl, KBr, KI, or RbCl.[121] Thus, higher photocatalytic activity of $BaNbO_2N$ is expected by controlling the morphology. $LaNbON_2$ can absorb visible light up to approximately 750 nm but is almost inactive for photocatalytic water oxidation. A plate-like $LaNbON_2$ acquired by the controlled nitridation of a unique $LaKNaNbO_5$ precursor exhibited enhanced photocatalytic O_2 evolution activity with an AQY of 0.82%.[122] As a typical Ti-based perovskite oxynitride, $LaTiO_2N$ with a bandgap of 2.1 eV could photocatalytically produced H_2 and O_2 using the appropriate sacrificial reagents.[118] Through refining the nitridation precursor by flux method and modifying suitable co-catalyst, single-crystalline $LaTiO_2N$ with meso and macroporous structure gave a high quantum efficiency of 27.1 ± 2.6% at 440 nm for O_2 evolution from $AgNO_3$ solution, which is the best values reported for 600 nm-class particulate photocatalysts.[119] In addition, surface treatment of $LaTiO_2N$ particles by the treatment with aqua regia solution evidently eliminated the defective surface reconstructed layers during the NH_3 nitridation, resulting in the two-fold enhancement of photocatalytic activity of $LaTiO_2N$ for H_2 evolution.[123]

3.3.3 Other Typical (Oxy)nitrides

In addition to the (oxy)nitrides with d^0 electronic configuration, there are some (oxy)nitride photocatalysts with d^{10} electronic configuration reported for water splitting. β-Ge_3N_4 was the first example of a non-oxide photocatalyst that decomposed water into H_2 and O_2 stoichiometrically. β-Ge_3N_4 was synthesized by nitriding GeO_2 powder at 1153 K for 10 h in an NH_3 atmosphere.[124] The overall water splitting activity on bare β-Ge_3N_4 is negligible. When loaded with RuO_2 as a co-catalyst, β-Ge_3N_4 became photocatalytically active for overall water splitting and its activity increased sharply with the RuO_2 loading, reaching a maximum at RuO_2 content of 1 wt%. Moreover, the treatment of β-Ge_3N_4 in a high-pressure NH_3 atmosphere at 823 K drastically improved the photocatalytic performance of overall water splitting. This can be attributed to the deduction of the defect density at the surface and in the bulk by the high-pressure treatment with NH_3.[125] It was also found that the reaction conditions and post-treatment greatly affect the photocatalytic performance of RuO_2-loaded β-Ge_3N_4.[126] The highest performance was achieved with 1 M H_2SO_4. However, the deactivation of RuO_2-loaded β-Ge_3N_4 occurred as the reaction progressed due to the liberation of Ge cations at the surface layer and the removal of the co-catalyst. Through the calcination of the used photocatalyst in air at 673 K and then reloading of the RuO_2 co-catalyst, the photocatalytic activity of β-Ge_3N_4 recovered up to 80% of the initial activity. However, the bandgap of β-Ge_3N_4 is *ca.* 3.8 eV, which is unable to absorb visible light. The wurtzite-type $(Zn_{1+x}Ge)(N_2O_x)$ solid solution with a similar lattice parameter as $ZnGeN_2$ and ZnO has been demonstrated to be an active photocatalyst for overall water splitting under ultraviolet and visible light.[127–129] Although $ZnGeN_2$ and ZnO have bandgap energies larger than *ca.* 3 eV, the $(Zn_{1+x}Ge)(N_2O_x)$ solid solution formed by nitridation of GeO_2 and ZnO (Zn content can be tuned) under an NH_3 flow at 1123 K for 15 h exhibited a bandgap of *ca.* 2.7 eV, and the bandgap decreased with increasing Zn concentration. The narrow bandgap of $(Zn_{1+x}Ge)(N_2O_x)$ solid solution is attributed to the p–d repulsion between N 2p + O 2p and Zn 3d orbitals leading to the upward shift of the top of valence band. When RuO_2 or $Rh_{2-x}Cr_xO_3$ was modified as the co-catalyst, $(Zn_{1.44}Ge)(N_{2.08}O_{0.38})$ achieved efficient overall water splitting with stoichiometric H_2 and O_2 production under visible light irradiation.

A solid solution of GaN and ZnO, $(Ga_{1-x}Zn_x)(N_{1-x}O_x)$, has also been reported as an important photocatalyst with d^{10} electronic configuration for one-step excitation overall water splitting under visible light.[130–133] Both GaN and ZnO are well-known materials having wurtzite structures with similar lattice parameters and bandgaps of *ca.* 3.4 and 3.2 eV, respectively. Similar to the $(Zn_{1+x}Ge)(N_2O_x)$ solid solution, the $(Ga_{1-x}Zn_x)(N_{1-x}O_x)$ showed a narrow bandgap responding to visible light because of the p–d repulsion between N 2p and Zn 3d orbitals. The $(Ga_{1-x}Zn_x)(N_{1-x}O_x)$ solid solution is conventionally synthesized from the mixture of Ga_2O_3 and ZnO under an NH_3 flow at the temperature higher than 1073 K.[130,133] In this reaction, $ZnGa_2O_4$ spinel is a key intermediate for the formation of $(Ga_{1-x}Zn_x)(N_{1-x}O_x)$ and limits the zinc concentration in the final solid solution oxynitride. Even if the Zn–Ga ratio in the ZnO–Ga_2O_3 mixtures is larger than 1/2, the excessive ZnO would be reduced to metallic Zn species and then vaporized in the high temperature NH_3 atmosphere. The bandgap of $(Ga_{1-x}Zn_x)(N_{1-x}O_x)$ is generally determined by the concentration of Zn species in solid solution. So significant attempts have been explored to increase the Zn content in $(Ga_{1-x}Zn_x)(N_{1-x}O_x)$ toward narrowing the bandgap of this solid solution for visible-light-driven solar H_2O splitting. For instance, $(Ga_{1-x}Zn_x)(N_{1-x}O_x)$ nanocrystals with x values from 0.3 to 0.87 and diameters of ~18 nm were synthesized by nitriding the mixture of nanocrystalline $ZnGa_2O_4$ and ZnO in varying ratios at 650 °C for 10 h under an NH_3 flow.[134] Because of the nanoscale precursors, the required nitridation temperature was lower than that for the preparation of the bulky zinc gallium oxynitride. At the milder synthetic conditions, the loss of Zn was inhibited and high-quality $(Ga_{1-x}Zn_x)(N_{1-x}O_x)$ nanocrystals were obtained. The bandgap of $(Ga_{1-x}Zn_x)(N_{1-x}O_x)$ nanocrystals decreased from 2.7 to 2.2 eV with increasing Zn content, corresponding to a 260% increase in the fraction of solar photons. The high-quality $(Ga_{1-x}Zn_x)(N_{1-x}O_x)$ with tunable bandgaps show the potential to achieve high photocatalytic performance. Subsequently, the composition of Zn in $(Ga_{1-x}Zn_x)(N_{1-x}O_x)$ nanocrystals further increased with an x value from 0.06 to 0.98.[135] A minimum bandgap of 2.25 eV was obtained at x = 0.87, whereas a blue shift occurred at x = 0.91 and 0.98. Another $(Ga_{1-x}Zn_x)(N_{1-x}O_x)$ solid solution was prepared by using quite different layered double hydroxide (LDH) precursors containing Ga^{3+}, Zn^{2+}, and CO_3^{2-} ions.[136] The Zn–Ga ratios in $(Ga_{1-x}Zn_x)(N_{1-x}O_x)$ were finely controlled by tuning the ratio in LDH, and the nitridation time for $(Ga_{1-x}Zn_x)(N_{1-x}O_x)$ significantly decreased because the Zn and Ga ions were homogeneously mixed at an atomic scale in LDH. The absorption edge of $(Ga_{1-x}Zn_x)(N_{1-x}O_x)$ exhibited a strong red shift and the bandgap decreased with the increase of x (from 2.6 eV at x = 0.46 to 2.37 eV at x = 0.81).

The photocatalytic performance of $(Ga_{1-x}Zn_x)(N_{1-x}O_x)$ depends on the co-catalyst modification, which will promote the migration of photogenerated charges, construct the catalytic reaction sites, and reduce the activation energy required for gas evolution. A series of transition metals and metal oxides are studied as the co-catalysts to enhance the overall water splitting activity of $(Ga_{1-x}Zn_x)(N_{1-x}O_x)$. It is reported that co-loading transition metals (*e.g.*, Fe, Ni, Co, Ru, Cu, Pd, Rh, Ir, Ag, or Pt) and Cr oxide as the co-catalysts significantly promote the photocatalytic performance of $(Ga_{1-x}Zn_x)(N_{1-x}O_x)$ for one-step excitation overall water splitting.[137] All dual co-catalysts produced H_2 and O_2 stoichiometrically from water except the Ag–Cr co-catalyst, and the activity was strongly dependent on the Cr loading. The highest photocatalytic activity was obtained by co-loading 1 wt% of Rh and 1.5 wt% of Cr co-catalyst on $(Ga_{1-x}Zn_x)(N_{1-x}O_x)$. Moreover, a new type of metal/Cr_2O_3 with a core–shell structure by photodeposition was developed as a co-catalyst for efficient overall water splitting.[131,138–140] Typically, noble metal nanoparticles were first loaded on $(Ga_{1-x}Zn_x)(N_{1-x}O_x)$ ($x = 0.12$) photocatalyst prepared from a noble metal complex salts such as Na_3RhCl_6.[130] Figure 3.11(a) shows a high-resolution transmission electron microscopy (TEM) image of Rh nanoparticulate co-catalyst loaded on the photocatalyst. The Cr_2O_3 shell layer (about 2 nm thickness) was subsequently photodeposited on the Rh nanoparticles using K_2CrO_4 as a precursor to form a core–shell nanostructure (Figure 3.11(b)). The Rh-loaded sample displayed a little photocatalytic activity, whereas the Rh/Cr_2O_3-co-loaded sample showed stable and stoichiometric gas evolutions of H_2 and O_2, indicating that the Cr_2O_3 shell played an indispensable role. As illustrated in Figure 3.11(c), the metallic Rh nanoparticles enable the extraction of photogenerated electrons from $(Ga_{1-x}Zn_x)(N_{1-x}O_x)$ and provide catalytic sites for proton reduction. During the photocatalytic overall water splitting, the bare Rh particles are also active for

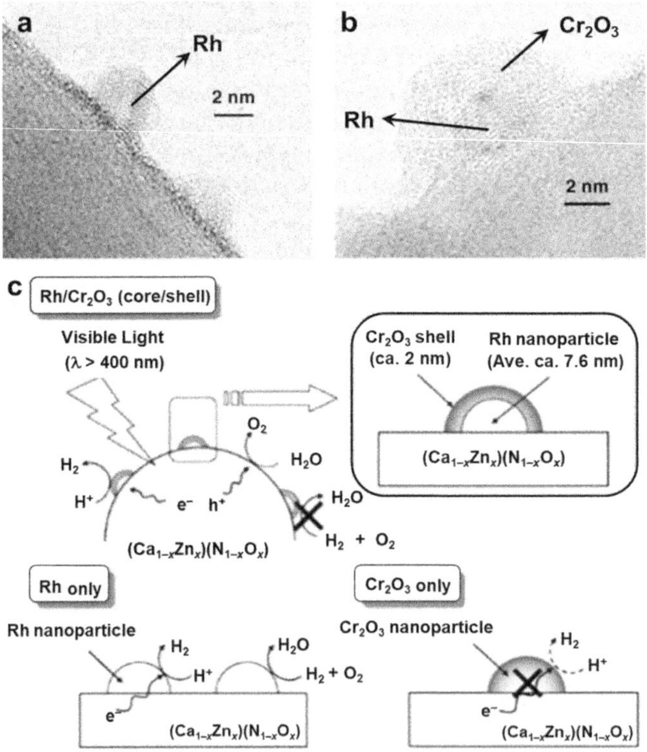

Figure 3.11 (a) and (b) TEM images of Rh-loaded $(Ga_{1-x}Zn_x)(N_{1-x}O_x)$ ($x = 0.12$) before and after photodeposition of a Cr_2O_3 shell. (c) Schematic reaction mechanism of $(Ga_{1-x}Zn_x)(N_{1-x}O_x)$ loaded with Rh/Cr_2O_3 for overall *water* splitting. Reproduced from ref. 131 with permission from John Wiley & Sons, Copyright © 2006 WILEY-VCH Verlag GmbH & Co. KGaA, Weinheim.

H_2O formation between H_2 and O_2. Cr_2O_3 itself does not work for the proton reduction reaction, but the Cr_2O_3 shell coated on Rh nanoparticles prevents the penetration of O_2 molecules, thereby suppressing the backward reaction from H_2 and O_2 and improving photocatalytic activity of overall water splitting. Similar results are observed when the Rh was replaced with other noble metals (*e.g.*, Pd, Pt) or metal oxides (*e.g.*, RuO_2, Rh_2O_3, NiO_x).[141]

Graphitic carbon nitride photocatalysts have attracted great attention, because of their composition of abundant carbon and nitrogen elements, leading to low cost and non-toxicity, high stability, easy synthesis and modification.[142] The valence band of graphitic carbon nitride is the highest occupied molecular orbital levels derived from nitrogen orbitals, and the conduction band the lowest unoccupied molecular orbital levels consisting predominantly of carbon orbitals, thus the bandgap structure of graphitic carbon nitride with a bandgap of 2.7–2.9 eV is suitable for water reduction and oxidation. In 2009, it was reported that polymeric carbon nitride evolved H_2 or O_2 in sacrificial half reactions under visible light irradiation.[143] The H_2 evolution ability for carbon nitride is much higher than the O_2 evolution ability, which is attributable to the sufficient reduction driving force of conduction band electrons and low oxidation driving force of valence band holes. A Z-scheme water splitting system, coupling the organic graphitic carbon nitride as H_2-evolution photocatalyst, was also successfully constructed. When combining with WO_3 as the O_2-evolution photocatalyst and IO_3^-/I^- as the redox mediator, the Pt-loaded g-C_3N_4 derived from an urea precursor rather than thiourea and dicyandiamide can achieve reproducible Z-scheme water splitting under visible light.[144] More importantly, it was discovered that evident one-step excitation water splitting activity was generated on graphitic carbon nitride nanosheets, when modifying Pt co-catalysts on the surface by photodeposition, in which Pt co-catalysts involved metallic Pt and PtO_x as promoters for H_2 evolution and O_2 evolution, respectively.[145] The water splitting activity with stoichiometric evolution of H_2 and O_2 gases on Pt-modified graphitic carbon nitride photocatalyst was significantly enhanced after further CoO_x modification as O_2 evolution co-catalysts. Polymeric carbon nitride usually has low crystallinity, resulting in low photocatalytic activity for overall water splitting. Therefore, many studies aim to prepare crystalline polymeric carbon nitride.[142,146,147] Among them, a highly-crystallized polytriazine imide intercalated with LiCl (PTI/Li^+Cl^-) was synthesized using an ionothermal method and achieved efficient photocatalytic overall water splitting by the loading of suitable co-catalysts.[148] The location of Pt or Co co-catalysts by site-selective photodeposition demonstrated that the reactive facets of PTI/Li^+Cl^- were $\{10\bar{1}0\}$ planes rather than $\{0001\}$ planes (Figure 3.12). As a result, the photocatalytic activity had an approximately linear correlation with the surface area ratio of $\{10\bar{1}0\}$ and $\{0001\}$ of the PTI/Li^+Cl^- crystals. Upon the optimization of exposed active facets, PTI/Li^+Cl^- exhibited excellent one-step excitation overall water splitting performance with an apparent quantum yield of 8% at 365 nm.

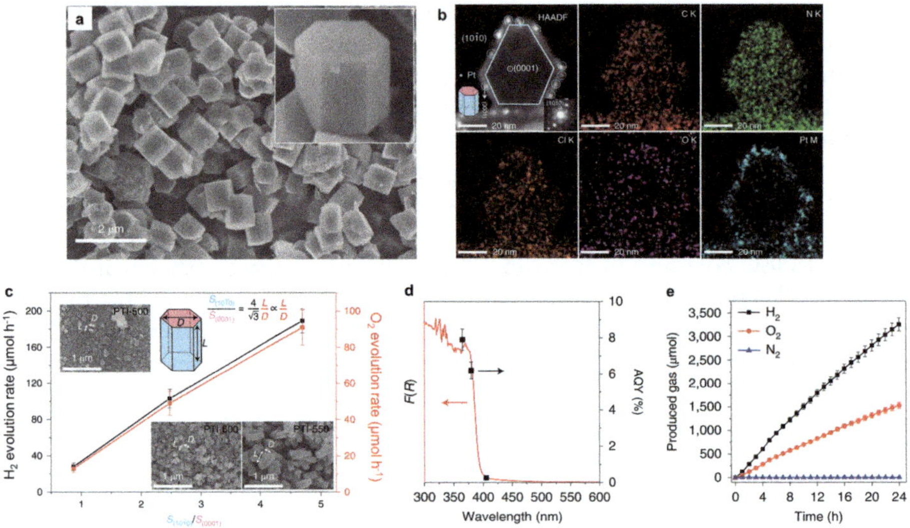

Figure 3.12 (a) SEM image showing the well-developed prism-like shape of PTI/Li^+Cl^- crystals. (b) The high-angle annular dark-field scanning transmission electron microscopy (HAADF-STEM) image of Pt-loaded PTI/Li^+Cl^- aligned close to the [0001] direction and EDS maps of the main constitutional elements. (c) The plot of maximum hydrogen and oxygen evolution rate as a function of the mean surface area ratio of the {100} and {0001} planes of the samples. (d) Wavelength-dependent apparent quantum yield for overall *water* splitting of PTI/Li^+Cl^-. (e) The accumulated amount of the gases produced by PTI/Li^+Cl^-. Reproduced from ref. 148 with permission from Springer Nature, Copyright 2020.

3.4 Strategies for Enhancing Activity

Many (oxy)nitride photocatalysts have been reported to exhibit excellent H_2 or O_2 evolution activity, while the efficiency of overall water splitting is still too low to satisfy the practical application. This is mainly because the crucial steps affecting the photocatalytic water splitting, namely, absorption of light, charge separation and migration, redox reaction on photocatalyst surface, still are not efficiently promoted. Many strategies have been proposed to enhance the photocatalytic performance of (oxy)nitride photocatalysts, such as doping, morphology control, heterostructures design, surface and co-catalyst modification. These strategies mainly focus on the improvement of charge separation, charge transfer, and surface reaction. In addition, the stability of photocatalyst and the suppression of backward reaction have an important influence on the performance, which can also be improved and controlled by the appropriate surface modification and co-catalyst design.

3.4.1 Doping

Doping is one of the most important strategies to promote activity, because the incorporation of heteroatoms in the semiconductor materials will change the original properties, such as electronic structure, band composition, defect concentration, and surface structure.[149–152] It has been well demonstrated that the fundamental properties of oxides could be modified by doping techniques. Especially, many doped oxides show visible-light-driven activities because of the introduction of new energy levels in the bandgap by doping foreign elements. The (oxy)nitrides are firstly derived from the doping of non-metal N element into the oxides, in which the hybridization of N 2p orbitals and O 2p orbitals results in the upward shift of the valence band. In addition to the non-metal elements (*e.g.*, N, S, P, B, Si), the doping of metal elements (*e.g.*, Rh, W, Mo, Cr, Zr, Mg, Cu, Zn, Al) also influences the band structure of the original materials.[153,154] When the original ions in the crystal structure are substituted by foreign ions with higher or lower valence states, it will form n-type or p-type doping. The formation of a p–n junction at the surface layer of the photocatalyst by non-homogeneous doping is desirable for the enhancement of water splitting. This is because the built-in field provides a supplementary driving force for charge separation and transfer. Moreover, the homogeneous doping, which occurs in the entire bulk of photocatalyst by selecting doping ions with ion radius close to the original ions, has been widely used to modify the electronic structure, reduce the defect density, and facilitate the surface reaction. Through appropriate doping techniques, the photocatalytic activity of (oxy)nitride photocatalysts was significantly enhanced and the overall water splitting on some (oxy) nitrides was also achieved. For instance, $LaTaON_2$ doped with Mg accomplished one-step excitation overall water splitting at wavelengths up to 600 nm, mainly because of the adjusted bandgap energy by the formation of $LaMg_xTa_{1-x}O_{1+3x}N_{2-3x}$ solid solution.[114] Zr-doped $LaTaON_2$ exhibited a superior photocatalytic activity for water splitting with an AQY of 0.93% at 420 ± 20 nm.[155] It is attributed that Zr dopants helped to decrease the defect density of reduced Ta^{4+} species, increase the surface hydrophilicity and induce the formation of mesoporous structures of $LaTaON_2$ (Figure 3.13).

3.4.2 Morphology Control

It is well-known that the morphological characteristics of photocatalysts, including particle size, surface area, exposed facets and so on, have important effects on photocatalytic performance. Generally, nanosized photocatalyst particles shorten the migration distance for photoexcited charge carries from interior to surface and exhibit a large surface area for the catalytic redox reaction. When compared to conventional bulky Ta_3N_5, nanostructured Ta_3N_5 has exhibited higher photocatalytic activity for H_2 evolution. Many efforts have been also employed to control the nanostructure of Ta_3N_5 crystals. For example, the porous Ta_3N_5 nanoplate

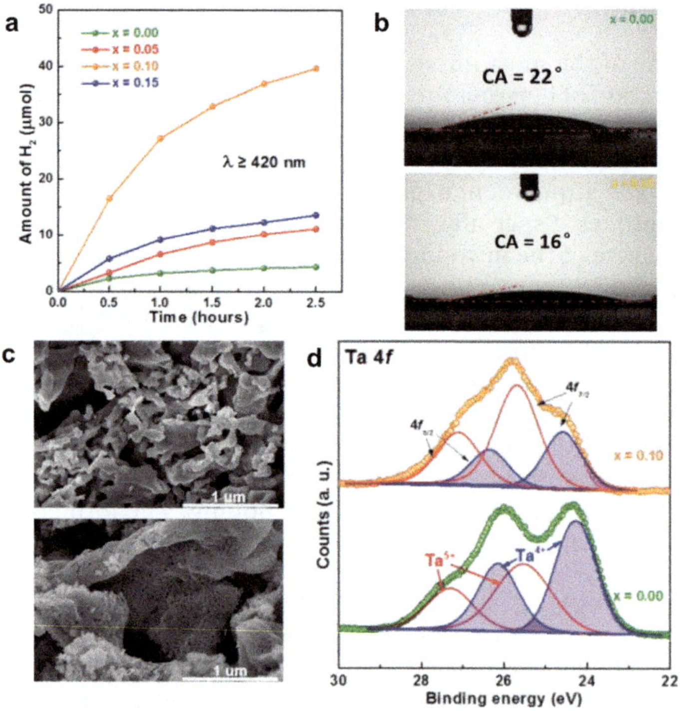

Figure 3.13 (a) Photocatalytic H_2 evolution on $LaTa_{1-x}Zr_xO_{1+y}N_{2-y}$ ($0 \leq x \leq 0.15$) samples. (b) Contact angles of stable *water* droplet on compressed pellets. (c) SEM images and (d) Ta 4f X-ray photoelectron spectra of $LaTa_{1-x}Zr_xO_{1+y}N_{2-y}$ ($x = 0.00$) and $LaTa_{1-x}Zr_xO_{1+y}N_{2-y}$ ($x = 0.10$). Reproduced from ref. 155 with permission from the Royal Society of Chemistry.

and octahedra Ta_3N_5 particles were synthesized through the nitridation of Ta_2O_5 templates derived from aerosol-assisted molten salt synthesis with different fluxes, as shown in Figure 3.14.[156] These special templates allowed Ta_3N_5 to have defined morphology, higher surface area, and lower defect states. Thus, the improved photocatalytic performance for H_2 or O_2 evolution was achieved on these shape defined Ta_3N_5. The ordered mesoporous Ta_3N_5 hollow spheres with high crystallinity and a high specific surface area were obtained by nitriding SiO_2-coated mesoporous Ta_2O_5 precursor.[157] After modification of Pt, the mesoporous Ta_3N_5 hollow spheres showed photocatalytic activity for H_2 production two times higher than the conventional bulk Ta_3N_5. The well-crystallized Ta_3N_5 with defined morphologies were also prepared by flux method.[158] The quality and morphology of the final products were mainly affected by the starting materials and nitridation conditions. Besides particle size and crystal shape, it has been demonstrated that the construction of spatial charge separation facets on photocatalysts is beneficial to photocatalytic performance. The [100] oriented Ta_3N_5 with co-exposed {001} and {010}

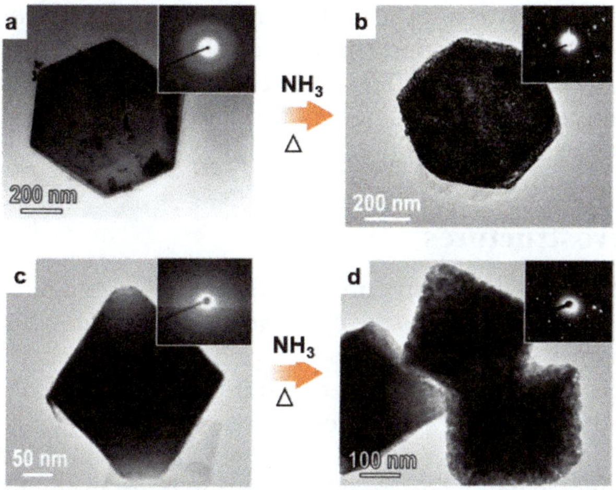

Figure 3.14 TEM images (top right insets: electron diffraction) of (a) Ta_2O_5 nanoplate, (b) Ta_3N_5 nanoplate, (c) Ta_2O_5 octahedra, and (d) Ta_3N_5 octahedra. Reproduced from ref. 156 with permission from the Royal Society of Chemistry.

facets were prepared by flux-assisted nitridation method and showed a significant improvement in photocatalytic activity.[159] Experimental and theoretical results indicated that the high activity was due to the efficient charge separation and transfer promoted by surface electric field between {010} and {001}. Similar results were observed on $BaTaO_2N$ with co-exposed {100} and {110} facets, in which photogenerated electrons and holes tended to migrate to the {100} and {110} facets, respectively, achieving a spatial charge separation to enhance the photocatalytic activity for H_2 evolution.[108]

Nevertheless, morphology control for (oxy)nitrides including particle size, shape, and exposed facets is still insufficient, mainly because of the harsh synthetic conditions of high temperature ammonolysis. Usually, (oxy)nitride materials with thermodynamically favourable shapes can be achieved if these materials show specific shape characteristics, other characteristics (*e.g.*, size, crystallinity, or composition) will change. In many cases, shape-defined oxides are obtained by simply adding the shape directing agents (*e.g.*, polyethylene glycol, polyvinylpyrrolidone and cetyltrimethylammonium bromide molecules), or by changing the synthetic conditions (*e.g.*, treatment temperature, treatment time, pressure, atmosphere). However, the morphologies of many (oxy)nitrides are determined by the corresponding metal oxide precursor with defined morphologies. Notably, porosity and defects will be introduced from this approach because of the unequal replacement of oxygen ions (O^{2-}) with nitrogen ions (N^{3-}) and phase transformation. At present, some (oxy)nitrides with smooth surface, high crystallinity, and specific morphology are obtained by flux-assisted method, which provides a direct and effective crystallization route for the products. The selection and collocation of flux reagents and the optimization of synthetic conditions are the key points of this method. To further enhance the photocatalytic performance *via* morphology control, more new synthetic approaches and techniques are needed to be developed.

3.4.3 Heterostructures

The construction of heterostructures to improve the charge separation and promote the light harvest is also the effective means to obtain higher photocatalytic performance. Numerous composite photocatalysts, such as TiO_2/Cu_2O, MoS_2/Ni_3S_2 and $BiVO_4/WO_3$, exhibited significantly promoted photocatalytic activities for water splitting.[160–163] The heterostructure is usually based on two types of semiconductors with suitable bandgap, matchable band position and crystal lattice, which will facilitate the formation of an intimate interface and create an effective internal electric field towards charge separation and transfer. The experimental techniques for the preparation of effective heterostructure remain difficult.[164] To date, many efforts have been made to construct heterostructure based on (oxy)nitrides. Conventionally, the heterostructure of (oxy) nitrides is formed through the surface decoration of the basic materials with foreign compounds. The efficient H_2-evolution photocatalyst $MgTa_2O_{6-x}N_y$/TaON heterostructure was prepared from MgO-modified Ta_2O_5 precursor.[90]

Considering the formation of (oxy)nitride from the transformation of oxide/oxysalt, the heterostructure can be directly obtained by the nitridation of a special oxide precursor with the controlled compositions. For instance, $Ta_3N_5/BaTaO_2N$ heterostructure with type II junction alignment was fabricated by one-step nitridation of single-phase $KBa_2Ta_5O_{15}$ precursor, in which the K element was evaporated and the O atoms were substituted by N atoms gradually under the ammonolysis process, as shown in Figure 3.15.[165] With the evaporation of K and the introduction of N atoms, the $KBa_2Ta_5O_{15}$ precursor first was transformed into Ta_3N_5, and the Ba atoms and the remaining Ta atoms were combined to form $BaTaO_2N$ on the surface of Ta_3N_5. As a result, the $Ta_3N_5/BaTaO_2N$ heterostructure with high crystallinity and intimate interface was achieved, exhibiting the promotion of the charge separation, and enhanced photocatalytic activity for Z-scheme water splitting. Moreover, $Sr_2Ta_2O_{7-x}N_x/SrTaO_2N$ heterostructure was constructed by employing $Sr_2Ta_2O_7$ as precursor *via* one-step ammonia thermal route.[166] The heterostructure exhibited a higher photocatalytic activity than the single $Sr_2Ta_2O_{7-x}N_x$ and $SrTaO_2N$ photocatalysts because of the matched energy band structure and favourable interface between $Sr_2Ta_2O_{7-x}N_x$ and $SrTaO_2N$.

3.4.4 Surface Modification

The photocatalytic reaction is an oxidation–reduction process that occurs on the surface of the photocatalyst materials, so that the structural properties of the surface and interface play a vital role in the photocatalytic activity. Constructing a well-crystallized surface layer and adsorption functional

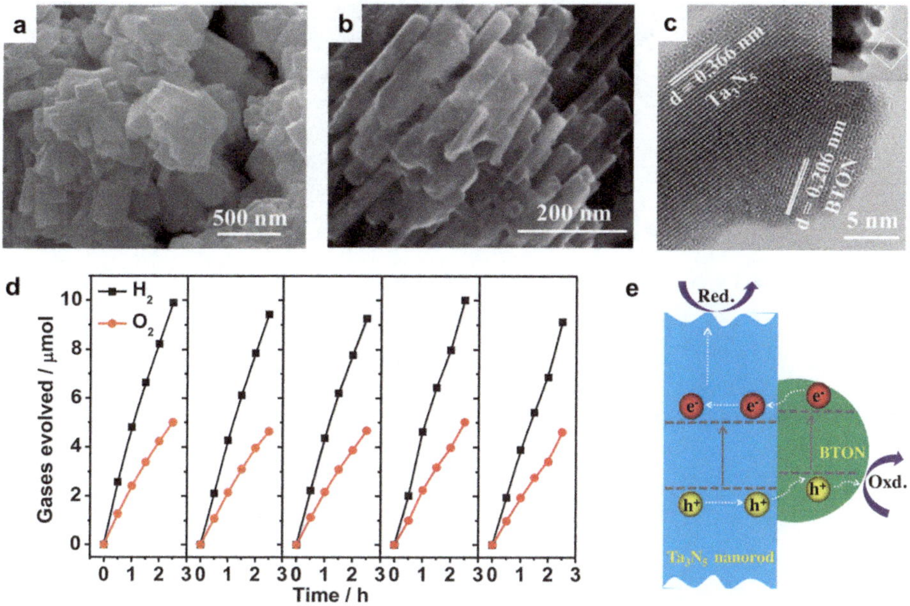

Figure 3.15 SEM images of (a) $KBa_2Ta_5O_{15}$ precursor and (b) $Ta_3N_5/BaTaO_2N$ sample. (c) HRTEM image of $Ta_3N_5/BaTaO_2N$ sample (top right inset: the enlarged view). (d) Photocatalytic activity of Z scheme water splitting with 0.3 wt% Pt-loaded $Ta_3N_5/BaTaO_2N$ (50 mg) as HEP and 0.45 wt% PtO_x-loaded WO_3 (100 mg) as OEP in NaI aqueous solution (150 mL, 2.0×10^{-3} M). (e) Schematic illustration of the photoexcited charge separation between Ta_3N_5 and $BaTaO_2N$ (Red., the reduction reaction; Oxd., the oxidation reaction.). Reproduced from ref. 165 with permission from John Wiley & Sons, Copyright © 2019 WILEY-VCH Verlag GmbH & Co. KGaA, Weinheim.

sites through surface modification is one of the effective means to improve photocatalytic efficiency.[167] Since the (oxy)nitrides are usually obtained from the NH_3 nitridation of the corresponding oxides, the formation of surface defects with reduced metal species is inevitable in such materials. ZrO_2 modification on Ta_3N_5 and TaON (oxy)nitrides has been demonstrated to be effective to reduce the surface defects during the nitridation process.[86,87,168–170] Moreover, the introduction of $BaZrO_3$ on the surface of $BaTaO_2N$ to form a solid solution passivated layer, not only prevented the generation of reduced tantalum species on the surface of $BaTaO_2N$, but also provided an increased driving force for redox reactions because of the slightly enlarged bandgap of $BaZrO_3$–$BaTaO_2N$ solid solution.[171] A facile surface modification method with alkaline metal (AM) salt is also employed to enhance the photocatalytic activity of (oxy)nitrides. For instance, Ta_3N_5 obtained from surface modified Ta_2O_5 with a small amount of AM showed an improvement in photocatalytic activity for O_2 evolution compared with conventional Ta_3N_5.[172] The addition of AM in the preparation of Ta_3N_5 promoted the partial dissolution of Ta_2O_5 and nucleation of $NaTaO_3$, therefore leading to the formation of well-defined

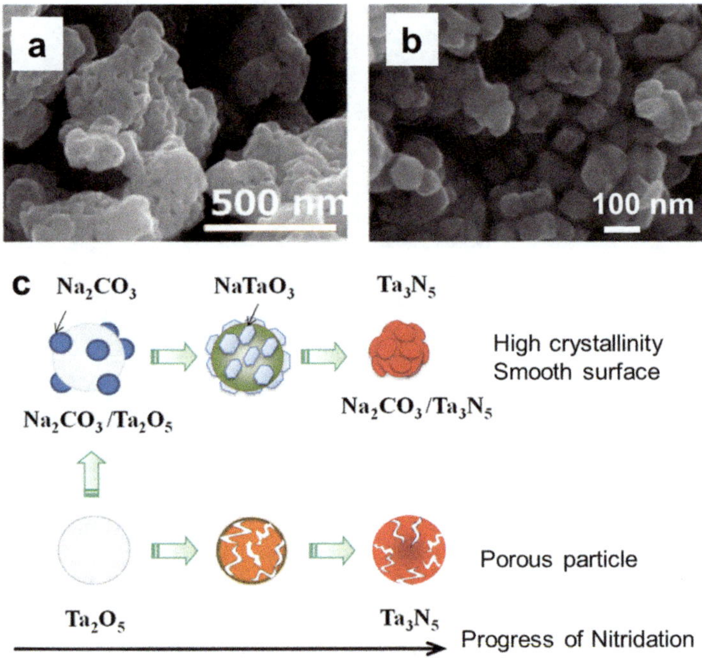

Figure 3.16 SEM images of Ta_3N_5 obtained from Ta_2O_5 (a) unmodified and (b) modified with Na_2CO_3. (c) Proposed scheme for evolution of conventional Ta_3N_5 (bottom) and Na_2CO_3/Ta_3N_5 (top). Reproduced from ref. 172 with permission from American Chemical Society, Copyright 2012.

Ta_3N_5 with better crystallinity and decreased surface defects, as shown in Figure 3.16.

Surface modification also shows a positive effect on the functionalization of photocatalyst materials for the forward reaction. Rutile titania (R-TiO_2) with a special surface adsorption property has been reported as a surface modifier for O_2-evolution photocatalyst in the Z scheme water splitting containing both I^- and IO_3^- ions. For example, Ta_3N_5 modified with Ir nanoparticle and R-TiO_2 successfully was applied as an O_2-evolution photocatalyst in the Z-scheme water splitting system with IO_3^-/I^- redox mediator under visible light irradiation.[173] The R-TiO_2 modifier enabled the selective adsorption of IO_3^- ions on the surface of Ta_3N_5, allowing the reduction of IO_3^- to I^- and water oxidation to O_2 to take place on Ta_3N_5 even in the presence of a considerable amount of I^-.

3.4.5 Co-catalysts Modification

Bare photocatalysts usually display low activities or no activities because of the rapid recombination of electrons and holes before migrating to the surface and the inefficient surface-active sites. Co-catalysts play indispensable roles in photocatalytic reaction, not only to promote the extraction of electrons and holes, but also to provide the reduction or oxidation sites, thereby dramatically increasing the photocatalytic efficiency.[174–176] The noble metals (Au, Pt, Ir, *etc.*) with larger work function than semiconductor, that is, lower Fermi level, which facilitate the migration of electrons from semiconductor to co-catalysts, are often used as the H_2-evolving co-catalysts. Among them, Pt has the largest work function and is considered as the most suitable co-catalyst for H_2 evolution. In recent years, some biomimetic metal–organic compounds have been found to be effective for H_2 evolution. For example, cobaloximes (Co(III) complexes) and hydrogenase mimic $[(\mu\text{-SPh-4-NH}_2)_2Fe_2(CO)_6]$ were employed as H_2-evolving co-catalysts on CdS and ZnS, respectively, and good photocatalytic performance was obtained in the presence of sacrificial agents.[177,178] The application of these biomimetic co-catalysts on (oxy)nitride is still rare. As for the O_2-evolving co-catalysts, there are mainly metal oxides such as IrO_x, CoO_x, RuO_x, MnO_x and so on. The modification of these co-catalysts on (oxy)nitride photocatalysts effectively improves the O_2 evolution activity.

Generally, the photocatalytic performance depends on the properties of photocatalysts and the functions of co-catalysts. An efficient co-catalyst should be compatible with the photocatalyst semiconductor in terms of energy level and electronic structure, so that the photogenerated charges are able to transfer through the interface from the semiconductor to co-catalyst. The most traditional co-catalyst modification is the impregnation method, which forms the random dispersion of co-catalyst nanoparticles on the surface of the photocatalyst. In contrast, *in-situ* photodeposition allows the co-catalyst to selectively locate at the charge-accumulating sites. In the overall water splitting reaction, noble metal co-catalysts also enable the backward reaction of H_2 and O_2 to form water. Thus, the core–shell-structured metal/Cr_2O_3 co-catalyst that is effective in the inhibition of water formation reaction is successfully developed, and the visible-light-driven overall water splitting has accomplished on a series of (oxy)nitrides with metal/Cr_2O_3 co-catalyst. To improve the photocatalytic efficiency and promote large-scale application, it is necessary to explore more low-cost and effective co-catalysts for the modification of (oxy)nitride.

3.5 Conclusion and Prospects

In summary, we present a brief introduction of the typical (oxy)nitrides and involve strategies for enhancing their photocatalytic performance. Apparently, sufficient light harvest, effective charge separation and migration, and suppression of backward reactions competing with the water-splitting reaction are crucial to realize high STH energy conversion efficiency. Although many (oxy)nitride photocatalysts have shown good activities for H_2 evolution or O_2 evolution in relative sacrificial agents, the present efficiency of overall water splitting based on (oxy)nitrides is still low. Therefore, it is highly desirable to develop new strategies to obtain high-quality (oxy)nitrides with minimized defects for efficiently utilizing photogenerated charges. In addition, facet engineering, shape-controlled synthesis, and nanostructure preparation for (oxy)nitride photocatalysts will be a promising approach to the improvement of photocatalytic activity. Furthermore, the mechanism of the charge separation and migration in photocatalysts and the catalytic reaction on surface are still worth investigating with assistance of advanced characterization technologies (*e.g.*, transient absorption spectroscopy (TAS), X-ray and neutron scattering) and theoretical studies (*e.g.*, DFT calculation). All the mentioned strategies and research directions are essential to achieving high-efficiency solar energy conversion based on (oxy)nitride photocatalysts with narrow bandgaps towards practical applications.

Acknowledgements

This work was financially supported by the Pioneer Initiative Action Project of Chinese Academy of Sciences and National Natural Science Foundation of China.

References

1. M. S. Dresselhaus and I. L. Thomas, *Nature*, 2001, **414**, 332.
2. J. Barber, *Chem. Soc. Rev.*, 2009, **38**, 185.
3. Z. Wang, C. Li and K. Domen, *Chem. Soc. Rev.*, 2019, **48**, 2109.
4. N. S. Lewis, *Science*, 2007, **315**, 798.
5. K. Maeda and K. Domen, *J. Phys. Chem. Lett.*, 2010, **1**, 2655.
6. A. Kudo and Y. Miseki, *Chem. Soc. Rev.*, 2009, **38**, 253.
7. M. G. Walter, E. L. Warren, J. R. McKone, S. W. Boettcher, Q. Mi, E. A. Santori and N. S. Lewis, *Chem. Rev.*, 2010, **110**, 6446.
8. T. Hisatomi, J. Kubota and K. Domen, *Chem. Soc. Rev.*, 2014, **43**, 7520.
9. J. Joy, J. Mathew and S. C. George, *Int. J. Hydrogen Energy*, 2018, **43**, 4804.
10. Y. Lin, G. Yuan, R. Liu, S. Zhou, S. W. Sheehan and D. Wang, *Chem. Phys. Lett.*, 2011, **507**, 209.
11. J. S. Lee, *Catal. Surv. Asia*, 2005, **9**, 217.
12. D. M. Fabian, S. Hu, N. Singh, F. A. Houle, T. Hisatomi, K. Domen, F. E. Osterloh and S. Ardo, *Science*, 2015, **8**, 2825.
13. F. F. Abdi, L. Han, A. H. M. Smets, M. Zeman, B. Dam and R. van de Krol, *Nat. Commun.*, 2013, **4**, 2195.
14. T. Takata, J. Jiang, Y. Sakata, M. Nakabayashi, N. Shibata, V. Nandal, K. Seki, T. Hisatomi and K. Domen, *Nature*, 2020, **581**, 411.
15. Y. Wei, J. Wang, R. Yu, J. Wan and D. Wang, *Angew. Chem., Int. Ed.*, 2019, **58**, 1422.
16. W. Chen, *Appl. Catal., B*, 2016, **192**, 145.
17. K. Maeda and K. Domen, *J. Phys. Chem. C*, 2007, **111**, 7851.

18. S. G. Ebbinghaus, H. P. Abicht, R. Dronskowski, T. Muller, A. Reller and A. Weidenkaff, *Prog. Solid State Chem.*, 2009, **37**, 173.
19. Y. Jun, J. Choi and J. Cheon, *Angew. Chem., Int. Ed.*, 2006, **45**, 3414.
20. H. Ahmad, S. K. Kamarudin, L. J. Minggu and M. Kassim, *Renewable Sustainable Energy Rev.*, 2015, **43**, 599.
21. J. Liu, Y. Li, X. Zhou, H. Jiang, H. Yang and C. Li, *J. Mater. Chem. A*, 2020, **8**, 17.
22. J. Ran, J. Zhang, J. Yu, M. Jaroniec and S. Qiao, *Chem. Soc. Rev.*, 2014, **43**, 7787.
23. T. Hisatomi and K. Domen, *Faraday Discuss.*, 2017, **198**, 11.
24. J. Seo, H. Nishiyama, T. Yamada and K. Domen, *Angew. Chem., Int. Ed.*, 2018, **57**, 8396.
25. M. Matsumura, Y. Saho and H. Tsubomura, *J. Phys. Chem.*, 1983, **87**, 3807.
26. J. R. Darwent and A. Mills, *J. Chem. Soc., Faraday Trans.*, 1982, **78**, 359.
27. J. Low, C. Jiang, B. Cheng, S. Wageh, A. A. Al-Ghamdi and J. Yu, *Small Methods*, 2017, **1**, 1700080.
28. K. Qi, B. Cheng, J. Yu and W. Ho, *Chin. J. Catal.*, 2017, **38**, 1936.
29. K. Sayama, K. Mukasa, R. Abe, Y. Abe and H. Arakawa, *Chem. Commun.*, 2001, 2416.
30. K. Sayama, K. Mukasa, R. Abe, Y. Abe and H. Arakawa, *J. Photochem. Photobiol., A*, 2002, **148**, 71.
31. R. Abe, K. Sayama and H. Sugihara, *J. Phys. Chem. B*, 2005, **109**, 16052.
32. H. Kato, M. Hori, R. Konta, Y. Shimodaira and A. Kudo, *Chem. Lett.*, 2004, **33**, 1348.
33. T. Hisatomi and K. Domen, *Nat. Catal.*, 2019, **2**, 387.
34. T. Hisatomi, K. Takanabe and K. Domen, *Catal. Lett.*, 2015, **145**, 95.
35. A. Fujishima and K. Honda, *Nature*, 1972, **238**, 37.
36. K. Domen, S. Naito, M. Soma, T. Onishi and K. Tamaru, *J. Chem. Soc., Chem. Commun.*, 1980, 543.
37. S. Sato and J. M. White, *Chem. Phys. Lett.*, 1980, **72**, 83.
38. K. Domen, S. Naito, T. Onishi and K. Tamaru, *Chem. Phys. Lett.*, 1982, **92**, 433.
39. M. Machida, J. Yabunaka and T. Kijima, *Chem. Commun.*, 1999, 1939.
40. J. Ye, Z. Zou and A. Matusshita, *Int. J. Hydrogen Energy*, 2003, **28**, 651.
41. H. Kato and A. Kudo, *Chem. Phys. Lett.*, 1998, **295**, 487.
42. H. Kato and A. Kudo, *Catal. Lett.*, 1999, **58**, 153.
43. A. Kudo and H. Kato, *Chem. Phys. Lett.*, 2000, **331**, 373.
44. A. Iwase, H. Kato, H. Okutomi and A. Kudo, *Chem. Lett.*, 2004, **33**, 1260.
45. H. Kato, K. Asakura and A. Kudo, *J. Am. Chem. Soc.*, 2003, **125**, 3082.
46. K. Sayama and H. Arakawa, *J. Photochem. Photobiol., A*, 1994, **77**, 243–247.
47. K. Sayama and H. Arakawa, *J. Phys. Chem.*, 1993, **97**, 531.
48. V. R. Reddy, D. W. Hwang and J. S. Lee, *Korean J. Chem. Eng.*, 2003, **20**, 1026.
49. J. J. Zou, C. J. Liu and Y. P. Zhang, *Langmuir*, 2006, **22**, 2334.

50. M. Uno, A. Kosuga, M. Okui, K. Horisaka, H. Muta, K. Kurosaki and S. Yamanaka, *J. Alloys Compd.*, 2006, **420**, 291.
51. K. Sayama, H. Arakawa and K. Domen, *Catal. Today*, 1996, **28**, 175.
52. X. Chen, T. Yu, X. Fan, H. Zhang, Z. Li, J. Ye and Z. Zou, *Appl. Surf. Sci.*, 2007, **253**, 8500.
53. K. Domen, A. Kudo, M. Shibata, A. Tanaka, K. Maruya and T. Onishi, *J. Chem. Soc., Chem. Commun.*, 1986, **23**, 1706.
54. K. Sayama, A. Tanaka, K. Domen, K. Maruya and T. Onishi, *J. Catal.*, 1990, **124**, 541.
55. G. Li, T. Kako, D. Wang, Z. Zou and J. Ye, *J. Phys. Chem. Solids*, 2008, **69**, 2487.
56. Q. P. Ding, Y. P. Yuan, X. Xiong, R. P. Li, H. B. Huang, Z. S. Li, T. Yu, Z. G. Zou and S. G. Yang, *J. Phys. Chem. C*, 2008, **112**, 18846.
57. B. Zielin´ ska, E. Borowiak-Palen and R. J. Kalenzuk, *J. Phys. Chem. Solids*, 2008, **69**, 236.
58. Y. Miseki, H. Kato and A. Kudo, *Chem. Lett.*, 2005, **34**, 54.
59. D. Chen and J. Ye, *Chem. Mater.*, 2009, **21**, 2327.
60. Y. Miseki, H. Kato and A. Kudo, *Energy Environ. Sci.*, 2009, **2**, 306.
61. H. G. Kim, D. W. Hwang, J. Kim, Y. G. Kim and J. S. Lee, *Chem. Commun.*, 1999, **12**, 1077.
62. K. Domen, J. Yoshimura, T. Sekine, A. Tanaka and T. Onishi, *Catal. Lett.*, 1990, **4**, 339.
63. T. Yanagida, Y. Sakata and H. Imamura, *Chem. Lett.*, 2004, **33**, 726.
64. Y. Sakata, Y. Matsuda, T. Yanagida, K. Hirata, H. Imamura and K. Teramura, *Catal. Lett.*, 2008, **125**, 22.
65. K. Ikarashi, J. Sato, H. Kobayashi, N. Saito, H. Nishiyama and Y. Inoue, *J. Phys. Chem. B*, 2002, **106**, 9048.
66. J. Sato, H. Kobayashi, K. Ikarashi, N. Saito, H. Nishiyama and Y. Inoue, *J. Phys. Chem. B*, 2004, **108**, 4369.
67. J. Sato, H. Kobayashi, S. Saito, H. Nishiyama and Y. Inoue, *J. Photochem. Photobiol., A*, 2003, **158**, 139.
68. J. Sato, S. Saito, H. Nishiyama and Y. Inoue, *J. Phys. Chem. B*, 2003, **107**, 7965.
69. J. Sato, H. Kobayashi and Y. Inoue, *J. Phys. Chem. B*, 2003, **107**, 7970.
70. J. Sato, N. Saito, H. Nishiyama and Y. Inoue, *J. Phys. Chem. B*, 2001, **105**, 6061.
71. J. Sato, N. Saito, H. Nishiyama and Y. Inoue, *Chem. Lett.*, 2001, **30**, 868.
72. K. Maeda and K. Domen, *J. Phys. Chem. C*, 2007, **111**(22), 7851.
73. A. Ishikawa, T. Takata, J. N. Kondo, M. Hara, H. Kobayashi and K. Domen, *J. Am. Chem. Soc.*, 2002, **124**, 13547.
74. A. Ishikawa, T. Takata, T. Matsumura, J. N. Kondo, M. Hara, H. Kobayashi and K. Domen, *J. Phys. Chem. B*, 2004, **108**, 2637.
75. M. Xiao, S. Wang, S. Thaweesak, B. Luo and L. Wang, *Engineering*, 2017, **3**, 365.
76. C. Zhen, R. Chen, L. Wang, G. Liu and H. Cheng, *J. Mater. Chem. A*, 2016, **4**, 2783.

77. M. Hara, E. Chiba, A. Ishikawa, T. Takata, J. N. Kondo and K. Domen, *J. Phys. Chem. B*, 2003, **107**, 13441.
78. Y. Moriya, T. Takata and K. Domen, *Coord. Chem. Rev.*, 2013, **257**, 1957.
79. Y. Wang, H. Suzuki, J. Xie, O. Tomita, D. J. Martin, M. Higashi, D. Kong, R. Abe and J. Tang, *Chem. Rev.*, 2018, **118**, 5201.
80. Z. Wang, C. Li and K. Domen, *Chem. Soc. Rev.*, 2019, **48**, 2109.
81. K. Maeda, *ACS Catal.*, 2013, **3**, 1486.
82. G. Hitoki, T. Takata, J. N. Kondo, M. Hara, H. Kobayashi and K. Domen, *Chem. Commun.*, 2002, 1698.
83. R. Abe, M. Higashi and K. Domen, *ChemSusChem*, 2011, **4**, 228.
84. M. Higashi, R. Abe, A. Ishikawa, T. Takata, B. Ohtani and K. Domen, *Chem. Lett.*, 2008, **37**, 138.
85. Y. Iwase, O. Tomita, M. Higashi and R. Abe, *Sustainable Energy Fuels*, 2017, **1**, 748.
86. K. Maeda, H. Terashima, K. Kase, M. Higashi, M. Tabata and K. Domen, *Bull. Chem. Soc. Jpn.*, 2008, **81**, 927.
87. K. Maeda, M. Higashi, D. Lu, R. Abe and K. Domen, *J. Am. Chem. Soc.*, 2010, **132**, 5858.
88. Y. Qi, Y. Zhao, Y. Gao, D. Li, Z. Li, F. Zhang and C. Li, *Joule*, 2018, **2**, 2393.
89. K. Maeda, D. Lu and K. Domen, *Chem. – Eur. J.*, 2013, **19**, 4986.
90. S. Chen, Y. Qi, T. Hisatomi, Q. Ding, T. Asai, Z. Li, S. S. Ma, F. Zhang, K. Domen and C. Li, *Angew. Chem., Int. Ed.*, 2015, **54**, 8498.
91. G. Hitoki, A. Ishikawa, T. Takata, J. Kondo, M. Hara and K. Domen, *Chem. Lett.*, 2002, **31**, 736.
92. Y. He, J. E. Thorne, C. H. Wu, P. Ma, C. Du, Q. Dong, J. Guo and D. Wang, *Chem*, 2016, **1**, 640.
93. Z. Wang, Y. Inoue, T. Hisatomi, R. Ishikawa, Q. Wang, T. Takata, S. Chen, N. Shibata, Y. Ikuhara and K. Domen, *Nat. Catal.*, 2018, **1**, 756.
94. Y. Qi, S. Chen, M. Li, Q. Ding, Z. Li, J. Cui, B. Dong, F. Zhang and C. Li, *Chem. Sci.*, 2017, **8**, 437.
95. M. Zhong, T. Hisatomi, Y. Sasaki, S. Suzuki, K. Teshima, M. Nakabayashi, N. Shibata, H. Nishiyama, M. Katayama, T. Yamada and K. Domen, *Angew. Chem., Int. Ed.*, 2017, **56**, 4739.
96. S. S. Ma, T. Hisatomi, K. Maeda, Y. Moriya and K. Domen, *J. Am. Chem. Soc.*, 2012, **134**, 19993.
97. S. Chen, S. Shen, G. Liu, Y. Qi, F. Zhang and C. Li, *Angew. Chem., Int. Ed.*, 2015, **54**, 3047.
98. J. Xiao, J. J. M. Vequizo, T. Hisatomi, J. Rabeah, M. Nakabayashi, Z. Wang, Q. Xiao, H. Li, Z. Pan, M. Krause, N. Yin, G. Smith, N. Shibata, A. Brückner, A. Yamakata, T. Takata and K. Domen, *J. Am. Chem. Soc.*, 2021, **143**, 10059.
99. Y. Qi, S. Chen, J. Cui, Z. Wang, F. Zhang and C. Li, *Appl. Catal., B*, 2018, **224**, 579.
100. S. S. Ma, K. Maeda, T. Hisatomi, M. Tabata, A. Kudo and K. Domen, *Chem. – Eur. J.*, 2013, **19**, 7480.

101. J. Cui, Y. Qi, B. Dong, L. Mu, Q. Ding, G. Liu, M. Jia, F. Zhang and C. Li, *Appl. Catal., B*, 2019, **241**, 1.
102. M. Higashi, R. Abe, T. Takata and K. Domen, *Chem. Mater.*, 2009, **21**, 1543.
103. Y. P. Luo, H. Li, Y. L. Luo, Z. Li, Y. Qi, F. Zhang and C. Li, *J. Energy Chem.*, 2022, **67**, 27.
104. J. Xu, C. Pan, T. Takata and K. Domen, *Chem. Commun.*, 2015, **51**, 7191.
105. T. Matoba, K. Maeda and K. Domen, *Chem. – Eur. J.*, 2011, **17**, 14731.
106. K. Maeda, D. Lu and K. Domen, *ACS Catal.*, 2013, **3**, 1026.
107. B. Dong, Y. Qi, J. Cui, B. Liu, F. Xiong, X. Jiang, Z. Li, Y. Xiao, F. Zhang and C. Li, *Dalton Trans.*, 2017, **46**, 10707.
108. Y. Luo, S. Suzuki, Z. Wang, K. Yubuta, J. J. M. Vequizo, A. Yamakata, H. Shiiba, T. Hisatomi, K. Domen and K. Teshima, *ACS Appl. Mater. Interfaces*, 2019, **11**, 22264.
109. Y. Luo, Z. Wang, T. Yamada, K. Yubuta, S. Suzuki, T. Hisatomi, K. Domen and K. Teshima, *ACS Appl. Energy Mater.*, 2020, **3**, 10669.
110. Y. Luo, Z. Wang, S. Suzuki, K. Yubuta, N. Kariya, T. Hisatomi, K. Domen and K. Teshima, *Cryst. Growth Des.*, 2020, **20**, 255.
111. Z. Wang, Y. Luo, T. Hisatomi, J. J. M. Vequizo, S. Suzuki, S. Chen, M. Nakabayashi, L. Lin, Z. Pan, N. Kariya, A. Yamakata, N. Shibata, T. Takata, K. Teshima and K. Domen, *Nat. Commun.*, 2021, **12**, 1005.
112. X. Wang, T. Hisatomi, Z. Wang, J. Song, J. Qu, T. Takata and K. Domen, *Angew. Chem., Int. Ed.*, 2019, **58**, 10666.
113. R. Wang, Y. Wang, S. Chang, S. Jin, Y. Shao and X. Xu, *J. Catal.*, 2020, **390**, 57.
114. C. Pan, T. Takata, M. Nakabayashi, T. Matsumoto, N. Shibata, Y. Ikuhara and K. Domen, *Angew. Chem., Int. Ed.*, 2015, **54**, 2955.
115. C. Pan, T. Takata, K. Kumamoto, S. S. Ma, K. Ueda, T. Minegishi, M. Nakabayashi, T. Matsumoto, N. Shibata, Y. Ikuhara and K. Domen, *J. Mater. Chem. A*, 2016, **4**, 4544.
116. J. Seo, T. Hisatomi, M. Nakabayashi, N. Shibata, T. Minegishi, M. Katayama and K. Domen, *Adv. Energy Mater.*, 2018, **8**, 1800094.
117. R. Wang, Q. Wang, J. Qian and X. Xu, *Inorg. Chem. Front.*, 2021, **8**, 2365.
118. A. Kasahara, K. Nukumizu, T. Takata, J. N. Kondo, M. Hara, H. Kobayashi and K. Domen, *J. Phys. Chem. B*, 2003, **107**, 791.
119. F. Zhang, A. Yamakata, K. Maeda, Y. Moriya, T. Takata, J. Kubota, K. Teshima, S. Oishi and K. Domen, *J. Am. Chem. Soc.*, 2012, **134**, 8348.
120. T. Hisatomi, C. Katayama, Y. Moriya, T. Minegishi, M. Katayama, H. Nishiyama, T. Yamada and K. Domen, *Energy Environ. Sci.*, 2013, **6**, 3595.
121. M. Kodera, M. Katayama, T. Hisatomi, T. Minegishi and K. Domen, *CrystEngComm*, 2016, **18**, 3186.
122. X. Wang, T. Hisatomi, J. Liang, Z. Wang, Y. Xiang, Y. Zhao, X. Dai, T. Takata and K. Domen, *J. Mater. Chem. A*, 2020, **8**, 11743.

123. M. Matsukawa, R. Ishikawa, T. Hisatomi, Y. Moriya, N. Shibata, J. Kubota, Y. Ikuhara and K. Domen, *Nano Lett.*, 2014, **14**, 1038.
124. J. Sato, N. Saito, Y. Yamada, K. Maeda, T. Takata, J. N. Kondo, M. Hara, H. Kobayashi, K. Domen and Y. Inoue, *J. Am. Chem. Soc.*, 2005, **127**, 4150.
125. Y. Lee, T. Watanabe, T. Takata, M. Hara, M. Yoshimura and K. Domen, *J. Phys. Chem. B*, 2006, **110**, 17563.
126. K. Maeda, N. Saito, D. Lu, Y. Inoue and K. Domen, *J. Phys. Chem. C*, 2007, **111**, 4749.
127. Y. Lee, H. Terashima, Y. Shimodaira, K. Teramura, M. Hara, H. Kobayashi, K. Domen and M. Yashima, *J. Phys. Chem. C*, 2007, **111**, 1042.
128. Y. Lee, K. Teramura, M. Hara and K. Domen, *Chem. Mater.*, 2007, **19**, 2120.
129. K. Takanabe, T. Uzawa, X. Wang, K. Maeda, M. Katayama, J. Kubota, A. Kudob and K. Domen, *Dalton Trans.*, 2009, 10055.
130. K. Maeda, T. Takata, M. Hara, N. Saito, Y. Inoue, H. Kobayashi and K. Domen, *J. Am. Chem. Soc.*, 2005, **127**, 8286.
131. K. Maeda, K. Teramura, D. Lu, N. Saito, Y. Inoue and K. Domen, *Angew. Chem., Int. Ed.*, 2006, **45**, 7806.
132. K. Maeda, K. Teramura, D. Lu, T. Takata, N. Saito, Y. Inoue and K. Domen, *Nature*, 2006, **440**, 295.
133. H. Chen, W. Wen, Q. Wng, J. C. Hanson, J. T. Muckerman, E. Fujita, A. I. Frenkel and J. A. Rodriguez, *J. Phys. Chem. C*, 2009, **113**, 3650.
134. K. Lee, B. M. Tienes, M. B. Wilker, K. J. Schnitzenbaumer and G. Dukovic, *Nano Lett.*, 2012, **12**, 3268.
135. K. Lee, Y. Lu, C. Chuang, J. Ciston and G. Dukovic, *J. Mater. Chem. A*, 2016, **4**, 2927.
136. J. Wang, B. Huang, Z. Wang, P. Wang, H. Cheng, Z. Zheng, X. Qin, X. Zhang, Y. Dai and M. Whangbo, *J. Mater. Chem.*, 2011, **21**, 4562.
137. K. Maeda, K. Teramura, N. Saito, Y. Inoue and K. Domen, *J. Catal.*, 2006, **243**, 303.
138. K. Maeda, H. Masuda and K. Domen, *Catal. Today*, 2009, **147**, 173.
139. K. Maeda, K. Teramura, D. Lu, N. Saito, Y. Inoue and K. Domen, *J. Phys. Chem. C*, 2007, **111**, 7554.
140. K. Maeda and K. Domen, *Bull. Chem. Soc. Jpn.*, 2016, **89**, 627.
141. K. Maeda, K. Teramura, T. Takata, M. Hara, N. Saito, K. Toda, Y. Inoue, H. Kobayashi and K. Domen, *J. Phys. Chem. B*, 2005, **109**, 20504.
142. L. Lin, Z. Yu and X. Wang, *Angew. Chem., Int. Ed.*, 2019, **58**, 6164.
143. X. Wang, K. Maeda, A. Thomas, K. Takanabe, G. Xin, J. M. Carlsson, K. Domen and M. Antonietti, *Nat. Mater.*, 2009, **8**, 76.
144. D. J. Martin, P. J. T. Reardon, S. J. A. Moniz and J. Tang, *J. Am. Chem. Soc.*, 2014, **136**, 12568.
145. G. Zhang, Z. Lan, L. Lin, S. Lin and X. Wang, *Chem. Sci.*, 2016, **7**, 3062.
146. Z. Zhang, K. Leinenweber, M. Bauer, L. A. J. Garvie, P. F. McMillan and G. H. Wolf, *J. Am. Chem. Soc.*, 2001, **123**, 7788.

147. E. Wirnhier, M. Döblinger, D. Gunzelmann, J. Senker, B. V. Lotsch and W. Schnick, *Chem. – Eur. J.*, 2011, **17**, 3213.
148. L. Lin, Z. Lin, J. Zhang, X. Cai, W. Lin, Z. Yu and X. Wang, *Nat. Catal.*, 2020, **3**, 649.
149. H. Ahn, K. Yoon, M. Kwak, J. Park and J. Jang, *ACS Catal.*, 2018, **8**, 11932.
150. Z. Zou, J. Ye, K. Sayama and H. Arakawa, *Nature*, 2000, **414**, 625.
151. Y. Ling, G. Wang, D. A. Wheeler, J. Z. Zhang and Y. Li, *Nano Lett.*, 2011, **11**, 2119.
152. R. Sasikala, A. R. Shirole, V. Sudarsan, C. Sudakar, R. Naik, R. Rao and S. R. Bharadwaj, *Appl. Catal., A*, 2010, **377**, 47.
153. Y. Li, G. Ma, S. Peng, G. Lu and S. Li, *Appl. Surf. Sci.*, 2008, **254**, 6831.
154. H. Chen, C. Huang, J. C. S. Wu and S. Lin, *J. Phys. Chem. C*, 2012, **116**, 7897.
155. Y. Wang, S. Jin, G. Pan, Z. Li, L. Chen, G. Liu and X. Xu, *J. Mater. Chem. A*, 2019, **7**, 5702.
156. J. Fu and S. E. Skrabalak, *J. Mater. Chem. A*, 2016, **4**, 8451.
157. T. Hisatomi, M. Otani, K. Nakajima, K. Teramura, Y. Kako, D. Lu, T. Takata, J. N. Kondo and K. Domen, *Chem. Mater.*, 2010, **22**, 3854.
158. T. Takata, D. Lu and K. Domen, *Cryst. Growth Des.*, 2011, **11**, 33.
159. C. Zhou, J. Zhou, L. Lu, J. Wang, Z. Shi, B. Wang, L. Pei, S. Yan, Z. Yu and Z. Zou, *Appl. Catal., B*, 2018, **237**, 742.
160. Y. Luo, S. Suzuki, Z. Wang, K. Yubuta, J. J. M. Vequizo, A. Yamakata, H. Shiiba, T. Hisatomi, K. Domen and K. Teshima, *ACS Appl. Mater. Interfaces*, 2019, **11**, 22264.
161. Y. Bessekhouad, D. Robert and J. V. Weber, *Catal. Today*, 2005, **101**, 315.
162. V. M. Daskalaki, M. Antoniadou, G. L. Puma, D. I. Kondarides and P. Lianos, *Environ. Sci. Technol.*, 2010, **44**, 7200.
163. L. J. Zhang, S. Li, B. K. Liu, D. J. Wang and T. F. Xie, *ACS Catal.*, 2014, **4**, 3724.
164. J. S. Jang, H. G. Kim, U. A. Joshi, J. W. Jang and J. S. Lee, *Int. J. Hydrogen Energy*, 2008, **33**, 5975.
165. B. Dong, J. Cui, Y. Gao, Y. Qi, F. Zhang and C. Li, *Adv. Mater.*, 2019, **31**, 1808185.
166. W. Zeng, Y. Bian, S. Cao, Y. Ma, Y. Liu, A. Zhu, P. Tan and J. Pan, *ACS Appl. Mater. Interfaces*, 2018, **10**, 21328.
167. H. Tong, S. Ouyang, Y. Bi, N. Umezawa, M. Oshikiri and J. Ye, *Adv. Mater.*, 2012, **24**, 229.
168. K. Maeda, H. Terashima, K. Kase, M. Higashi, M. Tabata and K. Domen, *Bull. Chem. Soc. Jpn.*, 2008, **81**, 927.
169. K. Maeda, M. Higashi, D. Lu, R. Abe and K. Domen, *J. Am. Chem. Soc.*, 2010, **132**, 5858.
170. S. S. K. Ma, K. Maeda and K. Domen, *Catal. Sci. Technol.*, 2012, **2**, 818.
171. T. Matoba, K. Maeda and K. Domen, *Chem*, 2011, **17**, 14731.
172. S. S. Ma, T. Hisatomi, K. Maeda, Y. Moriya and K. Domen, *J. Am. Chem. Soc.*, 2012, **134**, 19993.

173. M. Tabata, K. Maeda, M. Higashi, D. Lu, T. Takata, R. Abe and K. Domen, *Langmuir*, 2010, **26**, 9161.
174. A. Meng, J. Zhang, D. Xu, B. Cheng and J. Yu, *Appl. Catal., B*, 2016, **198**, 286.
175. R. Li, H. Han, F. Zhang, D. Wang and C. Li, *Energy Environ. Sci.*, 2014, **7**, 1369.
176. Y. Ma, R. Chong, F. Zhang, Q. Xu, S. Shen, H. Han and C. Li, *Phys. Chem. Chem. Phys.*, 2014, **16**, 17734.
177. F. Wen, J. Yang, X. Zong, B. Ma, D. Wang and C. Li, *J. Catal.*, 2011, **281**, 318.
178. F. Wen, X. Wang, L. Huang, G. Ma, J. Yang and C. Li, *ChemSusChem*, 2012, **5**, 849.

CHAPTER 4

Recent Advances in (Oxy)sulfide Photocatalysts for Solar-driven Water Splitting and CO_2 Reduction

AKIHIDE IWASE*[a]

[a]Department of Applied Chemistry, School of Science and Technology, Meiji University, Kanagawa 214-8571, Japan
*E-mail: a_iwase@meiji.ac.jp

4.1 General Introduction to Metal Sulfide Photocatalysts

Photocatalytic water splitting into H_2 and O_2, and CO_2 reduction accompanied by water oxidation to O_2 can be classified as artificial photosynthetic reactions in terms of photon to chemical energy conversion. Using a wide range of visible light in sunlight is required for photocatalyst materials to achieve highly efficient artificial photosynthesis. The most famous and traditional material group used as photocatalysts is metal oxides.[1–3] The band structure of a typical metal oxide is shown in Figure 4.1(a).[3,4] Metal oxides generally possess deep valence bands composed of O 2p orbitals at around +3.0 V *vs.* NHE at pH 0.[5] When the metal oxide possesses photocatalytic ability for water reduction to produce H_2, its conduction band should be located at equal to or more negative than 0 V *vs.* NHE at pH 0. The bandgaps of such

Sustainable Energy No. 1
Recent Developments in Functional Materials for Artificial Photosynthesis
Edited by Srabanti Ghosh and Qian Wang

Published by the Royal Society of Chemistry, www.rsc.org

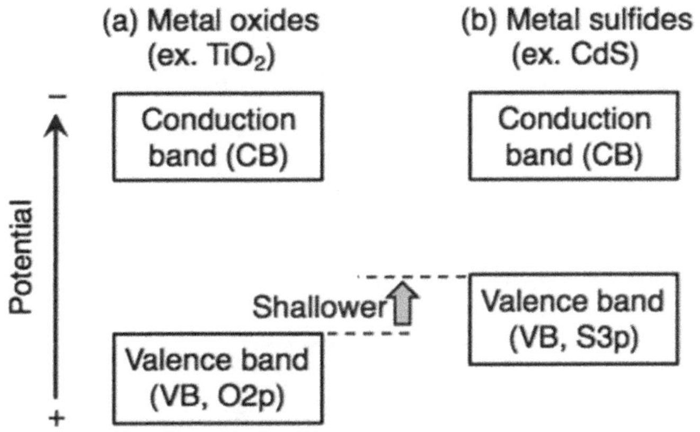

Figure 4.1 Band structures of (a) metal oxides and (b) metal sulfides.

metal oxides with photocatalytic H_2 evolution ability become larger than 3.0 eV, which corresponds to 410 nm due to the deep valence bands. Metal sulfides are also another famous material group used as photocatalysts.[3,6,7] The valence bands of the metal sulfides are formed by S 3p orbitals and are shallower than the valence bands composed of O 2p orbitals (Figure 4.1(b)). Hence, metal sulfides possess relatively narrow bandgaps, which are advantageous to visible light response. Therefore, metal sulfides are a favorable material group for efficient utilization of sunlight.

A disadvantage of non-metal oxide materials, such as metal sulfides, is poor chemical stability in water under photoirradiation. For example, a well-known CdS photocatalyst is oxidized by photogenerated holes in the CdS particles, as follows:[8,9] $CdS + 2h^+ \rightarrow Cd^{2+} + S$. This self-oxidation under photoirradiation is called photocorrosion. Thus, metal sulfides hardly oxidize water to O_2, and electron donors such as S^{2-} and SO_3^{2-} ions are used to work steadily, as shown in Figure 4.2.[10]

Various metal sulfides have been reported as photocatalysts, while metal cations that are contained in the metal sulfide photocatalyst as a main constituent element are limited.[3,6,7] For example, most of the metal sulfides contain metal cations with a d^{10} electron configuration, such as Zn^{2+}, Cd^{2+}, Ga^{3+}, In^{3+}, Ge^{4+} and Sn^{4+}. Recently, some metal sulfides that contain Nb^{5+}, Ta^{5+} and V^{5+} ions of a d^0 electron configuration have been found.[11,12]

4.2 Utilization of Metal Sulfides Toward Artificial Photosynthetic Reactions of Water Splitting and CO_2 Reduction

Figure 4.3 shows two ways to demonstrate water splitting and CO_2 reduction accompanied by water oxidation to O_2 using photocorrosive metal sulfides. The first way is the construction of a photoelectrode system (Figure 4.3(a)).

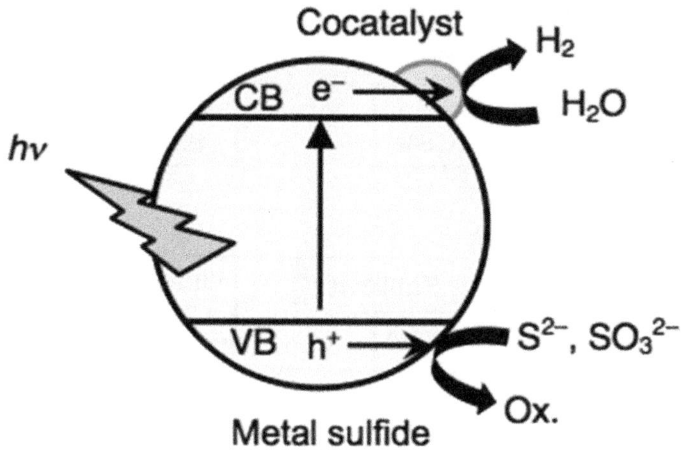

Figure 4.2 H_2 evolution over metal sulfide photocatalyst in the presence of sacrificial reagents such as S^{2-} and SO_3^{2-}.

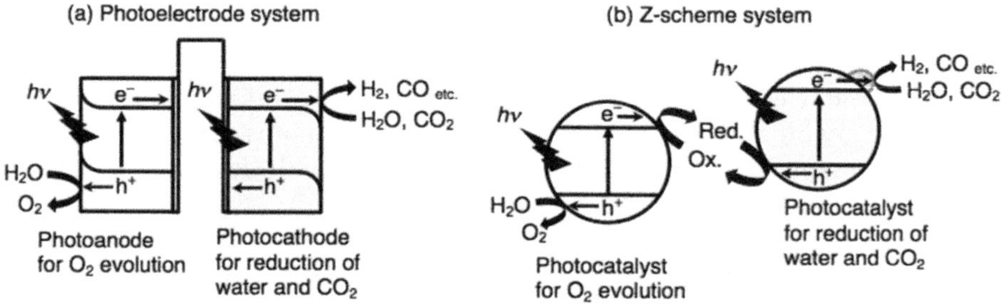

Figure 4.3 (a) Photoelectrode and (b) Z-scheme systems for water splitting and CO_2 reduction.

In the photoelectrode system, only a reduction reaction proceeds on the photocathode and photogenerated holes are removed to the photoanode. Therefore, even the photocorrosive metal sulfides would be applicable for the photocathode when the metal sulfides possess p-type semiconductor character (p-type metal sulfides). The second way is the construction of a Z-scheme system (Figure 4.3(b)). The Z-scheme system is composed of two different photocatalysts and an electron mediator. One photocatalyst reduces water and/or CO_2, and the other photocatalyst oxidizes water to O_2. The electron mediator carries the remaining electrons in the photocatalyst for water oxidation to the photocatalyst for reduction. Metal sulfides have the potential to work as the photocatalyst for the reduction of water and CO_2.

4.2.1 Metal Sulfide Photocathodes for Reduction of Water

The typical structure of recently developed metal sulfide photocathodes with high efficiencies is composed of p-type metal sulfide photocatalysts, n-type buffer layers, and co-catalysts for water reduction to H_2, as shown in

Figure 4.4.[13] The metal sulfide absorbs photons to generate electrons and holes. The holes migrate to conductive substrates such as Sn-doped In_2O_3 (ITO), F-doped SnO_2 (FTO), and metal layers. Photoexcited electrons transfer to n-type buffer through a p–n junction between p-type metal sulfide and n-type buffer layer, which promotes the electron transfer. The transferred electrons reduce water to H_2 over catalysts for H_2 evolution.

Typical metal sulfide photocathodes are summarized in Table 4.1. Cu(I)-containing metal sulfides, such as Cu_2S (BG 1.5 eV),[14] $CuInS_2$ (BG 1.5 eV),[15–19] $CuGaS_2$ (BG 2.3 eV),[20] Cu(In,Ga)S_2,[21,22] Cu_2ZnSnS_4 (BG 1.5 eV)[23–25] and Cu_2ZnGeS_4 (BG 2.2 eV),[26] are typical p-type metal sulfides. Moreover, $CuInS_2$, $CuGaS_2$, Cu_2ZnSnS_4 and Cu_2ZnGeS_4 function as a powdered photocatalyst for H_2 evolution in the presence of electron donors under visible light irradiation. Thus, Cu(I)-containing metal sulfides are suitable as the photocathode

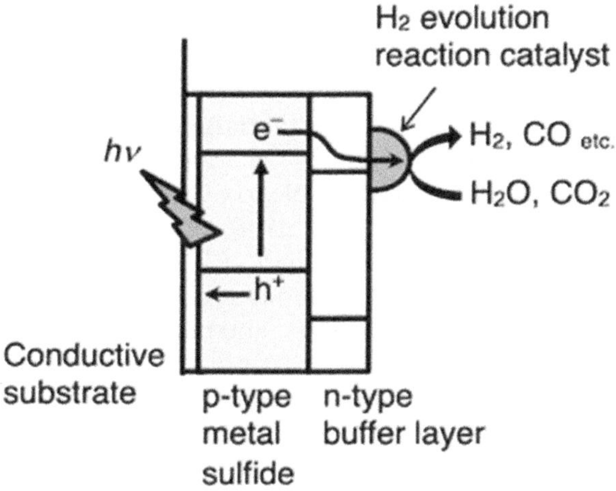

Figure 4.4 Typical structure of metal sulfide photocathodes for reduction of water and CO_2.

Table 4.1 Metal sulfide photocathodes for water reduction to produce H_2.

HER catalyst	Buffer layer	p-Type metal sulfide	Ref.
—	TiO_2/CdS	Cu_2S	14
Pt	TiO_2/CdS	$CuInS_2$	16
Pt	CdS	$CuInS_2$	17
Pt	In_2S_3	$CuInS_2$	17
—	C_{60}	SnS_2/$CuInS_2$	19
Pt	TiO_2/CdS	Cu(In,Ga)S_2	21
—	$C_3N_{4-x}S_{3x/2}$/CdS	Cu(In,Ga)S_2	22
Pt	In_2S_3/CdS	Cu_2ZnSnS_4	23
Pt	HfO_2/CdS/HfO_2	Cu_2ZnSnS_4	24
MoS_x	CdS	Cu_2ZnSnS_4	25
—	ZnS	Cu_2ZnGeS_4	26
No–Mo	TiO_2/CdS	Cu(In,Ga)(S,Se)$_2$	27

material. Due to their narrow bandgaps, $CuInS_2$ and Cu_2ZnSnS_4 are widely used as the photocathode for water reduction to H_2.

As n-type buffer layers, CdS and In_2S_3 are widely used.[14,16,17,21–25] The n-type buffer layers are deposited on the p-type metal sulfide layer to form p–n junction. The p–n junction promotes the electron transfer from the p-type metal sulfide to the n-type buffer layer, resulting in the suppression of charge recombination. To avoid the corrosion of the n-type buffer layer, TiO_2 is sometimes deposited on the n-type buffer layer as a protection layer.[14,22] Deposition of a HfO_2 layer is another efficient modification that promotes electron transfers not only from the p-type metal sulfide to the n-type buffer layer but also from the n-type buffer layer to the a H_2 evolution reaction catalyst, as seen in Pt-loaded HfO_2/CdS/HfO_2/Cu_2ZnSnS_4 photocathode.[24]

As H_2 evolution reaction (HER) catalysts, Pt with low overpotential for reduction of water to H_2 is widely used. Instead of the Pt, non-noble metal-based HER catalysts such as MoS_x[25] and Ni–Mo alloy[27] have been developed.

4.2.2 Metal Sulfide Photocathodes for the Reduction of CO_2

In CO_2 reduction, relatively negative potentials are required when compared to water reduction to H_2 (Figure 4.5).[28–30] Therefore, metal sulfides with high conduction band potential are applicable to CO_2 reduction.

Ru-Complex polymer-modified Cu_2ZnSnS_4 photocathodes reduce CO_2 to form HCOOH under visible light irradiation.[31] The Ru-complex polymer is known as an electrocatalyst for CO_2 reduction.[32] The photogenerated electrons migrate to the Ru-complex polymer and reduce CO_2 on the Ru-complex polymer.

The unmodified Cu_2ZnSnS_4 photocathode is also active for reduction of CO_2 to CO under visible light irradiation. The activity for CO_2 reduction is improved by modification with n-type buffer layers of In_2S_3 and CdS on the Cu_2ZnSnS_4 photocathode.[33] The modified Cu_2ZnSnS_4 photocathode produces HCOOH as well as CO. In addition to the Cu_2ZnSnS_4-based photocathodes,

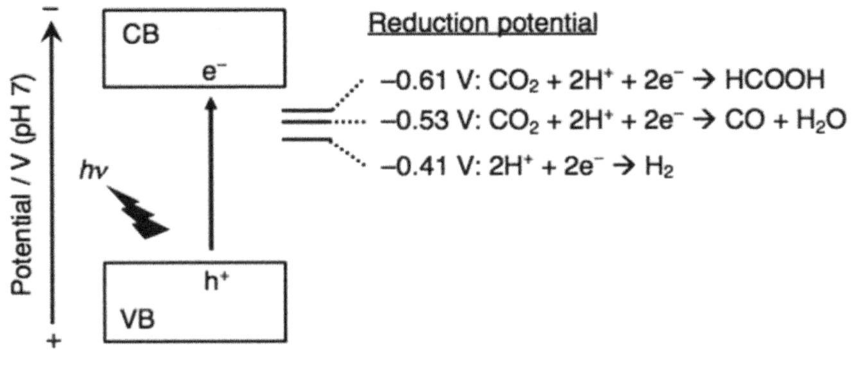

Figure 4.5 CO_2 reduction over photocatalysts.

$CuGaS_2$[34] and ZnS-modified Cu_2ZnGeS_4[26] photocathodes also produce CO by reduction of CO_2 under visible light irradiation.

4.2.3 Z-scheme Systems for Water Splitting and CO_2 Reduction

The key factor in constructing Z-scheme systems using metal sulfide powders lies in an electron mediator, because the photogenerated holes must smoothly be removed from metal sulfides before oxidation of metal sulfides (photocorrosion). There are three types of the Z-scheme system using metal sulfide powders: (1) powdered Z-scheme system with an ionic electron mediator; (2) powdered Z-scheme system with a solid-electron mediator; and (3) powder and photoelectrode-combined system with an ionic electron mediator (Figure 4.6).

Various redox couples such as Fe^{3+}/Fe^{2+},[35] I_3^-/I^-,[36,37] IO_3^-/I^-[38] and VO_2^+/VO^{2+}[39] have been reported as ionic electron mediators for powdered Z-scheme systems. Co complexes of $[Co(bpy)]^{3+/2+}$,[40] $[Co(phen)]^{3+/2+}$[40] and $[Co(terpy)]^{3+/2+}$[41] are successful examples as suitable electron mediators with metal sulfide powdered photocatalysts, while combination of usable photocatalysts are limited to a Ru co-catalyst-loaded $(CuGa)_{1-x}Zn_{2x}S_2$ H_2-evolving photocatalyst and a $BiVO_4$ O_2-evolving photocatalyst (Figure 4.6(a)). The Z-scheme systems with Ru-loaded $(CuGa)_{1-x}Zn_{2x}S_2$, $BiVO_4$, and Co complexes split water into

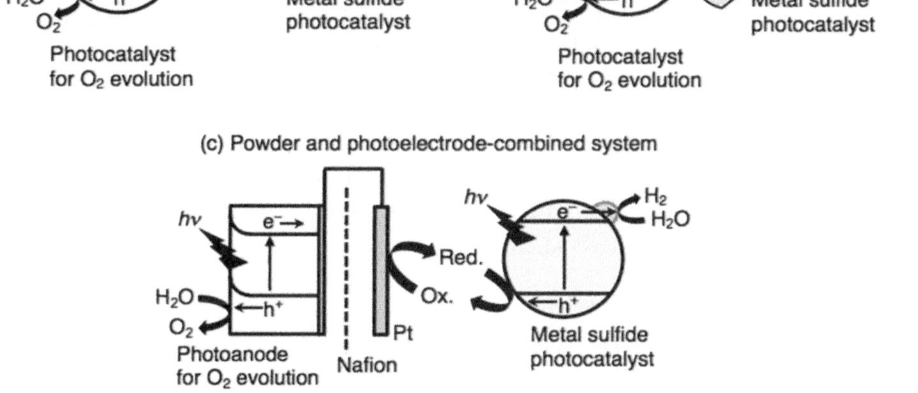

Figure 4.6 Various Z-scheme systems using powdered metal sulfide photocatalysts accompanied by oxidation of water to O_2.

H_2 and O_2 under visible light irradiation. The $(CuGa)_{1-x}Zn_{2x}S_2$ can be seen as a solid solution between $CuGaS_2$ and ZnS and shows higher photocatalytic activity for sacrificial H_2 evolution under visible light irradiation than $CuGaS_2$. Such formation of a solid solution is useful strategy to develop efficient metal sulfide photocatalysts. The Z-scheme system is also active for reduction of CO_2 to CO and HCOOH by loading of a [Ru(dpbpy)] complex on $(CuGa)_{1-x}Zn_{2x}S_2$ instead of the Ru co-catalyst.[42] The [Ru(dpbpy)] complex functions as CO_2 reduction sites.

Almost all electron mediators of powdered Z-scheme system are ionic redox couples as mentioned previously, while a reduced graphene oxide (RGO) works as a solid-state electron mediator (Figure 4.6(b)).[43] In the Z-scheme system with the RGO electron mediator, H_2-evolving photocatalysts and O_2 evolving photocatalysts need to possess p-type and n-type semiconductor properties, respectively.[44] Therefore, metal sulfides that work as a photocathode are usable as the H_2-evolving photocatalyst in the Z-scheme system. When rutile TiO_2 is used as the O_2 evolving photocatalyst, $CuInS_2$, $CuGaS_2$, Cu_2ZnSnS_4 and Cu_2ZnGeS_4 work as the H_2-evolving photocatalyst.[44] Note that the Z-scheme systems with rutile TiO_2 split water into H_2 and O_2 under only UV light irradiation due to the wide bandgap of rutile TiO_2 (BG: 3.0 eV). When visible-light-responsive $BiVO_4$ (BG: 2.4 eV) is used as the O_2 evolving photocatalyst, $CuGaS_2$, Cu_2ZnGeS_4 and $(CuGa)_{1-x}Zn_{2x}S_2$ work as the H_2-evolving photocatalyst.[45,46] The Z-scheme systems with $CuGaS_2$ with either TiO_2 or $BiVO_4$ are also active for CO_2 reduction to from CO.[45,47] It is important to note that RGO is oxidized and CO is generated due to the strong oxidation power of TiO_2. Therefore, the CO is generated by both reduction of CO_2 and oxidation of RGO when TiO_2 is used. In contrast, such oxidation of RGO is not happened when $BiVO_4$ is used, because the oxidation power of $BiVO_4$ is weaker than TiO_2. In other words, the CO is generated by only reduction of CO_2 when $BiVO_4$ is used.

The last example of the Z-scheme system using metal sulfide photocatalysts is a powder and photoelectrode-combined system. In more detail, the system is composed of $K_2[CdFe(CN)_6]$-modified CdS and $ZnIn_2S_4$ powders, a $[Fe(CN)_6]^{3-/4-}$ electron mediator, a Pt electrode and a CoO_x-loaded TaON photoelectrode (Figure 4.6(c)).[48] Photoexcited electrons in the metal sulfides reduce water to H_2 on the loaded Pt co-catalyst. Remaining holes are scavenged by $K_2[CdFe^{II}(CN)_6]$ to become $K[CdFe^{III}(CN)_6]$ and subsequently oxidize $[Fe(CN)_6]^{4-}$ to $[Fe(CN)_6]^{3-}$. O_2 is generated on the CoO_x-loaded TaON photoelectrode under visible light irradiation. Remained electrons migrate to the Pt electrode to reduce $[Fe(CN)_6]^{3-}$ to $[Fe(CN)_6]^{4-}$.

4.3 Oxysulfide Photocatalysts

The simplest way to demonstrate water splitting is a one-step photoexcitation system using a single photocatalyst. As mentioned above, metal sulfides are not suitable photocatalysts for oxidation of water to O_2 due to photocorrosion. In contrast to it, oxysulfides such as $Sm_2Ti_2S_2O_5$ (BG: 2.0–2.1 eV)[49] and $La_5Ti_2MS_5O_7$ (BG: 2.2 eV for M = Ag, 1.9 eV for M = Cu)[50] can oxidize water to O_2 under visible light irradiation when an IrO_2 co-catalyst for O_2 evolution is loaded. This is because the valence band maximum formed by the O 2p and S 3p hybridized orbitals stabilizes S^{2-} ions in the oxysulfides compared to those in the sulfides (Figure 4.7). $Sm_2Ti_2S_2O_5$ as well as $Y_2Ti_2S_2O_5$ (BG: 1.9 eV) shows photocatalytic activities not only for O_2 evolution but also for H_2 evolution in the presence of sacrificial reagents. The $Y_2Ti_2S_2O_5$ splits water into H_2 and O_2 under visible light irradiation upon loading with both IrO_2 and Cr_2O_3/Rh co-catalysts.[51]

Oxysulfides are also used as H_2-evolving photocatalysts for Z-schematic water splitting, as summarized in Table 4.2. The first four combinations in Table 4.2 are powdered systems with ionic redox couples of IO_3^-/I^-[52] and I_3^-/I^-,[53] as shown in Figure 4.6(a). The first system cannot function under visible light irradiation because of the use of the TiO_2 (rutile) with UV light response,[52] while the rutile TiO_2 possesses small enough bandgap to absorb sunlight. Instead of rutile TiO_2, visible-light-responsive WO_3 with a bandgap of 2.6–2.7 eV is used as O_2-evolving photocatalyst combined with various

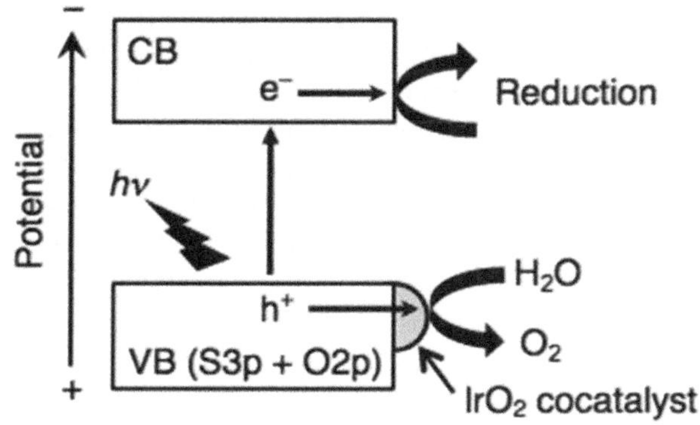

Figure 4.7 Oxidation of water to O_2 over a metal oxysulfide photocatalyst loaded with an IrO_2 co-catalyst.

Table 4.2 Z-scheme systems using oxysulfide photocatalysts for water splitting.

H_2-Evolving photocatalyst	Electron mediator	O_2-Evolving photocatalyst	Ref.
Pt/$Sm_2Ti_2S_2O_5$	IO_3^-/I^-	TiO_2 (rutile)	52
Pt,IrO_2/$Sm_2Ti_2S_2O_5$	I_3^-/I^-	PtO_x-loaded WO_3^a	53
Pt/$La_5Ti_2CuS_5O_7$	I_3^-/I^-	PtO_x-loaded WO_3^a	53
Rh/$La_6Ti_2S_8O_5$	I_3^-/I^-	PtO_x-loaded WO_3^a	53
Cr_2O_3/Rh/$La_5Ti_2CuS_5O_7$	Au layer	$BiVO_4$	54
Cr_2O_3/Rh/$La_5Ti_2Cu_{0.9}Ag_{0.1}S_5O_7^b$	Au layer	$BiVO_4$	54

[a] WO_3 treated with Cs^+ and H^+ ions.[55]
[b] $La_5Ti_2Cu_{0.9}Ag_{0.1}S_5O_7$ doped with Ga.[54]

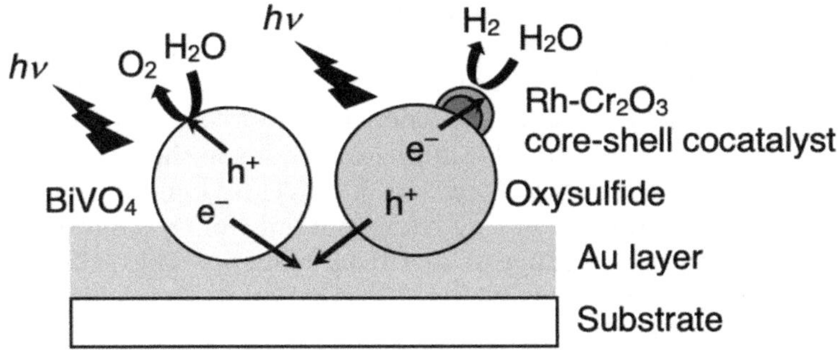

Figure 4.8 Sheet-type Z-scheme system using oxysulfide photocatalysts for water splitting under visible light irradiation.

oxysulfide photocatalysts of $Sm_2Ti_2S_2O_5$, $La_5Ti_2CuS_5O_7$ and $La_6Ti_2S_8O_5$ in the second to fourth systems.[53] Two different co-catalysts of Pt and IrO_2 loaded on $Sm_2Ti_2S_2O_5$ work as reduction sites for water to H_2 and oxidation sites for I^- to I_3^-, respectively. The last two combinations in Table 4.2 are a sheet-type system in which H_2-evolving and O_2-evolving photocatalysts are fixed on a conductive layer, as shown in Figure 4.8. In these systems, $BiVO_4$ with a bandgap of 2.4 eV, which is smaller than WO_3 is used as an O_2-evolving photocatalyst.[54] However, the usable light as the Z-scheme system is still limited by the O_2-evolving photocatalyst. Photogenerated holes in $BiVO_4$ oxidize water to O_2 and the photoexcited electrons transfer to the Au layer to combine with holes photogenerated holes in oxysulfide photocatalysts. The photoexcited electrons in the oxysulfide photocatalysts reduce H_2O to H_2 over a Rh–Cr_2O_3 core–shell co-catalyst. The Au layer contributes to efficient electron transfer because water splitting does not proceed in the absence of the Au layer.

References

1. F. E. Osterloh, *Chem. Mater.*, 2008, **20**, 35.
2. Y. Inoue, *Energy Environ. Sci.*, 2009, **2**, 364.
3. A. Kudo and Y. Miseki, *Chem. Soc. Rev.*, 2009, **38**, 253.
4. R. Abe, *J. Photochem. Photobiol., C*, 2010, **11**, 179.
5. D. E. Scaife, *Sol. Energy*, 1980, **25**, 41.
6. K. Zhang and L. Guo, *Catal. Sci. Technol.*, 2013, **3**, 1672.
7. S. Chandrasekaran, L. Yao, L. Deng, C. Bowen, Y. Zhang, S. Chen, Z. Lin, F. Peng and P. Zhang, *Chem. Soc. Rev.*, 2019, **48**, 4178.
8. H. Gerischer, *J. Electroanal. Chem. Interfacial Electrochem.*, 1975, **58**, 263.
9. A. J. Bard and M. S. Wrighton, *J. Electroanal. Chem.*, 1977, **124**, 1706.
10. J. F. Reber and K. Meier, *J. Phys. Chem.*, 1984, **88**, 5903.
11. T. Takayama, I. Tsuji, N. Aono, M. Harada, T. Okuda, A. Iwase, H. Kato and A. Kudo, *Chem. Lett.*, 2017, **46**, 616.
12. S. Ikeda, N. Aono, A. Iwase, H. Kobayashi and A. Kudo, *ChemSusChem*, 2019, **12**, 1977.

13. C. Y. Toe, S. Zhou, M. Gunawan, X. Lu, Y. H. Ng and R. Amal, *J. Mater. Chem. A*, 2021, **9**, 20277.
14. Y. X. Yu, L. Pan, M. K. Son, M. T. Mayer, W. D. Zhang, A. Hagfeldt, J. Luo and M. Grätzel, *ACS Energy Lett.*, 2018, **3**, 760.
15. G. Hodes, T. Engelhard, J. A. Turner and D. Cahen, *Sol. Energy Mater.*, 1985, **12**, 211.
16. J. Zhao, T. Minegishi, L. Zhang, M. Zhong, Gunawan, M. Nakabayashi, G. Ma, T. Hisatomi, M. Katayama, S. Ikeda, N. Shibata, T. Yamada and K. Domen, *Angew. Chem., Int. Ed.*, 2014, **53**, 11808.
17. Gunawan, W. Septina, T. Harada, Y. Nose and S. Ikeda, *ACS Appl. Mater. Interfaces*, 2015, **7**, 16086.
18. B. K. Patra, S. Khilari, D. Pradhan and N. Pradhan, *Chem. Mater.*, 2016, **28**, 4358.
19. F. Zhang, Y. Chen, W. Zhou, C. Ren, H. Gao and G. Tian, *ACS Appl. Mater. Interfaces*, 2019, **11**, 9093.
20. A. Iwase, Y. H. Ng, R. Amal and A. Kudo, *J. Mater. Chem. A*, 2015, **3**, 8566.
21. N. Guijarro, M. S. Prévot, X. Yu, X. A. Jeanbourquin, P. Bornoz, W. Bourée, M. Johnson, F. L. Formal and K. Sivula, *Adv. Energy Mater.*, 2016, **6**, 1501949.
22. D. Wang, C. Wang, F. P. G. de Arquer, J. Zhong, L. Qian, L. Fang, P. Liu, Y. Pang, M. Liu, M. Liu, G. Zheng, D. Sinton, E. H. Sargent, H. Yang and B. Zhang, *J. Mater. Chem. A*, 2017, **5**, 3167.
23. F. Jiang, T. Harada, Y. Kuang, T. Minegishi, K. Domen and S. Ikeda, *J. Am. Chem. Soc.*, 2015, **137**, 13691.
24. D. Huang, K. Wang, L. Li, K. Feng, N. An, S. Ikeda, Y. Kuang, Y. H. Ng and F. Jiang, *Energy Environ. Sci.*, 2021, **14**, 1480.
25. K. Feng, D. Huang, L. Li, T. Harada, S. Ikeda and F. Jiang, *Appl. Catal., B*, 2020, **268**, 118438.
26. S. Ikeda, S. Fujikawa, T. Harada, T. H. Nguyen, S. Nakanishi, T. Takayama, A. Iwase and A. Kudo, *ACS Appl. Energy Mater.*, 2019, **2**, 6911.
27. M. Baek, M. Zafar, S. Kim, D. H. kim, C. W. Jeon, J. Lee and K. Yong, *ChemSusChem*, 2018, **11**, 3679.
28. S. C. Roy, O. K. Varghese, M. Paulose and C. A. Grimes, *ACS Nano*, 2010, **4**, 1259.
29. S. N. Habisreutinger, L. Schmidt-Mende and J. K. Stolarczyk, *Angew. Chem., Int. Ed.*, 2013, **52**, 7372.
30. S. R. Lingampalli, M. M. Ayyub and C. N. R. Rao, *ACS Omega*, 2017, **2**, 2740.
31. T. Arai, S. Tajima, S. Sato, K. Uemura, T. Morikawa and T. Kajino, *Chem. Commun.*, 2011, **47**, 12664.
32. S. Chardon-Noblat, A. Deronzier, R. Ziessel and D. Zsoldos, *J. Electroanal. Chem.*, 1998, **444**, 253.
33. S. Kamimura, Y. Sasaki, M. Kanaya, T. Tsubota and T. Ohno, *RSC Adv.*, 2016, **6**, 112594.
34. S. Ikeda, Y. Tanaka, T. Kawaguchi, S. Fujikawa, T. Harada, T. Takayama, A. Iwase and A. Kudo, *Chem. Lett.*, 2018, **47**, 1424.

35. H. Kato, M. Hori, R. Konta, Y. Shimodaira and A. Kudo, *Chem. Lett.*, 2004, **33**, 1348.
36. R. Abe, K. Sayama, K. Domen and H. Arakawa, *Chem. Phys. Lett.*, 2001, **344**, 339.
37. K. Sayama, K. Mukasa, R. Abe, Y. Abe and H. Arakawa, *Chem. Commun.*, 2001, 2416.
38. R. Abe, K. Shinmei, K. Hara and B. Ohtani, *Chem. Commun.*, 2009, 3577.
39. Y. Miseki, S. Fujiyoshi, T. Gunji and K. Sayama, *J. Phys. Chem. C*, 2017, **121**, 9691.
40. Y. Sasaki, H. Kato and A. Kudo, *J. Am. Chem. Soc.*, 2013, **135**, 5441.
41. T. Kato, Y. Hakari, S. Ikeda, Q. Jia, A. Iwase and A. Kudo, *J. Phys. Chem. Lett.*, 2015, **6**, 1042.
42. T. M. Suzuki, S. Yoshino, T. Takayama, A. Iwase, A. Kudo and T. Morikawa, *Chem. Commun.*, 2018, **54**, 10199.
43. A. Iwase, Y. H. Ng, Y. Ishiguro, A. Kudo and R. Amal, *J. Am. Chem. Soc.*, 2011, **133**, 11054.
44. K. Iwashina, A. Iwase, Y. H. Ng, R. Amal and A. Kudo, *J. Am. Chem. Soc.*, 2015, **137**, 604.
45. A. Iwase, S. Yoshino, T. Takayama, Y. H. Ng, R. Amal and A. Kudo, *J. Am. Chem. Soc.*, 2016, **138**, 10260.
46. S. Yoshino, A. Iwase, Y. H. Ng, R. Amal and A. Kudo, *ACS Appl. Energy Mater.*, 2020, **3**, 5684.
47. T. Takayama, K. Sato, T. Fujimura, Y. Kojima, A. Iwase and A. Kudo, *Faraday Discuss.*, 2017, **198**, 397.
48. T. Shirakawa, M. Higashi, O. Tomita and R. Abe, *Sustainable Energy Fuels*, 2017, **1**, 1065.
49. A. Ishikawa, T. Takata, J. N. Kondo, M. Hara, H. Kobayashi and K. Domen, *J. Am. Chem. Soc.*, 2002, **124**, 13547.
50. T. Suzuki, T. Hisatomi, K. Teramura, Y. Shimodaira, H. Kobayashi and K. Domen, *Phys. Chem. Chem. Phys.*, 2012, **14**, 15475.
51. Q. Wang, M. Nakabayashi, T. Hisatomi, S. Sun, S. Akiyama, Z. Wang, Z. Pan, X. Xiao, T. Watanabe, T. Yamada, N. Shibata, T. Takata and K. Domen, *Nat. Mater.*, 2019, **18**, 827.
52. W. Zhao, K. Maeda, F. Zhang, T. Hisatomi and K. Domen, *Phys. Chem. Chem. Phys.*, 2014, **16**, 12051.
53. G. Ma, S. Chen, Y. Kuang, S. Akiyama, T. Hisatomi, M. Nakabayashi, N. Shibata, M. Katayama, T. Minegishi and K. Domen, *J. Phys. Chem. Lett.*, 2016, **7**, 3892.
54. S. Sun, T. Hisatomi, Q. Wang, S. Chen, G. Ma, J. Liu, S. Nandy, T. Minegishi, M. Katayama and K. Domen, *ACS Catal.*, 2018, **8**, 1690.
55. Y. Miseki, S. Fujiyoshi, T. Gunji and K. Sayama, *Catal. Sci. Technol.*, 2013, **3**, 1750.

CHAPTER 5

Efficacy of Bismuth Oxyhalides (BiOX-based) Materials for Enhanced Photocatalysis

SRABANTI GHOSH*[a,b], NEHA VERMA[c] AND RAJAKUMAR ANANTHAKRISHNAN[c]

[a]Energy Materials & Devices Division, CSIR – Central Glass and Ceramic Research Institute, Raja S. C. Mullick Road, Jadavpur, Kolkata-700032, India; [b]Academy of Scientific & Innovative Research (AcSIR), Ghaziabad, India; [c]Green Environmental Materials & Analytical Chemistry Laboratory, Department of Chemistry, Indian Institute of Technology, Kharagpur 721 302, India
*E-mail: srabanti@cgcri.res.in

5.1 Introduction

Oxyhalides, in particular, bismuth halides (BiOX, X = Cl, Br, I) based materials have emerged as a promising class in the area of photocatalysis.[1–3] Owing to their diverse nature, flexible properties, light-harvesting capabilities, they have driven enormous attention in the scientific community for a wide range of applications, such as ferroelectric materials, storage materials, and pigments.[4] The surface properties, morphologies, electronic structure, compositions of the materials are the most important parameters to tune their catalytic activities for photocatalytic waste water and indoor-gas purification,

Sustainable Energy No. 1
Recent Developments in Functional Materials for Artificial Photosynthesis
Edited by Srabanti Ghosh and Qian Wang

Published by the Royal Society of Chemistry, www.rsc.org

water splitting, organic synthesis, and selective oxidation of alcohol.[5–10] This chapter focuses on the strategies directed to modify the pristine oxyhalides nanostructured materials to boost their visible light photocatalytic efficiencies. In particular, the fundamental structural changes upon introducing the other materials, electronic properties, phase modification *via* doping, different types of heterojunctions, and the effect of non-metal interaction are outlined. The reaction conditions, reaction media, kinetics, types of catalysts employed in the methods used for the suitability of H_2 generation, water splitting, CO_2 reduction are discussed in detail. Furthermore, the scope for the rational design of oxyhalides-based materials and future outlook is proposed.

5.2 Pure Oxyhalides

Bismuth oxyhalides BiOX (X = Cl, Br, I), as another category of Bi-based semiconductors, have received immense attention for their outstanding optoelectronic properties and have shown promising applications in the field of the catalysis,[11] pigments,[12] and gas sensors.[13] All BiOX (X = Cl, Br, I) compounds are crystallized in a tetragonal matlockite structure.[14] The BiOX (X = Cl, Br, I) are belonging to the layered structure that consist of [Bi_2O_2] slabs with double halogen slabs along the direction [001]. An asymmetric decahedral symmetry obtained in presence of each bismuth centre is surrounded by four oxygen atoms and four halogen atoms in [X–Bi–O–Bi–X] unit (Figure 5.1).

BiOX demonstrates indirect transition bandgap (*i.e.*, electrons have to travel a certain *k*-spacing to be emitted to the VB) with low electron–hole recombination compared to direct bandgap semiconductors. It is well known that the bandgap values of BiOX (X = Cl, Br, I) decreases with increasing the atomic numbers of the halogen (*e.g.*, BiOCl (~3.2 eV), BiOBr (~2.7 eV), and BiOI (~1.7 eV)). This could allow for enhancing their visible light-harvesting capacity and, consequently, improved photocatalytic activities over a large

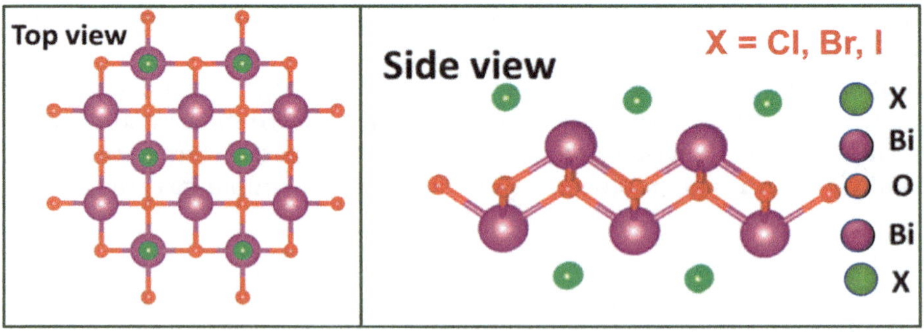

Figure 5.1 Top and side view of the geometric structure of single-layer BiOX. Reproduced from ref. 24 with permission from the Royal Society of Chemistry.

range.[15] Interestingly, the valence band maxima of BiOX is contributed by O 2p and X np orbitals (3p, 4p and 5p for Cl, Br and I, respectively) and the bottom of the conduction band is occupied by a Bi 6p state, which can facilitate the photogenerated electrons of BiOX to travel longer distances that promote the mobility of the charge carriers and effectively reduce the bandgap of the materials.[16] After a revolutionary discovery of work, exponential interest has been focused on the robust syntheses of BiOX (X = Cl, Br, I) nano/micro-structures to boost their photocatalytic efficiencies.[17] Different strategies such as thickness tuning, morphological control, heteroatom doping, surface modification, surface defect engineering, and engineering of heterojunctions have been developed to tailor the photocatalytic activity. Up to now, different BiOX nanostructures, *e.g.*, nanocrystals, porous nanospheres, nanosheets, nanowires, nanobelts, and flower-like structures *etc.*, have been tested for photocatalytic applications with enhanced light absorption and photo-induced charge transfer efficiency.[18–22] A summary of the electronic structure of oxyhalides, various modification strategies, and photocatalytic applications of BiOX-related nanomaterials follows.

5.2.1 The Electronic Structure of Oxyhalides

Bismuth oxyhalides (BiOX (X = Cl, Br, I)) are a class of V–VI–VII ternary semiconductor materials. These materials crystallize in a tetragonal matlockite structure that is a layered structure composed of [X–Bi–O–Bi–X] slabs combined *via* van der Waals interactions.[3] In this [X–Bi–O–Bi–X] motif, each bismuth center is surrounded by four oxygen atoms and four halogen atoms, which rise to an asymmetric decahedral symmetry. The combination of strong intralayer covalent bonding coupled with interlayer van der Waals attractions make these materials interesting for anisotropic structural, electrical, optical, and mechanical properties.[23]

The crystal structure also results in internal electric fields perpendicular to the [X–Bi–O–Bi–X] slabs. This can potentially promote effective charge separation along the [001] direction, thus further inhibiting the recombinations.[24] The electronic structures of single-layer BiOX have been calculated by HSE06 methods to overcome the limitation of DFT calculation.[25] Figure 5.2(a)–(c) illustrates that BiOX materials possess indirect bandgaps with the conduction band minimum (CBM) located at the Z points and the valence band maximum (VBM) located between the R and Z points. Remarkably, the single-layer BiOX have band edges situated in energetically favorable positions for water splitting as shown in Figure 5.2(d).[24]

Recently, Lu *et al.*[26] conducted a detailed study on electronic properties and lattice dynamics of BiOCl by using optical spectroscopy and first principles calculations. Notably, the large spin–orbit coupling effect of the Bi atom in the Brillouin zone, which lowered the bandgap by 230 meV from 3.59 eV to 3.36 eV and the high temperature and monolayer stability, make them suitable in photocatalysis.

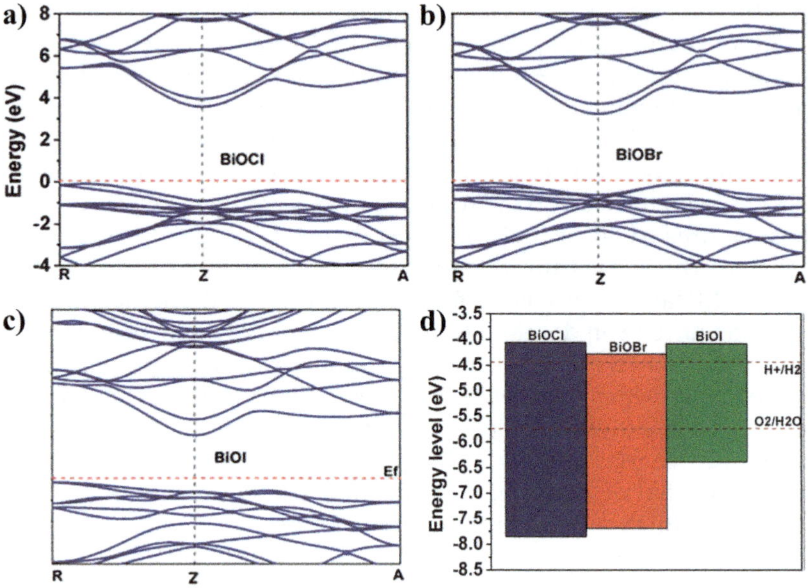

Figure 5.2 (a)–(c) The band structures of single-layer BiOCl, BiOBr, and BiOI at the Fermi level of 0 eV. (d) The band edge positions of single-layer BiOCl, BiOBr, and BiOI related to the vacuum level calculated with the HSE06 functional and the standard redox potentials for water splitting at pH = 0. Reproduced from ref. 24 with permission from the Royal Society of Chemistry.

5.2.2 Photocatalytic Applications of Oxyhalides

Single-layer BiOI possesses a favorable bandgap, suitable band edge positions, different orbital characteristics and different effective masses at the VBM and CBM, thus presenting excellent photocatalytic activity for water splitting.[5,27,28] BiOX-based materials are excellent candidates for a variety of photocatalytic applications which include: (a) photocatalytic oxygen evolution (OER), (b) photocatalytic hydrogen evolution (HER), (c) CO_2 photoreduction, (d) photocatalytic water splitting, (e) photocatalytic nitrogen fixation, (f) photocatalytic organic synthesis, (g) photocatalytic disinfection, and (f) photocatalytic removal of pollutants (Figure 5.3).[29,30]

Photocatalytic water splitting and CO_2 conversion into fuels through artificial photosynthesis is beneficial to renewable resource generation. For example, Lee *et al.*[31] fabricated flower-like hierarchical BiOX (X: Cl, Br, I) microspheres by a microwave-assisted solvothermal method that demonstrated a hydrogen evolution rate of 1316.9 μmol h^{-1} g^{-1} under visible light irradiation. It has been proposed that the conduction band level of BiOI catalysts provide a sufficient over-potential for water reduction and H_2 generation. BiOX photocatalysts with well-defined facets having different geometric and electronic structures stimulated significant research interest due

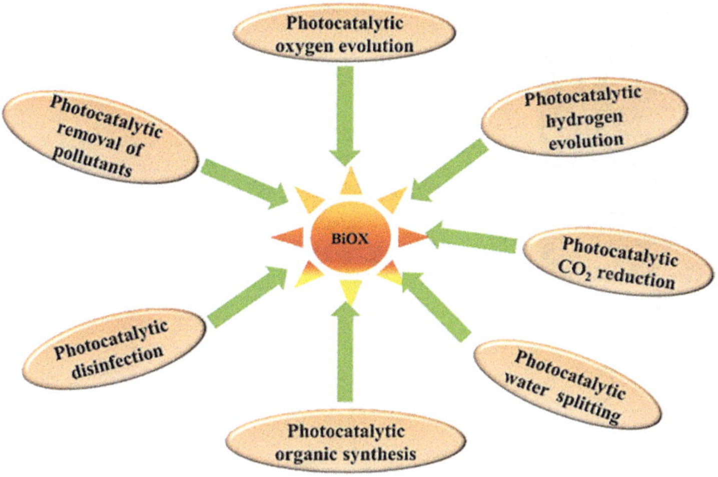

Figure 5.3 Various photocatalytic applications of BiOX photocatalysts.

to their intrinsic facet dependent photocatalytic properties.[32] Jiang reported the BiOCl single-crystalline nanosheets with exposed (001) and (010) facets that demonstrated facet-dependent photocatalytic activity for direct and indirect dye photosensitization degradation under UV light and visible light, respectively.[7] Recently, a series of BiOX photocatalysts have been developed for the facets effect, for example, porous BiOCl micro-flowers with (001) facets, BiOBr nanosheets with (001) facets,[33,34] highly symmetrical BiOI single-crystal nanosheets with dominant exposed (001) facets. Shi *et al.*[35] fabricated a well-defined BiOBr nanosheets dominating with respective facets, (001) and (010), to explore the effect of crystal facets for photocatalytic water splitting. Notably, the photoreactivity of BiOBr-(001) was shown to be higher than BiOBr-(010) for both hydrogen production (5.5 and 0.9 μmol m^{-2}) and oxygen evolution reactions (350 and 85 μmol m^{-2}) due to the highly exposed facets that lead to an efficient spatial separation of photogenerated charges. Liu and co-workers showed facet dependent photocatalytic activity and selectivity of CO_2 conversion to generate CH_4 and CO under light irradiation using photoinduced electrons of BiOI-001 and BiOI-100 nanosheets. BiOI-001 possesses efficient charge separation efficiency and a higher CB position, a large number O-vacancies that resulted in greater CO and CH_4 evolution (0.259 and 0.089 μmol h^{-1}, respectively) when compared to CO and CH_4 evolution over BiOI-100 (CO 0.076 μmol h^{-1} and CH_4 0.075 μmol h^{-1}).[36] Jiang *et al.*[37] fabricated BiOCl single-crystalline nanosheets with exposed {001} and {010} facets (Figure 5.4(a) and (b)) and BiOCl single-crystalline nanosheets with exposed {001} facets displayed high photocatalytic activity for pollutant degradation under UV light, but the counterpart with exposed {010} facets possessed superior activity for indirect dye photosensitization degradation under visible light. Notably, the internal electric field may promote charge

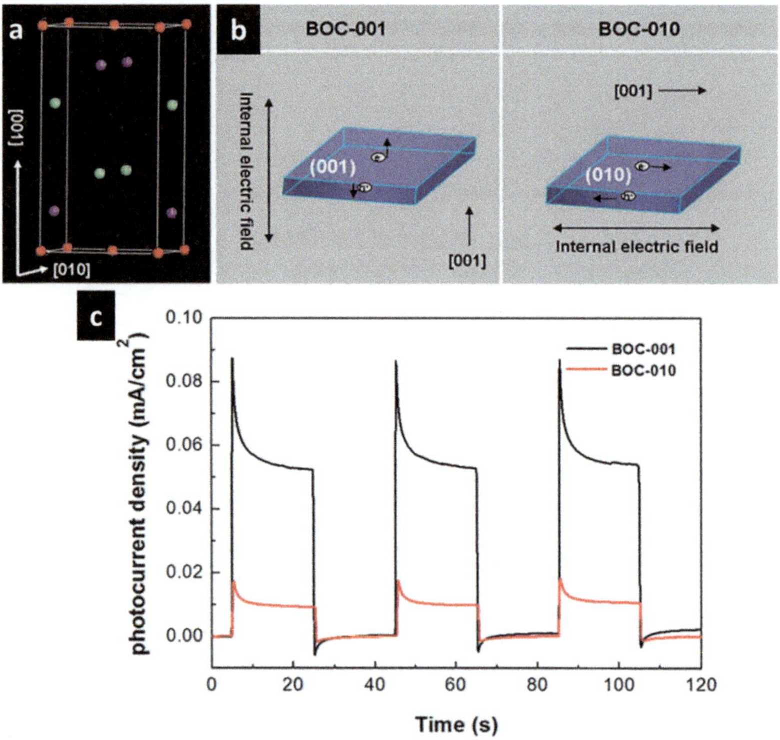

Figure 5.4 (a) Crystal structure of BiOCl. (b) Model showing the direction of the internal electric field in each of the BOC SCNSs. (c) Photocurrent responses of the BOC SCNSs under UV-vis irradiation. Reproduced from ref. 37 with permission from American Chemical Society, Copyright 2012.

transfer and the enhanced photocurrent obtained for the {001} facet-dominant nanosheets compared with that of the {010} facet-dominant counterparts (Figure 5.4(c)).

Very recently, Zou and co-workers[38] developed a new strategy to improve the catalytic efficiency of rose-like BiOCl with Bi vacancies rich assembled nanosheets having exposed active (001) facets that showed both the CO_2 capture and reduction with a higher value of CO generation (~22 μmol g^{-1} h^{-1}) under visible light. In another report, Ye *et al.*[39] have prepared bismuth-rich ultra-thin BiOBr for the photo-reduction of CO_2 into CO (~2.46 μmol g^{-1} h^{-1}) and CH_4 (~1.93 μmol g^{-1} h^{-1}). Notably, the CBM of a few-layers of BiOI illustrate an upward shift by ~0.17 eV, making the band alignment thermodynamically more favourable for the conversion of CO_2 into CO and CH_4.[40]

5.2.3 Modified Oxyhalides and Their Derivatives

In the case of semiconductor photocatalysts, fast photogenerated charge carrier recombination and poor photon absorption are the two major issues. Even though BiOX demonstrates strong photocatalytic performance under

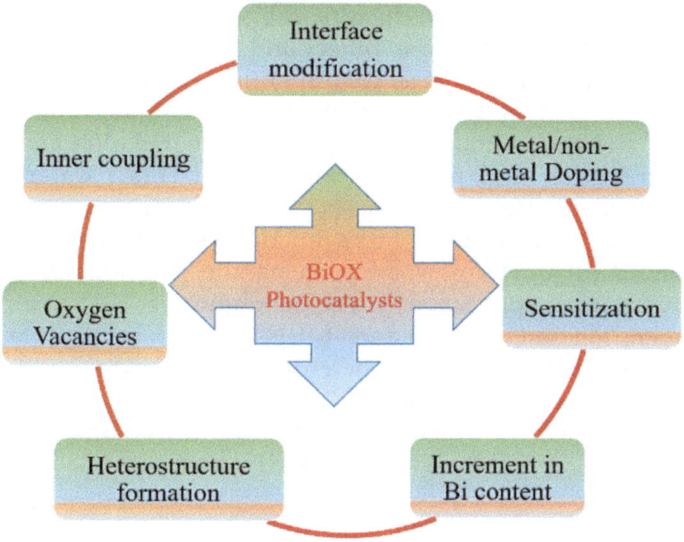

Figure 5.5 Various techniques for the increment of photocatalytic activity of BiOX photocatalysts.

visible light, it still lacks overall photocatalytic proficiency. Thus, it is difficult to implement for practical industrial applications. To achieve desirable redox ability for good photocatalytic activity, a large bandgap that relates to higher redox ability is desirable. However, for visible light photocatalysis, a small bandgap that helps in increasing the capability of light-harvesting is required. Therefore, to increase visible light activity and photocatalytic performance of BiOX, various strategies like heterostructure formation, enhancement in the Bi-content, metal/non-metal doping, interface modification, inner coupling between different bismuth oxyhalides, creation of oxygen vacancies, and use of sensitizers have been employed (Figure 5.5).[41–45]

5.3 Advancements in Oxyhalides Photocatalysts for Artificial Photosynthesis

5.3.1 Heterojunction Formation

Heterojunction formation is the most investigated method to change the bandgap positions that lead to work under visible light. A heterojunction formation between two semiconductors can efficiently enhance photocatalytic performance rather than single component materials because the coupling of two semiconductors is regarded as matching the energy of CB and VB, bandgap, and the crystal type. Therefore, a heterostructure is envisioned in designing a visible light photocatalyst having low surface charge recombination and improved redox properties.[46,47] There are several methods to design heterostructure photocatalysts that comprise heterojunction formation between two semiconductors. A brief outline of each method is discussed. A significant number of binary and ternary heterojunctions have been fabricated with BiOX, such as $TiO_2/Bi_4O_5Br_2$,[48] $SnWO_4$/BiOBr,[49] N-CQDs/OV-BiOBr,[50] BiOBr/ZnO/BiOI,[51] CdS/BiOI,[52] CuI/BiOI,[53] g-C_3N_4/BiOI,[54] g-C_3N_4/AgI/$Bi_4O_5I_2$,[55] and tested mainly for organic dye degradation and photocatalytic H_2 generation and photoelectrochemical (PEC) water oxidation (Table 5.1). Recently, He and co-workers[56] designed a new type of ternary layered materials of BiOBr/C Schottky junctions with two stack forms, the parallel and vertical junctions. Notably, the parallel junctions showed photocatalytic overall water splitting, with a H_2 generation rate of 240 μmol h^{-1} g^{-1} and O_2 generation rate of 110 μmol h^{-1} g^{-1}. Very recently, Wang *et al.*[57] designed an S-Scheme CdS/BiOCl heterojunction for the photoreduction of CO_2 that demonstrated enhanced CO_2 reduction activity due to the more effective charge transport, which is synergistically driven by the electric field at the heterojunction interface and the polarization electric field in the BiOCl phase.

5.3.2 Type II Heterojunctions

The constructed semiconductor heterojunctions are always fabricated by two semiconductors with overlapping band structure configurations and such semiconductors form a type-II photocatalytic mechanism.[58] When incident

Table 5.1 Comparison of the photocatalytic performance of BiOX-based materials.

Material	Photocatalytic H_2/O_2 generation, μmol h^{-1}	PEC activity, μA cm^{-2} (@1 V *vs.* Ag/AgCl)	Ref.
BiOX	1316.9(H_2)	—	31
BiOI	9200 (H_2)	2.4	61
BiOCl	3860(H_2)	11	
BiOBr	8500 (H_2)	0.76	
BiOI/$CuFe_2O_4$	11 230 (H_2)	36.67	61
BiOCl/$CuFe_2O_4$	22 200 (H_2)	13.63	
BiOBr/$CuFe_2O_4$	12 500 (H_2)	64.47	
TiO_{2-x}/BiOCl	—	700	78
Bi_4NbO_8Cl	25 (H_2) 12 (O_2)	—	79
Fe(III)-Modified BiOCl	30 (H_2)	350	80
BiOCl/β-FeOOH	13 650(H_2)	—	81
BiOCl/Cu(II)	16 (H_2)	40	82
{001}Facets BiOCl	58.6 (O_2)	—	83
Oxygen vacancy-mediated BiOCl	4 (H_2) 90(O_2)	—	84
BiOI	2730 (H_2)	—	85
g-C_3N_4/BiOI	—	800	86
BiOI nanoplatelet	—	1200	87
g-C_3N_4/Bi_4NbO_8Cl	287.71(H_2)	0.8	88
CeO_2/BiOI	323 (O_2)	—	89

light (visible light) falls on the semiconductor heterojunction to generate electron–hole pairs, electrons of semiconductor I (SM I) with a high conduction band edge position transfer to semiconductor II (SM II) with a low conduction band edge position. At the same time, photoinduced holes in both excited photocatalysts obey reverse transfer, *i.e.*, photo-induced holes and electrons are retained on SM II for the reduction reaction and SM I for the oxidation reaction. This is expected to improve the separation and recombination of charge carriers which considerably results in an enhancement in energy conversion efficiency of fabricated heterojunction (Figure 5.6).[59,60]

In the case of a type II photocatalyst, the redox ability of photogenerated charge carriers get weakened due to band-to-band migration and is not capable of effective photocatalytic reactions. Therefore, progress so far in the construction of the Type II heterojunction heterojunctions for oxyhalides is discussed. Recently, Bera *et al.*[61] reported a series of heterostructures using n-type bismuth oxyhalides and p-type spinel ferrites $CuFe_2O_4$ (CFO) by two step colloidal synthetic route for visible light-driven photocatalytic water splitting and H_2 generation. Interesting, a conventional Type-I and Type-II charge transfer mechanism has been proposed for BiOBr/CFO and BiOI/CFO heterostructures, respectively, which may effectively lower charge transfer resistance compared to bare BiOBr and BiOI suggests facile charge transfer (Figure 5.7).

5.3.3 Z-scheme Heterojunctions

The Z-scheme mechanism is the preferred type II mode photocatalyst (Figure 5.8). In the case of a Z-scheme photocatalyst, two semiconductor photocatalysts are interconnected. However, one is responsible for oxidation (SM II) while the other delivers a reduction reaction (SM I). Typically, the reduction photocatalyst holds usually a high conduction band position and displays a strong reducing ability whereas the oxidation photocatalyst has a lower valence band position and displays strong oxidative capacity.[62,63] Due to this exceptional charge transfer, Z-scheme photocatalyst not only displays excellent photocatalytic activity but also possesses other multiple features

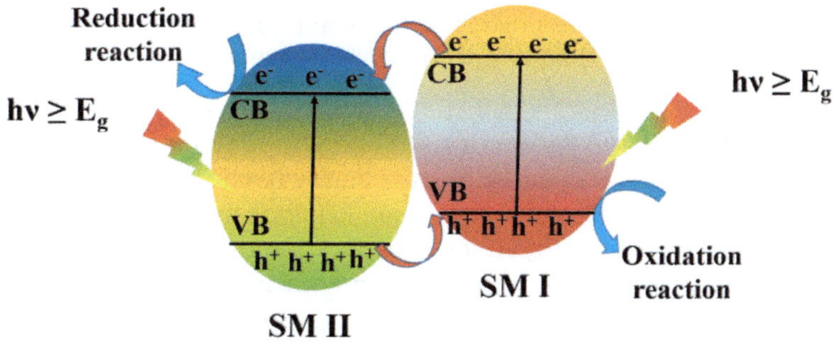

Figure 5.6 Type-II heterojunction photocatalytic mechanism.

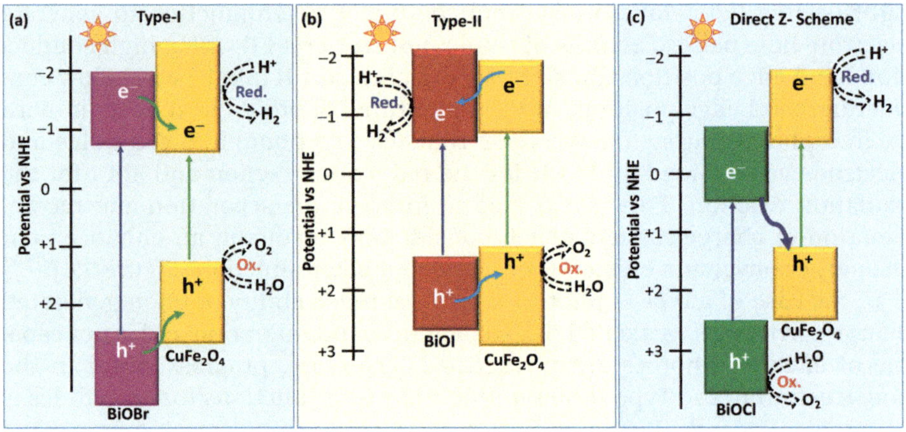

Figure 5.7 Band alignments of (a) BiOI/CFO, (b) BiOBr/CFO and (c) BiOCl/CFO heterostructures. Reproduced from ref. 61 with permission from American Chemical Society, Copyright 2022.

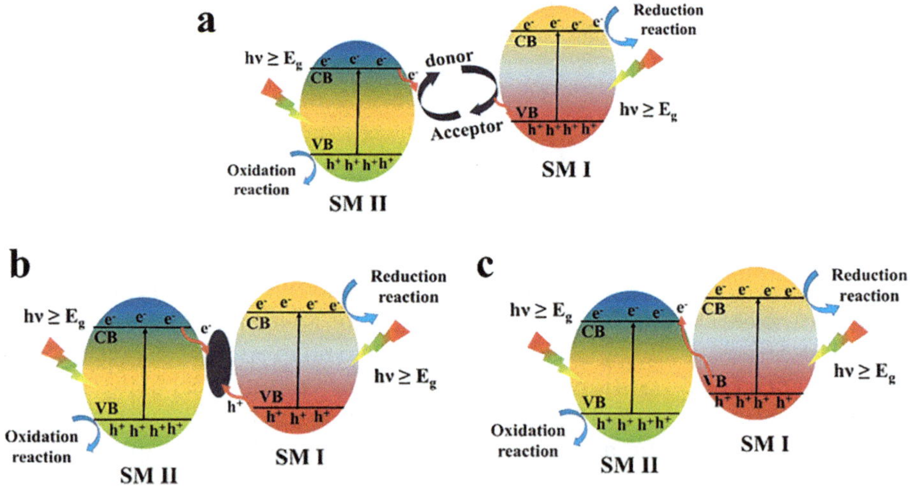

Figure 5.8 Z-scheme photocatalytic mechanism: (a) traditional Z-scheme, (b) all-solid-state Z-scheme, and (c) direct Z-scheme.

including: (1) spectra of various photocatalytic reactions, displaying that coupling of two photocatalysts of narrow bandgap can meet the necessity of improved redox ability for such reactions; (2) conservation of strong oxidation and reduction powers; (3) effective separation of oxidative and reductive active sites; and (4) enhanced light harvesting properties.

Z-scheme photocatalysts can be categorized into three main types: (1) traditional Z-scheme photocatalysts in which reversible redox ion pair is server phase transfer for charge carriers; (2) all-solid-state Z-scheme photocatalysts, in which an electron acts as a conductor to carry out charge transfer, and (3) direct Z-Scheme without using any charge transfer mediator. Z-scheme is widely used between two semiconductors. The semiconductor photocatalysts having a narrow bandgap can be used to fabricate direct Z-scheme photocatalysts to enlarge the visible light absorption range, hence increasing visible light active photocatalysis.[64–66] A 2D–2D Z-scheme heterojunction photocatalyst based on Bi_3O_4Cl and BiOCl nanosheets reported O_2 evolution in the presence of $FeCl_3$ as the electron scavenger reached to 58.6 μmol g^{-1} h^{-1} under visible light irradiation, which was three times higher than that of control Bi_3O_4Cl and BiOCl. It has been proposed that the rapid transfer of photogenerated electrons to the (001)-BiOCl facets trapped by Fe^{3+}, while the holes collected on the (001)-Bi_3O_4Cl facets contain plenty of active sites for the photocatalytic O_2 evolution.[67] The g-C_3N_4/$Bi_4O_5I_2$ heterostructure display higher photocatalytic performance for CO_2 reduction and shows the maximum production of CO (45.6 μmol g^{-1} h^{-1}) using I^{3-}/I^- redox mediator assisted Z-scheme mechanism.[68] Indirect Z-scheme BiOI/g-C_3N_4 photocatalysts have been reported with enhanced photoreduction CO_2 activity.[69] Another ternary heterostructure g-C_3N_4/BiOBr/Au showed promising photocatalytic activity for photoreduction of CO_2 and revealed simultaneously the highest CO (6.67 μmol g^{-1} h^{-1}) and CH_4 productivities (0.92 μmol g^{-1} h^{-1}) where the size of Au nanoparticles acted as the Z-scheme bridge and SPR centre during the photocatalytic process.[70] In another example, g-C_3N_4/BiOCl (defect-rich) demonstrated photoreduction of CO_2 and generated CO and CH_4 at a rate of 28.4 and 4.6 μmol g^{-1} h^{-1}, respectively, supported *via* interfacial oxygen vacancies that deliver a transport channel for the interfacial carriers, which in turn enhanced carrier transfer efficiency.[71] A direct Z-scheme BiOCl/CFO heterostructure has been formed between BiOCl and CFO with an intimate interfacial contact that demonstrated 5.7 times higher H_2 generation activity than pure BiOCl and two-fold improved catalytic efficiency compared to Type-II BiOI/CFO heterostructures (Figure 5.8(c)). Furthermore, a Z-scheme-based ternary heterostructure like BiOI-CD-CdS (BCC) has shown highly improved properties for 4-nitrophenol (4-NP) degradation and selective oxidation of benzyl alcohol under visible light illumination.[72] The morphology of the prepared ternary heterostructure (BCC) has been reported as in aggregated form (Figure 5.9(c) and (d)) while the BiOI-CD is spherical (Figure 5.9(a) and (b)). In this report, at approximately 4–6 nm carbon dots are essentially playing the role of electron transmission bridge between n-type CdS and p-type BiOI. The transient photocurrent response of ternary hybrid (BCC) has enhanced approximately seven times than bare CdS and BiOI-CD while 1.65 times greater as compared to 10 wt% BiOI-CdS (Figure 5.9(e)) indicating the predominant role of carbon dots in the effective separation of electron and hole. The Z-scheme-based BiOI-CD-CdS (denoted as 10 wt% BCC) exhibits a greater photocatalytic efficiency compared to the pristine BiOI and CdS. The results offer an enhanced degradation rate constant ($k = 12.67 \times 10^{-3}$ min^{-1}) that is almost 17.5 times and 6.5 times higher than the BiOI and CdS for 4-NP removal (Figure 5.9(f)).

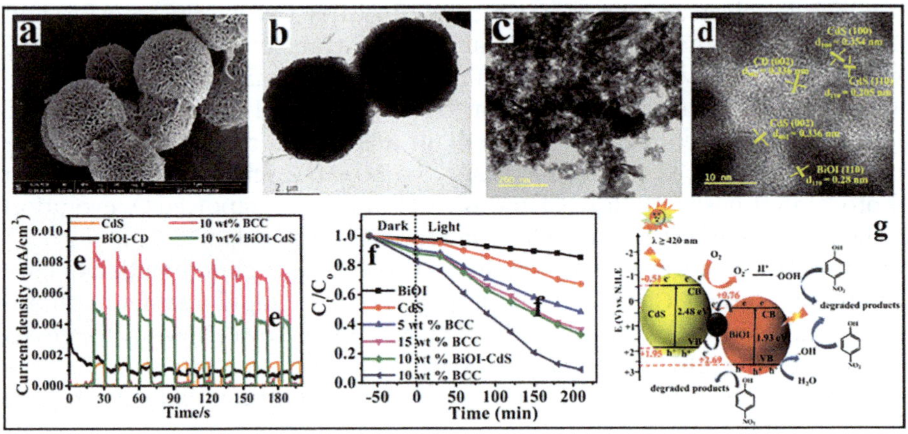

Figure 5.9 (a) and (b) FESEM and TEM image of BiOI-CD, (c) TEM image of BiOI-CD-CdS, (d) HRTEM image BCC at higher resolution indicating planes of CdS, BiOI, and CD, (e) transient photocurrent response of BiOI-CD, CdS, BiOI-CdS and BiOI-CD-CdS, (f) kinetic plots of photodegradation of 4-NP, (g) working principle of the hybrid (10 wt% BCC) for photodegradation of 4-NP in an aqueous medium. Reproduced from ref. 72 with permission from John Wiley & Sons, Copyright © 2021 Wiley-VCH GmbH.

In addition, the photocatalyst gave a 90% selective conversion of benzyl alcohol into benzaldehyde with high selectivity (98%). The working principle of the photocatalyst is proposed as Z-scheme pathways (Figure 5.9(g)). A series of novel Z-scheme visible-light photocatalysts based on the BiOCl-Au-CdS heterostructure,[73] $CoFe_2O_4$/BiOBr/graphene,[74] O-vacancy-BiOBr/$CoFe_2O_4$ heterojunction[75] have been reported for aueous pollutant removal such as degrading water dyes, Cr(VI) reduction and antibiotic degradation. Wang *et al.*[76] discussed the role of bridged Ag nanoparticles in the 2D/2D $Bi_5FeTi_3O_{15}$/g-C_3N_4 heterojunction extended the light absorption range and prolonged the lifetime of photogenerated electron–holes induced by $Bi_5FeTi_3O_{15}$ for efficient photocatalytic degradation of antibiotics. Kang *et al.*[77] explores the effect of dual direct Z-scheme MoS_2/g-C_3N_4/$Bi_{24}O_{31}Cl_{10}$ ternary heterojunction photocatalysts that may enhance the visible light absorption and favors excellent charge transfer and stronger redox capacity for photodegradation of antibiotic. However, it is important to mention that they have been rarely used for solar fuel production, such as water splitting and CO_2 photoreduction.

5.4 Conclusion and Future Prospects

For the development of a sustainable and cleaner environment, BiOX modified photocatalysts have captivated researchers for a prime role in applications of wastewater remediation, organic pollutant removal, CO_2 reduction,

N-fixation, disinfection, oxygen evolution, and hydrogen evolution. Since there are enormous research articles in evidence of BiOX-modified semiconductor photocatalysts, we herein mainly focus on current developments in BiOX-based (X = Cl, Br, I, and F) semiconductor photocatalysts for wastewater treatment, selective value-added transformations, and CO_2 reductions. For tuning the electronic structure of BiOX, various strategies have been applied to enhance the photocatalytic activity of BiOX-modified semiconductor photocatalysts. To date, extensive research is focused on utilizing BiOX-based photocatalysts owing to less toxicity, cost-effectiveness, photo-corrosion resistance, high chemical stability and high earth abundance. Properties such as tetragonal matlockite configuration, narrow bandgap because of their O 2p and Bi 6s hybridized valence band, high thermal firmness makes BiOX-based semiconductor photocatalysts an potential candidate for effective photocatalysis. For the enhanced photocatalytic performances of BiOX-based semiconductor photocatalysts, researchers are currently practicing several strategies like heterojunction construction, enhancement in bismuth content, metal/non-metal doping, interface modification by co-catalyst and metal nanoparticles, inner coupling between different bismuth oxyhalides, creation of oxygen vacancies and use of photosensitizers.[78–89] The co-catalyst supports a dynamic role in enhancing the photocatalytic activity of BiOX by acting as recombination sites and inciting charge separation. In addition, the Bi-rich strategy can raise the conduction band potential *via* fluctuating Bi content and subsequently increase the photocatalytic action of BiOX photocatalysts. Besides, BiOX construction is a tough task, mechanisms at their atomic level are much obligatory, and their revival from reaction solution exist as a challenge for recyclability. These shortcomings need to be understood *via* structural and mechanistic alteration to boost the photocatalytic efficacy of BiOX.

References

1. M. Guan, C. Xiao, J. Zhang, S. Fan, R. An, Q. Cheng, J. Xie, M. Zhou, B. Ye and Y. Xie, *J. Am. Chem. Soc.*, 2013, **135**, 10411–10417.
2. J. Di, J. Xia, H. Li, S. Guo and S. Dai, *Nano Energy*, 2017, **41**, 172–192.
3. J. Li, Y. Yu and L. Zhang, *Nanoscale*, 2014, **6**, 8473–8488.
4. J. Zhang, F. Shi, J. Lin, D. Chen, J. Gao, Z. Huang, X. Ding and C. Tang, *Chem. Mater.*, 2008, **20**, 2937–2941.
5. D. S. Bhachu, S. J. A. Moniz, S. Sathasivam, D. O. Scanlon, A. Walsh, S. M. Bawaked, M. Mokhtar, A. Y. Obaid, I. P. Parkin, J. Tang and C. J. Carmalt, *Chem. Sci.*, 2016, **7**, 4832–4841.
6. P. Gao, Y. Yang, Z. Yin, F. Kang, W. Fan, J. Sheng, L. Feng, Y. Liu, Z. Du and L. Zhang, *J. Hazard. Mater.*, 2021, **412**, 125186.
7. J. Jiang, K. Zhao, X. Xiao and L. Zhang, *J. Am. Chem. Soc.*, 2012, **134**, 4473–4476.
8. F. Yanfen, H. Yingping, Y. Jing, W. Pan and C. Genwei, *Environ. Sci. Technol.*, 2011, **45**, 1593–1600.

9. R. Yuan, S. Fan, Z. Huaxi, D. Zhengxin, S. Lin, Z. Li, Z. Zhang, C. Xu, L. Wu, X. Wang and X. Fu, *Angew. Chem., Int. Ed.*, 2013, **52**, 3222–3226.
10. N. Kijima, K. Matano, M. Saito, T. Oikawa, T. Konishi, H. Yasuda, T. Sato and Y. Yoshimura, *Appl. Catal., A*, 2001, **206**, 237–244.
11. X. Wei, M. U. Akbar, A. Raza and G. Li, *Nanoscale Adv.*, 2021, **3**, 3353–3372.
12. X. Jin, L. Ye, H. Xie and G. Chen, *Coord. Chem. Rev.*, 2017, **349**, 84–101.
13. C. R. Michel, N. L. L. Contreras and A. H. Martínez-Preciado, *Sens. Actuators, B*, 2011, **160**, 271–277.
14. L. Zhang, W. Wang, D. Jiang, E. Gao and S. Sun, *Nano Res.*, 2015, **8**, 821–831.
15. H. An, Y. Du, T. Wang, C. Wang, W. Hao and Z. Junying, *Rare Met.*, 2008, **27**, 243–250.
16. J. Xiong, P. Song, J. Di and H. Li, *J. Mater. Chem. A*, 2020, **8**, 21434–21454.
17. X. Zhang, Z. Ai, F. Jia and L. Zhang, *J. Phys. Chem. C*, 2008, **112**, 747–753.
18. G. Zhang, L. Zhang, Y. Liu, L. Liu, C.-P. Huang, H. Liu and J. Li, *ACS Appl. Mater. Interfaces*, 2016, **8**, 26783–26793.
19. D. Mao, S. Ding, L. Meng, Y. Dai, C. Sun, S. Yang and H. He, *Appl. Catal., B*, 2017, **207**, 153–165.
20. Y. Huang, H. Li, W. Fan, F. Zhao, W. Qiu, H. Ji and Y. Tong, *ACS Appl. Mater. Interfaces*, 2016, **8**, 27859–27867.
21. Q. Mu, Q. Zhang, H. Wang and Y. Li, *J. Mater. Chem.*, 2012, **22**, 16851–16857.
22. J. Bai, J. Sun, X. Zhu, J. Liu, H. Zhang, X. B. Yin and L. Liu, *Small*, 2020, **16**, e1904783.
23. M. Gondal, C. Xiaofeng and M. Dastageer, *Novel Bismuth-Oxyhalide-Based Materials and Their Applications*, 2017, p. 123.
24. J. Chen, M. Guan, W. Cai, J. Guo, C. Xiao and G. Zhang, *Phys. Chem. Chem. Phys.*, 2014, **16**, 20909–20914.
25. N. Seriani, W. Pompe and L. C. Ciacchi, *J. Phys. Chem. B*, 2006, **110**, 14860–14869.
26. J. Lu, W. Zhou, X. Zhang and G. Xiang, *J. Phys. Chem. Lett.*, 2020, **11**, 1038–1044.
27. L. Wang, L. Wang, Y. Du, X. Xu and S. X. Dou, *Mater. Today Phys.*, 2021, **16**, 100294.
28. N. Tian, C. Hu, J. Wang, Y. Zhang, T. Ma and H. Huang, *Coord. Chem. Rev.*, 2022, **463**, 214515.
29. Y. Yang, C. Zhang, C. Lai, G. Zeng, D. Huang, M. Cheng, J. Wang, F. Chen, C. Zhou and W. Xiong, *Adv. Colloid Interface Sci.*, 2018, **254**, 76–93.
30. I. Ahmad, S. Shukrullah, M. Yasin Naz, S. Ullah and M. Ali Assiri, *Ind. Eng. Chem. Res.*, 2022, **105**, 1–33.
31. G.-J. Lee, Y.-C. Zheng and J. J. Wu, *Catal. Today*, 2018, **307**, 197–204.
32. L. Ye, Y. Su, X. Jin, H. Xie and C. Zhang, *Environ. Sci.: Nano*, 2014, **1**, 90–112.
33. D.-H. Wang, G.-Q. Gao, Y.-W. Zhang, L.-S. Zhou, A.-W. Xu and W. Chen, *Nanoscale*, 2012, **4**, 7780–7785.
34. D. Zhang, J. Li, Q. Wang and Q. Wu, *J. Mater. Chem. A*, 2013, **1**, 8622–8629.

35. M. Shi, G. Li, J. Li, X. Jin, X. Tao, B. Zeng, E. A. Pidko, R. Li and C. Li, *Angew. Chem.*, 2020, **59**, 6590–6595.
36. L. Ye, X. Jin, X. Ji, C. Liu, Y. Su, H. Xie and C. Liu, *Chem. Eng. J.*, 2016, **291**, 39–46.
37. J. Jiang, K. Zhao, X. Xiao and L. Zhang, *J. Am. Chem. Soc.*, 2012, **10**, 4473–4476.
38. L. Wang, R. Wang, T. Qiu, L. Yang, Q. Han, Q. Shen, X. Zhou, Y. Zhou and Z. Zou, *Nano Lett.*, 2021, **21**, 10260–10266.
39. L. Ye, X. Jin, C. Liu, C. Ding, H. Xie, K. H. Chu and P. K. Wong, *Appl. Catal., B*, 2016, **187**, 281–290.
40. L. Ye, H. Wang, X. Jin, Y. Su, D. Wang, H. Xie, X. Liu and X. Liu, *Sol. Energy Mater. Sol. Cells*, 2016, **144**, 732–739.
41. H. Liu, W. Cao, Y. Su, Y. Wang and X. Wang, *Appl. Catal., B*, 2012, **111–112**, 271–279.
42. H. Li and L. Zhang, *Nanoscale*, 2014, **6**, 7805–7810.
43. J. Zhang, K. Zhu, Y. Zhu, C. Qin, L. Liu, D. Liu, Y. Wang, W. Gan, X. Fu and H. Hao, *Chem. Phys. Lett.*, 2020, **750**, 137483.
44. X. Liu, X. Jiang, Z. Chen, J. Yu and Y. He, *Mater. Lett.*, 2018, **210**, 194–198.
45. L. Ren, D. Zhang, X. Hao, X. Xiao, Y. Jiang, J. Gong, F. Zhang, X. Zhang and Z. Tong, *Mater. Res. Bull.*, 2017, **94**, 183–189.
46. C.-F. Fu, X. Li and J. Yang, *Chem. Sci.*, 2021, **12**, 2863–2869.
47. C. Hua, X. Dong, Y. Wang, N. Zheng, H. Ma and X. Zhang, *RSC Adv.*, 2018, **8**, 16513–16520.
48. S. Shafafi, A. Habibi-Yangjeh, S. Feizpoor, S. Ghosh and T. Maiyalagan, *Sep. Purif. Technol.*, 2020, **250**, 117179.
49. A. P. Chowdhury and B. H. Shambharkar, *Adv. Chem. Eng.*, 2020, **4**, 100040.
50. H. Yu, J. Huang, L. Jiang, Y. Shi, K. Yi, W. Zhang, J. Zhang, H. Chen and X. Yuan, *Chem. Eng. J.*, 2020, **402**, 126187.
51. S. Zarezadeh, A. Habibi-Yangjeh, M. Mousavi and S. Ghosh, *J. Photochem. Photobiol., A*, 2020, **389**, 112247.
52. C. Song, Y. Feng, W. Shi and C. Liu, *CrystEngComm*, 2016, **18**, 7796–7804.
53. M. Sun, J. Hu, C. Zhai, M. Zhu and J. Pan, *ACS Appl. Mater. Interfaces*, 2017, **9**, 13223–13230.
54. K. M. Alam, P. Kumar, P. Kar, U. K. Thakur, S. Zeng, K. Cui and K. Shankar, *Nanoscale Adv.*, 2019, **1**, 1460–1471.
55. A. Habibi-Yangjeh, S. Asadzadeh-Khaneghah and S. Ghosh, *Adv. Powder Technol.*, 2020, 31.
56. X. Zheng, L. Feng, Y. Dou, H. Guo, Y. Liang, G. Li, J. He, P. Liu and J. He, *ACS Nano*, 2021, **15**, 13209–13219.
57. S.-S. Wang, X. Liang, Y.-K. Lv, Y.-Y. Li, R.-H. Zhou, H.-C. Yao and Z.-J. Li, *ACS Appl. Energy Mater.*, 2022, **5**, 1149–1158.
58. M. Zhang, C. Lai, B. Li, F. Xu, D. Huang, S. Liu, L. Qin, Y. Fu, X. Liu, H. Yi, Y. Zhang, J. He and L. Chen, *Chem. Eng. Trans.*, 2020, **396**, 125343.
59. J. Li, Q. Zhou, F. Yang, L. Wu, W. Li, R. Ren and Y. Lv, *New J. Chem.*, 2019, **43**, 14829–14840.

60. T. Jiang, J. Li, Y. Gao, L. Li, T. Lu and L. Pan, *J. Colloid Interface Sci.*, 2017, **490**, 812–818.
61. S. Bera, S. Ghosh, T. Maiyalagan and R. N. Basu, *ACS Appl. Energy Mater.*, 2022, **5**, 3821–3833.
62. T. Zhang, X. Shao, D.-f. Zhang, X. Pu, Y. Tang, J. Yin, B. Ge and W. Li, *Sep. Purif. Technol.*, 2018, **195**, 332–338.
63. S. Ghosh, *Heterostructured Photocatalysts for Solar Energy Conversion*, Elsevier Inc., 20 October 2020, ISBN: 9780128200728.
64. H. Li, F. Deng, Y. Zheng, L. Hua, C. Qu and X. Luo, *Environ. Sci.: Nano*, 2019, **6**, 3670–3683.
65. M. Zhang, H.-f. Yin, J.-c. Yao, M. Arif, B. Qiu, P.-f. Li and X.-h. Liu, *Colloids Surf., A*, 2020, **602**, 124778.
66. A. Raza, Z. Qin, S. O. A. Ahmad, M. Ikram and G. Li, *J. Environ. Chem. Eng.*, 2021, **9**, 106569.
67. S. Ning, X. Shi, H. Zhang, H. Lin, Z. Zhang, J. Long, Y. Li and X. Wang, *Sol. RRL*, 2019, **3**, 1900059–1900067.
68. Y. Bai, L. Ye, L. Wang, X. Shi, P. Wang, W. Bai and P. K. Wong, *Appl. Catal., B*, 2016, **194**, 98–104.
69. J.-C. Wang, H.-C. Yao, Z.-Y. Fan, L. Zhang, J.-S. Wang, S.-Q. Zang and Z.-J. Li, *ACS Appl. Mater. Interfaces*, 2016, **8**, 3765–3775.
70. Y. Bai, T. Chen, P. Wang, L. Wang, L. Ye, X. Shi and W. Bai, *Sol. Energy Mater. Sol. Cells*, 2016, **157**, 406–414.
71. Y. Chen, F. Wang, Y. Cao, F. Zhang, Y. Zou, Z. Huang, L. Ye and Y. Zhou, *ACS Appl. Energy Mater.*, 2020, **3**, 4610–4618.
72. N. Verma and R. Ananthakrishnan, *ChemPhotoChem*, 2021, **5**, 1–15.
73. Q. Li, Z. Guan, D. Wu, X. Zhao, S. Bao, B. Tian and J. Zhang, *ACS Sustainable Chem. Eng.*, 2017, **8**, 6958–6968.
74. M. Li, C. Song, Y. Wu, M. Wang, Z. Pan, Y. Sun, L. Meng, S. Han, L. Xu and L. Gan, *Appl. Surf. Sci.*, 2019, **478**, 744–753.
75. P. f. Zhu, J. Xu, L. Xie, M. Duan, X. Wu, X. Xiao and M. Liu, *Colloids Surf., A*, 2022, **645**, 128810.
76. K. Wang, J. Li and G. Zhang, *ACS Appl. Mater. Interfaces*, 2019, **31**, 27686–27696.
77. J. Kang, C. Jin, Z. Li, M. Wang, Z. Chen and Y. Wang, *J. Alloys Compd.*, 2020, **825**, 153975.
78. R. Fu, X. Zeng, L. Ma, S. Gao, Q. Wang, Z. Wang, B. Huang, Y. Dai and J. Lu, *J. Power Sources*, 2016, **312**, 12–22.
79. H. Fujito, H. Kunioku, D. Kato, H. Suzuki, M. Higashi, H. Kageyam and R. Abe, *J. Am. Chem. Soc.*, 2016, **138**, 2082–2085.
80. Y. Mi, L. Wen, Z. Wang, D. Cao, R. Xu, Y. Fang, Y. Zhou and Y. Lei, *Nano Energy*, 2016, **30**, 109–117.
81. W. Li, S. He, X. Wang, Q. Ma and C. Zhao, *Int. J. Energy Res.*, 2019, **43**, 1–10.
82. L. Zhang, W. Wang, S. Sun, Y. Sun, E. Gao and J. Xu, *Appl. Catal., B*, 2013, **132–133**, 315–320.

83. M. Shi, G. Li, J. Li, X. Jin, X. Tao, B. Zeng, E. A. Pidko, R. Li and C. Li, *Angew. Chem.*, 2020, **59**, 6590–6595.
84. T.-H. Chen, M. Yoshida, S. Tsunekawa, J.-H. Wu, K.-Y. A. Lind and C. Hu, *Catal. Sci. Technol.*, 2020, **10**, 3223–3231.
85. K. M. Alam, P. Kumar, P. Kar, U. K. Thakur, S. Zeng, K. Cui and K. Shankar, *Nanoscale Adv.*, 2019, **1**, 1460–1471.
86. N. T. Hahn, S. Hoang, J. L. Self and C. B. Mullins, *ACS Nano*, 2012, **60**, 7712–7722.
87. Y. You, S. Wang, K. Xiao, T. Ma, Y. Zhang and H. Huang, *ACS Sustainable Chem. Eng.*, 2018, **6**, 16219–16227.
88. S. Sultana, S. Mansingh and K. M. Parida, *J. Phys. Chem. C*, 2018, **122**, 808–819.
89. M. Du, S. Zhang, Z. Xing, Z. Li, J. Yin, J. Zou, Q. Zhu and W. Zhou, *Langmuir*, 2019, **35**, 7887–7895.

CHAPTER 6

Molecular Catalysts Immobilised on Photocathodes for Solar Fuel Generation

C. E. CREISSEN*[a]

[a]School of Chemical and Physical Sciences, Keele University, Keele ST5 5BG, UK
*E-mail: c.e.creissen@keele.ac.uk

6.1 Introduction

Our energy supply needs to drastically change to alleviate the catastrophic effects of climate change.[1,2] Reliance on fossil fuels protrudes beyond electricity and gasoline production, extending into the commodity industry as the source for plastics, pharmaceuticals, and other chemical products. The ultimate objective to reduce the global carbon footprint is countered by a growing need for improved energy access in developing communities where the most readily available, portable, and inexpensive sources are derived from crude oils.[3] Modern sustainable development solutions are beginning to remove electricity dependence on these supply chains through the growth of renewable energy sectors such as wind and solar power.[4] However, to enhance access and limit energy inequities, grid-independent standalone solutions are required. One research area capable of addressing this issue is the sunlight-driven conversion of H_2O and CO_2 into chemical energy carriers— so called 'solar fuels'.[5]

Sustainable Energy No. 1
Recent Developments in Functional Materials for Artificial Photosynthesis
Edited by Srabanti Ghosh and Qian Wang

Published by the Royal Society of Chemistry, www.rsc.org

Inspiration for synthetic approaches derives from the core biological processes involved in natural photosynthesis: light harvesting, charge separation, and catalysis. Billions of years of evolution have supplied natural systems with the complex molecular tools required to sustain life. Consequently, these pathways do not provide considerable efficiency for the direct synthesis of fuels for human consumption. In artificial systems, these sunlight-driven electron transfer pathways can be redirected to form solar fuels. With directives from enzymes, tuneable molecular catalysts that facilitate high rates of reaction with minimal energy input have been developed. These molecular complexes, designed through rational modification of the metal(s) and ligand(s), add benefits of low light absorption and high single-site activity. Furthermore, spectroscopic insights and performance evaluations over the past decades have provided operative guidelines for the synthesis and integration of effective catalysts. Molecular catalysts are therefore ideal candidates for visible-light-driven reduction reactions when coupled with light-absorbing components.

One of the most interesting approaches to integrate molecular catalysts is through the assembly of photoelectrochemical (PEC) devices, which have the potential to operate without an externally applied bias, thereby offering a sustainable solution to the generation of chemical energy carriers. Such devices face challenges in terms of energy efficient operation, which in part arises from the difficulty of forming effective photocathodes (PCs) for reduction reactions. However, analysis of functional 'hybrid photocathodes' featuring surface-immobilised molecular catalysts has deepened our understanding of specific limitations and drawbacks. This in turn has resulted in targeted approaches that can bypass significant issues and effectively generate chemical products.

This chapter will discuss combined developments in the fields of molecular catalysis and semiconductor (SC) photoelectrochemistry to provide practical guidelines for solar fuels research. A specific focus is given to molecular catalysts immobilised on PCs for CO_2 reduction (CO_2R) and H_2 evolution. Separate discussions of SC properties and molecular catalysts are first provided with requirements and practical considerations, followed by specific examples of hybrid PCs to highlight interesting developments in the field.

6.2 Photoelectrochemical Cells

The process of converting water and CO_2 into value-added chemicals requires a significant energy input. Although the cathodic half-reactions do not have significant thermodynamic requirements (E_{red}), when coupled with the oxygen evolution reaction (OER) at the anode, the minimum energy requirement (E°_{cell}) is much larger (Table 6.1).

Despite the challenges involved in finding suitable methods to supply the large amount of electrical input with solar light, functional solar-fuel devices have been developed using several different technological approaches. In general, solar-driven electrochemical systems can be divided into two main

Table 6.1 Standard cathode ($E^\circ_{cathode}$) and full-cell (E°_{cell}) potentials for HER and CO_2R products from molecular catalysts coupled to *water* oxidation.

Reaction	$E^\circ_{cathode}$/V *vs.* RHE	$\lvert E^\circ_{cell} \rvert$ V
$2H_2O \rightarrow 2H_2 + O_2$	0.00	1.23
$2CO_2 \rightarrow 2CO + O_2$	−0.10	1.34
$CO_2 + 2H_2O \rightarrow CH_4 + 2O_2$	0.17	1.06
$2CO_2 + 2H_2O \rightarrow 2HCOOH + O_2$	−0.25	1.48
$CO_2 + H_2O \rightarrow HCOO^- + O_2$	−0.02	1.25

subsets: PEC and photovoltaic-electrochemical (PV-EC).[6] PV-EC devices do not deal with the direct coupling of light absorption with catalysis and the engineering requirements are different from PEC devices, which can theoretically operate without the need for external wiring and additional framing which can be costly.[7] These differences make direct comparison of their performance difficult, as although the same metrics can in principle be used, the two forms of technology are not at the same stage of development. The differentiation between PV-EC and integrated PEC (^{i}P) systems that incorporate buried junctions is part of an ongoing debate in the field. However, with respect to molecular catalysis, most systems are ^{i}P or PEC-based and are therefore the only types of system discussed here. Terminologies and different taxonomies to describe devices can be found in several reports,[6,8,9] and more recently, in a dedicated benchmarking database that pools examples of existing solar-driven devices that operate without applied bias.[10]

6.2.1 Tandem PEC Cells

Single light-absorbing components often fail to meet the high energetic requirements for solar fuel production. Consequently, research has steered towards the development of tandem PEC cells composed of stacked photoelectrodes (Figure 6.1(a) and (b)). This tactic enhances the attainable efficiency whilst simultaneously extending the choice of SCs available for each half reaction. In terms of photoanode (PA) materials, there are several relatively stable PA systems that can operate with high reaction rates, however corresponding PCs suffer from multiple drawbacks that often result in low currents or instability. It is important to note that the photocurrent for a PEC cell is limited by the lowest current-generating half-cell, therefore PCs are often the bottleneck to achieving high efficiency devices. Here, three main types of PC are discussed: photocathodes (PC), integrated photocathodes (^{i}PC), and dye-sensitised photocathodes (DSPCs), as shown in Figure 6.1(c). A more detailed explanation of SC properties and their behaviour in aqueous solutions is provided in the next section.

The electrical input required to drive a full reaction is best described by eqn (6.1), in which E_{cell} is the total cell voltage, E_{anode} and $E_{cathode}$ are the thermodynamic potentials for the individual half reactions, η_c is the cathode

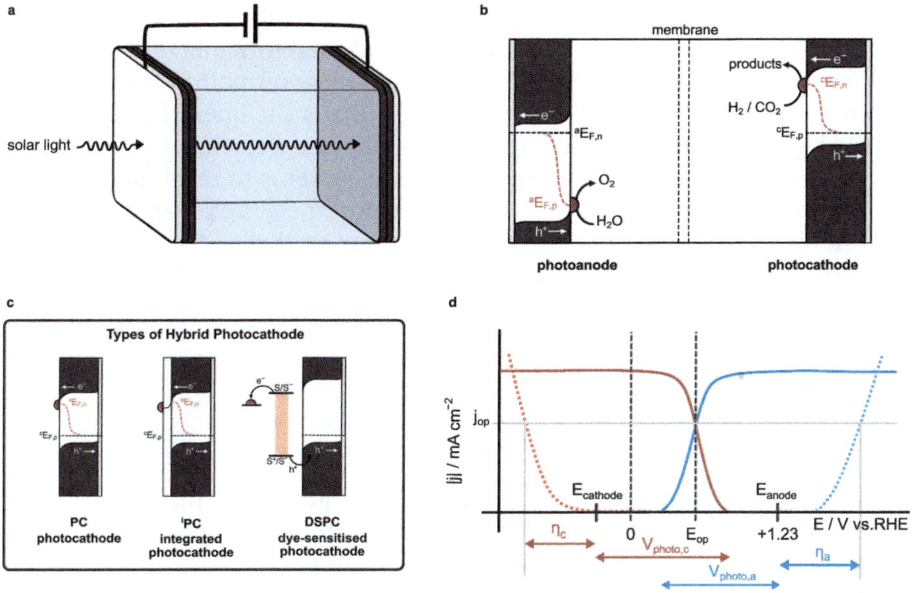

Figure 6.1 (a) Schematic of a standalone tandem PEC cell. (b) Internal charge transfer and catalysis in a functional PEC cell. (c) The three types of hybrid photocathodes discussed. (d) The current–voltage response of a tandem PEC cell in dark (dotted lines) and under illumination (solid lines), in which $V_{photo,a}$ and $V_{photo,c}$ are the photovoltages generated by the photoanode and photocathode, respectively, j_{op} is the operating current, E_{op} is the operating voltage, E_{anode} and $E_{cathode}$, are the thermodynamic potentials for each half reaction, and η_a and η_c are the anode and cathode overpotentials.

overpotential, η_a is the anode overpotential, and E_σ includes all terms relating to various system-specific losses (for example ohmic diffusion, membrane resistance).[11] Here, E_{anode} and $E_{cathode}$ are derived from the Nernst equation so the pH in each compartment is factored in. The pH terms can be separated to give eqn (6.2), in which E°_{anode} and $E^\circ_{cathode}$ are standard half-cell redox potentials, and $\Delta pH = pH_{anode} - pH_{cathode}$. This shows that maintaining a pH gradient is beneficial to cell potential and therefore must be factored in when calculating solar-to-fuel efficiency.[6]

$$E_{cell} = [E_{anode} + \eta_a] - [E_{cathode} + \eta_c] + E_\sigma \tag{6.1}$$

$$E_{cell} = E^\circ_{anode} - E^\circ_{cathode} - 0.059\Delta pH + \eta_a - \eta_c + E\sigma \tag{6.2}$$

To achieve bias-free solar fuel production, the voltage supplied by the combination of photoelectrodes must exceed E_{cell}.[12] Evaluation of a full cell is best exemplified by looking at the current–voltage plots for a tandem configuration under illumination compared with dark conditions (Figure 6.1(d)). The photovoltages (V_{photo}) at the anode and cathode represent the difference between the onset potential under illumination and the thermodynamic potential for the reaction ($E_{cathode}$ or E_{anode}) and therefore dictate the operating potential (E_{op}). By plotting $|j|$ against E, a crossover point for two photoelectrodes under illumination can be seen—this is the operating point of a coupled system and gives the highest possible bias-free current density for the tandem cell. PC improvements can enable systems to reach higher bias-free currents. Two parameters are of high importance: the photovoltage and the photocurrent. These two parameters are controlled by the SC, the catalyst, and the interfacial connection between these two components, all of which are discussed in the following sections.

6.3 Photocathodes

6.3.1 Semiconductor Electrochemistry

Hybrid PCs for solar fuel production rely on the transport of photo-generated electrons to an immobilised catalyst to drive a reduction reaction. Certain properties of SC materials in aqueous solution enable light absorption and directional charge transport to achieve effective solar-driven catalysis. These properties are described here to show why the choice of SC is extremely important when developing a molecular-catalyst-based PC.

6.3.1.1 Band Structure

SC electrodes exhibit different behaviour from metallic electrodes due to the presence of a bandgap. In crystalline materials, the multiple bonds between atoms create a periodic lattice. The formation of molecular orbitals through the combination of atoms creates discrete electron energy levels, as described by the tight binding model (linear combination of atomic orbitals). In the simplest case, a two-atom system, this results in the formation of bonding and anti-bonding orbitals (Figure 6.2(a)). In solid materials, the large numbers of atoms combining to form molecular orbitals eventually leads to such a low energy difference between states that a 'band' is created which is

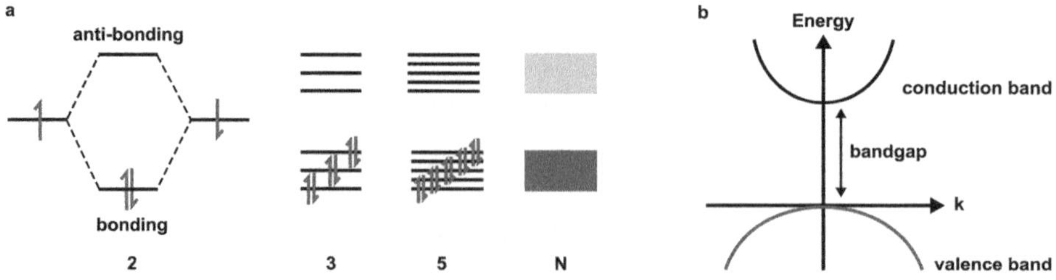

Figure 6.2 (a) Formation of bands in semiconductors through the combination of atomic orbitals. (b) Dispersion relations for conduction and valence bands of a semiconductor, in which k is the wave vector.

essentially a continuum of allowed energy levels (Figure 6.2(a)). When thinking of bands and electron transport, it is more useful to use Bloch's theorem, in which electrons are treated as travelling waves but within a periodic lattice. The periodicity leads to a periodic potential energy profile with two regions of allowed wavefunctions: the valence band (filled with electrons) and the conduction band (empty of electrons) (Figure 6.2(b)).[13,14]

The direction and the curvature of each band has an associated effective electron mass, where a negative mass corresponds to holes in the valence band (movement of positive charge) and a positive electron mass corresponds to electrons in the conduction band (movement of negative charge). The transport of holes and electrons to catalytic sites enables SC electrodes to conduct redox reactions. Therefore, the charge carrier mobility, which is inversely proportional to the effective electron mass, is a key parameter in SC electrochemistry. The difference between the unoccupied and occupied bands is the bandgap—this controls the wavelength of light for which excitation of an electron from the valence band to the conduction band can occur. The bandgap can be direct (k-vectors aligned over $k = 0$) or indirect (k vectors not aligned), in which case the absorption or emission of a phonon is required to conserve momentum. Equally important is the differentiation between n-type and p-type SCs within the context of electrochemistry. The majority charge carriers are holes in the p-type case and electrons in the n-type case. This distinction alters the band behaviour in aqueous conditions and facilitates effective charge separation.

The band structure defines the light-absorption wavelength, the charge transfer kinetics, and the valence band and conduction band positions. Therefore, material properties can control the photoelectrochemical properties of specific SCs. Although disparity exists in the surface structure and stability, which are particularly important for electrochemical functionality, the correct band structure is a vital requirement for effective PEC devices. This simplified understanding of the role band structure plays in SCs is a necessary step in the development of PEC systems.

6.3.1.2 Band Bending and Charge Transport in p-SCs

PCs are constructed using p-type SCs due to the beneficial effects arising from band bending when a SC is interfaced with an electrolyte solution. This effect stems from the variation of electrochemical potential within a SC arising from charge transfer between the material and solution-phase species. Upon contact with a redox species in solution, the SC acts to balance charge in the system by accepting electrons from the higher-energy solution species (Figure 6.3(a)).

Ultimately, the energy of the bands in electrochemical terms differs depending on the SC–liquid junction as well as the applied potential and illumination conditions. To understand the effect of applying a potential to a p-SC, a simplified diagram that ignores the potential drop across the Helmholtz layer can be used. In the case where a cathodic bias (negative) is applied

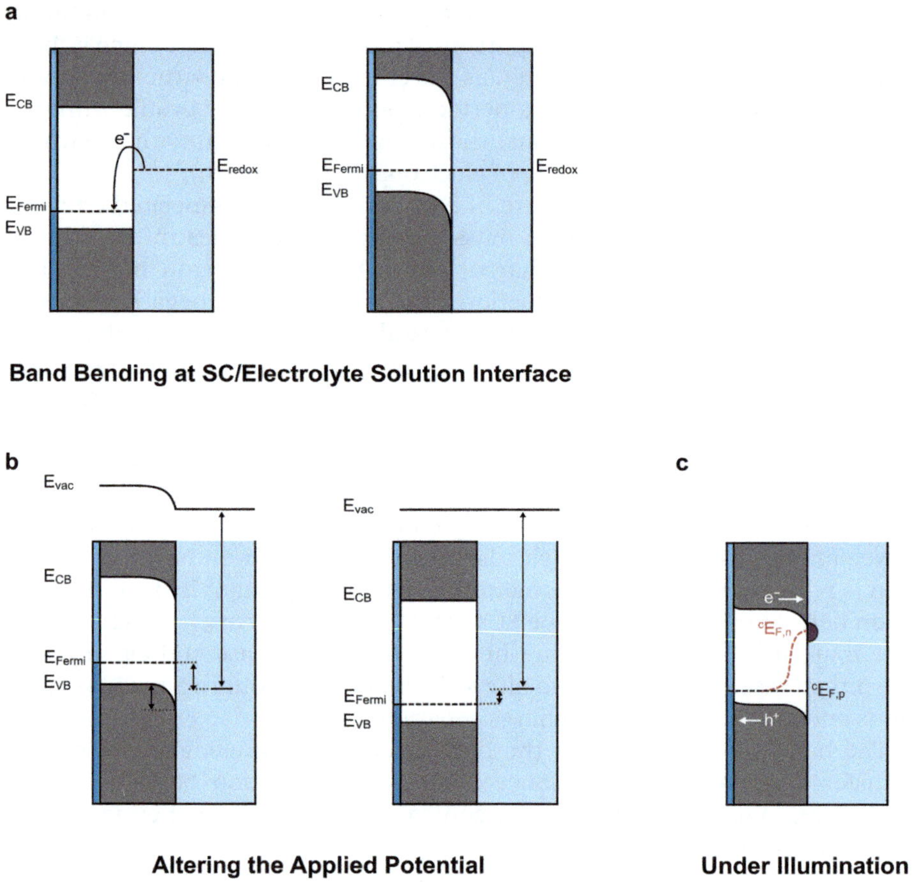

Figure 6.3 Representation of the behaviour of bands in a p-type semiconductor when (a) interfaced with an electrolyte solution, (b) applying a potential (left to right represents moving to more positive potentials), and (c) under illumination, in which h^+ represent holes, e^- are electrons, and ${}^{c}E_{F,p}$ is the quasi-Fermi level for electrons.

to the p-SC, a change in potential occurs in the space charge region of the SC. This is because the charge carrier density in the electrolyte solution tends to be much higher than that in the SC. As a result, the bulk of the SC is shifted cathodically (negative in potential) with respect to the band edges, which are now depleted of minority carriers (electrons in this case) (Figure 6.3(b)). If the potential is pushed more positive (anodically) then the p-SC will reach a point at which no net band banding is observed, which is called flatband conditions (Figure 6.3(b)). The flatband potential, which can be determined from Mott–Schottky plots using electrochemical impedance spectroscopy, is a good representation of the conduction band position in n-type SCs or the valence band position in p-type SCs.

Absorption of photons with sufficient energy promotes electrons from the valence band to the conduction band, creating electron–hole pairs. For a p-SC, the holes and electrons are separated due to the electric field (drift current) arising from band bending in the space charge region. In p-SCs, holes (majority carriers) are transported toward the metallic contact and subsequently to the photoanode, whereas electrons (minority carriers) are shuttled to the SC–liquid junction to carry out a reduction reaction. Under illumination, a p-SC has an excess concentration of electrons at the SC–liquid interface and the quasi-Fermi level for electrons (${}^{c}E_{F,n}$), which denotes the concentration at each point, varies across the space charge region (Figure 6.3(c)). A p-SC with a ${}^{c}E_{F,n}$ level more negative than the redox potential of the reduction reaction has sufficient thermodynamic driving force for the reaction. However, reaction kinetics for bare SC surfaces tend to be low and therefore additional catalysts are required to facilitate reactions at low overpotentials.

6.3.2 Semiconductor Choice

6.3.2.1 Stability Issues

The need for visible light absorption and a compatible conduction band position imposes several restrictions on the type of p-SC that can be used for reduction reactions. These requirements are also complicated by commonly observed problems with instability under working conditions. PCs need to incorporate SCs that: (1) absorb a large portion of the visible spectrum; and (2) have a sufficiently cathodic conduction band position. Figure 6.4 shows

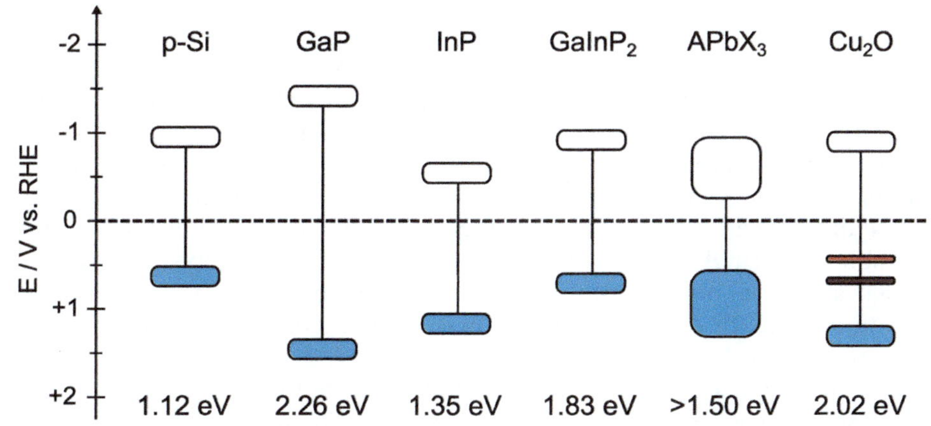

Figure 6.4 Common p-type semiconductors with their valence band (blue) and conduction band (white) positions along with their bandgaps. For Cu_2O, the positions of the different redox potentials within the bandgap that lead to degradation are highlighted.

the band structures and bandgaps of p-SC materials that have been coupled with molecular catalysts for solar-driven CO_2R and H_2 evolution.

Bare p-SC surfaces exhibiting favourable band structure tend to suffer from instability in aqueous conditions. For example, p-Si has a bandgap of 1.12 eV, allowing it to absorb a large portion of the solar spectrum. As an extrinsic SC, p-Si gains its p-type conductivity through doping with an acceptor impurity such as boron, which introduces holes with high mobility in the valence band. Despite exhibiting favourable band structure and being effective in solar cells, Si is susceptible to formation of an insulating SiO_2 layer on its surface, which blocks electron transfer, eventually rendering the PC inactive. For Cu_2O, again the bandgap (2.0 eV) is well-suited for visible absorption, but it possesses two accessible redox potentials that lead to degradation (Figure 6.4).[15] The first is the photoreduction of Cu_2O to form metallic Cu at +0.47 V *vs.* RHE (eqn (6.3)), and the second is the photooxidation of Cu_2O to form CuO at +0.67 V *vs.* RHE (eqn (6.4)).

$$Cu_2O + 2H^+ + 2e^- \rightleftharpoons 2Cu + H_2O \tag{6.3}$$

$$2CuO + 2H^+ + 2e^- \rightleftharpoons Cu_2O + H_2O \tag{6.4}$$

With other SCs the susceptibility for decomposition or time-dependent deactivation can vary depending on the conditions used or the composition. However, such cathode materials can be protected from degradation through the use of protective layers that enable charge transfer and block water access.

6.3.2.2 Protective Layers and Buried Junctions

Due to the instability issues discussed for common p-SCs, deposition of protective layers to form PV-integrated photocathodes (^{i}PC) is often required. Although ^{i}PCs operate under light conditions, the catalyst itself is not always exposed to the light such as when back-side illumination is used—this can be beneficial in terms of avoiding light instability with certain molecular catalysts. To protect PCs, there are a number of materials and methods that result in stable electrodes.[16–22] One of the most effective methods for forming a stable conformal coating on the surface is atomic layer deposition (ALD), which avoids significant pinhole formation, enabling otherwise unstable SCs to operate for significant amounts of time under illumination in aqueous conditions.

For heterogeneous catalysts, SCs can be protected with thin layers of the catalysts themselves. This is well-illustrated with MoS_2 protected Si PC, which is highly active and stable for the HER.[21,22] However, molecular catalysts require a scaffold support that is compatible with typical anchoring groups as well as being highly porous to accommodate sufficient loadings. Therefore, additional support layers or deposition of porous materials that serve both functions are highly desirable. An example of a protective layer that can also act as a porous scaffold is mesoporous TiO_2. With SCs such as p-Si, for which the aerobic growth of an insulating SiO_x layer is relatively slow, rapid coating with TiO_2 after etching of the oxide layer in HF can enable stable operation for several hours in aqueous conditions.[23] However, with more sensitive materials such as lead halide perovskites ($APbX_3$), the extreme sensitivity prohibits such methods. Instead, innovative protection methods such as the low temperature deposition of an InBiSn alloy (Field's metal) can avoid rapid degradation issues.[24,25] The scaffold for immobilisation of the molecular catalyst then requires formation or electrical connection on top of the protective layer, such as through growth on a separate metallic support that is then electrically integrated with the protected PC.[26]

The presence of the protective layer can have a profound effect on the electrochemical properties of the SC. In most cases the deposition of a passivating layer leads to the formation of a buried junction. The formation of a buried junction means that a photovoltage is unaffected by the catalyst, however the built-in potential due to band bending may be diminished by such layers.[27] In cases where ion-permeable catalyst layers are used, so-called 'adaptive' junctions can be formed where the photovoltage is modified by the layer and the catalyst deposited on the surface.[28,29] In both cases, the effective potential supplied by the SC under illumination may vary from that expected and therefore careful choice of an appropriate protection method should be made when designing SC systems for incorporation of molecular catalysts.

6.4 Molecular Catalysts

Reduction reactions to generate solar fuels require catalysts that lower the activation barrier for the reaction and provide the ideal environment to obtain high selectivity. The two main catalyst branches are molecular and heterogeneous, with catalysts such as metal–organic frameworks (MOFs), covalent organic frameworks (COFs), and single-atom catalysts (SACs) bridging the gap between these two extremes. Heterogeneous catalysts (*i.e.*, metallic, or semiconducting materials) can provide high reaction rates but there is often a lack of clarity in defining active sites and differentiating between inactive or active subcomponents. Identification of specific active oxidation states, oxide formation and modification, catalyst reconstruction, and the role of intercalated ions and/or substrate poisoning, all complicate the elucidation of structure-performance relationships that would enable systematic improvements to be made. Furthermore, even with the information in hand, modifications are often not easily controllable with interdependent processes occurring simultaneously that interfere with the projected performance enhancement. However, the high current densities at low overpotentials that are achievable with heterogeneous catalysts make such systems desirable for integration in demonstrator devices and commercial-scale reactors in the near future.

Molecular catalysts offer advantages over their heterogeneous counterparts, as they are well-defined, modifiable, and typically do not limit light absorption. By altering the metal species and the primary and secondary coordination spheres around the metal centre, they can be rationally designed to provide a desired function. The structure-function relationships defined for molecular complexes are easily understood using a range of spectroscopic, chemical, and electrochemical techniques, therefore a huge number of examples highlighting improvements have paved the way for integration with SC electrodes. Developments in homogeneous (solution-based) electrocatalysts for reduction reactions have revealed the most active and selective classes of catalysts. Understanding the important performance-enhancing features has enabled development of effective immobilisation strategies to integrate them in high-performing solar fuel generating devices. This section will give an overview of different classes of molecular catalysts that have been modified with anchoring groups for immobilisation on different types of electrodes.

6.4.1 Performance Metrics

To understand and compare the performance of electrocatalysts immobilised on PCs, several figures of merit can be used. These vary in the individual parameters they describe so a combination of multiple metrics is often required to show how effective a catalyst is. In general, lower electrical input with maximum rates of reaction are desired if the product selectivity is high.

In photoelectrocatalysis, a fundamental challenge exists in lowering the electrical input to drive a reaction. This is best described by the overpotential (η), which is defined as the additional potential applied beyond the thermodynamic potential for a reaction at a specific current density (eqn (6.5)). In molecular catalysis, the onset potential, E_{onset}, (most commonly the second reduction) is used as an indicator of the minimum voltage required to drive the reaction and is simply the point at which catalytic current can be observed.

$$\eta = |E_{\text{onset}} - E^{\circ}| \tag{6.5}$$

The η value is usually assessed using cyclic voltammetry of the complex immobilised on a conductive support or dissolved in solution. However, with p-type SCs used for PCs, the reduction of a catalyst in dark conditions is not usually possible because the conduction band sits at a too negative position for it to be reduced. This also means that the redox properties of a surface-bound species are not easily determined in dark conditions, limiting the amount of information that can be obtained from otherwise straightforward voltammetry techniques. Instead, assessment in solution phase can be used to extract values to understand if coupling with a p-SC is electrochemically viable. One problem with this method arises from the solubility of certain complexes, which warrants dissolution in organic solvents. Although there are good techniques to convert between references used in organic solvents and H_2O, alterations to the catalytic activity and onset potential arising from such conditions make it difficult to translate values to aqueous conditions as required for sustainable catalysis.

Alternatively, the catalyst can be immobilised on a degenerate SC such as indium tin oxide (ITO) or fluorine tin oxide (FTO) as these possess metal oxide surfaces that are compatible with commonly used anchoring groups and are similar in surface structure to protection layers used for passivation of p-SCs. These degenerate SCs behave like metallic materials so are useful in extracting information about the redox properties of the catalyst. Values can be easily obtained to give a general idea of the characteristics when immobilised on a PC. It is important to note that there are ranges of stability for such materials and often cathodic stability is lower than anodic stability therefore researchers should ensure a compatible region is used.[30]

PEC systems aim to provide the highest photocurrent density (j), which is typically reported for a geometric area, at the lowest overpotential. There are several options for assessment. Linear sweep voltammograms (LSVs) under intermittent light irradiation show how much current a PC generates at different applied potentials, allowing for observation of dark processes that occur as well as showing the position of the onset potential. This can also be conducted under continuous illumination and compared with a dark scan. For product detection and full analysis, electrolysis under illumination is required. This involves holding a potential (chronoamperometry) or a current (chronopotentiometry) across the electrode and sampling gas and liquid products for analysis whilst illuminating with solar light. The number of products can be used to determine other related values discussed below.

With molecular catalysts, good indicators of the stability and rate of catalysis are the turnover number (TON) and turnover frequency (TOF), given in eqn (6.6) and (6.7), in which n_{product} is the mol of product and n_{catalyst} is the mol of catalyst used.

$$\text{TON} = \frac{n_{\text{product}}}{n_{\text{catalyst}}} \tag{6.6}$$

$$\text{TOF} = \frac{\text{TON}}{\text{time}} \tag{6.7}$$

The TON is a measure of how effective and stable the catalysis is. The TOF is simply the time-dependent TON and therefore gives a good indicator of the rate of reaction. However, difficulty in comparing these numbers arises from differences in experimental conditions. For example, a catalyst that operates for a short time but with high rates of reaction will give a high TOF but low TON, therefore when evaluating the performance care should be taken to check the conditions used to determine each value.

The selectivity of a catalyst is a measure of the product distribution. If one specific product is targeted, then a selectivity of 100% is desired. For the case of CO_2R, the selectivity is often different for certain products as H_2 evolution is competitive and multiple products can be obtained. A good measure of selectivity is the faradaic efficiency (FE) (eqn (6.8)).

$$\mathrm{FE} = \frac{n_{\mathrm{product}} \times F \times z}{Q} \times 100 \tag{6.8}$$

The FE is usually determined after bulk photoelectrolysis and reflects the number of electrons going into product generation. If the FE for all products is not 100% then it can be assumed that electrochemical degradation, side reactions, bubble interference, undetected products, or time-variant production is occurring, seeing as some of the current is not going into products. In photoelectrochemistry, there are often problems with lower than unity FE due to photo-induced corrosion, where some of the current goes into degradation of the SC and/or catalyst and therefore must be considered. Using these performance metrics, a comparison between different classes of catalysts is possible.

6.4.2 Immobilisation Methods for Molecular Catalysts

Strategies to anchor molecular components on surfaces rely on an assortment of interactions that are specific to individual materials. The method is dictated by conditions and surface structure of the SC. Effective anchoring groups must be durable under operating conditions and enable electrical communication between the SC surface and the catalytic site. The anchor itself can play a role in altering activity and selectivity, as certain groups are non-innocent. For example, immobilisation on the SC surface can remove surface states to promote or inhibit catalysis and the anchor can act as an electron withdrawing or donating group to alter the redox potential of the catalyst. Specific anchoring groups and linkers can also modify the local electric field, which can alter catalytic behaviour due to interactions with intermediates or substrates within the electrochemical double layer.[31] Therefore choosing an appropriate anchoring group and understanding the possible interactions that can occur on different surface types is important for surface-bound molecular catalysis. Here, immobilisation strategies are presented for two predominant surface types: metal oxides and carbon-based materials. Other surfaces are grouped together as examples are more limited.

6.4.2.1 Anchoring on Metal Oxides

Metal oxide surfaces are often used for integrated PCs as they act as good protective layers for SCs that are susceptible to degradation. The most common anchoring groups for MO_x surfaces are carboxylates, phosphonates, and silanes. The different possible binding modes of each are displayed in Figure 6.5(a). Although hydroxamates, silatranes and acetylacetonates have also been incorporated in some systems, their usage as catalysts in PCs is

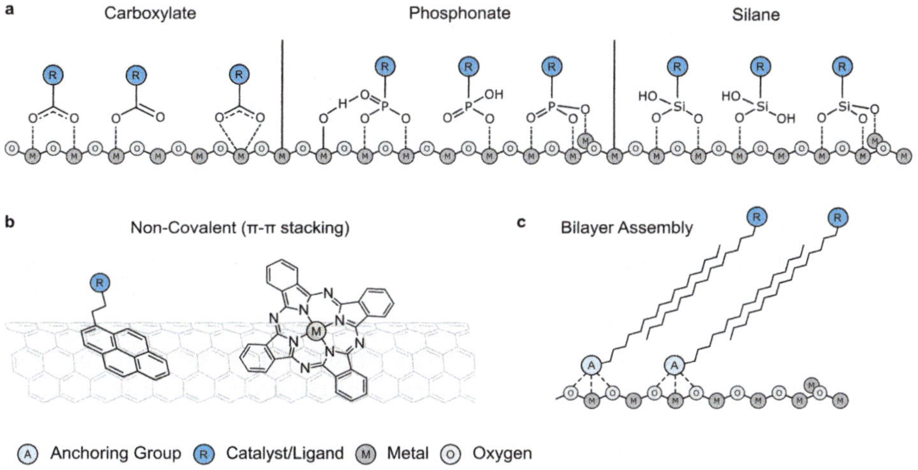

Figure 6.5 (a) Common anchoring groups for metal oxide surfaces. (b) Anchoring examples on carbon nanotubes. (c) Bilayer assembly method for immobilisation.

limited. For carboxylates and phosphonates, layers of catalyst can be covalently anchored on the surface of the metal oxide through self-assembly. Typically, this involves soaking in solutions of the corresponding carboxylic acid or phosphonic acid-bearing complexes, followed by rinsing in the deposition solvent to remove any unanchored catalyst. For silanes, the process is similar but often involves mild heating in inert conditions.

The pH limit for carboxylates restricts their use in CO_2R systems due to the need for CO_2-saturated bicarbonate electrolyte solution (pH ~6) in H-cell configurations, therefore switching to phosphonate or silane anchored catalysts that are more stable under these conditions is required. However, it should be noted that for both H_2 evolution and CO_2R, the consumption of protons under operating conditions results in a high pH close to the electrode surface, even with effective buffers.[32–34] Therefore, if high currents are achieved, instability of such anchoring groups, and consequently desorption may be observed.

6.4.2.2 *Anchoring on Carbon Surfaces*

Although carbon-based electrodes are not ideal for permitting light transport, they can be employed when backside illumination is used or if appropriately thin layers are used. One of the most effective methods is non-covalent grafting of catalysts through π–π stacking.[5,35,36] These non-covalent interactions are strong enough to stabilise species on the surface and the conjugation ensures good electrical coupling between the electrode and the catalyst. Examples of such systems include pyrenes, porphyrins, and phthalocyanines, all of which have a high affinity for carbon nanotubes (CNTs) (Figure 6.5(b)).

6.4.2.3 Anchoring on Other Surface Types

Thiols can be self-assembled on metal sites (Cu, Ag, Au) and chalcogenides but are susceptible to reductive desorption. This is because the application of a cathodic potential results in electron transfer to the M–S bond, which is reduced and the thiolate can dissociate.[37–39] Therefore, potential windows compatible with such species must be carefully chosen or novel devices that limit overpotentials are required for high-current operation.[37]

Self-assembled bilayers can also be used to non-covalently attach molecules (Figure 6.5(c)).[40] These use non-covalent interactions between long alkyl chains to retain molecules close to the electrode. With such systems, a chain length must be found that does not significantly limit water access, due to the high hydrophobicity, but also enables electronic communication with the electrode without the molecules being lost through desorption. An interesting approach is to use host–guest interactions for example through the immobilisation of cyclodextrin on a surface and using aromatic rings as the guest molecules.[41,42] However, these two approaches still require anchoring groups for the alkyl chain or the host molecule.

6.5 H_2 Evolution Catalysts

A vast range of complexes that are active for H_2 evolution exist in the literature. These include multiple ligand types and metal centres, which have been reviewed elsewhere.[5,35,43] Here, we concentrate on several interesting examples employing non-precious metals that have been functionalised with anchoring groups, and describe some key features that enable them to effectively generate H_2. This includes complexes that have been anchored on metal oxide nanoparticles, which are directly applicable to metal-oxide-coated SCs. Although these examples are in no way extensive, they highlight some of the issues associated with photoelectrochemical hydrogen generation with molecular catalysts.

Molecular catalysts for H_2 evolution have drawn inspiration from the active sites of natural enzymes such as [FeFe] hydrogenases (Figure 6.6(a)). Although direct structural mimics have not proven highly active, some lessons learnt from enzymatic systems have assisted development of molecular catalysts. One example of such functionality is the incorporation of pendant amines as proton relays. With an [FeFe] hydrogenase, it was shown through reconstitution with synthetic bridging dithiolate cofactors, that a pendant amine group acts as a proton acceptor to facilitate effective H_2 generation (Figure 6.6(a)). This trend translated to synthetic bimetallic Fe–Fe catalysts, where complexes incorporating similar pendant amine functionality outperformed those without (Figure 6.6(a)).[44] Performance-enhancing structural motifs such as proton relays have also been applied to other classes of molecular catalysts.

One class of catalysts taking advantage of such effects are Ni bis(diphosphine) catalysts (Figure 6.6(b)).[43,45,46] The non-coordinated pendant amine acts as a proton relay and the flexibility enables intra and intermolecular

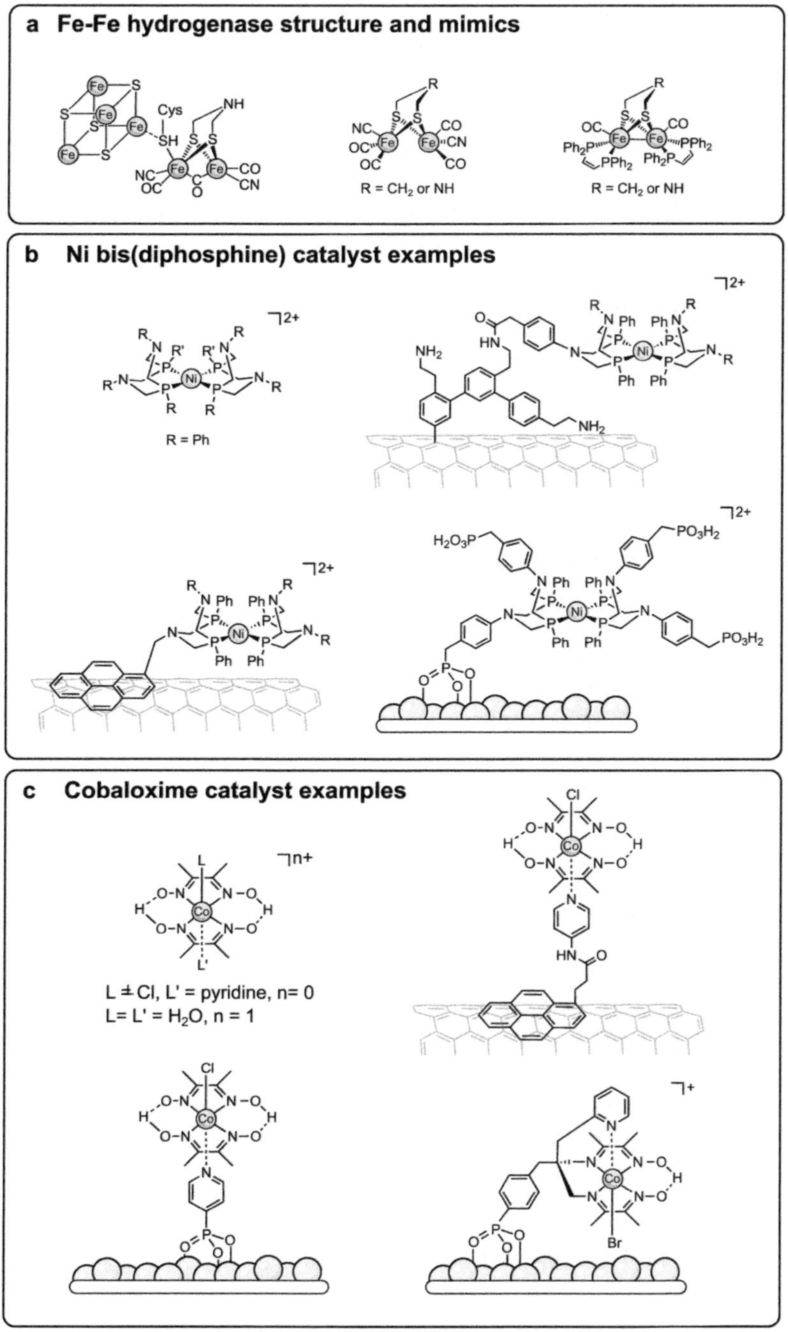

Figure 6.6 (a) [FeFe] hydrogenase active site, synthetic cofactors, and molecular catalysts highlighting the importance of pendant amines in H_2 production. (b) Ni bis(diphosphine) electrocatalyst structure along with examples of immobilisation on different surfaces. (c) Cobaloxime electrocatalyst structures with examples of immobilisation on different surfaces.

proton or hydride transfer.[47] Modifications with anchoring groups at the N position have been effective in generating highly active electrocatalytic systems. For example, electroreduction of a diazonium salt on CNTs generated 4-(2-aminoethyl)phenyl that was used for peptide coupling with an ester on the para position of the N-bound phenyl of the ligand, enabling high activity for H_2 evolution (Figure 6.6(b)).[48] An alternative method attached pyrene anchoring groups to the N positions of all four diphosphine ligands, enabling direct anchoring on CNTs without additional surface modification (Figure 6.6(b)).[49] Modification with phosphonic acid anchoring groups on all four N groups of the diphosphine ligand promoted anchoring to TiO_2 for electrocatalytic H_2 generation in aqueous conditions (Figure 6.6(b)).[50] Interestingly, modifications at different positions can also impact activity. For example, a recent report highlighted how the attachment of the phosphonate group on the P position could free-up one pendant amine to enhance H_2 evolution, showing that care should be taken in deciding how to link the immobilising moiety without compromising activity.[51]

Cobalt complexes with tetradentate bis(dimethylglyoximato) $(dmg)_2$ ligands in the equatorial position, termed cobaloximes, are among the most explored catalysts for H_2 evolution (Figure 6.6(c)).[52–56] Their high activity in aqueous conditions, low onset potential, and typically straightforward preparation and modification procedures make them appealing candidates for immobilisation on photoelectrodes. For example, functionalisation with a pyrene anchor enabled immobilisation on a CNT-based electrode (Figure 6.6(c)), whereas phosphonic acid functionalisation enabled immobilisation on metal oxide surfaces.[57,58] Although electron donation from the axial pyridine moiety improves activity,[54] it is labile under catalytic conditions and therefore the stability is low. A study highlighting the versatility of these catalysts offered a route to bypass this issue whilst maintaining high activity through placement of the anchoring group on the equatorial ligand and retention of a pendant pyridine moiety in the axial position (Figure 6.6(c)).[59] These catalysts have been used for photoelectrochemical reduction reactions due to their accessibility and high activity, however the stability issues require careful assessment of the true active species.

These examples highlight how a variety of strategies can be used to immobilise effective catalysts on different surfaces. Examples incorporating similar functionality amongst others for coupling with solar light are outlined in the next section.

6.5.1 CO_2 Reduction Catalysts

As with H_2 evolution, types of complexes and comparative performance analysis for CO_2R catalysts have been covered elsewhere.[5,35,36,60–63] A great deal of work on the development of molecular catalysts for CO_2R has highlighted how surface modification can have a significant impact on selectivity, overpotential, and activity. Here, specific examples are used to highlight some core catalysts available for immobilisation in solar-driven CO_2R systems.

Detailed evaluation of performance is not provided but this rather serves to highlight some complexes that can be employed in PEC systems.

Polypyridine-based catalysts have proven highly effective for CO_2R and a number of different structures have been explored.[62,64] Catalysts based on [Re(bpy)(CO)$_3$X] complexes initially reported by Lehn and colleagues, have been extremely proficient for CO_2R to CO.[65–67] Methods for immobilisation using phosphonate (Figure 6.7(a)) and pyrene (Figure 6.7(b)) anchoring groups have enabled immobilisation on metal oxide (TiO_2) and carbon-based (carbon black) surfaces, respectively.[68,69] Analogous catalysts based on similar structures but replacing the Re centre with non-precious metals have yielded impressive results. [Mn(bpy)(CO)$_3$Br] complexes in particular exhibit high activity in homogeneous conditions.[70] Immobilisation of such complexes has shown some interesting phenomena that are related to the susceptibility for dimerisation, which can alter product selectivity.[71] When immobilised on TiO_2 films through the use of phosphonic acid groups (Figure 6.7(c)), UV-vis spectroelectrochemistry showed the formation of a dimer species that was responsible for CO generation.[72] In a subsequent study, immobilisation on CNTs using a pyrene anchoring group permitted control over the surface concentration to show that with low amounts of catalyst on the surface, dimerisation could be inhibited to direct selectivity towards formate rather than CO (Figure 6.7(d)).[73] The two methods gave rise to mechanistic alterations, highlighting the importance of evaluating molecular catalysts on surfaces prior to coupling in PEC devices.

Terpyridines have also been used for CO_2R, with the most selective examples being based on Co(II) centres.[74,75] One example used an electrografting method to functionalise a glassy carbon surface with a terpyridine ligand that was subsequently metallated to form a surface-bound Co-terpyridine catalyst (Figure 6.7(e)).[76] Although this worked in aqueous conditions, the selectivity for CO was low. An interesting example using phosphonic acid groups on the terpyridine ligand showed that selectivity for formate and CO was altered by the amount of H_2O present in the electrolyte solution—this was due to involvement of the phosphonic acid group in the mechanism (Figure 6.7(f)).[77] These examples highlight how the anchoring group can alter catalysis.

Macrocycles such as Ni cyclams have proven effective catalysts for CO_2R to CO, as Earth-abundant alternatives to precious-metal-based catalysts.[78–82] An example using a carboxylate anchoring group (Figure 6.7(g)) showed that it was possible to enhance light-activated charge transfer from TiO_2 to the immobilised complex, however the poor stability in protic solvents showed that more stable anchoring groups are required to couple such catalysts with photoelectrodes.[83] Porphyrins and phthalcoyanine derivatives have been immobilised on metal oxides using carboxylate and phosphonate anchoring groups, and on carbon surfaces through pyrene modification and 'click' chemistry.[5] However, one of the most straightforward strategies to immobilise such catalysts on CNT surfaces is self-assembly of the unmodified complexes, where π–π stacking occurs without the need for additional anchoring

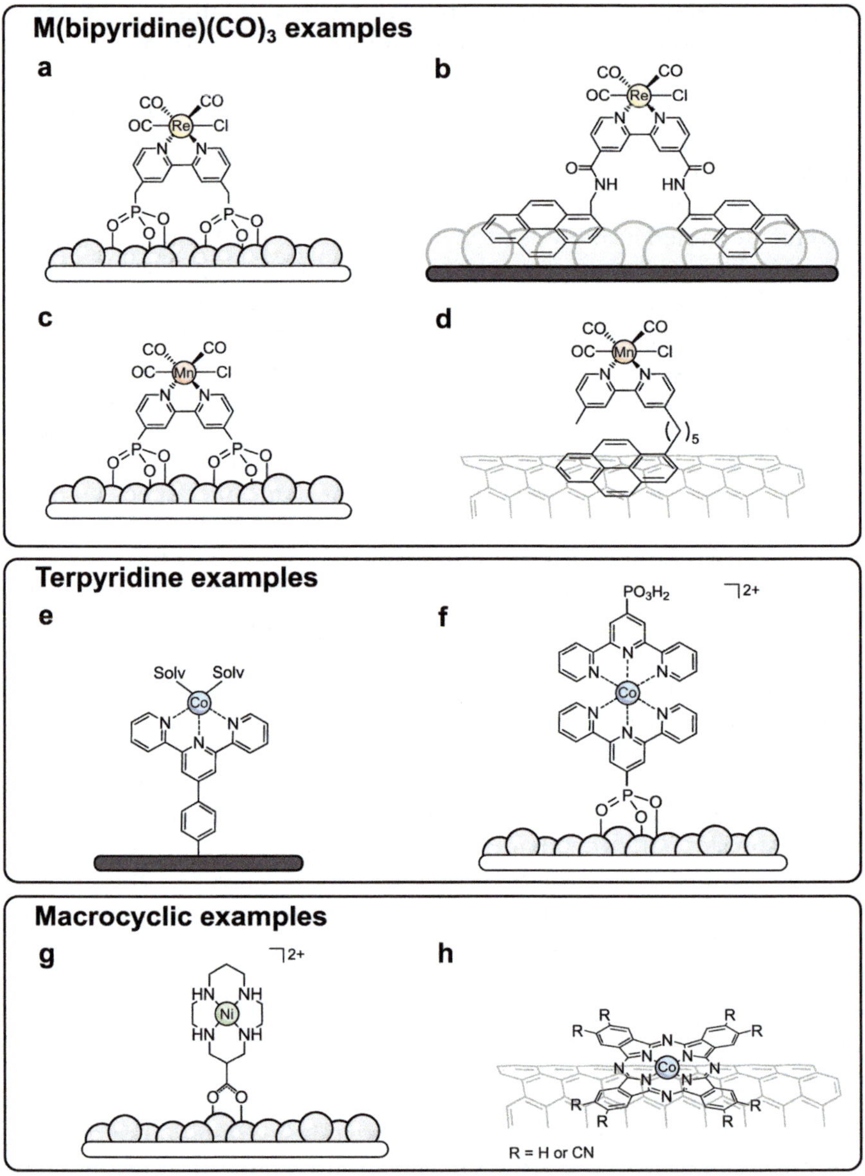

Figure 6.7 (a) and (b) Examples of [Re(bpy)$(CO)_3$X], and (c) and (d) [Mn(bpy)$(CO)_3$X] catalysts immobilised on different electrode types. (e) and (f) Co(terpyridine)$_n$ examples on different surface types. (g) Ni-cyclam immobilised on a TiO_2 surface. (h) Co-phthalocyanine catalyst immobilised on CNTs.

components (Figure 6.7(h)).[84] The simplicity of this route makes such commercially available catalysts accessible to a wide range of researchers and methods to coat the surface of PCs with compatible layers are discussed in the next section.

6.5.2 Catalyst Stability and Experimental Issues

As well as being able to desorb from the surface, electrocatalysts also exhibit several other pathways that can result in decomposition or structural modifications. Reductive decomposition of catalysts can occur in certain cases. For example, cobaloxime catalysts have been shown to decompose under reductive potentials resulting in ligand dissociation and deposition of a heterogeneous catalyst on the surface of the electrode.[85] This was further exemplified for cyclams, which undergo irreversible decomposition under cathodic applied potentials.[86] Interestingly, recent studies have identified that in some cases, for example with copper phthalocyanines and porphyrins as well as polymeric forms, reversible formation of clusters can occur.[87,88] With *ex-situ* analysis, such alterations are not observed as the structure pre- and post-electrolysis remains identical. Only using *in-operando* X-ray absorption spectroscopy can the reversible process be correctly assigned.[89] This is significant when assessing stability of such catalysts because structures based on pre-catalyst structure obtained from *ex situ* analysis fail to provide correct routes to rational modification. Certain catalysts are also susceptible to decomposition or molecular alterations under light illumination. For example [Mn(bpy)$(CO)_3$X] complexes lose CO following illumination in the visible region, which can give false product quantification if low yields are obtained. Methods to confirm the presence of nanoparticles or spectroscopic techniques to ensure structural features are retained should be commonplace when assessing molecular catalysts.[90]

Additional issues can arise from secondary contributions from metals or contaminants. Examples include reference electrodes that can leach Ag, impure electrolytes that can contain metal ions, metallic crocodile clips that can corrode if contacted with solution, and light-induced decomposition of epoxy or other glues. In these cases, ^{13}C labelling can be used to verify the source of products for CO_2R and careful tests such as post-electrolysis XPS, IR spectroscopy, and ^{1}H NMR can confirm the structural integrity of molecular catalysts.

6.6 Photocathodes with Molecular Catalysts

PCs are built on SCs that can absorb visible light to drive electrochemical reduction reactions. As discussed, types of visible-light-absorbing SCs differ in the way they absorb light, separate charge, and transport electrons and holes. Understanding how to control these specific characteristics is key to controlling catalytic reactions when p-SCs feature immobilised molecular catalysts. In this section, specific examples using a range of different SC materials are used to highlight effective strategies to integrate molecular catalysts with narrow bandgap SCs. These are divided into PC systems and ^{i}PC systems, which feature protective layers.

6.6.1 Photocathodes (PCs) with Molecular Catalysts

Although stability issues are commonly observed with p-SCs under illumination in aqueous conditions, several PCs that directly incorporate molecular catalysts have been reported for operation over short time periods. An early example used p-Si coated with Mo_3S_4 clusters, deposited through drop casting.[91] With Si pillars formed through photolithography, currents up to -9 mA cm^{-2} were obtained at 0.0 V *vs.* RHE. However, a large portion of the clusters were lost to the electrolyte solution following PEC testing, therefore anchoring strategies to immobilise molecular catalysts on the surface would be required to enhance stability.

The most effective immobilisation strategy for direct anchoring of molecular catalysts on p-SCs takes advantage of UV-induced grafting. A study showed that this grafting route was possible through the immobilisation of Ni (bis) diphosphine catalyst on GaP (100) and Si (111) surfaces, but no photocurrent studies were conducted in this case (Figure 6.8(a)).[92] However, a similar approach was used to incorporate Co and Fe porphyrin catalysts on a freshly etched GaP (100) surface to form functional PCs (Figure 6.8(b)).[93] These assemblies could reach current densities of approximately -1.3 mA cm^{-2} at 0.0 V *vs.* RHE, with the more effective Co catalyst giving a TOF of 3.9 s^{-1}. These systems were further developed using UV-induced surface grafting to first deposit a polypyridyl layer on GaP (100) to which a catalyst could be coordinated. Initially used for a cobaloxime catalyst (Figure 6.8(c)),[94] the method was employed for a more active Co-porphyrin (Figure 6.8(d)).[95] The initial layer provides coordination sites for the Co-porphyrin to bind through soaking in solution, enabling the catalyst to reach -1.2 mA cm^{-2} at 0.0 V *vs.* RHE. A follow-up study on this immobilised catalyst confirmed that the light absorption from the porphyrin itself had a negligible influence on H_2 generation, emphasising the need for electronic connection between the p-SC and the

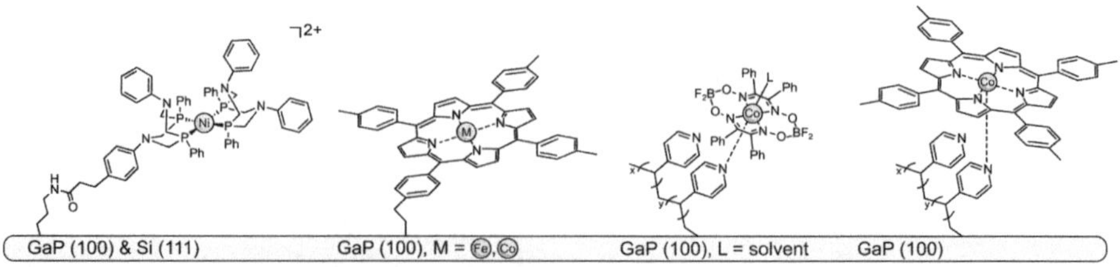

Figure 6.8 (a–d) Examples of immobilisation of molecular catalysts on light-active p-SCs using UV-induced grafting.

catalyst.[96] This method was also extended to an imidazolium-based polymer, highlighting the diversity of modular assembly in cobaloxime-based PCs.[97]

Electropolymerised molecular catalysts have also been effective in functionalising bare p-SC surfaces. A first example incorporated a Ru-complex polymer with a Zn-doped InP electrode to enable CO_2R to HCOOH.[98] In a subsequent study, a series of analogues were assessed for CO_2R on InP, GaP, and N-Ta_2O_4 electrodes.[99] The addition of a phosphonate anchoring group to the Ru complex enhanced CO_2R performance.

Although effective PCs have been produced using a range of interesting immobilisation procedures and p-SC electrodes, the overriding issue of material stability has led research towards protective layers in ^{i}PC examples. Several important studies are highlighted in the next section to show how ^{i}PCs can be used to form effective solar fuel platforms.

6.6.2 Integrated Photocathodes (^{i}PC) with Molecular Catalysts

There are many more examples of ^{i}PCs than PCs seeing as common instability issues can be partially bypassed through introduction of a protective layer. This layer serves two purposes: it prevents rapid degradation of the SC, and it acts as a support to which the catalyst can anchor. Unlike PCs, where deposition techniques are dependent on the surface of the light absorber, here the common usage of metal oxide protective coatings means that the wide range of appropriate anchoring groups discussed in the previous section can be used.

Construction of effective systems is reliant on catalyst/SC assemblies that are compatible with aqueous conditions. This has proven problematic for several ^{i}PCs. With Cu_2O, prevention of short-term photodegradation was made possible due to pioneering work that demonstrated the protective ability of deposited metal oxide layers.[100] The layers not only serve as stabilising materials but also enable immobilisation of molecular catalysts. Initially exemplified with a solution-phase Re catalyst,[101] a protected Cu_2O electrode was modified to incorporate a nanostructured TiO_2 surface capable of hosting a phosphonic-acid-modified Re(bpy)$(CO)_3$ catalyst.[102] However, in both cases organic electrolyte solutions were used, which are not compatible with sustainable full-cell CO_2R.

The most productive examples have been based on TiO_2-protected p-Si. Studies using heterogeneous catalysts highlighted how coatings formed through ALD on Si can enable effective electron transport through the so-called 'leaky' metal oxide for water oxidation.[16] The conduction band position of TiO_2 is sufficiently positive to enable electron transfer from p-Si, which results in the formation of highly mobile electrons in the conduction band of TiO_2. Interestingly, mesoporous TiO_2 (*meso*-TiO_2) layers can provide sufficient protective properties for short term operation whilst simultaneously acting as a high-surface-area support for molecular species. In one example, a p-Si|*meso*-TiO_2 electrode was functionalised with either a

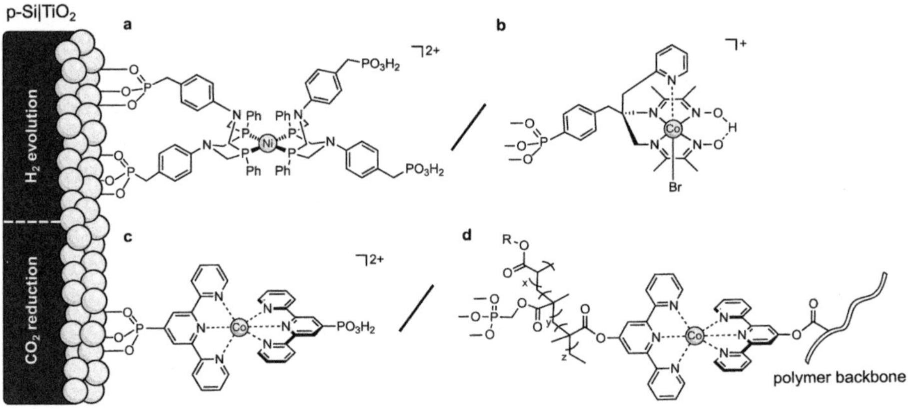

Figure 6.9 Examples of mesoporous TiO_2-protected p-Si ^{i}PCs with catalysts for: (a) and (b) H_2 evolution, and (c) and (d) CO_2 reduction.

phosphonic-acid-bearing Ni (bis)diphosphine (Figure 6.9(a)) or cobaloxime catalyst (Figure 6.9(b)) for H_2 evolution.[103] With the Ni catalyst, the system consistently produced H_2 over the course of 24 h with an approximate stable photocurrent density of −1 mA cm^{-2} at 0.0 V *vs.* RHE. These p-Si|*meso*-TiO_2 electrodes were also used for CO_2 reduction to CO with a Co-phthalocyanine,[104] and to HCOOH and CO with a phosphonated Co bis(terpyridine) catalyst (Figure 6.9(c)).[77] In this case, involvement of a possible protonated phosphonic acid intermediate was suggested to be the cause behind a sufficiently lowered catalytic onset potential, enabling the p-Si|*meso*-TiO_2 ^{i}PCs to operate. The knowledge gained from integration in an ^{i}PC architecture was further developed through the use of a polymer-incorporated Co-terpyridine (Figure 6.9(d)).[105] Using a hydrophobic unit, and by optimising the ratio of bis- and mono-tpy, solar-driven CO production could be enhanced. These systems exemplify how immobilisation and photoelectrocatalytic testing can lead to insights that guide rational catalyst modifications.

Systems employing a dual protection strategy have successfully incorporated ALD_{TiO_2} coatings with additional *meso*-TiO_2 supports for enhanced molecular loading. A good example of this effective approach used a phosphonated cobaloxime catalyst, which was further stabilised by an additional ALD_{TiO_2} coating on top of the immobilised catalyst, preventing short-term desorption which was apparent with purely covalent immobilisation (Figure 6.10(a)).[106] Alternative anchoring groups have also been used for similar systems. A recent example highlighted how hydroxamate groups could anchor N_5-chelated Ni and Co complexes to TiO_2-protected p-Si, which outperformed an analogous cobaloxime catalyst.[107] The Ni catalyst gave a photocurrent of approximately −1.0 mA cm^{-2} at 0.0 V *vs.* RHE, enabling bias-free operation when coupled with a Co_4O_4 cubane modified $BiVO_4$ photoanode (Figure 6.10(b)).

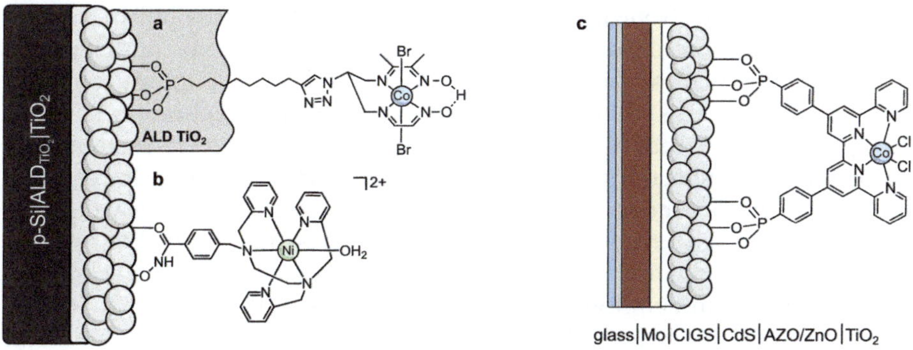

Figure 6.10 ^{i}PCs for H_2 evolution and CO_2R. (a) p-Si|ALD$_{TiO_2}$ with an additional ALD$_{TiO_2}$ layer on top of the Co catalyst, and (b) p-Si|ALD$_{TiO_2}$ with a molecular Ni catalyst bound using a hydroxamate anchoring group. (c) CIGS-based ^{i}PC with a Co quaterpyridine catalyst for CO_2R.

Examples of different SC surfaces coated with TiO_2 layers can facilitate molecular modification for aqueous solar fuel production. For example, a metal oxide, $CuFe_xO_y$|ALD$_{TiO_2}$, was modified with a carboxylate-bearing cobaloxime catalyst,[108] and an oxysulfide, $La_5Ti_2Cu_{0.9}Ag_{0.1}S_5O_7$|$TiO_{2(sputtered)}$, was functionalised with a phosphonate-appended Ni bis(disphosphine) catalyst, both proving effective for H_2 evolution. TiO_2 modification was recently extended to a $Cu(In_{0.1}Ga_{0.9}Se_2)$ (CIGS) to enable immobilisation of a Co-quaterpyridine catalyst that is proficient for CO generation from CO_2 (Figure 6.10(c)).[109] In this example, the catalyst produced CO with a FE of 97% with a photocurrent of −0.8 mA cm^{-2} at −0.06 V *vs.* RHE with frontside illumination.

Although TiO_2 is an effective metal oxide for permitting frontside illumination in certain cases, non-transparent protective layers can be employed when backside illumination is used. This approach is well-demonstrated with metal halide perovskites ($APbX_3$) protected from moisture sensitivity with InBiSn alloys (Field's metal).[26,110] Through deposition of CNT sheets on the surface of a Field's-metal-protected lead halide perovskite, a Co-porphyrin could be non-covalently anchored to catalytically convert CO_2 to CO, however with a low FE_{CO} of 16% under 1 Sun illumination for a current density of −5.6 mA cm^{-2}.[111] With such modifications, immobilisation of more selective non-covalently grafted CO_2R catalysts could give rise to effective solar CO_2 conversion in future work.

6.6.3 Outlook

Hybrid PCs have made great advances in the past years due to a deeper understanding of the design rules for integration of effective catalysts. These advances have encompassed stability, selectivity, and activity, due to effective anchoring strategies, catalyst alterations, and good protection layers among other contributors. In CO_2R catalysis, device designs such as gas-fed flow cells and membrane electrode assemblies have enhanced the possibility of reaching high currents at low overpotentials.[11,37,112–116] Recently, this has extended to novel methods for implementation of PCs,[117–120] but has not yet extended to molecular systems, therefore there is room for implementation of interesting phenomena that can occur when molecular catalysts are used in such devices. In particular, the susceptibility for SC degradation may be lower when humidified gas is used and therefore promising systems previously deemed unstable could see a resurgence. Additionally, the search for stable p-SCs is continually growing, and novel materials may prove fruitful in the search for effective PCs.[121] Integration of molecular catalysts in PCs or ^{i}PCs that have long term stability can be realised through scientific advances in this domain.

6.7 Dye-sensitised Photocathodes with Molecular Catalysts

The combination of visible-light-absorbing chromophores with wide-band-gap SCs to drive reduction reactions is an interesting approach to generate solar fuels. The extensive range of photosensitisers (PSs), such as molecular dyes and quantum dots, means that tailor-made devices can be constructed in a modular fashion using an array of rational immobilisation approaches. DSPCs also bypass stability issues of PCs as the p-SCs used for building such systems tend to be less susceptible to degradation as they do not absorb the incoming solar light and therefore do not photodegrade under operating conditions. Although the development of DSPCs has been slower than corresponding photoanodes, a deep understanding of the limiting points has been derived from combined experimental and theoretical analysis.[122–125]

6.7.1 Operating Principles

Molecular DSPCs are typically composed of a p-SC, a dye, and a catalyst, all of which participate in charge transport. Following initial light excitation of the dye, two charge transfer processes occur. First, the photoexcited dye injects a hole into the p-SC, to generate a reduced dye state (PS^-), and then the PS^- species can then transfer an electron to reduce the molecular catalyst (Figure 6.11(a)). The cycle must repeat two times (for a two-electron reduction) for the system to generate products.

Although each step sounds simple in theory, there are several energetic criteria for effective charge transfer and catalysis. The first step of hole injection relies on two interacting components: the p-SC and the PS. The photoexcited PS must have a sufficiently positive enough electrochemical potential to inject an electron into the valence band of the p-SC. The second step of electron transfer to the immobilised catalyst requires a sufficiently negative electrochemical potential of the reduced PS^- to reduce the catalyst. The thermodynamic properties of each component can be obtained using several techniques. The dye $E_{(PS/PS-)}$ value can be obtained from the reduction

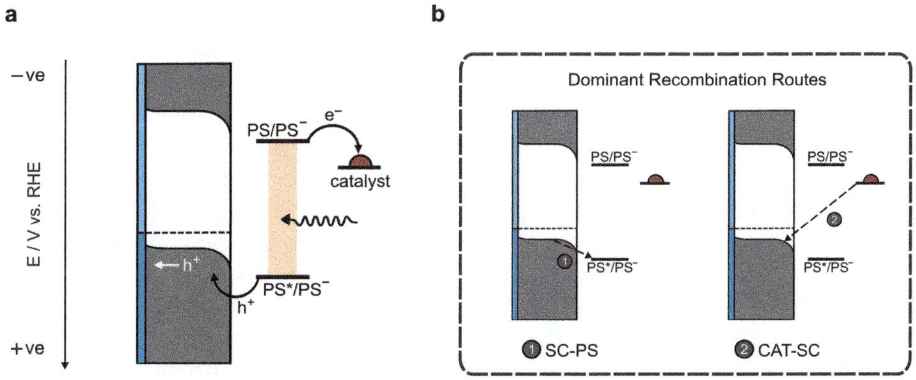

Figure 6.11 (a) Operation of a DSPC, where h^+ and e^- are holes and electrons, respectively. (b) Two major recombination routes that limit performance in DSPCs.

potential, either in organic solution or immobilised on a conductive support and added to the E_{00} values obtained from the intersection of the normalised absorption and fluorescence spectra to obtain the $E_{(PS*/PS-)}$ values. For the p-SC, Mott–Schottky analysis using electrochemical impedance spectroscopy can be used to obtain a flatband position, corresponding roughly to the thermodynamic potential for hole injection. Finally, for the molecular catalyst the reduction potential (typically the second reduction) can be obtained from conventional cyclic voltammetry techniques.

Nonetheless, even with correct electrochemical properties and high functionality with a sacrificial electron acceptor (SEA), some systems fail to generate appreciable photocurrents. This is because the photocurrent, although controlled by the rates of each charge transfer step, is restricted by the presence of numerous recombination pathways that exist between the components. Figure 6.11(b) depicts the two major recombination routes for DSPCs. Here, the problems associated with these recombination routes and methods to bypass them are discussed with respect to DSPCs featuring molecular catalysts. Specific cases are highlighted with reference to the recombination routes using abbreviated forms (*i.e.*, CAT-SC or SC-PS).

6.7.2 Photosensitisers

Advances in the development of DSSCs have shown that a wide range of dyes can be tailored to provide high photocurrents when transfer of an electron to an n-type SC is required. However, the design rules for p-DSSCs are quite different, as the hole injection step and subsequent charge separation requires specific characteristics to ensure long-lived reduced states to enable effective electron transport to the catalyst.

Differences between p-SC and n-SC systems are well-demonstrated by attempts to integrate derivatives of ruthenium tris-2,2′-bipyridine ($[Ru(bpy)_3]^{2+}$)

dyes with p-SCs. $[Ru(bpy)_3]^{2+}$ dyes can generate long-lived excited states when immobilised on TiO_2, which arise from broad metal-to-ligand charge transfer bands, where the electron is located on the aromatic ligand and the hole situated at the metal centre.[126–128] However, when immobilised on p-SCs, these dyes exhibit fast charge recombination between holes in the p-SC and the reduced dye (SC-PS) leading to much lower photocurrents.[129] One way to limit SC-PS recombination is to spatially separate the electron in the PS^- species (corresponding to the ground state LUMO) from the p-SC whilst simultaneously placing the HOMO close to the anchoring group to facilitate fast hole injection. Such an effect was demonstrated in an SEA system through the replacement of one bipyridine unit of a phosphonic-acid-functionalised $[Ru(bpy)_3]^{2+}$ PS with dipyridophenazine (dppz).[130] The modification enabled removal of electron density away from the phosphonate anchoring groups, limiting SC-PS recombination.

A good method to assess the ability of a PS/p-SC assembly to function effectively is using a water-soluble SEA. Common examples include $[Co(NH_3)_5Cl]^{2+}$ and 4,4′-dithiodipyridine, which undergo rapid irreversible reduction. With an SEA, the kinetics of catalysis or reduction of the catalyst are not the limiting step and therefore only the dye-to-SEA electron transfer is observed as the photocurrent. This also removes a recombination pathway from the catalyst to the SC, which as described below is one of the major limiting points in DSPCs. However, such experiments are not a true measure of the maximum observable photocurrent because the enhanced kinetics of electron transfer to an SEA exaggerate the ability of a system to function for catalysis. Nonetheless, these experiments are extremely useful when probing the different ability of dyes to separate charge from the SC and therefore highlight ways to modify dye structure to avoid SC-PS recombination.

The development of p-type dye-sensitised solar cells (p-DSSCs) has highlighted several modifications that enable efficient charge transfer to p-type metal oxides. Organic dyes have been particularly effective in DSPCs.[131–135] As well as exhibiting high extinction coefficients due to strong π–π^* transitions, introduction of push–pull functionality and understanding of the factors that control the bandgap and HOMO and LUMO positions have enabled effective organic dye synthesis and integration.[132,136] Organic dyes therefore not only enhance performance but reduce the large costs associated with precious-metal-based PSs.

6.7.3 Dyads and Supramolecular Assemblies

Spectroscopic characterisation of DSPCs has shown that a dominant limitation is fast recombination between the reduced catalyst and the SC (CAT-SC).[137] To circumvent this issue, dyads or supramolecular assemblies can be used to spatially separate the catalyst from the SC. This has been well-demonstrated in several systems using different approaches.

One approach is to form supramolecular assemblies of phosphonic acid-bearing complexes with Zr^{IV} linkers. Layer-by-layer modification was used to form a NiO-based DSPC featuring a hexaphosphonated $[Ru(bpy)_3]^{2+}$ dye, a Zr^{IV} linker, and a tetraphosphonated Ni (bis)diphosphine catalyst (Figure 6.12).[138] The full assembly displayed photocurrents of −8.8 μA cm^{-2} at +0.3 V *vs.* RHE, whereas the co-assembled electrode only showed maximal photocurrents of −0.9 μA cm^{-2}, highlighting that catalyst separation is an effective method to prevent CAT-SC recombination. This assembly method was extended in another study to incorporate a dianiline donor species on an ITO-based electrode with a Ru dye and the same Ni catalyst.[139] The donor was shown to direct holes to the ITO surface and enable effective H_2 production.

Alternatively, dyad species incorporating dye and catalyst components can be used to the same effect.[140–146] The efficacy of the dyad approach was demonstrated through the synthesis of a dyad featuring a Co diamine-dioxime catalyst, linked to a push–pull organic dye, in which the triphenylamine component acts as a donor and the ethyl cyanoacetate as an acceptor (Figure 6.13(a)).[140] However, the high performance with this dyad immobilised on NiO was similar to a co-grafted counterpart showing that the major recombination routes can still exist with such complex architectures.[141] One extremely stable example of a dyad system used a $[Ru(bpy)_3]^{2+}$ dye immobilised on NiO through a silane anchoring group, and linked to a $[Re(bpy)(CO)_3Cl]$ catalyst for CO_2 reduction (Figure 6.13(b)).[143] The DSPC was capable of generating CO

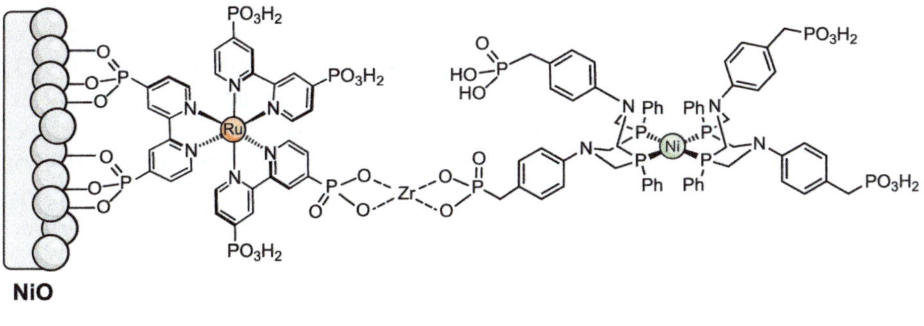

Figure 6.12 Supramolecular assembly of a DSPC for H_2 evolution using Zr^{IV} linkers to connect phosphonic acid groups of the dye with those of the catalyst.

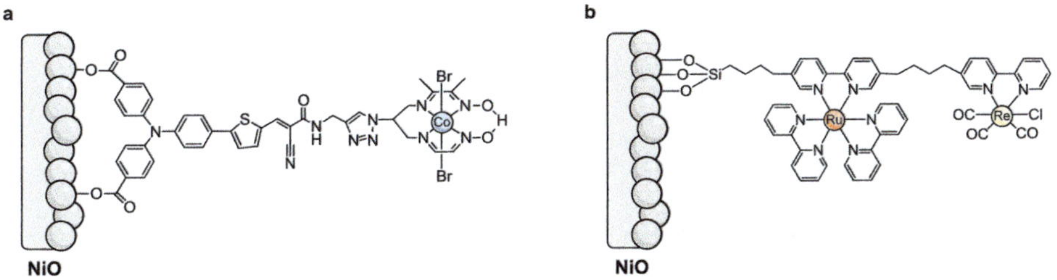

Figure 6.13 Dyad assemblies for (a) H_2 evolution with a Co catalyst linked with an organic push–pull dye, and (b) CO_2R with a Ru–Re complex.

with a photocurrent of approximately −15 μA cm^{-2} at ∼0 V *vs.* RHE over the course of 10 h.

6.7.4 Alternative p-Type Semiconductors

Since the first reported p-DSSC, p-type NiO has been the predominant SC used for DSPCs.[147] Spectroscopic techniques have been used to identify weak points in the operation of DSPCs and recent work has shown that although the hole injection process is often ultra-fast giving rise to the PS^- species, hole trapping in NiO can be detrimental to activity.[148] Surface trap states facilitate fast SC-PS and CAT-SC charge recombination, therefore the most effective way of avoiding these recombination routes is through the development of alternative p-SCs.

Delafossite structures (Figure 6.14(a)) have emerged as high performance p-SCs in DSPCs.[149–153] These materials, initially explored for p-DSSCs,[154,155] tend to exhibit higher hole mobility than NiO and can therefore transport holes from the surface more effectively to limit recombination. $CuCrO_2$ electrodes synthesised using a sol–gel procedure were coupled with a diketopyrrolopyrrole dye and a Ni bis(disphosphine) catalyst to generate relatively high yields of H_2.[151] The advantages of $CuCrO_2$ were further highlighted through the construction of inverse opal $CuCrO_2$ electrodes, on which the former diketopyrrolopyrrole dye was compared with a perylene monoimide dye (Figure 6.14(b)).[149] Analysis of the final DSPCs with the immobilised Ni catalyst confirmed that the perylene monoimide dye structure on $CuCrO_2$ inverse opal electrodes showed high performance (−25 μA cm^{-2} at +0.0 V *vs.* RHE) despite the two components being co-immobilised on the surface. With $CuGaO_2$, p-SC advantages were highlighted through immobilisation of a Ru–Re dyad for CO_2R, in which the early onset potential enabled coupling with a TaON/CoO_x water oxidation photoanode.[152] The DSPC outperformed

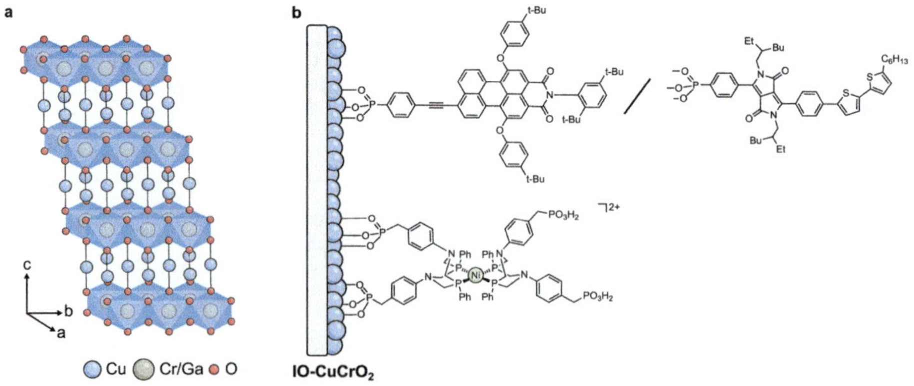

Figure 6.14 (a) $CuMO_2$ delafossite structure. (b) Examples of $CuCrO_2$ DSPCs based on perylene monoimide and diketopyrrolopyrrole dyes coupled with a molecular Ni catalyst.

an analogous system based on NiO, further highlighting how SC replacement can enhance activity.[156]

6.7.5 Outlook

Further work to establish performance-limiting aspects of DSPCs, and novel routes to avoid dominant recombination routed will be key in the development of effective systems. There are additional issues that also have to be tackled, for example problems with Ni dissociation which can play a role in catalysis for NiO-based systems.[157] However, alternative p-SCs can avoid these issues and enhance performance, but so far the exploration has been limited. Approaches combining the benefits of directional charge transport in DSPCs with the higher ability to generate photocurrent of PCs have shown some promise in producing high currents for CO_2R.[158] Furthermore, alternative reactions such as oxygen reduction have also been effective using DSPCs.[159] The progress made in this field is encouraging to researchers aiming to understand the photochemical processes occurring at p-SC/dye/catalyst interfaces, and is important in the design of selective systems for directional charge transport and catalysis.

References

1. IPCC, *AR6 Climate Change 2021*: The Physical Science Basis, 2021.
2. P. U. Clark, J. D. Shakun, S. A. Marcott, A. C. Mix, M. Eby, S. Kulp, A. Levermann, G. A. Milne, P. L. Pfister, B. D. Santer, D. P. Schrag, S. Solomon, T. F. Stocker, B. H. Strauss, A. J. Weaver, R. Winkelmann, D. Archer, E. Bard, A. Goldner, K. Lambeck, R. T. Pierrehumbert and G. K. Plattner, *Nat. Clim. Change*, 2016, **6**, 360–369.
3. K. B. Debnath and M. Mourshed, *Nat. Energy*, 2018, **3**, 172–184.
4. BP p.l.c, *BP Statistical Review of World Energy*, London, 2019.
5. K. E. Dalle, J. Warnan, J. J. Leung, B. Reuillard, I. S. Karmel and E. Reisner, *Chem. Rev.*, 2019, **119**, 2752–2875.
6. C. E. Creissen and M. Fontecave, *Adv. Energy Mater.*, 2021, **11**, 2002652.
7. S. Ardo, D. Fernandez Rivas, M. A. Modestino, V. Schulze Greiving, F. F. Abdi, E. Alarcon Llado, V. Artero, K. Ayers, C. Battaglia, J. P. Becker, D. Bederak, A. Berger, F. Buda, E. Chinello, B. Dam, V. Di Palma, T. Edvinsson, K. Fujii, H. Gardeniers, H. Geerlings, S. M. Hashemi, S. Haussener, F. Houle, J. Huskens, B. D. James, K. Konrad, A. Kudo, P. P. Kunturu, D. Lohse, B. Mei, E. L. Miller, G. F. Moore, J. Muller, K. L. Orchard, T. E. Rosser, F. H. Saadi, J. W. Schüttauf, B. Seger, S. W. Sheehan, W. A. Smith, J. Spurgeon, M. H. Tang, R. Van De Krol, P. C. K. Vesborg and P. Westerik, *Energy Environ. Sci.*, 2018, **11**, 2768–2783.
8. A. C. Nielander, M. R. Shaner, K. M. Papadantonakis, S. A. Francis and N. S. Lewis, *Energy Environ. Sci.*, 2015, **8**, 16–25.

9. J. W. Ager, M. R. Shaner, K. A. Walczak, I. D. Sharp and S. Ardo, *Energy Environ. Sci.*, 2015, **8**, 2811–2824.
10. I. Holmes-Gentle, *et.al.*, SolarFuelsDB, https://solarfuelsdb.epfl.ch/.
11. A. Peugeot, C. E. Creissen, M. W. Schreiber and M. Fontecave, *ChemElectroChem*, 2021, **8**, 2726–2736.
12. M. T. Mayer, *Curr. Opin. Electrochem.*, 2017, **2**, 104–110.
13. S. H. Simon, *The Oxford Solid State Basics*, Oxford University Press, Oxford, 1st edn, 2013.
14. R. Memming, *Semiconductor Electrochemistry*, Wiley-VCH, Weinheim, 2nd edn, 2015.
15. R. Wick and S. D. Tilley, *J. Phys. Chem. C*, 2015, **119**, 26243–26257.
16. S. Hu, M. R. Shaner, J. A. Beardslee, M. Lichterman, B. S. Brunschwig and N. S. Lewis, *Science*, 2014, **344**, 1005–1009.
17. T. Wang, Z. Luo, C. Li and J. Gong, *Chem. Soc. Rev.*, 2014, **43**, 7469–7484.
18. R. Liu, Z. Zheng, J. Spurgeon and X. Yang, *Energy Environ. Sci.*, 2014, **7**, 2504–2517.
19. A. B. Laursen, T. Pedersen, P. Malacrida, B. Seger, O. Hansen, P. C. K. Vesborg and I. Chorkendorff, *Phys. Chem. Chem. Phys.*, 2013, **15**, 20000–20004.
20. B. Seger, T. Pedersen, A. B. Laursen, P. C. K. Vesborg, O. Hansen and I. Chorkendorff, *J. Am. Chem. Soc.*, 2013, **135**, 1057–1064.
21. J. D. Benck, S. C. Lee, K. D. Fong, J. Kibsgaard, R. Sinclair and T. F. Jaramillo, *Adv. Energy Mater.*, 2014, **4**, 1400739.
22. L. A. King, T. R. Hellstern, J. Park, R. Sinclair and T. F. Jaramillo, *ACS Appl. Mater. Interfaces*, 2017, **9**, 36792–36798.
23. J. J. Leung, J. Warnan, D. H. Nam, J. Z. Zhang, J. Willkomm and E. Reisner, *Chem. Sci.*, 2017, **8**, 5172–5180.
24. V. Andrei, R. L. Z Hoye, M. Crespo-Quesada, M. Bajada, S. Ahmad, M. De Volder, R. Friend, E. Reisner, V. Andrei, M. Crespo-Quesada, M. Bajada, E. Reisner, R. L. Z Hoye, R. Friend, S. Ahmad and M. De Volder, *Adv. Energy Mater.*, 2018, **8**, 1801403.
25. M. Crespo-Quesada, L. M. Pazos-Outón, J. Warnan, M. F. Kuehnel, R. H. Friend and E. Reisner, *Nat. Commun.*, 2016, **7**, 1–7.
26. E. Edwardes Moore, V. Andrei, S. Zacarias, I. A. C. Pereira and E. Reisner, *ACS Energy Lett.*, 2020, **5**, 232–237.
27. S. W. Boettcher, E. L. Warren, M. C. Putnam, E. A. Santori, D. Turner-Evans, M. D. Kelzenberg, M. G. Walter, J. R. McKone, B. S. Brunschwig, H. A. Atwater and N. S. Lewis, *J. Am. Chem. Soc.*, 2011, **133**, 1216–1219.
28. F. Lin and S. W. Boettcher, *Nat. Mater.*, 2014, **13**, 81–86.
29. M. J. Kenney, M. Gong, Y. Li, J. Z. Wu, J. Feng, M. Lanza and H. Dai, *Science*, 2013, **342**, 836–840.
30. J. D. Benck, B. A. Pinaud, Y. Gorlin and T. F. Jaramillo, *PLoS One*, 2014, **9**, e107942.
31. D. Bhattacharyya, P. E. Videla, M. Cattaneo, V. S. Batista, T. Lian and C. P. Kubiak, *Chem. Sci.*, 2021, **12**, 10131–10149.
32. E. E. Moore, S. J. Cobb, A. M. Coito, A. R. Oliveira, I. A. C. Pereira and E. Reisner, *Proc. Natl. Acad. Sci. U. S. A.*, 2022, **119**, e2114097119.

33. Z. Zhang, L. Melo, R. P. Jansonius, F. Habibzadeh, E. R. Grant and C. P. Berlinguette, *ACS Energy Lett.*, 2020, **5**, 3101–3107.
34. K. Yang, R. Kas and W. A. Smith, *J. Am. Chem. Soc.*, 2019, **141**, 15891–15900.
35. R. M. Bullock, A. K. Das and A. M. Appel, *Chem. – Eur. J.*, 2017, **23**, 7626–7641.
36. A. Perazio, G. Lowe, R. Gobetto, J. Bonin and M. Robert, *Coord. Chem. Rev.*, 2021, **443**, 214018.
37. C. E. Creissen, J. G. Rivera de la Cruz, D. Karapinar, D. Taverna, M. W. Schreiber and M. Fontecave, *Angew. Chem.*, 2022, **61**, e202206279.
38. J. C. Love, L. A. Estroff, J. K. Kriebel, R. G. Nuzzo and G. M. Whitesides, *Chem. Rev.*, 2005, **105**, 1103–1169.
39. J. R. Pankhurst, P. Iyengar, A. Loiudice, M. Mensi and R. Buonsanti, *Chem. Sci.*, 2020, **11**, 9296–9302.
40. L. Wang, D. E. Polyansky and J. J. Concepcion, *J. Am. Chem. Soc.*, 2020, **141**, 8020–8027.
41. L. Sévery, J. Szczerbiński, M. Taskin, I. Tuncay, F. Brandalise Nunes, C. Cignarella, G. Tocci, O. Blacque, J. Osterwalder, R. Zenobi, M. Iannuzzi and S. D. Tilley, *Nat. Chem.*, 2021, **13**, 523–529.
42. H. Li, F. Li, Y. Wang, L. Bai, F. Yu and L. Sun, *ChemPlusChem*, 2016, **81**, 1056–1059.
43. D. L. Dubois, *Inorg. Chem.*, 2014, **53**, 3935–3960.
44. M. E. Carroll, B. E. Barton, T. B. Rauchfuss and P. J. Carroll, *J. Am. Chem. Soc.*, 2012, **134**, 18843–18852.
45. M. Rakowski Dubois and D. L. Dubois, *Acc. Chem. Res.*, 2009, **42**, 1974–1982.
46. A. D. Wilson, R. H. Newell, M. J. McNevin, J. T. Muckerman, M. R. DuBois and D. L. DuBois, *J. Am. Chem. Soc.*, 2006, **128**, 358–366.
47. E. S. Rountree and J. L. Dempsey, *J. Am. Chem. Soc.*, 2015, **137**, 13371–13380.
48. A. Le Goff, V. Artero, B. Jousselme, P. D. Tran, N. Guillet, R. Métayé, A. Fihri, S. Palacin and M. Fontecave, *Science*, 2009, **326**, 1384–1387.
49. P. D. Tran, A. Le Goff, J. Heidkamp, B. Jousselme, N. Guillet, S. Palacin, H. Dau, M. Fontecave and V. Artero, *Angew. Chem., Int. Ed.*, 2011, **50**, 1371–1374.
50. M. A. Gross, A. Reynal, J. R. Durrant and E. Reisner, *J. Am. Chem. Soc.*, 2014, **136**, 356–366.
51. F. M. Brunner, M. L. Neville and C. P. Kubiak, *Inorg. Chem.*, 2020, **59**, 16872–16881.
52. P.-A. Jacques, V. Artero, J. Pecaut and M. Fontecave, *Proc. Natl. Acad. Sci. U. S. A.*, 2009, **106**, 20627–20632.
53. J. L. Dempsey, B. S. Brunschwig, J. R. Winkler and H. B. Gray, *Acc. Chem. Res.*, 2009, **42**, 1995–2004.
54. M. Razavet, V. Artero and M. Fontecave, *Inorg. Chem.*, 2005, **44**, 4786–4795.
55. C. C. L. McCrory, C. Uyeda and J. C. Peters, *J. Am. Chem. Soc.*, 2012, **134**, 3164–3170.

56. V. Artero, M. Chavarot-Kerlidou and M. Fontecave, *Angew. Chem., Int. Ed.*, 2011, **50**, 7238–7266.
57. B. Reuillard, J. Warnan, J. J. Leung, D. W. Wakerley and E. Reisner, *Angew. Chem.*, 2016, **128**, 4020–4025.
58. A. Reynal, J. Willkomm, N. M. Muresan, F. Lakadamyali, M. Planells, E. Reisner and J. R. Durrant, *Chem. Commun.*, 2014, **50**, 12768–12771.
59. J. Willkomm, N. M. Muresan and E. Reisner, *Chem. Sci.*, 2015, **6**, 2727–2736.
60. B. Kumar, M. Llorente, J. Froehlich, T. Dang, A. Sathrum and C. P. Kubiak, *Annu. Rev. Phys. Chem.*, 2012, **63**, 541–569.
61. K. A. Grice, *Coord. Chem. Rev.*, 2017, **336**, 78–95.
62. E. Boutin, L. Merakeb, B. Ma, B. Boudy, M. Wang, J. Bonin, E. Anxolabéhère-Mallart and M. Robert, *Chem. Soc. Rev.*, 2020, **49**, 5772–5809.
63. B. Das, A. Thapper, S. Ott and S. B. Colbran, *Sustainable Energy Fuels*, 2019, **3**, 2159–2175.
64. N. Elgrishi, M. B. Chambers, X. Wang and M. Fontecave, *Chem. Soc. Rev.*, 2017, **46**, 761–796.
65. J. Hawecker, J. M. Lehn and R. Ziessel, *J. Chem. Soc., Chem. Commun.*, 1983, 536–538.
66. J. Hawecker, J. M. Lehn and R. Ziessel, *J. Chem. Soc., Chem. Commun.*, 1984, 328–330.
67. J. Hawecker, J. -M Lehn and R. Ziessel, *Helv. Chim. Acta*, 1986, **69**, 1990–2012.
68. J. D. Blakemore, A. Gupta, J. J. Warren, B. S. Brunschwig and H. B. Gray, *J. Am. Chem. Soc.*, 2013, **135**, 18288–18291.
69. C. D. Windle, E. Pastor, A. Reynal, A. C. Whitwood, Y. Vaynzof, J. R. Durrant, R. N. Perutz and E. Reisner, *Chem. – Eur. J.*, 2015, **21**, 3746–3754.
70. M. Bourrez, F. Molton, S. Chardon-Noblat and A. Deronzier, *Angew. Chem., Int. Ed.*, 2011, **50**, 9903–9906.
71. M. D. Sampson, A. D. Nguyen, K. A. Grice, C. E. Moore, A. L. Rheingold and C. P. Kubiak, *J. Am. Chem. Soc.*, 2014, **136**, 5460–5471.
72. T. E. Rosser, C. D. Windle and E. Reisner, *Angew. Chem., Int. Ed.*, 2016, **128**, 7514–7518.
73. B. Reuillard, K. H. Ly, T. E. Rosser, M. F. Kuehnel, I. Zebger and E. Reisner, *J. Am. Chem. Soc.*, 2017, **139**, 14425–14435.
74. N. Elgrishi, M. B. Chambers and M. Fontecave, *Chem. Sci.*, 2015, **6**, 2522–2531.
75. N. Elgrishi, M. B. Chambers, V. Artero and M. Fontecave, *Phys. Chem. Chem. Phys.*, 2014, **16**, 13635–13644.
76. N. Elgrishi, S. Griveau, M. B. Chambers, F. Bedioui and M. Fontecave, *Chem. Commun.*, 2015, **51**, 2995–2998.
77. J. J. Leung, J. Warnan, K. H. Ly, N. Heidary, D. H. Nam, M. F. Kuehnel and E. Reisner, *Nat. Catal.*, 2019, **2**, 354–365.
78. J. D. Froehlich and C. P. Kubiak, *Inorg. Chem.*, 2012, **51**, 3932–3934.

79. S. Pugliese, N. T. Huan, J. Forte, D. Grammatico, S. Zanna, B. L. Su, Y. Li and M. Fontecave, *ChemSusChem*, 2020, **13**, 6449–6456.
80. A. Zhanaidarova, C. E. Moore, M. Gembicky and C. P. Kubiak, *Chem. Commun.*, 2018, **54**, 4116–4119.
81. J. Song, E. L. Klein, F. Neese and S. Ye, *Inorg. Chem.*, 2014, **53**, 7500–7507.
82. M. F. Kuehnel, C. D. Sahm, G. Neri, J. R. Lee, K. L. Orchard, A. J. Cowan and E. Reisner, *Chem. Sci.*, 2018, **9**, 2501–2509.
83. G. Neri, J. J. Walsh, C. Wilson, A. Reynal, J. Y. C. Lim, X. Li, A. J. P. White, N. J. Long, J. R. Durrant and A. J. Cowan, *Phys. Chem. Chem. Phys.*, 2015, **17**, 1562–1566.
84. X. Zhang, Z. Wu, X. Zhang, L. Li, Y. Li, H. Xu, X. Li, X. Yu, Z. Zhang, Y. Liang and H. Wang, *Nat. Commun.*, 2017, **8**, 14675.
85. S. Cobo, J. Heidkamp, P.-A. Jacques, J. Fize, V. Fourmond, L. Guetaz, B. Jousselme, V. Ivanova, H. Dau, S. Palacin, M. Fontecave and V. Artero, *Nat. Mater.*, 2012, **11**, 802–807.
86. T. N. Huan, E. S. Andreiadis, J. Heidkamp, P. Simon, E. Derat, S. Cobo, G. Royal, A. Bergmann, P. Strasser, H. Dau, V. Artero and M. Fontecave, *J. Mater. Chem. A*, 2015, **3**, 3901–3907.
87. Z. Weng, Y. Wu, M. Wang, J. Jiang, K. Yang, S. Huo, X. F. Wang, Q. Ma, G. W. Brudvig, V. S. Batista, Y. Liang, Z. Feng and H. Wang, *Nat. Commun.*, 2018, **9**, 415.
88. D. Karapinar, A. Zitolo, T. N. Huan, S. Zanna, D. Taverna, L. H. Galvão Tizei, D. Giaume, P. Marcus, V. Mougel and M. Fontecave, *ChemSusChem*, 2020, **13**, 173–179.
89. C. E. Creissen and M. Fontecave, *Nat. Commun.*, 2022, **13**, 2280.
90. V. Artero and M. Fontecave, *Chem. Soc. Rev.*, 2013, **42**, 2338–2356.
91. Y. Hou, B. L. Abrams, P. C. K. Vesborg, M. E. Björketun, K. Herbst, L. Bech, A. M. Setti, C. D. Damsgaard, T. Pedersen, O. Hansen, J. Rossmeisl, S. Dahl, J. K. Nørskov and I. Chorkendorff, *Nat. Mater.*, 2011, **10**, 434–438.
92. G. F. Moore and I. D. Sharp, *J. Phys. Chem. Lett.*, 2013, **4**, 568–572.
93. D. Khusnutdinova, A. M. Beiler, B. L. Wadsworth, S. I. Jacob and G. F. Moore, *Chem. Sci.*, 2016, **8**, 253–259.
94. D. Cedeno, A. Krawicz, P. Doak, M. Yu, J. B. Neaton and G. F. Moore, *J. Phys. Chem. Lett.*, 2014, **5**, 3222–3226.
95. A. M. Beiler, D. Khusnutdinova, B. L. Wadsworth and G. F. Moore, *Inorg. Chem.*, 2017, **56**, 12178–12185.
96. B. L. Wadsworth, N. P. Nguyen, D. Nishiori, A. M. Beiler and G. F. Moore, *ACS Appl. Energy Mater.*, 2020, **3**, 7512–7519.
97. A. M. Beiler, D. Khusnutdinova, S. I. Jacob and G. F. Moore, *ACS Appl. Mater. Interfaces*, 2016, **8**, 10038–10047.
98. T. Arai, S. Sato, K. Uemura, T. Morikawa, T. Kajino and T. Motohiro, *Chem. Commun.*, 2010, **46**, 6944–6946.
99. S. Sato, T. Arai, T. Morikawa, K. Uemura, T. M. Suzuki, H. Tanaka and T. Kajino, *J. Am. Chem. Soc.*, 2011, **133**, 15240–15243.

100. A. Paracchino, V. Laporte, K. Sivula, M. Grätzel and E. Thimsen, *Nat. Mater.*, 2011, **10**, 456–461.
101. M. Schreier, P. Gao, M. T. Mayer, J. Luo, T. Moehl, M. K. Nazeeruddin, S. D. Tilley and M. Grätzel, *Energy Environ. Sci.*, 2015, **8**, 855–861.
102. M. Schreier, J. Luo, P. Gao, T. Moehl, M. T. Mayer and M. Grätzel, *J. Am. Chem. Soc.*, 2016, **138**, 1938–1946.
103. J. J. Leung, J. Warnan, D. H. Nam, J. Z. Zhang, J. Willkomm and E. Reisner, *Chem. Sci.*, 2017, **8**, 5172–5180.
104. S. Roy, M. Miller, J. Warnan, J. J. Leung, C. D. Sahm and E. Reisner, *ACS Catal.*, 2021, **11**, 1868–1876.
105. J. J. Leung, J. A. Vigil, J. Warnan, E. Edwardes Moore and E. Reisner, *Angew. Chem., Int. Ed.*, 2019, **58**, 7697–7701.
106. S. Chandrasekaran, N. Kaeffer, L. Cagnon, D. Aldakov, J. Fize, G. Nonglaton, F. Baleras, P. Mailley and V. Artero, *Chem. Sci.*, 2019, **10**, 4469–4475.
107. L. Gong, P. Zhang, G. Liu, Y. Shan and M. Wang, *J. Mater. Chem. A*, 2021, **9**, 12140–12151.
108. C. Tapia, E. Bellet-Amalric, D. Aldakov, F. Boudoire, K. Sivula, L. Cagnon and V. Artero, *Green Chem.*, 2020, **22**, 3141–3149.
109. P. B. Pati, R. Wang, E. Boutin, S. Diring, S. Jobic, N. Barreau, F. Odobel and M. Robert, *Nat. Commun.*, 2020, **11**, 1–9.
110. V. Andrei, R. L. Z. Hoye, M. Crespo-Quesada, M. Bajada, S. Ahmad, M. De Volder, R. Friend and E. Reisner, *Adv. Energy Mater.*, 2018, **8**, 1801403.
111. V. Andrei, B. Reuillard and E. Reisner, *Nat. Mater.*, 2020, **19**, 189–194.
112. C. M. Gabardo, C. P. O'Brien, J. P. Edwards, C. McCallum, Y. Xu, C. T. Dinh, J. Li, E. H. Sargent and D. Sinton, *Joule*, 2019, **3**, 2777–2791.
113. J. E. Huang, F. Li, A. Ozden, A. S. Rasouli, F. P. G. de Arquer, S. Liu, S. Zhang, M. Luo, X. Wang, Y. Lum, Y. Xu, K. Bertens, R. K. Miao, C. T. Dinh, D. Sinton and E. H. Sargent, *Science*, 2021, **372**, 1074–1078.
114. F. P. García de Arquer, C. T. Dinh, A. Ozden, J. Wicks, C. McCallum, A. R. Kirmani, D. H. Nam, C. Gabardo, A. Seifitokaldani, X. Wang, Y. C. Li, F. Li, J. Edwards, L. J. Richter, S. J. Thorpe, D. Sinton and E. H. Sargent, *Science*, 2020, **367**, 661–666.
115. T. Burdyny and W. A. Smith, *Energy Environ. Sci.*, 2019, **12**, 1442–1453.
116. S. Ren, D. Joulié, D. Salvatore, K. Torbensen, M. Wang, M. Robert and C. P. Berlinguette, *Science*, 2019, **365**, 367–369.
117. T. A. Kistler, N. Danilovic and P. Agbo, *J. Electrochem. Soc.*, 2019, **166**, H656–H661.
118. T. A. Kistler, D. Larson, K. Walczak, P. Agbo, I. D. Sharp, A. Z. Weber and N. Danilovic, *J. Electrochem. Soc.*, 2019, **166**, H3020–H3028.
119. T. A. Kistler, M. Y. Um and P. Agbo, *J. Electrochem. Soc.*, 2020, **167**, 066502.
120. T. A. Kistler, M. Y. Um, J. K. Cooper, I. D. Sharp and P. Agbo, *Adv. Energy Mater.*, 2021, **11**, 2100070.
121. J. H. Montoya, L. C. Seitz, P. Chakthranont, A. Vojvodic, T. F. Jaramillo and J. K. Nørskov, *Nat. Mater.*, 2017, **16**, 70–81.
122. H. L. Wu, X. B. Li, C. H. Tung and L. Z. Wu, *Adv. Sci.*, 2018, **5**, 1700684.

123. V. Nikolaou, A. Charisiadis, G. Charalambidis, A. G. Coutsolelos and F. Odobel, *J. Mater. Chem. A*, 2017, **5**, 21077–21113.
124. E. A. Gibson, *Chem. Soc. Rev.*, 2017, **46**, 6194–6209.
125. B. Zhang and L. Sun, *Chem. Soc. Rev.*, 2019, **48**, 2216–2264.
126. J. V. Caspar and T. J. Meyer, *J. Am. Chem. Soc.*, 1983, **105**, 5583–5590.
127. B. Durham, J. V. Caspar, J. K. Nagle and T. J. Meyer, *J. Am. Chem. Soc.*, 1982, **104**, 4803–4810.
128. K. Kalyanasundaram, *Coord. Chem. Rev.*, 1982, **46**, 159–244.
129. R. J. Dillon, L. Alibabaei, T. J. Meyer and J. M. Papanikolas, *ACS Appl. Mater. Interfaces*, 2017, **9**, 26786–26796.
130. N. Queyriaux, R. A. Wahyuono, J. Fize, C. Gablin, M. Wächtler, E. Martinez, D. Léonard, B. Dietzek, V. Artero and M. Chavarot-Kerlidou, *J. Phys. Chem. C*, 2017, **121**, 5891–5904.
131. F. Li, K. Fan, B. Xu, E. Gabrielsson, Q. Daniel, L. Li and L. Sun, *J. Am. Chem. Soc.*, 2015, **137**, 9153–9159.
132. J. R. Swierk, D. D. Méndez-Hernández, N. S. McCool, P. Liddell, Y. Terazono, I. Pahk, J. J. Tomlin, N. V. Oster, T. A. Moore, A. L. Moore, D. Gust and T. E. Mallouk, *Proc. Natl. Acad. Sci. U. S. A.*, 2015, **112**, 1681–1686.
133. A. M. Brown, L. J. Antila, M. Mirmohades, S. Pullen, S. Ott and L. Hammarström, *J. Am. Chem. Soc.*, 2016, **138**, 8060–8063.
134. R. J. Kamire, M. B. Majewski, W. L. Hoffeditz, B. T. Phelan, O. K. Farha, J. T. Hupp and M. R. Wasielewski, *Chem. Sci.*, 2017, **22**, 32–57.
135. D. A. Hoogeveen, M. Fournier, S. A. Bonke, X. Y. Fang, A. J. Mozer, A. Mishra, P. Bäuerle, A. N. Simonov and L. Spiccia, *Electrochim. Acta*, 2016, **219**, 773–780.
136. J. T. Kirner and R. G. Finke, *J. Mater. Chem. A*, 2017, **5**, 19560–19592.
137. J. M. Gardner, M. Beyler, M. Karnahl, S. Tschierlei, S. Ott and L. Hammarström, *J. Am. Chem. Soc.*, 2012, **134**, 19322–19325.
138. M. A. Gross, C. E. Creissen, K. L. Orchard and E. Reisner, *Chem. Sci.*, 2016, **7**, 242–247.
139. B. Shan, A. Das, S. L. Marquard, B. H. Farnum, D. Wang, R. M. Bullock and T. J. Meyer, *Energy Environ. Sci.*, 2016, **9**, 3693–3697.
140. N. Kaeffer, J. Massin, C. Lebrun, O. Renault, M. Chavarot-Kerlidou and V. Artero, *J. Am. Chem. Soc.*, 2016, **138**, 12308–12311.
141. N. Kaeffer, C. D. Windle, R. Brisse, C. Gablin, D. Leonard, B. Jousselme, M. Chavarot-Kerlidou and V. Artero, *Chem. Sci.*, 2018, **9**, 6721–6738.
142. Z. Ji, M. He, Z. Huang, U. Ozkan and Y. Wu, *J. Am. Chem. Soc.*, 2013, **135**, 11696–11699.
143. T.-T. Li, B. Shan and T. J. Meyer, *ACS Energy Lett.*, 2019, **4**, 629–636.
144. N. Põldme, L. O'Reilly, I. Fletcher, J. Portoles, I. V. Sazanovich, M. Towrie, C. Long, J. G. Vos, M. T. Pryce and E. A. Gibson, *Chem. Sci.*, 2019, **10**, 99–112.
145. E. Giannoudis, S. Bold, C. Müller, A. Schwab, J. Bruhnke, N. Queyriaux, C. Gablin, D. Leonard, C. Saint-Pierre, D. Gasparutto, D. Aldakov, S. Kupfer, V. Artero, B. Dietzek and M. Chavarot-Kerlidou, *ACS Appl. Mater. Interfaces*, 2021, **13**, 49802–49815.

146. S. Lyu, J. Massin, M. Pavone, A. B. Muñoz-García, C. Labrugère, T. Toupance, M. Chavarot-Kerlidou, V. Artero and C. Olivier, *ACS Appl. Energy Mater.*, 2019, **2**, 4971–4980.
147. J. He, H. Lindström, A. Hagfeldt and S.-E. Lindquist, *J. Phys. Chem. B*, 1999, **103**, 8940–8943.
148. L. D'Amario, J. Föhlinger, G. Boschloo and L. Hammarström, *Chem. Sci.*, 2017, **9**, 223–230.
149. C. E. Creissen, J. Warnan, D. Antón-García, Y. Farré, F. Odobel and E. Reisner, *ACS Catal.*, 2019, **9**, 9530–9538.
150. M. F. Kuehnel, C. E. Creissen, C. D. Sahm, D. Wielend, A. Schlosser, K. L. Orchard and E. Reisner, *Angew. Chem., Int. Ed.*, 2019, **58**, 5059–5063.
151. C. E. Creissen, J. Warnan and E. Reisner, *Chem. Sci.*, 2018, **9**, 1439–1447.
152. H. Kumagai, G. Sahara, K. Maeda, M. Higashi, R. Abe and O. Ishitani, *Chem. Sci.*, 2017, **8**, 4242–4249.
153. C. D. Windle, H. Kumagai, M. Higashi, R. Brisse, S. Bold, B. Jousselme, M. Chavarot-Kerlidou, K. Maeda, R. Abe, O. Ishitani and V. Artero, *J. Am. Chem. Soc.*, 2019, **141**, 9593–9602.
154. D. Xiong, W. Zhang, X. Zeng, Z. Xu, W. Chen, J. Cui, M. Wang, L. Sun and Y. B. Cheng, *ChemSusChem*, 2013, **6**, 1432–1437.
155. Z. Xu, D. Xiong, H. Wang, W. Zhang, X. Zeng, L. Ming, W. Chen, X. Xu, J. Cui, M. Wang, S. Powar, U. Bach and Y. B. Cheng, *J. Mater. Chem. A*, 2014, **2**, 2968–2976.
156. G. Sahara, H. Kumagai, K. Maeda, N. Kaeffer, V. Artero, M. Higashi, R. Abe and O. Ishitani, *J. Am. Chem. Soc.*, 2016, **138**, 14152–14158.
157. D. A. Hoogeveen, M. Fournier, S. A. Bonke, A. Nattestad, A. Mishra, P. Bäuerle, L. Spiccia, A. J. Mozer and A. N. Simonov, *J. Phys. Chem. C*, 2017, **121**, 25836–25846.
158. B. Shan, S. Vanka, T. T. Li, L. Troian-Gautier, M. K. Brennaman, Z. Mi and T. J. Meyer, *Nat. Energy*, 2019, **4**, 290–299.
159. O. Jung, M. L. Pegis, Z. Wang, G. Banerjee, C. T. Nemes, W. L. Hoffeditz, J. T. Hupp, C. A. Schmuttenmaer, G. W. Brudvig and J. M. Mayer, *J. Am. Chem. Soc.*, 2018, **140**, 4079–4084.

CHAPTER 7

Recent Advances in the Development of Metal–Organic Frameworks-based Photocatalysts for Water Splitting and CO_2 Reduction

EHIAGHE AGBOVHIMEN ELIMIAN[†a, b], AYAT N. EL-SHAZLY[†c], MAHMOUD A. HAMZA[†d, e], RAMADAN A. GEIOUSHY*[c], JAFAR ALI[f, g], AYMAN N. SABER[h], PEIDONG SU[i], OSAMA A. FOUAD[c], WAHEED IQBAL[j], PHUONG NGUYEN-TRIK[k] AND RIDHA DJELLABI*[l]

[a]Department of Chemical Engineering, Faculty of Science and Engineering, University of Nottingham, Ningbo 315100, China; [b]Department of Plant Biology and Biotechnology, University of Benin, Nigeria; [c]Central Metallurgical Research and Development Institute, P.O. Box 87, Helwan, Cairo, Egypt; [d]Chemistry Department, Faculty of Science, Ain Shams University, Abbasia, Cairo, Egypt; [e]School of Chemical Engineering and Advanced Materials, The University of Adelaide, Adelaide, South Australia 5005, Australia; [f]Department of Biochemistry, University of Sialkot, Pakistan; [g]College of New Energy and Environment, Jilin University, Changchun, 130021, China; [h]Pesticide Residues and Environmental Pollution Department, Central Agricultural Pesticide Laboratory, Agricultural Research Center, Dokki, Giza 12618, Egypt; [i]School of Chemical & Environmental Engineering, China University of Mining & Technology (Beijing), Beijing 100083, China;

[†]These authors contributed equally to this work.

Sustainable Energy No. 1
Recent Developments in Functional Materials for Artificial Photosynthesis
Edited by Srabanti Ghosh and Qian Wang

Published by the Royal Society of Chemistry, www.rsc.org

[j]MIIT Key Laboratory of Special Functional and Intelligent Polymer Materials, MOE Key Laboratory of Supernormal Material Physics and Chemistry, Shaanxi Key Laboratory of Polymer Science and Technology, Department of Chemistry, School of Chemistry and Chemical Engineering, Northwestern Polytechnical University, Xi'an, 710129, PR China; [k]Laboratory of Advanced Materials for Energy and Environment, Université Du Québec à Trois-Rivières (UQTR), 3351, Boul. des Forges, C.P. 500, Trois-Rivières, Québec, G9A 5H7, Canada; [l]Departament d'Enginyeria Química, Universitat Rovira i Virgili, Av Països Catalans 26, 43007, Tarragona, Spain
*E-mail: r_gushy@yahoo.com, ridha.djellabi@yahoo.com

7.1 Introduction

In recent decades we have witnessed a surge in air pollution due to widespread industrial emissions. This poses a serious threat to individual health and natural ecosystems both at a global and regional level. The emission of various industrial and greenhouse gasses, *i.e.*, NO, NO_2, CO, CO_2 and CH_4, from fossil fuel consumption has long-lasting impacts on climate change and global warming.[1,2] It has been estimated that CO_2 is responsible for about 76% of the total greenhouse gasses (GHGs) released by anthropogenic activities. Recently, it was reported that the annual average concentration of CO_2 has been reported to be more than 400 ppm.[3] During the last two decades, CO_2 yield in the atmosphere has been increased by two orders of magnitude due to anthropogenic actions, *i.e.*, combustion of fossil fuels.[4] The constant rise in atmospheric CO_2 disturbs the carbon cycle and raises concerns about climate change. The impacts of global warming can be reduced by mitigating the CO_2 emission and increasing the sequestration.[2,5]

The Paris climate agreement (2015) and the G20 summit committed to lowering the average increase in global temperature by 2 °C after achieving net-zero carbon emissions.[6] A wide range of technologies is being applied to fight CO_2 pollution. Capture and storage is a classical method used to reduce net CO_2 emission.[2,7] Generally, CO_2 can be converted into various fuels through electrochemical, photocatalytic, biochemical, and mineralization processes.[8] Transforming GHGs into organic fuels can surely help in alleviating global warming and energy crises. As CO_2 can quickly react with metal and metal oxides to form carbonates it can be stored with lime within coastlines.[7] Alternatively, geological/glacier traps, polymeric foams, and metal–organic frameworks (MOFs) can store a sufficient amount of CO_2.

Chemical fixation of CO_2 into organic molecules is an energy-intensive process. Direct hydrogenation can produce CH_3OH, CH_4, and $CHOO^-$ using metal/hybrid oxides under moderate to high temperatures.[5] Moreover, some organic/inorganic reagents can also react with CO_2 to form carbonates, such as in the industrial production of urea. Electrochemical reduction of CO_2 is also an attractive approach due to ambient operating conditions.[9] High overpotential and complicated steps involved in the electrochemical reduction of carbon limit the long-term stability and large-scale applications.[10] Photosynthesis is another biochemical pathway that converts photon energy into chemical energy (carbohydrates). During photosynthesis, H_2O is oxidized, and CO_2 is reduced through biocatalysis. Similarly, photon energy can be converted into chemical energy *via* photocatalysis, where light excitation creates electron–hole pairs. Photocatalysis has the advantage of plentiful light (solar) source, facile preparation of catalyst materials, and easy operation.[11,12] The inertness of CO_2 and positive Gibbs free energy ($\Delta G° = -400$ kJ mol^{-1}) demands efficient catalytic materials for the activation and transformation of CO_2.[2]

Previous studies have shown that H_2 is needed for the whole scavenging during the photoreduction of CO_2 into renewable hydrocarbons. Numerous reducing substances, *i.e.*, CH_4 and alcohols, can act as the source of H_2 to promote CO_2 photoconversion. Hydrogenation of CO_2 with H_2 is considered more feasible compared with water.[5] Thus, photocatalytic water splitting and simultaneous reduction of CO_2 will be an eco-friendly and sustainable approach.[12] The foremost challenge in mimicking natural photosynthesis is designing efficient photocatalytic reduction system with high adsorption and stability. In recent years, a lot of research has been reported on developing a new class of (photo)-catalytic materials, known as MOFs, which connecting metals species can fabricate with multi-dentate organic ligands.[13] Extremely high porosity and large surface area make MOFs ideal materials for acid–base catalysis, redox reactions, gas sorption, and asymmetric reactions.[14]

MOFs are promising photocatalysts because of their ability to be tailored toward specific properties, *i.e.*, large catalytic surface, excellent light absorption, and charges separation catalytic.[15–17] MOFs-based photocatalysts have received much attention from the scientific community in recent years for the photocatalytic conversion of CO_2[18,19] and water-splitting.[20,21] This chapter is a systematic and in-depth discussion in terms of designing and optimizing materials for CO_2 photoreduction and water-splitting processes. The mechanistic pathways for photocatalytic and photoelectrochemical CO_2 photoreduction and water-splitting are also addressed.

7.2 Principle of Photocatalytic Reduction of CO_2 and H_2 Generation

7.2.1 Photoreduction of CO_2

Rapid industrialization has resulted in severe environmental problems such as a rise in CO_2 emissions associated with climate change. The conversion of CO_2 emission into valuable organic fuels has gained considerable attention as a practical approach to combat the issue of carbon emission and energy crises.[22,23] Carbon dioxide is a linear chemical stable molecule characterized by poor electron affinity. The gaseous molecules consist of two C=O bonds, each with dissociation energy of 750 kJ mol^{-1}, which is higher than that of the C–C (336 kJ mol^{-1}), C–O (327 kJ mol^{-1}), or C–H (441 kJ mol^{-1}) bond due to the strong O–H bond interaction.[24,25] In the thermocatalysis process, high temperatures (500 K) and pressures (10 bar) are required to activate and promote CO_2 reduction. For example, the production of CO and H_2 (the syngas) is carried out at around 800 to 1000 °C *via* the dry reforming process.[26,27] However, the thermal processes are not energy-efficient due to the dependence on additional thermal energy input. Therefore, other carbon-efficient techniques have still been developed or optimized. Table 7.1 summarizes details of current techniques for CO_2 reduction.

In a general photocatalytic reaction, the energy of the conduction band must be equal to or higher than the redox potential of the given reaction to proceed. Due to the high endothermicity of the CO_2 reduction process, several wide bandgap semiconductor materials such as TiO_2, $BiVO_4$, $SrTiO_3$, ZnO, SiC, ZnS are widely applied. They can provide sufficient thermodynamic

Table 7.1 Summary of the principles, advantages, and disadvantages of the current techniques for CO_2 reduction.

Techniques	Principles	Advantages	Disadvantages	Ref.
Thermal catalysis	The reduction of CO_2 under high temperature and pressure	High conversion efficiency	Low product selectivity Large energy consumption	22
Photocatalysis	CO_2 can be reduced directed by the photoproduced electrons on the conduction band to form hydrocarbon fuels	Sunlight to power the reaction mild operating conditions	Low selectivity of photocatalysts Low conversion rates in some cases	28
Electrocatalysis	CO_2 reduction is initiated by the electrochemically generated electrons	Mild operating condition controllable reaction process, the catalysts can be recycled	Poor selectivity of electrocatalyst Low catalytic activity Poor stability	29
Photo-electrocatalysis	The combined application of photoelectrons and electric field to catalytically reduce CO_2	High solar to fuel efficiency High product selectivity	Low solubility of CO_2 in the electrolyte solution Requires a large amount of photovoltage	30
Photo-enzyme catalysis	Enzymes, including enzyme cascades, are explored for the *in vitro* CO_2 reduction into chemical fuels	High product selectivity High conversion efficiency	Inactivation of the enzyme and light corrosion The reaction is readily reversible	

force to drive the reduction of CO_2 and H_2O oxidation.[26,31] Figure 7.1 shows a simplified scheme of the photocatalytic reduction of CO_2 under solar light. The adsorption of CO_2 is the initial step during the photocatalytic reaction. Interactions between CO_2 molecules and the photocatalyst surface can boost the conversion of CO_2 by photogenerated electrons. The surface-bound CO_2 (carbonate or $CO_2^{\bullet-}$) species exhibit a bent structure and lower LUMO levels than gaseous CO_2 are more photoactive.[26]

The light activation of chemically inert CO_2 by photogenerated electrons triggers the cascade of reactions for CO_2 reduction, which involves the generation of $CO_2^{\bullet-}$ anion firstly.[32] It is essential to point out that the transfer of a single electron from the conduction band of photocatalyst to CO_2 is thermodynamically unfavorable, as compared to multi-electrons transfer since it requires lower energy. Enhanced separation of electron–hole pairs boosts furthers the activation and conversion of CO_2. The semiconductor should have a highly negative conduction band edge for better reduction. In contrast, its valence band energy should be lesser than the redox potential of water oxidation (0.817 V *vs.* SHE in pH 7.0 aqueous solution).[31] In addition, to accelerate the photocatalytic conversion of CO_2, increasing the lifetime of photogenerated charges is highly recommended. Recombination of charges typically takes 10^{-9} s, which is faster than the reduction reaction (~10^{-3}–10^{-8} s); therefore, decreasing the time gap will result in a more effective reduction.[32,33] Photocatalytic CO_2 reduction has been studied on several semiconductor materials (such as TiO_2, Bi_2MoO_6, In_2O_3, and Ta_2O_5). The formation of heterojunction systems (type II or Z-scheme), doping or introduced defects, *i.e.*, O_2 vacancies, were found to be very effective for enhancing the photoconversion of CO_2.[31,34] After the migration of separated electrons to the surface of the photocatalyst, the reduction of CO_2 takes place to form organic products, namely CO, CH_4, and CH_3OH, *etc.* However, positive holes oxidize water into H^+ and O_2.[33] The selectivity of the generated products can be controlled by the properties of the photocatalyst and the conduction of the medium.[32]

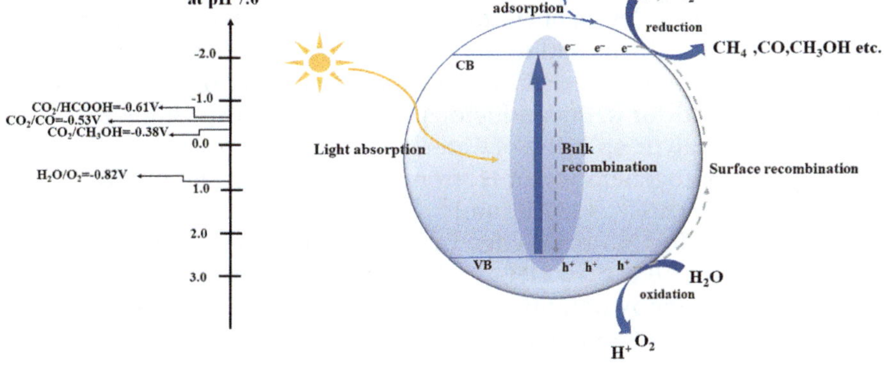

Figure 7.1 Simplified diagram showing the photoreduction process of CO_2.

However, based on the literature, CO and CH_4 are mainly produced in the gas phase system, while CH_3OH, CH_3COOH, and HCOOH are in the liquid phase.[26,35] The following equations (eqn (7.1)–(7.5)) show the possible CO_2 conversion reactions into different products.

$$CO_2 + e^- \rightarrow CO_2^{\bullet -}, \quad E^\circ = -1.90 \tag{7.1}$$

$$CO_2 + 2H^+ + 2e^- \rightarrow HCOOH, \quad E^\circ = -0.61\ V \tag{7.2}$$

$$CO_2 + 4H^+ + 4e^- \rightarrow HCHO + H_2O, \quad E^\circ = 0.43\ V \tag{7.3}$$

$$CO_2 + 6H^+ + 6e^- \rightarrow CH_3OH + H_2O, \quad E^\circ = -0.38\ V \tag{7.4}$$

$$CO_2 + 8H^+ + 8e^- \rightarrow CH_4 + 2H_2O, \quad E^\circ = -0.24\ V \tag{7.5}$$

7.2.2 Photocatalytic Hydrogen Production

Recent research efforts have been focused on developing solar hydrogen as an alternative fuel, prompted by environmental concerns and a desire to replace nuclear and fossil energy utilization with regenerative energy.[36] Although limited in volume produced, the photocatalytic generation of hydrogen through water splitting is considered an alternative and renewable method for simultaneous generation of hydrogen and wastewater treatment.[37] Several semiconductor catalysts such as TiO_2,[38,39] CdS,[40] $BiVO_4$,[41] g-C_3N_4,[38,42,43] *etc.*, have been applied for photocatalytic hydrogen generation. The modification of photocatalysts by doping, photosensitizing and heterojunction systems has been widely reported to enhance the yield of produced H_2. The photocatalytic formation of H_2 is realized by the reduction of protons *via* the photogenerated electrons, while the positive holes oxidize water molecules.[44,45] Figure 7.2 shows a simple scheme for the photocatalytic generation of H_2. The bandgap and the levels of conduction and valence bands play an essential role in successfully initiating the photogeneration of H_2. Therefore, due to the additional potential required to transfer an electron, the theoretical bandgap proposed for semiconductor photocatalysts is around 2.0 eV for optimal photocatalytic splitting of water.[46]

During the photogeneration of H_2 from water splitting, an uphill thermodynamic reaction ($\Delta G^\circ = +237$ kJ mol^{-1}) involves a four-electron transfer. In addition, the conduction band edge must be more negative than the redox potential of H^+/H_2 (0.0 V *vs.* NHE), while the valence band must be more positive than the redox potential O_2/H_2O (1.23 V *vs.* NHE). Therefore, it is recommended to use semiconductors with bandgaps of around 1.23 eV. In addition, unwanted reactions, *i.e.*, the generation of H_2O molecules from H_2 and O_2, should be avoided to obtain a stable H_2 production.

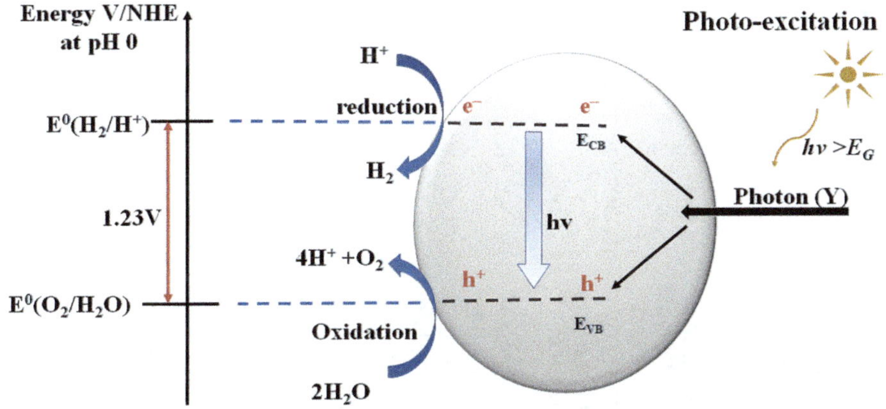

Figure 7.2 The scheme shows the photocatalytic generation of H_2 from water splitting.

7.3 MOF-based Materials for Photocatalytic CO_2 Reduction

The application of solar energy for the catalytic reduction of CO_2 into high value-added fuels such as $HCOO^-$, CO, CH_3OH, and CH_4 under mild reaction conditions is a promising route to address energy and environmental issues.[47,48] The process, however, remains challenging due to the linear structure and stability of the oxidation state of CO_2 ($\Delta_f G_{298}{}^{\circ} = -394.36$ kJ mol^{-1}). Typically, pristine semiconductor materials are characterized by high-rate e^-/h^+ pairs recombination, resulting in a lower CO_2 reduction process.[49,50] It is highly desirable to develop a photocatalyst that promotes photogenerated charge separation, reduces the activation energy, and accelerates the reduction rate. MOFs have emerged as two-or three-dimensional porous crystalline photocatalyst materials fabricated by metal ions/clusters coordinated with organic bridging ligands.[49,51] The diversity and functionality of various organic species, ligands, inorganic metal ions, and metal clusters applied in the fabrication process provide MOF materials with exceptional textural properties. A large number of MOF-based materials have been reported (around 20 000), each having its own set of physical, optical, and catalytic properties.[52–54] MOFs possess an incredibly high surface area (>1000 m^2 g^{-1}) and tunable pore sizes (≤9.8 nm).[55] The tailorable porosity promotes the transfer and proximity of reactants molecules to the active sites throughout the framework.[56–59] The high specific surface area promotes the adsorption of reactants on the catalytic surface, favoring the acceleration of subsequent reactions steps.[60] Therefore, MOFs promise porous materials, especially in catalysis. Table 7.2 lists the recent advancements on the synthesis of MOFs with active metal clusters for CO_2 reduction.

Table 7.2 MOFs with active metal clusters for photocatalytic CO_2 reduction.[a]

MOF photocatalyst	Active metal cluster	Organic metal linker	CO_2 adsorption capacity/ $cm^3\ g^{-1}$	Sacrificial agent	Synthetic method	Reaction conditions (catalyst weight, solvent, light source, irradiation, time)	Product	Yield/ μmol $g^{-1}\ h^{-1}$	Ref.
Fe-Modified MOFs									
MIL-101(Fe)	Fe–O	BDC	26.4[b]	TEOA	Hydrothermal	50 mg, MeCN–TEOA (5 : 1), 300 W Xe lamp ($\lambda > 300$ nm), 8 h	$HCOO^-$	59	72
MIL-53(Fe)	Fe–O	BDC	13.5[b]	TEOA	Hydrothermal	50 mg, MeCN–TEOA (5 : 1), 300 W Xe lamp ($\lambda > 300$ nm), 8 h	$HCOO^-$	29.7	72
NH_2-modified MOFs									
NH_2-MIL-125(Ti)	Ti–O cluster	NH_2-BDC	132.2[b]	TEOA	Post-synthetic ion-exchange	50 mg, MeCN–TEOA (5:1), Xe lamp, 10 h	$HCOO^-$	8.14	73
NH_2-Uio-66(Zr)	Zr–O cluster	DTA	68[b]	TEOA	Post-synthetic ion-exchange	50 mg, MeCN–TEOA (5 : 1), 300 W Xe lamp ($\lambda > 420$ nm),10 h	$HCOO^-$	3.4	78
NH_2-MIL-101(Fe)	Fe–O	NH_2-BDC	34.0[b]	TEOA	Post-synthetic ion-exchange	50 mg, MeCN–TEOA (5 : 1), 300 W Xe lamp ($\lambda > 420$ nm), 8 h	$HCOO^-$	178	72
NH_2-MIL-53(Fe)	Fe–O	NH_2-BDC	20.0[b]	TEOA	Post-synthetic ion-exchange	50 mg, MeCN–TEOA (5 : 1), 300 W Xe lamp ($\lambda > 420$ nm), 8 h	$HCOO^-$	46.5	72
NH_2-UiO-66(Zr/Ti)-120-16	Zr–O cluster Ti–O cluster	NH_2-BCD	80[b]	TEOA	Post-synthetic ion-exchange	50 mg, MeCN–TEOA (5 : 1), 300 W Xe lamp ($\lambda > 420$ nm), 10 h	$HCOO^-$	5.8	78

Carboxylate porphyrinic MOFs									
PCN-222	Zr–O	H2TCPP ligand	58[b]	TEOA	Solvothermal	50 mg in the MeCN–TEOA Xe 420–800 nm, 10 h	$HCOO^-$	30	79
PCN-601	Ni–O cluster	*N*,*N*-Diethyl-1,4-phenylene-diamine (DPD)	82.4[b]	TEOA	Solvothermal	5 mg, simulated solar irradiation (AM 1.5 G)	CH_4	92.0	80
Photosensitizer-MOFs composite									
Co-ZIF-9	Co–O cluster	$[Ru(bpy)_3]Cl_2·6H_2O$ (bpy = 2′2-bipyridine)	—	TEOA	—	1 mg, acetonitrile–H_2O (4 : 1), 300 W Xe lamp (λ > 420 nm), 2.5 h	CO	41.8	81
Co-MOF	Co–O cluster	HL^{3-} ligand	—	BNAH	Solvothermal	4 mg, 5 mL DMAc, 300 W Xe lamp (λ > 400 nm), 6 h	$HCOO^-$	456	82
Co-ZIF-67	Co–O cluster	$[Ru(bpy)_3]Cl_2·6H_2O$ (bpy = 2′2-bipyridine)	29[b]	TEOA	Co-precipitation	8 mg, TEOA–water–acetonitrile (1 : 2: 3), 300 W Xe lamp (λ > 400 nm), 0.5 h	CO	37.4	83
Co_6-MOF	Co 6 (μ3 -OH) 6	$[Ru(bpy)_3]Cl_2·6H_2O$ (bpy = 4,4′-bipyridine) NTB	55.24[b] 38.17[c]	TEOA	Solvothermal	TEOA–water–acetonitrile (1 : 1 : 4), 150 W Xe lamp at 420 ≤ λ ≤ 780) 3 h	CO	39.36	84

[a]BDC: 1,4-benzenediacarboxylate; NH_2-BDC: 2-amino-benzene-1,4-dicarboxylate; $(NH_2)_2$-BDC: 2,5-diaminobenzene-1,4-dicarboxylic acid; BPDC: 4,4′-biphenyldicarboxylic acid; DTA: 2,5-diaminoterephthalate; DPD: *N*,*N*-diethyl-1,4-phenylene-diamine (DPD); H_2TCPP: tetrakis(4-carboxyphenyl)porphyrin (H_2TCPP).

[b]CO_2 adsorption capacity of MOFs measured at 273 K and 1 atm.

[c]CO_2 adsorption capacities of MOFs measured at 298 K and 1 atm.

Unlike porous materials, *i.e.*, zeolites, that lack the capacity to be excited upon light irradiation, MOFs possess light absorption ability and generate photo charge carriers by exciting the metal nodes or enhancing charge separation *via* the ligand-to-metal charge transfer (LMCT).[61] In 2007 it was discovered that MOF-5 could be excited under light and underwent charge separation to generate electrons and holes in microseconds.[62] However, when compared to common semiconductor materials, the photoactivity of MOFs is less efficient. Thus, significant efforts have been made to develop novel procedures for the fabrication and functionalization of MOF materials for enhanced photoactivity. Metal complexes as visible-light sensitizers have been widely connected with MOFs materials to activate MOFs materials or/and act as reductive sites for CO_2 conversion.[63–65] By varying the organic ligand composition/metals and synthesis methods, MOFs can be tailored for efficient solar energy utilization, which is superior to semiconductor-based photocatalysts. Li and co-workers reported the enhanced performance of $Cu_3(BTC)_2$@TiO_2 under UV illumination. The $Cu_3(BTC)_2$@TiO_2 achieved a CH_4 yield over five-folds superior to pure TiO_2 and $Cu_3(BTC)_2$. Although the bare $Cu_3(BTC)_2$ could absorb UV light, its conjugated structure suppressed charge separation, resulting in low photocatalytic reduction efficiency.[64] Crake and colleagues reported that MOFs based on NH_2-UiO-66 coated with TiO_2 nanosheets effectively capture CO_2 photoreduction under UV-visible light irradiation. The strategy permitted the formation of a Type II heterojunction between TiO_2 and NH_2-UiO-66, which in turn boosts the separation of e^-/h^+ charges, resulting in enhanced CO_2 photoreduction of 1.9-fold superior to TiO_2.[66] Another study reported that, upon visible light irradiation, O-ZnO/rGO/UiO-66-NH_2 Z-scheme heterojunction promotes the production of a high yield of CH_3OH and HCOOH compared to O-ZnO/UiO-66-NH_2 and ZnO/rGO/UiO-66-NH_2.[67] For enhanced charge separation and CO_2 capturing, Kong *et al.*[68] have fabricated a Core@Shell $CsPbBr_3$ ZIF-67-based photocatalyst (Figure 7.3). The authors report that Co centers in ZIF-67 can boost the generation of photoproduced charges and improves the adsorption of CO_2 for enhanced CO_2 reduction into CO and CH_4. Idris *et al.*[69] have functionalized Fe-MOF nanosheets with visible-light-responsive $[Ru(bpy)]_3^{2+}$. A high photoactivity CO_2-to-CO conversion of 1120 μmol g^{-1} h^{-1} was recorded due to the high yield of electrons generation and separation, as shown in Figure 7.4.

The exceptional textural property of MOFs makes them very effective for CO_2 photoreduction. To develop MOF-based photocatalysts with high efficiency, strong light absorption ability is essential. The CO_2 reduction phenomena imitate the natural photosynthesis of green plants. The ability of photocatalyst to undergo electronic excitation under solar light irradiation is highly desirable. To achieve efficient photocatalytic CO_2 reduction, the MOF-based photocatalyst should possess the three critical requirements: strong light absorption ability, effective charges separation and surface transfer of photogenerated charge carriers.[70]

The redox-active metal clusters are promising targets for enhanced visible light-harvesting of MOF-based photocatalysts. It was reported that MOFs

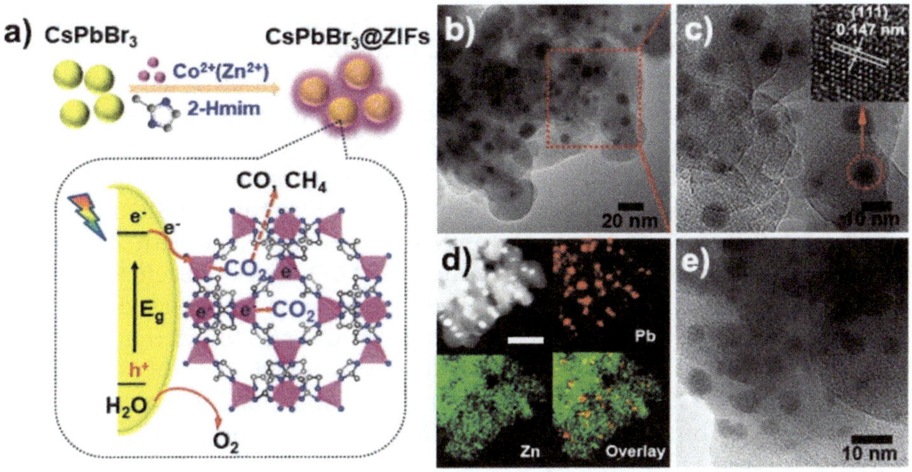

Figure 7.3 Synthesis process and CO_2 photoreduction process of $CsPbBr_3$/ZIFs. (b,c) Low- and high-magnification TEM images of $CsPbBr_3$@ZIF-8. The inset of (c) is the high-resolution image of a single $CsPbBr_3$ QD marked with a red circle. (d) High-angle annular dark-field STEM image and elemental mapping of $CsPbBr_3$@ZIF-8. Scale bar: 50 nm. (e) TEM image of $CsPbBr_3$@ZIF-67. Reproduced from ref. 68 with permission from American Chemical Society, Copyright 2018.

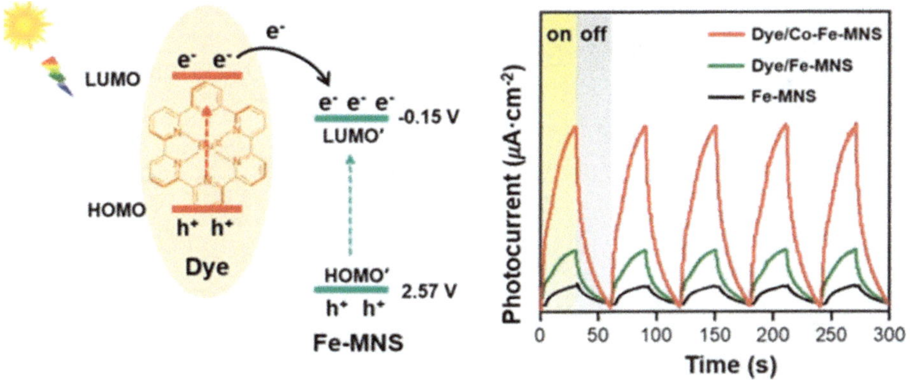

Figure 7.4 Diagram of Dye/Fe-MNS and photocurrent of samples. Reproduced from ref. 69 with permission from Elsevier, Copyright 2021.

prepared by Ti^{4+}, Zr^{4+}, Zn^{2+}, and Cu^{2+} metallic cations could reduce CO_2 under UV light.[71] The Fe-based MOFs photocatalyst consists of the iron oxo clusters (Fe–O clusters), excitable under visible light. The Fe-MOF series exhibited good photoreduction of CO_2 into $HCOO^-$ with yields of 59.0, 29.7, and 9.0 μmol for MIL-101(Fe), MIL-53(Fe), and MIL-88B(Fe), respectively.[72] It was claimed that Fe–O clusters can be excited effectively by visible light, then it initiates the transfer of electrons from O^{2-} to Fe^{3+}, resulting in the reduction of Fe^{3+} to Fe^{2+} to participate in the CO_2 reduction.

The electron-rich organic linkers of MOF-based photocatalyst serve as photon-absorption centers for the absorption of visible light. For example, it was reported that amine-functionalized MOF NH_2-MIL-125(Ti) with titanium metal nodes could significantly improve the conversion of CO_2 because of the LMCT mechanism as proved by electron spin resonance (ESR) analysis.[59] A similar tendency was observed for NH_2-UiO-66(Zr) MOF with Zr^{3+} redox centers, which can boost the photoreduction of CO_2 in the presence of triethanolamine as a hole scavenger.[73] Defect-engineering of Zr(IV)-based MOF was reported recently by Wang and co-workers,[74] and it was found that the MOF with ligand vacancy showed the best photocatalytic reduction of CO_2 (Figure 7.5). The highest CO yield of 30.5 μmol g^{-1} h^{-1} was obtained with the best sample that exhibits 9.2 times better efficiency than missing-cluster-based

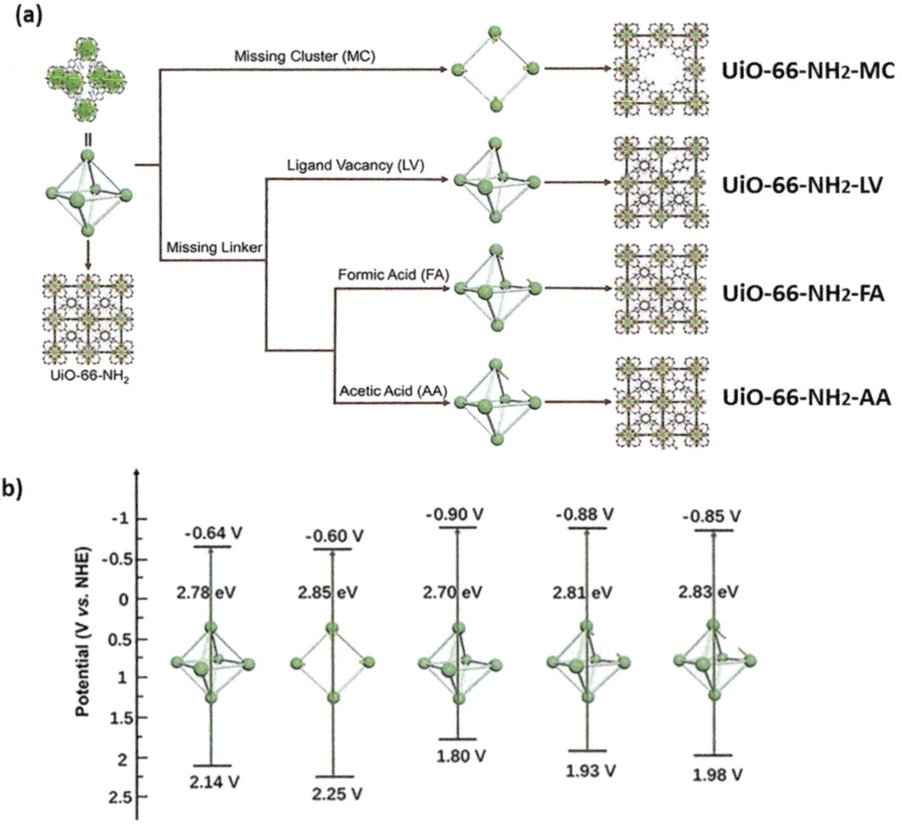

Figure 7.5 (a) A schematic diagram showing the different defective structures of Zr(IV)-based MOFs and (b) their corresponding bandgaps. Reproduced from ref. 74 with permission from Elsevier, Copyright 2021.

MOF. Inter-clusters synergy to enhance the ability of Fe-MOF towards the photoreduction of CO_2 to HCOOH was reported.[75]

Modifying MOF-based photocatalysts with plasmonic noble metals, photosensitizers, and other semiconductor materials can improve light response and photocatalyst activity.[76] The incorporation of noble metals into MOFs results in strong light absorbance due to the localized surface plasmon resonance (LSPR). The coating of Re_3-MOF onto Ag nanocubes (Ag⊂Re_3-MOF) resulted in a seven times improvement of CO_2 to CO photoreduction with long-term stability.[27] Similarly, the thin porphyrin paddle-wheel framework-3 (PPF-3) nanosheets (PPF-3_1) prepared by an electrostatic interaction display strong light absorption capacity due to the anchoring of Au nanoparticles onto the surface of the nanosheets. The Au anchored PPF-3_1 exhibits enhanced CO_2 photoreduction into HCOOH using acetonitrile–ethanol as solvent due to the LSPR caused by AuNPs. The integration of photosensitizers for the extension of light response of MOF with metalloporphyrin for enhanced visible light response has been reported to improve the photoreduction of CO_2.[77]

The charge carriers' efficient charge separation and reaction efficiency are prerequisites for effective CO_2 photoreduction. The crystalline nature of MOFs boosts the charge transfer of the photoexcited organic linkers to the metal clusters, thereby enhancing charge separation and promotion of photocatalytic efficiency as represented.[60]

The organic linker and inorganic metal ions/clusters are catalytically active and uniformly distributed throughout the framework, reducing transfer distance between the charge carriers and pore structure. The tunable pore with varying topology promotes the mass transfer during the reaction. MOFs-based photocatalyst can concentrate the yield of CO_2 gas for further photocatalytic conversion.[85–87] In addition, the organic linkers in MOF are electron-rich aromatic polycarboxylates capable of absorbing photons upon light irradiation.[88]

As illustrated in eqn (7.6), the bandgap energy (E_g) of a semiconductor material is defined as the potential difference between the conduction band (CB) and valence band (VB). However, in MOF-based photocatalyst, the VB and CB energy levels are referred to as highest occupied molecular orbitals (HOMO) and the lowest unoccupied molecular orbitals (LUMO), respectively.

$$E_g = E_C - E_V \tag{7.6}$$

Density functional theory calculations (DFT) suggest that the bandgap of MOFs is created by the difference between the energy level of the organic linkers regarded as the HOMO and the metal cluster regarded as the LUMO. It indicates that the alignment of both the organic linkers and metal clusters influences MOFs' electronic properties.[89] Based on these alignments, the oxidation reactions occur on organic linkers, while reduction reactions occur on the metal clusters. After attaining the charge-separation stare, the photogenerated electrons diffuse from the HOMO to the catalytic centers to interact with the adsorbed CO_2 and initiate the photocatalytic reaction.[90]

Based on the above requirements regarding visible-light absorption and efficient charge separation, the single component photocatalyst such as TiO_2 displays photoconversion efficiency, thereby limiting practical applications. The rod-like Co-ZIF-9 doped with Ag nanoparticles (≤5 nm) displayed a high photocatalytic reduction of CO_2 into CO under visible light. The Ag nanoparticles efficiently enhanced the photocatalyst reactivity by more than two-fold.[91] The integration of MOFs with inorganic semiconductors has been demonstrated to significantly boost the photocatalytic efficiency for CO_2 reduction. The integration of TiO_2 into MOF materials has reportedly enhanced photocatalytic performance. The hybrid material $Cu_3(BTC)_2$@TiO_2 achieves a high photocatalytic reduction of CO_2 to CH_4 under UV irradiation. The integration of TiO_2 with the HKUST-1 resulted in enhanced CO_2 selectivity and reactivity *via* the generation and transfer of photoelectrons for the TiO_2 to the MOF microporous core.[64] Table 7.3 showed the recent devolvement of MOFs-based composites for photocatalytic CO_2 reduction.

MOF-based materials are still advantageous due to the high reusability of the composites compared to transition metal oxides. Despite the extensive efforts reported in this area, the activities of MOF materials suffer from several limitations. First, the facile synthesis of highly visible light-responsive MOF-based for CO_2 reduction is still a challenge.[28,105,106] The poor stability of MOFs-based materials under high temperature or relative humidity is another factor that limits practical application. The structural collapse and decarboxylation of organic linkers have been observed under prolonged irradiation.[107] MOFs structure under long-time stability tests has been reported. Moreover, under humid conditions, water molecules interact with the metal-coordination sites and break the oxo-cluster bond, resulting in the framework's collapse.[108–110] MOFs should be prepared by both the visible light response and the hydro-thermal stability.

7.4 MOF-based Materials for Photocatalytic Water Splitting

Semiconductor-assisted photocatalysis is an ideal economic approach for water splitting *via* permanently accessible sunlight *via* photoelectrochemical (PEC) or photochemical (PC) approaches.[111–115] Numerous MOF-based photocatalysts have been investigated as potential photoactive materials towards water-splitting.[116–118] In addition, the commonly used semiconductors, such as TiO_2 and ZnO, have significant disadvantages. These include their considerable bandgap energy, excluding their photoactivity under UV only and the rapid e–h recombination causing a significant reduction of their photocatalytic activities.[119,120] However, MOFs showed excellent photoactivity due to their characteristic high surface area, rich porosity, inherited structural versatility and diverse chemical compositions, which facilitate the photogenerated charges separation.[121] However, the photocatalytic performance of MOFs could be boosted by editing the metallic cluster and/or

Table 7.3 MOF composites for photocatalytic CO_2 reduction.

MOF	Active metal	Synthetic method	Product	Yield/μmol g^{-1} h^{-1}	Ref.
Pt/NH_2-MIL125(Ti)	Pt	Impregnation	Formate	12.96	92
Au/NH_2-MIL-125(Ti)	Au	Impregnation	Formate	9.06	92
Au@PtAg@ZIF-8	Au@PtAg	*In situ* growth	CH_4, CO	1.2, 14.5	93
Au@Ag@ZIF-8	Au@Ag	*In situ* growth	CH_4, CO	0.95, 13.4	93
Au@ZIF-8	Au	*In situ* growth	CH_4, CO	0.75, 13.2	93
Au@Pd@MOF-74	Au@Pd	Ultrasonic synthesis	CO	12.31	94
Pt/MOF-74	Pt	Ultrasonic synthesis	CH_4, CO	9.04, 8.85	94
Pt/Au@Pd@MOF-74	Pt/Au@Pd	Ultrasonic synthesis	CH_4, CO	12.35, 2.42	94
CdS/ZIF-8	CdS	*In situ* deposition	CO	32.13	95
Co-ZIF-9/TiO_2	TiO_2	*In situ* growth	CH_4, CO	1.30, 0.99	96
ZIF-67@a-TiO_2	Amorphous TiO_2	Solvothermal	CO	43.8	97
TiO_2/HKUST-1	TiO_2 NPs	Step-by-step growth	CH_4	1.05	98
TiO_2/ZIF-8	Grid-like TiO_2	*In situ* growth	CH_4, CO	108, 530	99
BIF-20@g-C_3N_4	g-C_3N_4	Electrostatic assembly	CH_4	53.869	100
2DT-AuCu-ZIF-8	TiO_2, AuCu	Spin-coating, *in situ* growth	CO, CH_4	86.9	101
Fe–N-TiO_2/CPO-Cu-27	Fe,N co-doped TiO_2	*In situ* method	CH_4, CH_3OH	48.23, 2.19	102
ZnO/rGO/UiO-66-NH_2	ZnO, reduced graphene	Solvothermal	CH_3OH, formate	14.86, 4.20	67
O-ZnO/rGO/UiO-66-NH_2	ZnO, reduced graphene	Solvothermal	CH_3OH, formate	34.83, 6.41	67
O-ZnO/UiO-66-NH_2	ZnO, graphene	Solvothermal	CH_3OH, formate	9.67, 4.94	67
N-CNDs/NH_2-UiO-66	N-Carbon nanodots	Solvothermal	Formate HCOOH	~10.5	103
rGO@NH_2-MIL-125	rGO	Solvothermal	Methyl formate	1.116	104

the organic linkers or being combined with other metals, semiconductors, and/or carbon materials. Consequently, MOFs could substantially enhance the charge separation efficiency and improve and boost the photocatalytic activity toward water-splitting.[118,122] Therefore, this section will demonstrate the latest studies in applying MOF-based photocatalysts in water splitting. Furthermore, the different mechanisms of photocatalytic enhancement will also be discussed.

Generally, the construction of a hybrid structure or heterojunction can be considered as one of the most practical strategies to improve the charge separation efficiency *via* providing various charge transfer pathways rather than the recombination pathway. These can be categorized into four types: (1) Type I heterojunction; (2) Type II heterojunction; (3) Z-scheme heterojunction, and (4) Schottky junction, as demonstrated in Figure 7.6.[123] Several studies investigated the mixing of MOFs with other semiconductors and/or metals to form such heterojunction to enhance separation of charges and improve the photocatalytic performance toward water-splitting. Some of the recent attempts are listed in Table 7.4.

To begin with, L. Yu *et al.*[124] proposed a facile hydrothermal fabrication of a robust photocatalyst of TiO_2-MOF nanoribbons (Figure 7.7(a)) with unsaturated metal sites for PEC water splitting. The constructed Ni-MOF/TiO_2 binary nanocomposite displayed a double improvement in the photocurrent density compared to the bare TiO_2, as displayed in Figure 7.7(b). Besides, they assumed that the superior photocatalytic performance of MOFs toward water-splitting could be ascribed to the abundance of the unsaturated metal sites at the edge faces of MOFs. Thus, they proposed a secondary coordination strategy to prove the selective quenching of the unsaturated metallic sites on a Ni-MOF with small ligands such as terephthalic acid (TPA), acetate (CH_3COO^-), phosphate (PO_4^{3-}), oxalate ($C_2O_4^{2-}$), and $EDTA^{2-}$ established

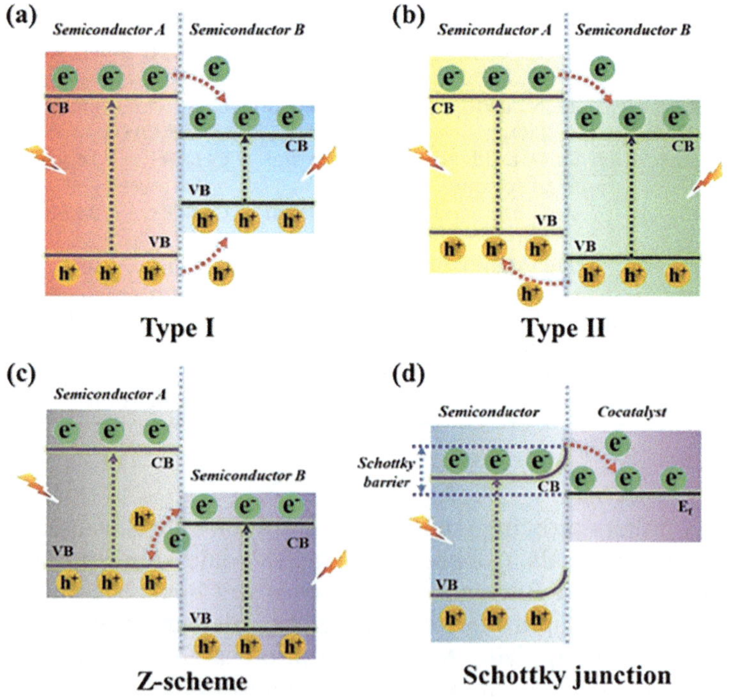

Figure 7.6 Schematic illustration of four heterojunction mechanisms [(a) Type I, (b) Type II, (c) Z-scheme, and (d) Schottky Junction] for enhancing the charge separation efficiency. Reproduced from ref. 123 with permission from Elsevier, Copyright 2020.

Table 7.4 Comparison of the photoactivity of MOF-based materials towards water splitting, expressed in photocurrent or photocatalytic hydrogen production rates.

Photocatalyst	Photocatalytic performance	Light source	Medium/ sacrificial agents	Ref.
TiO_2	0.42 mA cm^{-2}	AM 1.5 G (100 mW cm^{-2})	0.5 M Na_2SO_4 (pH = 6.8)	124
Ni-MOF/TiO_2	0.82 mA cm^{-2} at 1.23 V *vs.* RHE			
TiO_2	0.23 mA cm^{-2}	AM 1.5 G (100 mW cm^{-2})	0.5 M Na_2SO_4 (pH = 6.8)	125
NiFe-MOF/TiO_2	0.77 mA cm^{-2} at 1.23 V *vs.* RHE at 1.23 V *vs.* RHE			
Ti-MOF	1.40 mA cm^{-2}	300 W Xe lamp (336 mW cm^{-2})	Simulated seawater including 0.03 M Na_2SO_4 (pH = 6.82)	122
Ti-Nanotube arrays (TNTAs)	2.44 mA cm^{-2}			
Ti-MOF@TNTAs	3.04 mA cm^{-2} at 1.21 V *vs.* RHE			
ZnO NRs	~0.08 mA cm^{-2}	AM 1.5 G (300 mW cm^{-2})	pH = 13	126
Co-Doped ZnO (ZnO:Co)	~0.11 mA cm^{-2}			
ZnO/ZIF-8	~0.12 mA cm^{-2}			
ZnO:Co/ZIF-8	~0.16 mA cm^{-2} at 1.33 V *vs.* RHE			
ZnO	~0.22 mA cm^{-2}	AM 1.5 G (100 mW cm^{-2})	0.5 M Na_2SO_4 (pH = 6.82)	127
ZnO@ZIF-8	~0.29 mA cm^{-2}			
N-Doped C dots/ ZnO/ZIF-8	~0.46 mA cm^{-2} at 0.5 V *vs.* Ag/AgCl			
Nickel phosphonate MOF	2200 and 1700 μmol g^{-1} under UV-Vis and AM 1.5 G	150 W Xe lamp	Water–methanol	128
Cu^{II}-MOF	1.36 mmol h^{-1} g^{-1}	300 W Xe lamp (>420 nm)	5% triethylamine–water, (pH = 10)	129
Pt(4.3 wt%)/ Cu^{II}-MOF	2.51 mmol h^{-1} g^{-1}			
MOF (UIO-66)	0.025 mA cm^{-2}	400 W Xe lamp (AM 1.5 G)	0.2 M Na_2SO_4 (pH = 6.8)	130
MOF/S, N GQDs	0.050 mA cm^{-2}			
MOF/Pd	0.100 mA cm^{-2}			
MOF/Pd/S, N GQDs	0.3 mA cm^{-2}			
UiO-66 (Zr)	~30 μmol g^{-1}	Xe lamp (150 mW cm^{-2}), >420 nm	Water	131
UiO-66 (Ce)	~60 μmol g^{-1}			
UiO-66 (Zr/Ce)	~120 μmol g^{-1}			
UiO-66 (Zr/Ti)	~140 μmol g^{-1}			
UiO-66 (Zr/Ce/Ti)	~200 μmol g^{-1} after 24 h			
Fe_2O_3	0.27 mA cm^{-2}	AM 1.5 G (100 mW cm^{-2})	1 M NaOH (pH = 13.6)	132
Fe_2O_3/ZIF-67	1.75 mA cm^{-2}			
Fe_2O_3/MIL-88B	1.52 mA cm^{-2}			
Fe_2O_3/MIL-88B@ ZIF-6	2.52 mA cm^{-2} at 1.23 V *vs.* RHE			
Fe_2O_3	0.90 mA cm^{-2}	AM 1.5 G (100 mW cm^{-2})	1 M NaOH (pH = 13.6)	133
Ti:Fe_2O_3	1.41 mA cm^{-2}			
Co-MOF/Ti:Fe_2O_3	2.24 mA cm^{-2} at 1.23 V *vs.* RHE			

(*continued*)

Table 7.4 (*continued*)

Photocatalyst	Photocatalytic performance	Light source	Medium/ sacrificial agents	Ref.
Ti-Hematite (Ti-H)	0.65 mA cm^{-2}	AM 1.5 G (100 mW cm^{-2})	1 M NaOH (pH = 13.6)	134
Ti-Porpous hematite (Ti-PH)	1.42 mA cm^{-2}			
Ti-PH/Ni-F MOF	2.00 mA cm^{-2}			
Ti-PH/Ni-F MOF/ cobaloxime	2.45 mA cm^{-2} at 1.23 V *vs.* RHE			

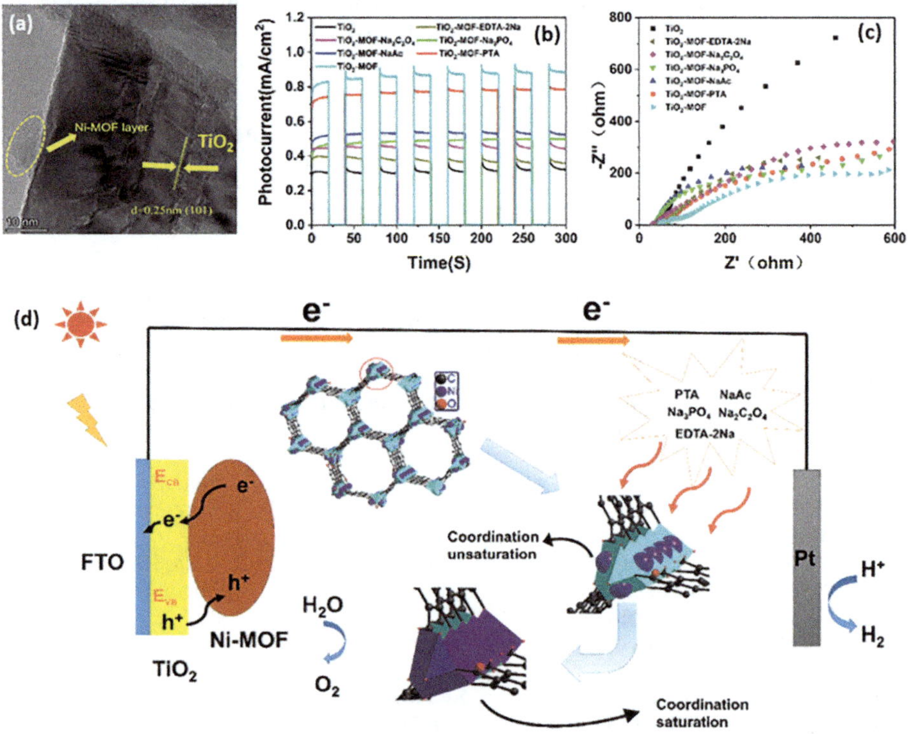

Figure 7.7 (a) HR-TEM image of TiO_2-MOF photoanode. (b) *J–t* curves under chopped illumination at 1.23 V *vs.* RHE under. (c) PEIS under illumination. (d) The proposed mechanism of enhanced performance of TiO_2-MOF photoanode toward water splitting. Reproduced from ref. 124 with permission from Elsevier, Copyright 2021.

secondary coordination toward the metallic sites of Ni. Figure 7.7(b) shows the reduction of the photocurrent densities indicating the removal of unsaturated Ni 2p sites leading to the depression of the water splitting reactions. However, the size and coordination of the various guest ligands affected the photocurrent due to their different modes of secondary coordination with the unsaturated Ni 2p sites. They found that Ni-MOF/TiO_2 displayed a superior photocatalytic performance than the pristine TiO_2 based on the enhanced charge separation as deduced from the photoelectrochemical impedance spectra (Figure 7.7(c)). Finally, they attributed the enhanced performance of Ni-MOF/TiO_2 composites due to narrowing the bandgap energies and enhancing the charge separation owing to the constructed Type-II heterojunction, as displayed in Figure 7.7(d).

Similarly, W. Cui *et al.*[125] stated the synthesis of a hybrid-photoanode composed of TiO_2 and NiFe-MOF using a sacrifice template procedure to fabricate unsaturated metallic sites on NiFe-MOF for boosting the photocatalytic efficiency toward water splitting. It was found that the photocurrent density of the hybrid photoanode was about 3.35 times that of pristine TiO_2, as demonstrated in Table 7.3. Besides, it was found that photoconversion efficiency was doubled to be 0.19% at 0.85 V *vs.* RHE (Figure 7.8(a)). The incident photon to current conversion efficiency (IPCE) was enhanced to 42% at 390 nm (Figure 7.8(b)) in addition to the substantial increase in the charge

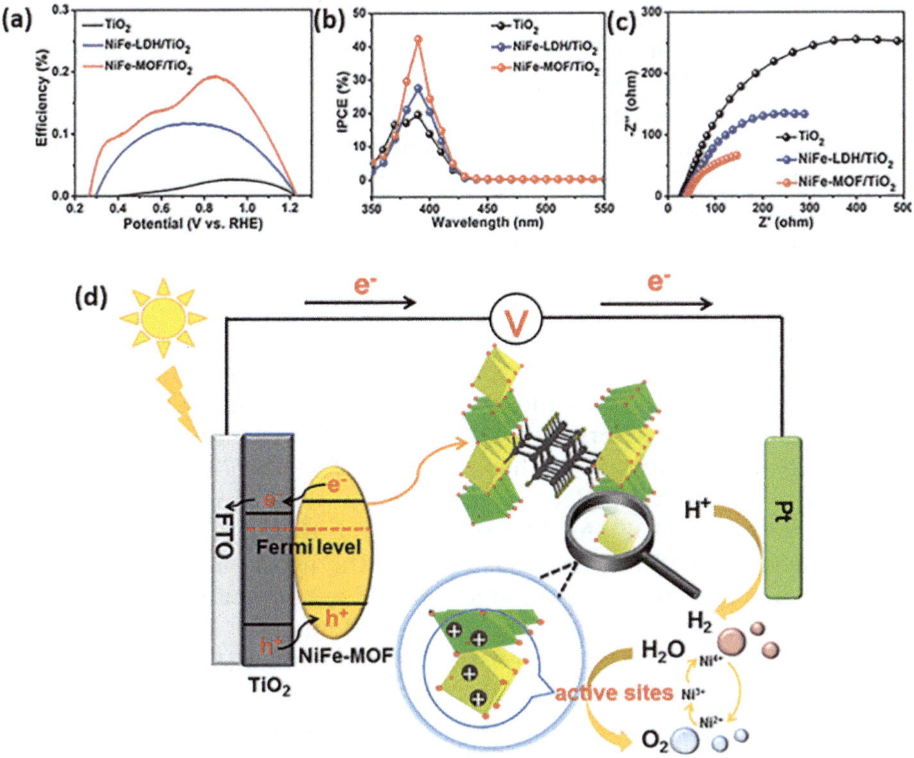

Figure 7.8 (a) The photoconversion efficiency. (b) IPCE spectra. (c) The PEIS under illumination. (d) The proposed mechanism of enhanced performance of TiO_2/NiFe-MOF photoanode toward water splitting. Reproduced from ref. 125 with permission from Elsevier, Copyright 2020.

injection to 92.8% and the separation of charges ability to 37.4%, as deduced from the PEIS (Figure 7.8(c)). They attributed the enhancement construction of a Type-II heterojunction (Figure 7.8(d)), in addition to the outstanding long-term stability of the fabricated NiFe-MOF/TiO_2 hybrid photoanode over 4 h of illumination.

In another study, Galán-González *et al.*[126] improved the photocatalytic properties of ZnO NRs by combining doped Co and functionalization of the surface *via* coupling with ZIF-8, where a superior IPCE% of 75% was achieved. The unique and exciting aspect of the work in this study is the controllable growth deposition of nanoscale ZIF-8 shell on ZnO surface by spin coating technique. They synthesized a series of four samples denoted as ZnO, ZnO:Co, ZnO@ZIF-8, and ZnO:Co@ZIF-8, were chosen to be compared for photoelectrochemical water splitting. The results reveal that both cobalt doping and combing with ZIF-8 boost the current density to be doubled compared to bare ZnO NRs. Furthermore, the passivation of ZnO with the ZIF-8 outer layer causes the elimination of some cathodic peaks that appear in the low potential region and indicate undesired reactions (*i.e.*, the water molecules are not fully oxidized). The enhancement in the photo-induced electrons and holes. The performance enhancement mechanism is displayed as follows. Under irradiation, e^{-}–h^{+} charges are induced at the Co-doped ZnO core and ZIF-8 shell. The excitation of electrons can be carried out in two ways in the Co-doped ZnO core, from the conduction band or sp–d related energy level provided by the cobalt doping. However, h^{+} can be transferred to the ZIF-8 shell. Therefore, the shell behaved as an h^{+} mediator layer between the electrolyte and the ZnO core, improving the charge separation. Besides, the Type-II band alignment of ZnO and ZIF-8 enables the efficient electron transfer of the ZIF-8 and enhances the charge separation efficiency.

In addition to this, Song *et al.*[122] reported the PEC water splitting by a novel hybrid photoanode Ti-MOF functionalized TNTAs. They employed a facile hydro-thermal method to fabricate a hybrid electrode of NH_2-MIL-125(Ti), a titanium-based MOF, decorated on titanium nanotube arrays (TNTAs) (Figure 7.9(a)). The as-fabricated TNTAs@Ti-MOF was employed as a photoanode for H_2O splitting. The hybrid anode of TNTAs@Ti-MOF exhibited the highest photocurrent density compared to the former corresponding single catalysts, Ti-MOF and TNTAs, where the hybrid structure TNTAs@Ti-MOF, as presented in Table 7.4. Besides, the hybrid anode of TNTAs@Ti-MOF achieved a high H_2 production rate of 132.86 μmol h^{-1} cm^{-2}, which was about 14-fold superior to bare TNTAs (9.44 μmol h^{-1} cm^{-2}). In addition to the long-term stability of the constructed photoanode (Figure 7.9(b)). The mechanism of this significant enhancement was attributed to boosting the light absorption capability as well as promoting the charge separation based on the constructed Type-II heterojunction, as displayed in Figure 7.9(c).

Moreover, the metal doping/decoration strategy effectively boosted the photocatalytic performance of various photocatalysts towards various photocatalytic applications.[114,120,135,136] Generally, metal doping of the semiconductor provides mid-state levels between CB and VB of the semiconductor. These

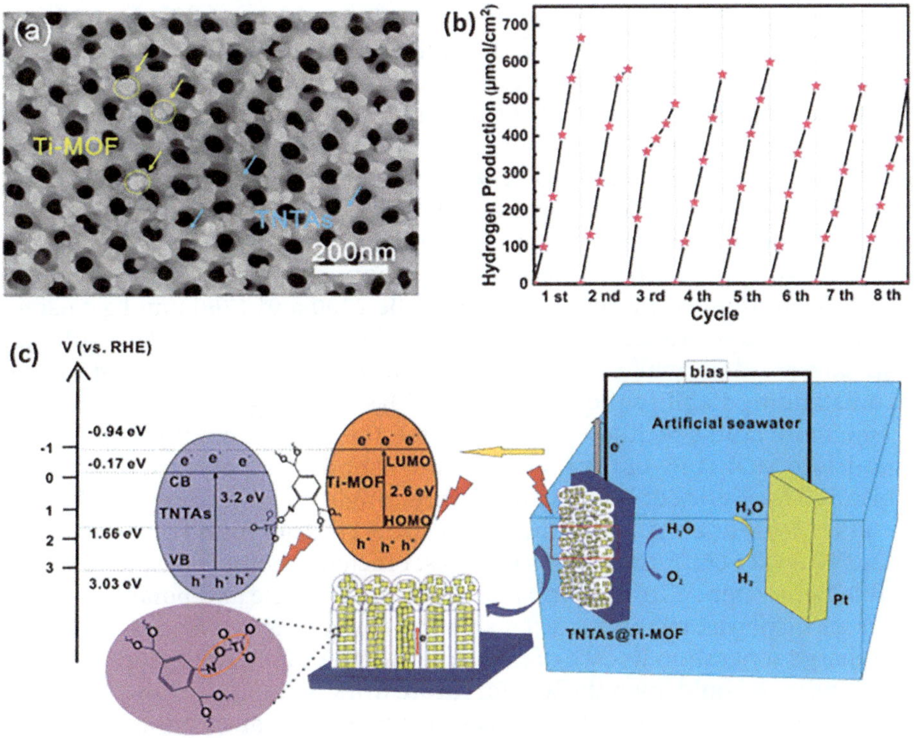

Figure 7.9 (a) HR-TEM image. (b) Cyclic tests of PEC activity *vs.* Pt system. (c) The proposed mechanism of the superior performance of TNTAs@Ti-MOF (20/800/160) photoanode toward water splitting using artificial seawater. Reproduced from ref. 122 with permission from Elsevier, Copyright 2019.

levels can act as an alternative CB causing the reduction of bandgap energy and/or act as a trapping center leading to boost the separation of charges and, consequently, improving the photoactivity.[136] As a prime example, very recently, Li *et al.*[129] reported the fabrication of a crystalline microporous Cu^{II}-based MOF and investigated the influence of Pt loading percentages. They reported that the optimized Pt loading percentage was 4.38 wt%, which achieved enhancement of Pt/Cu^{II}-MOF nanoribbon when compared to the bare Cu^{II}-MOF nanoribbon, where the estimated hydrogen production rate was about 2.51 mmol g^{-1} h^{-1}, which is almost double that of the bare Cu^{II}-MOF nanoribbon and about 5-fold of the referential Pt NP. This significant improvement is due to the synergism between the catalytic active Cu^{II} and Pt centers that favorably dominate the CB's potential and enhance the charge separation based on the constructed metallic Schottky junction.

Also, scientists are still innovating novel classes of MOFs with new structures and compositions. For instance, Salcedo-Abraira *et al.*[128] reported the facile hydrothermal synthesis of an innovative microporous two-dimensional nickel phosphonate MOF, represented as IEF-13, which exhibited a triclinic structure with the space group $P\bar{1}$: bi-octahedral nickel nodes and photoelectroactive tritopic phosphonate ligands. It was found that the IEF-13 shows coordinatively unsaturated Ni sites indicating their superior photocatalytic performance along with exceptional thermal and chemical stability. Besides, IEF-13 exhibited narrow bandgap energy of 2.83 eV, signifying its photocatalytic activity under visible irradiation. The photocatalytic activity of the novel nickel phosphate MOF, IEF-13 was examined toward overall water splitting and hydrogen production without the aid of any co-catalysts, where the rates of hydrogen production were estimated to be 2200 and 1700 μmol g^{-1} using methanol as a sacrificial under UV-visible light and stimulated sunlight, respectively. Additionally, the constructed novel IEF-13 photocatalyst exhibits outstanding stability and reachability with sustaining its crystal structure.

One more study, Melillo *et al.*[131] reported the enhancing of photoactivity of Zr-based MOF (UiO-66) by modifying metal node composition, where they fabricated five series of UiO-66 MOFs using single metallic (Zr or Ce), binary metallic (Zr/Ce or Zr/Ti), and trimetallic (Zr/Ce/Ti). They found that the trimetallic MOF, UiO-66(Zr/Ce/Ti), was the most efficient photocatalyst toward the visible-lightwater-splitting compared to the other single and binary counterparts, as displayed in Table 7.3. This improvement was ascribed to the effective charge separation in case of trimetallic MOF. Finally, they proposed that their findings would open the way for future work on multi-metallic MOFs as visible-light-driven photocatalysts by shifting the light absorption capability toward the visible region and enhancing the charge separation efficiency.

Furthermore, some scientists attempted to enhance the photoactivity of MOFs toward water splitting with different simultaneous approaches, such as metal doping with semiconductor coupling or semiconductor coupling with carbon-based materials, *etc.* As a prime example, Zargazi *et al.*[130] reported the fabrication of an innovative design of photoanode composed of an inverse opal-like UiO-66 MOF modified by combining metal sensitization using Pd and carbon-based materials decoration using like S, N graphene QDs. It displayed superior photocatalytic performance toward water splitting, as displayed in Figure 7.10.

Also, Han *et al.*[127] reported the synthesis of highly ordered nitrogen-doped carbon dots (N-CD) as photosensitizers on MOF-decorated ZnO nanotubes for enhanced PEC water splitting. They reported the photoelectrochemical hydrogen production by three types of electrodes of ZnO nanotubes, ZnO@ZIF-8, and N-CD/ZnO@ZIF-8 (Figure 7.11(a) and (b)). In this work, ZnO-based MOF was employed as the host (ZnO@ZIF-8), and N-CD was employed as the guest. The N-CD embedded ZnO@ZIF-8 exhibited a significant increase in the current photo density of ~0.46 mA cm^{-2} at 0.5 V *vs.* Ag/AgCl, compared to other photoanodes of ZnO and ZnO @ ZIF-8 of ~0.22 mA cm^{-2} and ~0.29 mA cm^{-2}, respectively (Figure 7.11(c)). This was attributed to the significant improvement of charge separation as deduced from electrochemical impedance spectra (EIS), as displayed in Figure 7.11(d).

The suffering of Fe_2O_3 from the shortness in the diffusion length (2–4 nm) and lifetime (picosecond time scale), resulting in increasing the

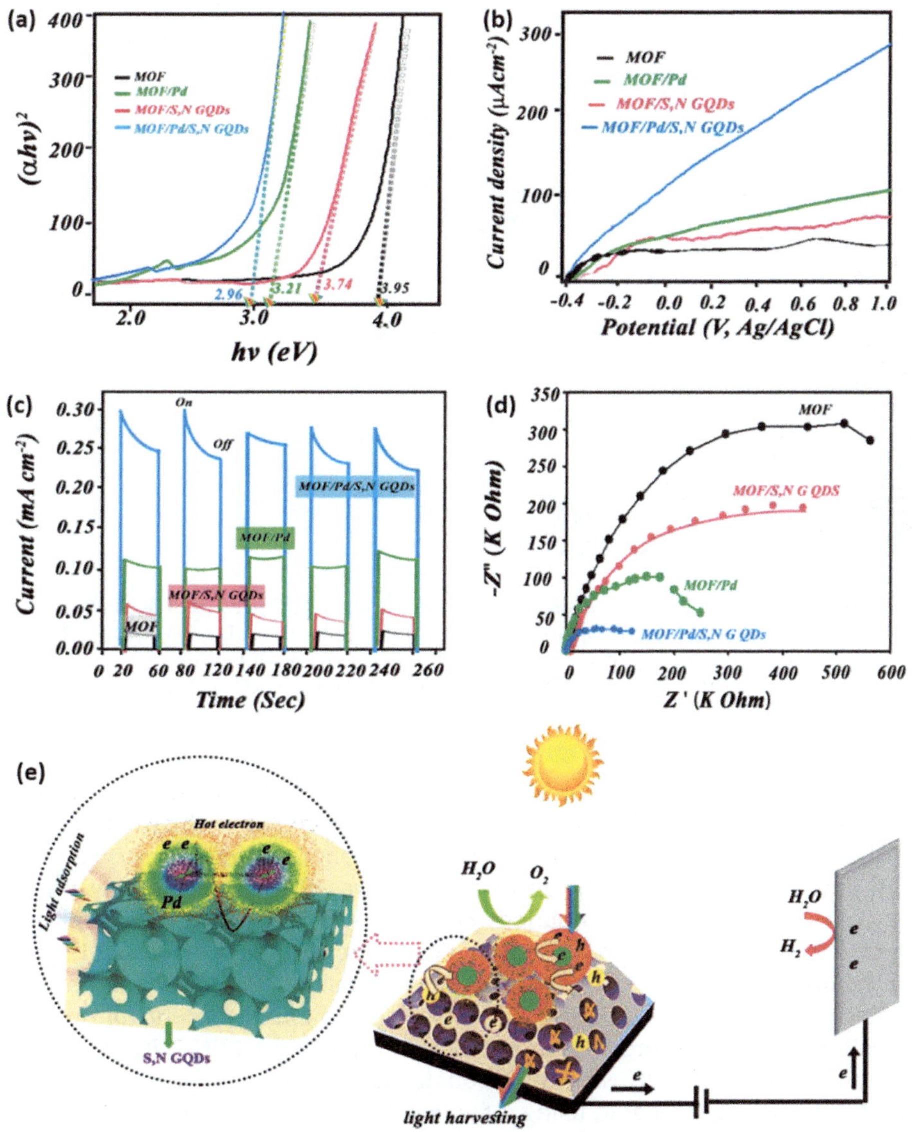

Figure 7.10 (a) Tauc plots. (b) Photocurrent densities. (c) *J*–*t* curves with chopped illumination. (d) Nyquist curve under light irradiation. (e) The proposed mechanism of enhanced performance of MOF/Pd/S,N-GQDs photoanode toward water splitting. Reproduced from ref. 130 with permission from Elsevier, Copyright 2021.

recombination rate and limitation of its photocatalytic activity toward water splitting. Hong *et al.*[132] overcame these drawbacks by coating the as-fabricated Fe_2O_3 photoanodes with dual MOFs (MIL-88B@ZIF-67), which enhanced the charge transfer and afforded photocurrent density of 2.52 mA cm^{-2} at 1.23 V *vs.* RHE, under illumination. Notably, the higher value in the photocurrent

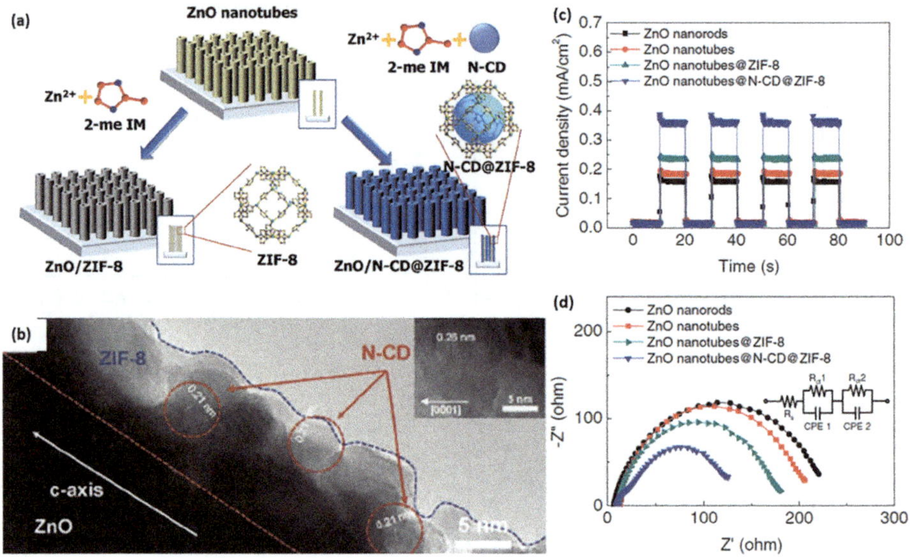

Figure 7.11 (a) Schematic illustrations of synthesis pathway. (b) HR-TEM. (c) *J–t* curves with chopped illumination. (d) EIS of N-CD@ZIF-8/ZnO. Reproduced from ref. 127 with permission from John Wiley & Sons, Copyright © 2019 WILEY-VCH Verlag GmbH & Co. KGaA, Weinheim.

than that of bare Fe_2O_3 photoanodes (0.27 mA cm^{-2}) is related to the significant reduction in the recombination rate compared to the modification with a single MOF. Along with the enhanced photocurrent density, a large cathodic shift (119 mV) in the onset potential was observed for the optimized Fe_2O_3/MIL-88B@ZIF-67 photoanode compared to bare Fe_2O_3. Higher photocurrent densities, which represented a cathodic shift, enhanced the IPCE% of the optimized Fe_2O_3/MIL-88B@ZIF-67 and reached 30.85% at 330 nm.

Recently, Wang and co-workers[134] have made significant contributions to developing organo-inorganic photocatalytic anodes for PEC water splitting. They fabricated a Ti-PH/NiFe MOF/cobaloxime hybrid structure. Ti-PH/NiFe MOF was initially prepared using a facile solvothermal method followed by a drop-coating of cobaloxime complex. The Ti-doped porous hematite (Ti-PH) was used for light-harvesting, and NiFe-MOF acted as a hole-transport material for charge transport. At the same time, molecular cobaloxime was employed as the catalyst in the oxygen evolution reaction. The hybrid photoanode displayed a remarkable enhancement in the photocurrent density, Table 7.3, along with the outstanding stability over 6 h of illumination. Additionally, a large cathodic shift (100 mV) in the photocurrent onset potential was observed after incorporating cobaloxime to the Ti-PH/NiFe MOF. They attributed the enhancement in the photoactivity of the hybrid photoanode, Ti-PH/Ni-F MOF/Cobaloxime, to the significant improvement of the effective charge separation.

To sum up, the studies mentioned above in this section highlighted the significant trends in improving the activity of MOF-based photocatalysts toward photocatalytic water splitting, including semiconductor coupling, MOF–MOF coupling, carbon-based material coupling, hybrid modifications, ligand functionalization, and metal doping/decoration, design of novel MOFs, *etc.* In addition to the possibility of mixing between these approaches to achieve much better performance toward water splitting.

7.5 Conclusions

This report reviews the application of MOFs-based materials for CO_2 photoconversion and H_2 generation. In general, MOFs-based materials have proved to be very effective for CO_2 conversion and H_2 generation. In addition, MOFs could be modified and combined easily with a wide range of (photo)-materials for enhanced photocatalytic activity. MOFs-based photocatalysts showed great electrochemical activity as well. The following points discuss some future trends and recommendations.

- Further investigations for the hybridization of MOFs and their combination with other materials for synergism or/and novel unique functionals are still challenging for CO_2 conversion and H_2 generation. Combining high storage capacity and photoactivity may lead to the best performance of MOFs-based photocatalysts.
- Based on the recent literature, the defect-engineering of the MOFs approach successfully enhanced MOFs' ability in terms of light response and photoactivity. This topic would be interesting for the scientific community regarding CO_2 conversion and H_2 generation. It is challenging to achieve better selectivity of MOF-based photocatalysts in CO_2 conversion to valuable products through self-designing or surface modification.
- A deactivation step is expected during the use of MOFs-based photocatalysts, similar to all heterogeneous photocatalytic processes. Therefore, approaches to limiting fast deactivation or/and carrying out a quick regeneration are recommended.
- A MOF-based composite with enhanced performance requires more precise characterization than pure materials, especially at the micro-level, which continues to be a challenge. An additional method for understanding the interface microstructure of materials is density functional theory (DFT) which better reveals the effects of composition and structural modification.
- Green synthesis of MOFs-based materials is required. Many types of MOFs require the use of huge amounts of toxic solvents, *e.g.*, dimethylformamide (DMF). However, it is recommended to use green methods to prepare MOFs. On top of that, to prepare MOFs, metals with less toxicity, *e.g.*, Fe, Cu, Zn, would be applied.

- Despite reduced costs when compared with commercial catalysts such as Pt and Ru, MOF-based materials still require process optimization to achieve large-scale production with controlled costs.

References

1. J. Ali, A. Sohail, L. Wang, M. Rizwan Haider, S. Mulk and G. Pan, Electromicrobiology as a promising approach towards renewable energy and environmental sustainability, *Energies*, 2018, **11**(7), 1822.
2. V.-H. Nguyen and J. C. Wu, Recent developments in the design of photoreactors for solar energy conversion from water splitting and CO_2 reduction, *Appl. Catal., A*, 2018, **550**, 122–141.
3. H. Mikulčić, J. J. Klemeš, M. Vujanović, K. Urbaniec and N. Duić, Reducing greenhouse gasses emissions by fostering the deployment of alternative raw materials and energy sources in the cleaner cement manufacturing process, *J. Cleaner Prod.*, 2016, **136**, 119–132.
4. R. Lal, R. F. Follett, B. A. Stewart and J. M. Kimble, Soil carbon sequestration to mitigate climate change and advance food security, *Soil Sci.*, 2007, **172**(12), 943–956.
5. V.-H. Nguyen, B.-S. Nguyen, Z. Jin, M. Shokouhimehr, H. W. Jang, C. Hu, P. Singh, P. Raizada, W. Peng and S. S. Lam, Towards artificial photosynthesis: Sustainable hydrogen utilization for photocatalytic reduction of CO_2 to high-value renewable fuels, *Chem. Eng. J.*, 2020, 126184.
6. N. Höhne, M. den Elzen, J. Rogelj, B. Metz, T. Fransen, T. Kuramochi, A. Olhoff, J. Alcamo, H. Winkler, and S. Fu, *Emissions: World Has Four Times the Work or One-Third of the Time*, Nature Publishing Group, 2020.
7. N. Roy, N. Suzuki, C. Terashima and A. Fujishima, Recent improvements in the production of solar fuels: from CO_2 reduction to water splitting and artificial photosynthesis, *Bull. Chem. Soc. Jpn.*, 2019, **92**(1), 178–192.
8. Z. Hu, Z. Shen and J. C. Yu, Converting carbohydrates to carbon-based photocatalysts for environmental treatment, *Environ. Sci. Technol.*, 2017, **51**(12), 7076–7083.
9. K. O. Yoro, and M. O. Daramola, CO_2 emission sources, greenhouse gases, and the global warming effect, in *Advances in Carbon Capture*, Elsevier, 2020, pp. 3–28.
10. J. Ali, L. Wang, H. Waseem, R. Djellabi, N. Oladoja and G. Pan, FeS@ rGO nanocomposites as electrocatalysts for enhanced chromium removal and clean energy generation by microbial fuel cell, *Chem. Eng. J.*, 2020, **384**, 123335.
11. R. Djellabi, B. Yang, H. M. A. Sharif, J. Zhang, J. Ali and X. Zhao, Sustainable and easy recoverable magnetic TiO_2-Lignocellulosic Biomass@ Fe_3O_4 for solar photocatalytic water remediation, *J. Cleaner Prod.*, 2019, **233**, 841–847.
12. L. Wang, J. Ali, C. Zhang, G. Mailhot and G. Pan, Simultaneously enhanced photocatalytic and antibacterial activities of TiO_2/Ag

composite nanofibers for wastewater purification, *J. Environ. Chem. Eng.*, 2020, **8**(1), 102104.
13. Y. Li, H. Xu, S. Ouyang and J. Ye, Metal–organic frameworks for photocatalysis, *Phys. Chem. Chem. Phys.*, 2016, **18**(11), 7563–7572.
14. P. Verma, D. J Stewart and R. Raja, Recent Advances in Photocatalytic CO_2 Utilisation Over Multifunctional Metal–Organic Frameworks, *Catalysts*, 2020, **10**(10), 1176.
15. H. N. Abdelhamid and A. P. Mathew, Cellulose–metal organic frameworks (CelloMOFs) hybrid materials and their multifaceted Applications: A review, *Coord. Chem. Rev.*, 2022, **451**, 214263.
16. Y. Zhou, R. Abazari, J. Chen, M. Tahir, A. Kumar, R. R. Ikreedeegh, E. Rani, H. Singh and A. M. Kirillov, Bimetallic metal–organic frameworks and MOF-derived composites: Recent progress on electro-and photoelectrocatalytic applications, *Coord. Chem. Rev.*, 2022, **451**, 214264.
17. G. Luciani, C. Imparato and G. Vitiello, Photosensitive hybrid nanostructured materials: The big challenges for sunlight capture, *Catalysts*, 2020, **10**(1), 103.
18. R. R. Ikreedeegh and M. Tahir, A critical review in recent developments of metal-organic-frameworks (MOFs) with band engineering alteration for photocatalytic CO_2 reduction to solar fuels, *J. CO2 Util.*, 2021, **43**, 101381.
19. I. I. Alkhatib, C. Garlisi, M. Pagliaro, K. Al-Ali and G. Palmisano, Metal-organic frameworks for photocatalytic CO_2 reduction under visible radiation: A review of strategies and applications, *Catal. Today*, 2020, **340**, 209–224.
20. R. Dutta, R. Shrivastav, M. Srivastava, A. Verma, S. Saxena, N. K. Biswas, V. R. Satsangi and S. Dass, MOFs in photoelectrochemical water splitting: New horizons and challenges, *Int. J. Hydrogen Energy*, 2021, **47**(8), 5192–5210.
21. H. Luo, Z. Zeng, G. Zeng, C. Zhang, R. Xiao, D. Huang, C. Lai, M. Cheng, W. Wang and W. Xiong, Recent progress on metal-organic frameworks based-and derived-photocatalysts for water splitting, *Chem. Eng. J.*, 2020, **383**, 123196.
22. C. Chauvier and T. Cantat, *A Viewpoint on Chemical Reductions of Carbon–Oxygen Bonds in Renewable Feedstocks Including CO_2 and Biomass*, ACS Publications, 2017.
23. Z. a. Kovačič, B. Likozar and M. Huš, Photocatalytic CO_2 reduction: A review of ab initio mechanism, kinetics, and multiscale modeling simulations, *ACS Catal.*, 2020, **10**(24), 14984–15007.
24. S. Lingampalli, M. M. Ayyub and C. Rao, Recent progress in the photocatalytic reduction of carbon dioxide, *ACS Omega*, 2017, **2**(6), 2740–2748.
25. T. P. Nguyen, D. L. T. Nguyen, V.-H. Nguyen, T.-H. Le, D.-V. N. Vo, Q. T. Trinh, S.-R. Bae, S. Y. Chae, S. Y. Kim and Q. V. Le, Recent advances in TiO_2-based photocatalysts for reduction of CO_2 to fuels, *Nanomaterials*, 2020, **10**(2), 337.

26. A. Corma and H. Garcia, Photocatalytic reduction of CO_2 for fuel production: Possibilities and challenges, *J. Catal.*, 2013, **308**, 168–175.
27. K. M. Choi, D. Kim, B. Rungtaweevoranit, C. A. Trickett, J. T. D Barmanbek, A. S. Alshammari and O. M. Yaghi, Plasmon-enhanced photocatalytic CO_2 conversion within metal–organic frameworks under visible light, *J. Am. Chem. Soc.*, 2017, **139**(1), 356–362.
28. Z. Fu, Q. Yang, Z. Liu, F. Chen, F. Yao, T. Xie, Y. Zhong, D. Wang, J. Li and X. Li, Photocatalytic conversion of carbon dioxide: From products to design the catalysts, *J. CO_2 Util.*, 2019, **34**, 63–73.
29. R. B. Song, W. Zhu, J. Fu, Y. Chen, L. Liu, J. R. Zhang, Y. Lin and J. J. Zhu, Electrode materials engineering in electrocatalytic CO_2 reduction: Energy input and conversion efficiency, *Adv. Mater.*, 2020, **32**(27), 1903796.
30. X. Chang, T. Wang, P. Yang, G. Zhang and J. Gong, The development of co-catalysts for photoelectrochemical CO_2 reduction, *Adv. Mater.*, 2019, **31**(31), 1804710.
31. J. Mao, T. Peng, X. Zhang, K. Li and L. Zan, Selective methanol production from photocatalytic reduction of CO_2 on $BiVO_4$ under visible light irradiation, *Catal. Commun.*, 2012, **28**, 38–41.
32. C.-C. Yang, J. Vernimmen, V. Meynen, P. Cool and G. Mul, Mechanistic study of hydrocarbon formation in photocatalytic CO_2 reduction over Ti-SBA-15, *J. Catal.*, 2011, **284**(1), 1–8.
33. Y. Zhang, B. Xia, J. Ran, K. Davey and S. Z. Qiao, Atomic-level reactive sites for semiconductor-based photocatalytic CO_2 reduction, *Adv. Energy Mater.*, 2020, **10**(9), 1903879.
34. J. Wang, S. Lin, N. Tian, T. Ma, Y. Zhang and H. Huang, Nanostructured Metal Sulfides: Classification, Modification Strategy, and Solar-Driven CO_2 Reduction Application, *Adv. Funct. Mater.*, 2021, **31**(9), 2008008.
35. T. Nakajima, Y. Tamaki, K. Ueno, E. Kato, T. Nishikawa, K. Ohkubo, Y. Yamazaki, T. Morimoto and O. Ishitani, Photocatalytic reduction of low concentration of CO_2, *J. Am. Chem. Soc.*, 2016, **138**(42), 13818–13821.
36. X. Liu and H. Zhuang, Recent progresses in photocatalytic hydrogen production: design and construction of Ni-based co-catalysts, *Int. J. Energy Res.*, 2021, **45**(2), 1480–1495.
37. J. Zhang, W. Hu, S. Cao and L. Piao, Recent progress for hydrogen production by photocatalytic natural or simulated seawater splitting, *Nano Res.*, 2020, **13**(9), 2313–2322.
38. D. C. Hoekstra, K. Nickmans, J. Lub, M. G. Debije and A. P. Schenning, Air-curable, high-resolution patternable oxetane-based liquid crystalline photonic films via flexographic printing, *ACS Appl. Mater. Interfaces*, 2019, **11**(7), 7423–7430.
39. K. Dai, J. Lv, J. Zhang, G. Zhu, L. Geng and C. Liang, Efficient visible-light-driven splitting of water into hydrogen over surface-fluorinated anatase TiO_2 nanosheets with exposed {001} facets/layered CdS–diethylenetriamine nanobelts, *ACS Sustainable Chem. Eng.*, 2018, **6**(10), 12817–12826.

40. Y. Xu, W. Zhao, R. Xu, Y. Shi and B. Zhang, Synthesis of ultrathin CdS nanosheets as efficient visible-light-driven water splitting photocatalysts for hydrogen evolution, *Chem. Commun.*, 2013, **49**(84), 9803–9805.
41. J. Wang, M. Kuo, P. Zeng, L. Xu, S. Chen and T. Peng, Few-layer $BiVO_4$ nanosheets decorated with $SrTiO_3$: Rh nanoparticles for highly efficient visible-light-driven overall water splitting, *Appl. Catal., B*, 2020, **279**, 119377.
42. W. Iqbal, B. Yang, X. Zhao, M. Rauf, M. Waqas, Y. Gong, J. Zhang and Y. Mao, Controllable synthesis of graphitic carbon nitride nanomaterials for solar energy conversion and environmental remediation: the road travelled and the way forward, *Catal. Sci. Technol.*, 2018, **8**(18), 4576–4599.
43. W. Iqbal, B. Qiu, Q. Zhu, M. Xing and J. Zhang, Self-modified breaking hydrogen bonds to highly crystalline graphitic carbon nitrides nanosheets for drastically enhanced hydrogen production, *Appl. Catal., B*, 2018, **232**, 306–313.
44. K. Maeda and K. J. Domen, New non-oxide photocatalysts designed for overall water splitting under visible light, *J. Phys. Chem. C*, 2007, **111**, 7851.
45. X. Chen, S. Shen, L. Guo and S. S. Mao, Semiconductor-based photocatalytic hydrogen generation, *Chem. Rev.*, 2010, **110**(11), 6503–6570.
46. H. M. El-Bery and H. N. Abdelhamid, Photocatalytic Hydrogen Generation via Water Splitting using ZIF-67 derived Co_3O_4@ C/TiO_2, *J. Environ. Chem. Eng.*, **2021**, 105702.
47. S. N. Habisreutinger, L. Schmidt-Mende and J. K. Stolarczyk, Photocatalytic reduction of CO_2 on TiO_2 and other semiconductors, *Angew. Chem., Int. Ed.*, 2013, **52**(29), 7372–7408.
48. X. Chang, T. Wang and J. Gong, CO 2 photo-reduction: insights into CO_2 activation and reaction on surfaces of photocatalysts, *Energy Environ. Sci.*, 2016, **9**(7), 2177–2196.
49. M. Ding, R. W. Flaig, H.-L. Jiang and O. M. Yaghi, Carbon capture and conversion using metal–organic frameworks and MOF-based materials, *Chem. Soc. Rev.*, 2019, **48**(10), 2783–2828.
50. V. Humphrey, J. Zscheischler, P. Ciais, L. Gudmundsson, S. Sitch and S. I. Seneviratne, Sensitivity of atmospheric CO_2 growth rate to observed changes in terrestrial water storage, *Nature*, 2018, **560**(7720), 628–631.
51. H. Furukawa, K. E. Cordova, M. O'Keeffe and O. M. Yaghi, The chemistry and applications of metal-organic frameworks, *Science*, 2013, **341**, 1230444.
52. S. Xu, S. Chansai, Y. Shao, S. Xu, Y.-c. Wang, S. Haigh, Y. Mu, Y. Jiao, C. E. Stere and H. Chen, Mechanistic study of non-thermal plasma assisted CO_2 hydrogenation over Ru supported on MgAl layered double hydroxide, *Appl. Catal., B*, 2020, **268**, 118752.
53. L. Ma, C. Abney and W. Lin, Enantioselective catalysis with homochiral metal–organic frameworks, *Chem. Soc. Rev.*, 2009, **38**(5), 1248–1256.

54. M. Ramyashree, S. S. Priya, N. C. Freudenberg, K. Sudhakar and M. Tahir, Metal-organic framework-based photocatalysts for carbon dioxide reduction to methanol: A review on progress and application, *J. CO_2 Util.*, 2020, 101374.
55. K. Sumida, D. L. Rogow, J. A. Mason, T. M. McDonald, E. D. Bloch, Z. R. Herm, T.-H. Bae and J. R. Long, Carbon dioxide capture in metal–organic frameworks, *Chem. Rev.*, 2012, **112**(2), 724–781.
56. M. Ding, X. Cai and H.-L. Jiang, Improving MOF stability: approaches and applications, *Chem. Sci.*, 2019, **10**(44), 10209–10230.
57. S. Kitagawa, Metal–organic frameworks (MOFs), *Chem. Soc. Rev.*, 2014, **43**(16), 5415–5418.
58. B. Li, H. M. Wen, Y. Cui, W. Zhou, G. Qian and B. Chen, Emerging multifunctional metal–organic framework materials, *Adv. Mater.*, 2016, **28**(40), 8819–8860.
59. L. Jiao, J. Y. R. Seow, W. S. Skinner, Z. U. Wang and H.-L. Jiang, Metal–organic frameworks: Structures and functional applications, *Mater. Today*, 2019, **27**, 43–68.
60. L. Jiao, Y. Wang, H. L. Jiang and Q. Xu, Metal–organic frameworks as platforms for catalytic applications, *Adv. Mater.*, 2018, **30**(37), 1703663.
61. D. Sun and Z. Li, Robust Ti-and Zr-based metal-organic frameworks for photocatalysis, *Chin. J. Chem.*, 2017, **35**(2), 135–147.
62. M. Alvaro, E. Carbonell, B. Ferrer, F. X. Llabrés i Xamena and H. Garcia, Semiconductor behavior of a metal-organic framework (MOF), *Chem. - Eur. J.*, 2007, **13**(18), 5106–5112.
63. S. Wang and X. Wang, Photocatalytic CO_2 reduction by CdS promoted with a zeolitic imidazolate framework, *Appl. Catal., B*, 2015, **162**, 494–500.
64. R. Li, J. Hu, M. Deng, H. Wang, X. Wang, Y. Hu, H. L. Jiang, J. Jiang, Q. Zhang and Y. Xie, Integration of an inorganic semiconductor with a metal–organic framework: a platform for enhanced gaseous photocatalytic reactions, *Adv. Mater.*, 2014, **26**(28), 4783–4788.
65. M. Wang, D. Wang and Z. Li, Self-assembly of CPO-27-Mg/TiO_2 nanocomposite with enhanced performance for photocatalytic CO_2 reduction, *Appl. Catal., B*, 2016, **183**, 47–52.
66. A. Crake, K. C. Christoforidis, A. Kafizas, S. Zafeiratos and C. Petit, CO_2 capture and photocatalytic reduction using bifunctional TiO_2/MOF nanocomposites under UV–vis irradiation, *Appl. Catal., B*, 2017, **210**, 131–140.
67. J. Meng, Q. Chen, J. Lu and H. Liu, Z-scheme photocatalytic CO_2 reduction on a heterostructure of oxygen-defective ZnO/reduced graphene oxide/UiO-66-NH_2 under visible light, *ACS Appl. Mater. Interfaces*, 2018, **11**(1), 550–562.
68. Z.-C. Kong, J.-F. Liao, Y.-J. Dong, Y.-F. Xu, H.-Y. Chen, D.-B. Kuang and C.-Y. Su, Core@shell $CsPbBr_3$@zeolitic imidazolate framework nanocomposite for efficient photocatalytic CO_2 reduction, *ACS Energy Lett.*, 2018, **3**(11), 2656–2662.

69. A. M. Idris, X. Jiang, J. Tan, Z. Cai, X. Lou, J. Wang and Z. Li, Dye-Sensitized Fe-MOF nanosheets as Visible-Light driven photocatalyst for high efficient photocatalytic CO_2 reduction, *J. Colloid Interface Sci.*, 2022, **607**, 1180–1188.
70. R. Li, W. Zhang and K. Zhou, Metal–organic-framework-based catalysts for photoreduction of CO_2, *Adv. Mater.*, 2018, **30**(35), 1705512.
71. X. Li and Q.-L. Zhu, MOF-based materials for photo-and electrocatalytic CO_2 reduction, *EnergyChem*, 2020, **2**(3), 100033.
72. D. Wang, R. Huang, W. Liu, D. Sun and Z. Li, Fe-based MOFs for photocatalytic CO_2 reduction: role of coordination unsaturated sites and dual excitation pathways, *ACS Catal.*, 2014, **4**(12), 4254–4260.
73. D. Sun, Y. Fu, W. Liu, L. Ye, D. Wang, L. Yang, X. Fu and Z. Li, Studies on Photocatalytic CO_2 Reduction over NH_2-Uio-66 (Zr) and Its Derivatives: Towards a Better Understanding of Photocatalysis on Metal–Organic Frameworks, *Chem. – Eur. J.*, 2013, **19**(42), 14279–14285.
74. S.-Q. Wang, X. Gu, X. Wang, X.-Y. Zhang, X.-Y. Dao, X.-M. Cheng, J. Ma and W.-Y. Sun, Defect-engineering of Zr(IV)-based metal-organic frameworks for regulating CO_2 photoreduction, *Chem. Eng. J.*, 2022, **429**, 132157.
75. Y.-J. Wang, H.-J. Wang, F. Luo, S. Yao, T.-B. Lu and Z.-M. Zhang, Inter-clusters synergy in iron-organic frameworks for efficient CO_2 photoreduction, *Appl. Catal., B*, 2022, **300**, 120487.
76. R. Djellabi, L. Noureen, V.-D. Dao, D. Meroni, E. Falletta, D. D. Dionysiou and C. L. Bianchi, Recent advances and challenges of emerging solar-driven steam and the contribution of photocatalytic effect, *Chem. Eng. J.*, 2022, **431**, 134024.
77. L. Wang, P. Jin, S. Duan, H. She, J. Huang and Q. Wang, In-situ incorporation of Copper (II) porphyrin functionalized zirconium MOF and TiO_2 for efficient photocatalytic CO_2 reduction, *Sci. Bull.*, 2019, **64**(13), 926–933.
78. D. Sun, W. Liu, M. Qiu, Y. Zhang and Z. Li, Introduction of a mediator for enhancing photocatalytic performance via post-synthetic metal exchange in metal–organic frameworks (MOFs), *Chem. Commun.*, 2015, **51**(11), 2056–2059.
79. H.-Q. Xu, J. Hu, D. Wang, Z. Li, Q. Zhang, Y. Luo, S.-H. Yu and H.-L. Jiang, Visible-light photoreduction of CO_2 in a metal–organic framework: boosting electron–hole separation via electron trap states, *J. Am. Chem. Soc.*, 2015, **137**(42), 13440–13443.
80. Z.-B. Fang, T.-T. Liu, J. Liu, S. Jin, X.-P. Wu, X.-Q. Gong, K. Wang, Q. Yin, T.-F. Liu and R. Cao, Boosting Interfacial Charge-Transfer Kinetics for Efficient Overall CO_2 Photoreduction via Rational Design of Coordination Spheres on Metal–Organic Frameworks, *J. Am. Chem. Soc.*, 2020, **142**(28), 12515–12523.
81. S. Wang, W. Yao, J. Lin, Z. Ding and X. Wang, Cobalt imidazolate metal–organic frameworks photosplit CO_2 under mild reaction conditions, *Angew. Chem.*, 2014, **126**(4), 1052–1056.

82. W.-M. Liao, J.-H. Zhang, Z. Wang, Y.-L. Lu, S.-Y. Yin, H.-P. Wang, Y.-N. Fan, M. Pan and C.-Y. Su, Semiconductive amine-functionalized Co(II)-MOF for visible-light-driven hydrogen evolution and CO_2 reduction, *Inorg. Chem.*, 2018, **57**(18), 11436–11442.
83. J. Qin, S. Wang and X. Wang, Visible-light reduction CO_2 with dodecahedral zeolitic imidazolate framework ZIF-67 as an efficient co-catalyst, *Appl. Catal., B*, 2017, **209**, 476–482.
84. J. Zhao, Q. Wang, C. Sun, T. Zheng, L. Yan, M. Li, K. Shao, X. Wang and Z. Su, A hexanuclear cobalt metal–organic framework for efficient CO_2 reduction under visible light, *J. Mater. Chem. A*, 2017, **5**(24), 12498–12505.
85. L. Jiao, Y. Wang, H.-L. Jiang and Q. Xu, Metal-organic Frameworks as Platforms for Catalytic Applications, *Adv. Mater.*, 2018, **30**, 1703663.
86. Y.-z. Li, Z.-h. Fu and G. Xu, Metal-organic framework nanosheets: Preparation and applications, *Coord. Chem. Rev.*, 2019, **388**, 79–106.
87. B. Li, M. Chrzanowski, Y. Zhang and S. Ma, Applications of metal-organic frameworks featuring multi-functional sites, *Coord. Chem. Rev.*, 2016, **307**, 106–129.
88. E. Carbonell, B. Ferrer, i. Llabrés, F. X. Xamena and H. Garcia, *Chem. – Eur. J.*, 2007, **13**, 5106.
89. Y. Horiuchi, T. Toyao, M. Saito, K. Mochizuki, M. Iwata, H. Higashimura, M. Anpo and M. Matsuoka, Visible-light-promoted photocatalytic hydrogen production by using an amino-functionalized Ti (IV) metal–organic framework, *J. Phys. Chem. C*, 2012, **116**(39), 20848–20853.
90. J. Liou, American Mineralogist, *Am. Mineral.*, 1971, **56**(3–4), 507–531.
91. M. Chen, L. Han, J. Zhou, C. Sun, C. Hu, X. Wang and Z. Su, Photoreduction of carbon dioxide under visible light by ultra-small Ag nanoparticles doped into Co-ZIF-9, *Nanotechnology*, 2018, **29**(28), 284003.
92. D. Sun, W. Liu, Y. Fu, Z. Fang, F. Sun, X. Fu, Y. Zhang and Z. Li, Noble metals can have different effects on photocatalysis over metal–organic frameworks (MOFs): a case study on M/NH_2-MIL-125 (Ti)(M= Pt and Au), *Chem. – Eur. J.*, 2014, **20**(16), 4780–4788.
93. Y. Su, H. Xu, J. Wang, X. Luo, Z.-l. Xu, K. Wang and W. Wang, Nanorattle Au@ PtAg encapsulated in ZIF-8 for enhancing CO_2 photoreduction to CO, *Nano Res.*, 2019, **12**(3), 625–630.
94. Y. Han, H. Xu, Y. Su, Z.-l. Xu, K. Wang and W. Wang, Noble metal (Pt, Au@ Pd) nanoparticles supported on metal organic framework (MOF-74) nanoshuttles as high-selectivity CO_2 conversion catalysts, *J. Catal.*, 2019, **370**, 70–78.
95. M. Wang, G. Tan, H. Ren, A. Xia and Y. Liu, Direct double Z-scheme Og-C_3N_4/Zn_2SnO_4N/ZnO ternary heterojunction photocatalyst with enhanced visible photocatalytic activity, *Appl. Surf. Sci.*, 2019, **492**, 690–702.
96. S. Yan, S. Ouyang, H. Xu, M. Zhao, X. Zhang and J. Ye, Co-ZIF-9/TiO_2 nanostructure for superior CO_2 photoreduction activity, *J. Mater. Chem. A*, 2016, **4**(39), 15126–15133.

97. H. Wang, D. Wu, C. Yang, H. Lu, Z. Gao, F. Xu and K. Jiang, Multi-functional amorphous TiO_2 layer on ZIF-67 for enhanced CO_2 photoreduction performances under visible light, *J. CO_2 Util.*, 2019, **34**, 411–421.
98. B. Di Credico, M. Redaelli, M. Bellardita, M. Calamante, C. Cepek, E. Cobani, M. D'Arienzo, C. Evangelisti, M. Marelli and M. Moret, Step-by-step growth of hkust-1 on functionalized TiO_2 surface: An efficient material for CO_2 capture and solar photoreduction, *Catalysts*, 2018, **8**(9), 353.
99. Z. Huang, P. Dong, Y. Zhang, X. Nie, X. Wang and X. Zhang, A ZIF-8 decorated TiO_2 grid-like film with high CO_2 adsorption for CO_2 photoreduction, *J. CO_2 Util.*, 2018, **24**, 369–375.
100. G. Xu, H. Zhang, J. Wei, H.-X. Zhang, X. Wu, Y. Li, C. Li, J. Zhang and J. Ye, Integrating the g-C_3N_4 nanosheet with B–H bonding decorated metal–organic framework for CO_2 activation and photoreduction, *ACS Nano*, 2018, **12**(6), 5333–5340.
101. T. Butburee, Z. Sun, A. Centeno, F. Xie, Z. Zhao, D. Wu, P. Peerakiatkhajohn, S. Thaweesak, H. Wang and L. Wang, Improved CO_2 photocatalytic reduction using a novel 3-component heterojunction, *Nano Energy*, 2019, **62**, 426–433.
102. A. Shariati, Fe-N-TiO_2/CPO-Cu-27 nanocomposite for superior CO_2 photoreduction performance under visible light irradiation, *Sol. Energy*, 2019, **186**, 166–174.
103. L. Chen, F. Yu, X. Shen and C. Duan, N-CND modified NH_2-UiO-66 for photocatalytic CO_2 conversion under visible light by a photo-induced electron transfer process, *Chem. Commun.*, 2019, **55**(33), 4845–4848.
104. Y. Zhao, W. Cai, J. Chen, Y. Miao and Y. Bu, A highly efficient composite catalyst constructed from NH2-MIL-125 (Ti) and reduced graphene oxide for CO_2 photoreduction, *Front. Chem.*, 2019, **7**, 789.
105. W. Tu, Y. Zhou and Z. Zou, Photocatalytic conversion of CO_2 into renewable hydrocarbon fuels: state-of-the-art accomplishment, challenges, and prospects, *Adv. Mater.*, 2014, **26**(27), 4607–4626.
106. C. Liu, W. Wang, B. Liu, J. Qiao, L. Lv, X. Gao, X. Zhang, D. Xu, W. Liu and J. Liu, Recent advances in MOF-based nanocatalysts for photo-promoted CO_2 reduction applications, *Catalysts*, 2019, **9**(8), 658.
107. S. Zhang, L. Li, S. Zhao, Z. Sun and J. Luo, Construction of interpenetrated ruthenium metal–organic frameworks as stable photocatalysts for CO_2 reduction, *Inorg. Chem.*, 2015, **54**(17), 8375–8379.
108. Y. Liu, C. Chen, J. Valdez, D. M. Meira, W. He, Y. Wang, C. Harnagea, Q. Lu, T. Guner and H. Wang, Phase-enabled metal-organic framework homojunction for highly selective CO_2 photoreduction, *Nat. Commun.*, 2021, **12**(1), 1–13.
109. B. Liu, K. Vikrant, K.-H. Kim, V. Kumar and S. K. Kailasa, Critical role of water stability in metal–organic frameworks and advanced modification strategies for the extension of their applicability, *Environ. Sci.: Nano*, 2020, **7**(5), 1319–1347.

110. M. A. Tekalgne, H. Do, A. Hasani, Q. Van Le, H. Jang, S. Ahn and S. Kim, Two-dimensional materials and metal-organic frameworks for the CO_2 reduction reaction, *Mater. Today Adv.*, 2020, **5**, 100038.
111. M. Hamza, A. El-Shazly, S. A. Tolba and N. K. Allam, Novel Bi-based photocatalysts with unprecedented visible light-driven hydrogen production rate: Experimental and DFT insights, *Chem. Eng. J.*, 2020, **384**, 123351.
112. A. N. El-Shazly, M. A. Hamza and N. K. Allam, Enhanced photoelectrochemical water splitting via engineered surface defects of $BiPO_4$ nanorod photoanodes, *Int. J. Hydrogen Energy*, 2021, **46**, 23214–23224.
113. A. N. El-Shazly, A. H. Hegazy, M. Rashad, M. El-Shahat and N. K. Allam, Ultrathin ALD TiO_2 shells for enhanced photoelectrochemical solar fuel generation, *J. Alloys Compd.*, 2018, **739**, 178–183.
114. A. N. El-Shazly, A. H. Hegazy, E. El Shenawy, M. A. Hamza and N. K. Allam, Novel facet-engineered multi-doped TiO_2 mesocrystals with unprecedented visible light photocatalytic hydrogen production, *Sol. Energy Mater. Sol. Cells*, 2021, **220**, 110825.
115. M. A. Hamza, A. N. El-Shazly and N. K. Allam, Facile template-free one-pot room-temperature synthesis of novel m-Bi(OH)CrO_4 microspheres, *Mater. Lett.*, 2020, **262**, 127188.
116. W. Wang, X. Xu, W. Zhou and Z. Shao, Recent progress in metal-organic frameworks for applications in electrocatalytic and photocatalytic water splitting, *Adv. Sci.*, 2017, **4**(4), 1600371.
117. M. Ali, E. Pervaiz, T. Noor, O. Rabi, R. Zahra and M. Yang, Recent advancements in MOF-based catalysts for applications in electrochemical and photoelectrochemical water splitting: A review, *Int. J. Energy Res.*, 2021, **45**(2), 1190–1226.
118. N. J. Vickers, Animal communication: when i'm calling you, will you answer too? *Curr. Biol.*, 2017, **27**(14), R713–R715.
119. M. Hamza, Z. Abou-Gamra, M. Ahmed and H. Medien, The critical role of Tween 80 as a 'green'template on the physical properties and photocatalytic performance of TiO_2 nanoparticles for Rhodamine B photodegradation, *J. Mater. Sci.: Mater. Electron.*, 2020, **31**(6), 4650–4661.
120. E. M. Hashem, M. A. Hamza, A. N. El-Shazly, M. F. Sanad, M. M. Hassan and S. O. Abdellatif, Investigating the UV absorption capabilities in novel Ag@RGO/ZnO ternary nanocomposite for optoelectronic devices, *Nanotechnology*, 2020, **32**(8), 085701.
121. A. Černý, Solutions to the multi-dimensional Prouhet–Tarry–Escott problem resulting from composition of balanced morphisms, *Inf. Comput.*, 2017, **253**, 424–435.
122. H. Song, Z. Sun, Y. Xu, Y. Han, J. Xu, J. Wu, T. Sun, H. Meng and X. Zhang, Fabrication of NH_2-MIL-125 (Ti) incorporated TiO_2 nanotube arrays composite anodes for highly efficient PEC water splitting, *Sep. Purif. Technol.*, 2019, **228**, 115764.
123. H. Hou and X. Zhang, Rational design of 1D/2D heterostructured photocatalyst for energy and environmental applications, *Chem. Eng. J.*, 2020, **395**, 125030.

124. L. Yu, W. Fan, N. He, Y. Liu, X. Han, F. Qin, J. Ding, G. Zhu, H. Bai and W. Shi, Effect of unsaturated coordination on photoelectrochemical properties of Ni-MOF/TiO_2 photoanode for water splitting, *Int. J. Hydrogen Energy*, 2021, **46**(34), 17741–17750.
125. W. Cui, H. Bai, J. Shang, F. Wang, D. Xu, J. Ding, W. Fan and W. Shi, Organic-inorganic hybrid-photoanode built from NiFe-MOF and TiO_2 for efficient PEC water splitting, *Electrochim. Acta*, 2020, **349**, 136383.
126. A. Galán-González, A. K. Sivan, J. Hernández-Ferrer, L. Bowen, L. Di Mario, F. Martelli, A. M. Benito, W. K. Maser, M. U. Chaudhry and A. Gallant, *Cobalt-doped ZnO Nanorods Coated with Nanoscale Metal-Organic Framework Shells for Water-Splitting Photoanodes*, 2020.
127. H. Han, F. Karlicky, S. Pitchaimuthu, S. H. R. Shin and A. Chen, Highly Ordered N-Doped Carbon Dots Photosensitizer on Metal–Organic Framework-Decorated ZnO Nanotubes for Improved Photoelectrochemical Water Splitting, *Small*, 2019, **15**(40), 1902771.
128. P. Salcedo-Abraira, S. M. Vilela, A. A. Babaryk, M. Cabrero-Antonino, P. Gregorio, F. Salles, S. Navalón, H. García and P. Horcajada, Nickel phosphonate MOF as efficient water splitting photocatalyst, *Nano Res.*, 2021, **14**(2), 450–457.
129. L. Li, Y. Zhao, Q. Wang, Z.-Y. Liu, X.-G. Wang, E.-C. Yang and X.-J. Zhao, BoostingphotocatalytichydrogenproductionactivitybymicroporousCuII-MOFnanoribbondecoratedwithPtnanoparticles,*Inorg.Chem.Front.*,2021, **8**, 3556–3565.
130. M. Zargazi and M. H. Entezari, Photoelectrochemical water splitting by a novel design of photo-anode: inverse opal-like UiO-66 sensitized by Pd and decorated with S, N graphene QDs, *Electrochim. Acta*, 2021, **391**, 138926.
131. A. Melillo, M. Cabrero-Antonino, S. Navalón, M. Álvaro, B. Ferrer and H. García, Enhancing visible-light photocatalytic activity for overall water splitting in UiO-66 by controlling metal node composition, *Appl. Catal., B*, 2020, **278**, 119345.
132. D. H. Hong, D. A. Reddy, K. A. J. Reddy, M. Gopannagari, D. P. Kumar and T. K. Kim, Synergetic catalytic behavior of dual metal-organic framework coated hematite photoanode for photoelectrochemical water splitting performance, *J. Catal.*, 2020, **391**, 471–479.
133. F. Wu, J. Xie, Y. You, Z. Zhao, L. Wang, X. Chen, P. Yang and Y. Huang, Cobalt Metal–Organic Framework Ultrathin Cocatalyst Overlayer for Improved Photoelectrochemical Activity of Ti-Doped Hematite, *ACS Appl. Energy Mater.*, 2020, **3**(5), 4867–4876.
134. T. Wang, H. Cao, M. Huang and P. Du, Ultrathin MOF Coupling with Molecular Cobaloxime to Construct an Efficient Hybrid Hematite Photoanode for Photocatalytic Water Splitting, *J. Phys. Chem. C*, 2021, **125**(42), 23153–23161.
135. A. N. El-Shazly, G. S. El-Sayyad, A. H. Hegazy, M. A. Hamza, R. M. Fathy, E. El Shenawy and N. K. Allam, Superior visible light antimicrobial performance of facet engineered cobalt doped TiO_2 mesocrystals in pathogenic bacterium and fungi, *Sci. Rep.*, 2021, **11**(1), 1–14.

136. M. A. Hamza, S. A. Abd El-Rahman, A. N. El-Shazly, E. M. Hashem, R. T. Mohamed, E. M. El-Tanany and M. G. Elmahgary, Facile one-pot ultrasonic-assisted synthesis of novel Ag@ RGO/g-C_3N_4 ternary 0D@ 2D/2D nanocomposite with enhanced synergetic tandem adsorption-photocatalytic degradation of recalcitrant organic dyes: Kinetic and mechanistic insights, *Mater. Res. Bull.*, 2021, **142**, 111386.

CHAPTER 8

2D Nanomaterials-based Heterostructures for H_2O Splitting and CO_2 Reduction

RESHMA BHOSALE[a], DONG JIN KIM[b], SURESH GOSAVI[c] AND SURENDRA TONDA*[b]

[a]Department of Environmental Science, Savitribai Phule Pune University, Pune 411007, Maharashtra, India; [b]School of Architecture, Civil, Environmental and Energy Engineering, Kyungpook National University, Daegu 41566, Republic of Korea; [c]Department of Physics, Savitribai Phule Pune University, Pune 411007, Maharashtra, India

*E-mail: surendar.t86@gmail.com, surendart@knu.ac.kr

8.1 Introduction

We are all aware that augmented energy demand due to population growth and rapid industrialization is a threat on natural reservoirs as well as the environment. Thus, to address the issues of energy shortage and global warming, the scientific community has emanated green and sustainable technologies for the long-term development of human beings. Among several possible technologies, photocatalysis has been widely explored recently since it turns abundant solar energy into chemical energy (hydrogen or hydrocarbons). This effective approach not only produces green fuels to solve the issue of energy crises, but also mitigates the greenhouse effect caused by carbon

Sustainable Energy No. 1
Recent Developments in Functional Materials for Artificial Photosynthesis
Edited by Srabanti Ghosh and Qian Wang

Published by the Royal Society of Chemistry, www.rsc.org

dioxide (CO_2) emissions. There has been substantial progress in the quest for novel photocatalysts since Fujishima made a breakthrough in the realm of photocatalysis in 1972.[1] To date, however, the focus has been on modifying or improving the performance of existing photocatalysts through the exploration of new strategies. Among the newly emerged photocatalysts, two-dimensional (2D) materials have garnered exponentially increased attention due to their astonishing structural, physiochemical, and electronic properties.[2–4] However, there is still room to boost performance through coupling (heterostructure formation) or hybridization, as this strategy provides a useful and robust way for not only controlling the physicochemical properties of 2D materials, but also for exploring new nanocomposites with novel functionalities. The sheet-like structure of 2D materials act as a building block for the design and construction of novel hybrid catalysts.[5] In this chapter, we review and summarize the development of a series of 2D photocatalysts and their distinct heterostructures for the application of photocatalytic water splitting and CO_2 reduction. Additionally, it emphasizes on future research directions for exploring new strategies for developing 2D material-based hybrid systems.

8.1.1 Rise of 2D Nanomaterials as Photocatalysts

Both, photocatalytic water splitting for hydrogen generation and CO_2 reduction to value added fuels are solar driven reactions achieved with the aid of nanomaterials as photocatalyst of appropriate bandgap and energy level positions. When the incident sunlight is greater than the bandgap energy, the nanomaterial-based catalyst gets excited and an electron–hole pair is generated. The electron migrates to the conduction band (CB) to perform the function of the reduction reaction. The holes migrate to the valence band (VB) for the oxidation reaction. Efficiency of the reaction depends on the positioning of energy levels with respect to the redox potentials of H_2O or CO_2 and how efficiently the electron–hole pair are separated and consumed for their respective reactions.[5,6] Therefore, it is essential to have robust and efficient photocatalyst for water splitting. Some semiconductors, such as metal oxides absorb in the UV region and fail to utilize the major visible portion of the solar spectrum whereas in some visible light absorbing semiconductors, the energy levels do not overlap with the redox potential. Hence, light harvesting, and band alignment are the two stringent requirements to be fulfilled by a single photocatalyst. Unfortunately, even though it is satisfied, a photocatalyst is not able to deliver efficient photocatalytic performance due to recombination of photogenerated charges caused by defects and traps available on the bulk of the surface.[7] This is a major bottleneck in the field of photocatalysis and restricts the practical application of the photocatalyst. Therefore, search of new and more competent photocatalyst with rational design of surface to address the above-mentioned challenges is the present requisite.

Among the newly explored photoactive materials, 2D materials have recently received immense attention because they represent an emerging class of materials that have a sheet-like configuration. Ever since the discovery of graphene in 2014 by Novoselov and co-workers,[8] 2D photocatalysts have become a topical subject in the area of photocatalysis. Because of their unprecedented structural, physical, and chemical characteristics, 2D photocatalysts have quickly risen to the top of the list of most popular research topics in the recent years.

In comparison to one dimensional (1D) and zero dimensional (0D) photocatalysts, 2D photocatalysts possess extraordinary properties and advantages that endow them with the potential for photocatalytic applications (Figure 8.1). Theses merits are: (1) the bandgap energy and optical response of 2D materials can be modified by simply altering the number of layers; (2) they have high surface-to-volume ratios, which means that most of the active sights are exposed on the surface of the photocatalyst boosting the photocatalytic activity; (3) the ultra-thin nature of 2D materials minimizes the charge migration distance and thus the recombination of charges, increasing the overall efficiency; (4) the high fraction of coordinated unsaturated surface sites can act as active site for photocatalytic reactions; (5) high exposure of surface atoms in 2D materials provides easy regulation of properties by surface modification or functionalization, elemental doping and by strain or defect phase engineering; (6) nano-sized 2D materials are excellent catalyst support which provides platform for the catalyst dispersion; and (7) unique structural features of 2D materials offer accessibility for hybridization or for composite formation with other catalysts that can perform synergistically to enhance the photocatalytic activity.[9,10]

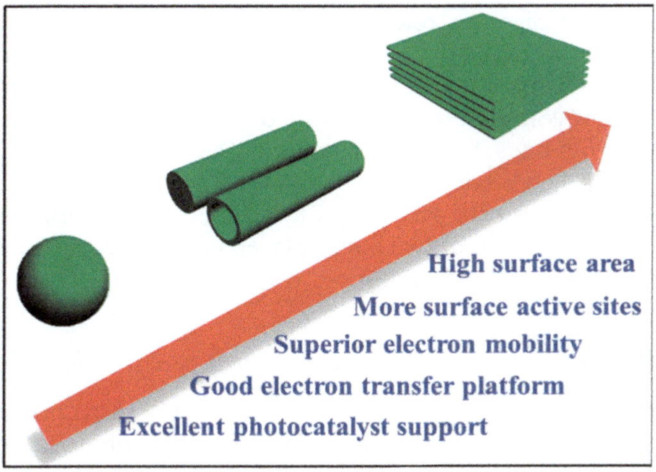

Figure 8.1 Schematic representation of the merits of 2D materials over 0D and 1D materials. Reproduced from ref. 9 with permission from the Royal Society of Chemistry.

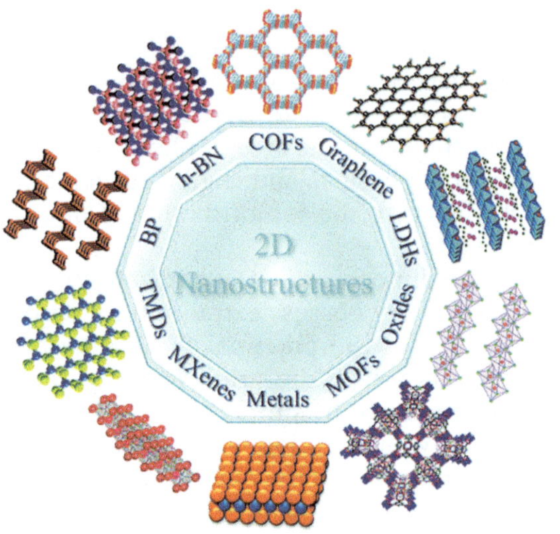

Figure 8.2 Different class of 2D materials. Reproduced from ref. 10 with permission from the Royal Society of Chemistry.

Hence, further, in this chapter we are going to discuss the different classes of renowned 2D photocatalysts and their exceptional characteristics along with their novel heterostructures employed for the application of water splitting and CO_2 reduction with several intriguing examples (Figure 8.2).[10]

8.1.2 Engineering of 2D Materials by Formation of Heterostructures (Hybridization)

In chemistry, hybridization refers to the concept of an innovative multicomponent material that works synergistically to improve physicochemical properties through nano-mixing or coupling of two materials. This results in the creation of completely new atomic orbitals with the same electronic potential as the old ones. Hybridization with 2D materials not only allows for the tailoring of the physicochemical characteristics of the hybridized materials, but it also allows for the creation of unpredicted functionalities through the synergistic integration of two materials. Hence, 2D layered structures act as building blocks for hybridization.

8.1.2.1 Requirements of Heterostructure or Hybridization

Even though 2D photocatalysts are considered as the most potential candidates for water splitting and CO_2 photoreduction, their photocatalytic performance is still constrained by a number of obstacles, such as: (1) the exciton binding energy within the 2D layered material is very high due to the lower electron screening effect than in the bulk, which is not ideal for photocatalytic activities because electron–hole pair separation becomes challenging;

(2) if exposed to air or water, many 2D semiconductors undergo oxidation, decreasing their overall performance; and (3) the redox potentials of some 2D materials are unsatisfactory for CO_2 conversion and overall water splitting.[7]

To address the aforementioned complications, several strategies have been adopted, including incorporation with metals or non-metals, creation of defects or strain, and heterostructure formation with metals or semiconductors. The photocatalytic activity of a photocatalyst is determined not only by its electron affinity, band positioning, and structure, but also by the interface built between the photocatalyst and a co-catalyst, or between two photocatalysts. Therefore, for an effective photocatalytic conversion process, the spatial integration of two photocatalysts to build a strong interface is critical as it provides a hybridized electronic structure. Consequently, this enables a sufficient charge separation by accelerating photoexcited charge transfer processes, which in turn enhances the efficiency of a photocatalytic process. Additionally, band potentials can be favorably matched to the reduction and oxidation potentials of target molecules, increasing the overall efficiency of system. Furthermore, the unique optoelectronic and catalytic properties of the two photocatalysts act synergistically to boost photoactivity as well as stability by alleviating corrosion and agglomeration, when compared to a single photocatalyst. Hence, hybridization or heterostructure formation has proven to be an excellent strategy for engineering high class 2D photocatalysts with enhanced performance. As compared to 0D–0D, 1D–0D, 2D–0D, 1D–1D, 2D–1D heterostructures, 2D–2D shows higher photocatalytic performance and long-term stability because of large contact area for efficient interfacial charge migration and extended visible light absorption (Figure 8.3).[7]

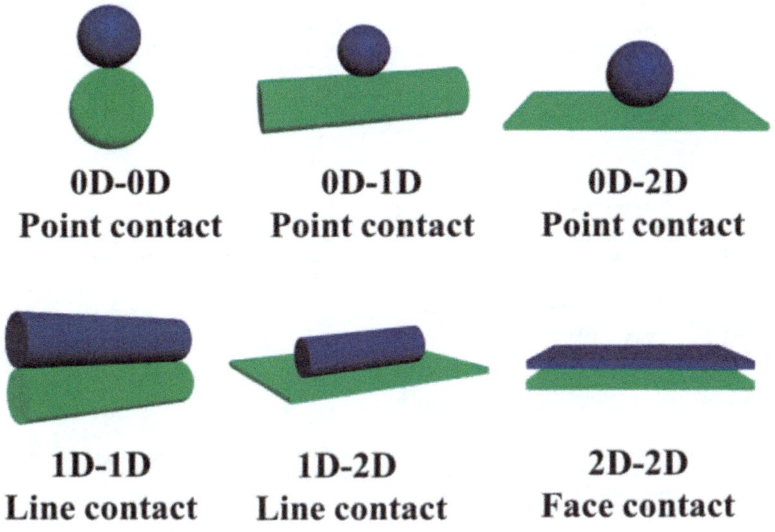

Figure 8.3 Schematic illustration of 2D based composites. Reproduced from ref. 9 with permission from the Royal Society of Chemistry.

8.1.2.2 Merits of 2D–2D Heterostructures

The various merits of 2D–2D hybrid heterostructures are as follows (Figure 8.4).[11]

(1) Exfoliated 2D nanosheets, particularly those of thin thickness, have all of their components exposed to the surface, resulting in strong interaction with incoming hybridized species. Such strong interactions are much more common in 2D–2D hybrid heterostructures than in low-dimensional heterostructures such as quantum dots (0D), and nanowires/tubes (1D). This type of opportunity allows for much better control over the electronic structure as well as the physicochemical properties of heterostructure species.

(2) The exfoliated 2D nanosheets resemble graphene nanosheets in terms of morphology, which aids in obtaining a strong interaction between the two 2D material and improves functionality.

(3) Often, 2D–2D nanostructures lead to stacked structures resembling a house of cards, resulting in micropores created by the intercalation of hybrid structure and mesopores created by the restacking of 2D crystallites. This categorized porous configuration enhances electrochemical activity, catalytic activity, and absorptivity of the hybrid material.

(4) The majority of 2D ultrathin sheets are produced through the exfoliation of pristine or bulk 2D material. During exfoliation, cationic or

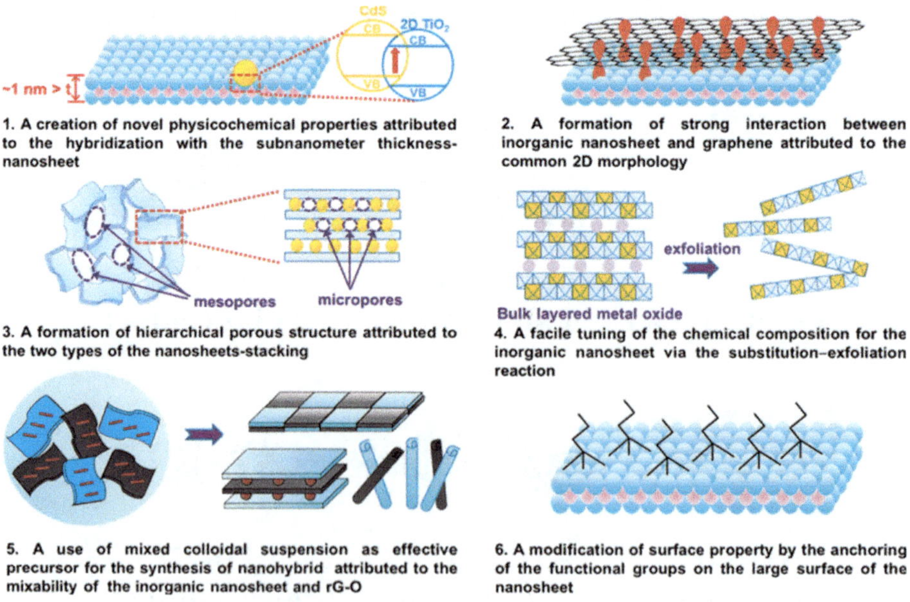

Figure 8.4 Advantage of 2D heterostructure by hybridization. Reproduced from ref. 11 with permission from American Chemical Society, Copyright 2014.

anionic substitution can occur while the host layer remains intact. This degree of adaptability in crystal structure and the chemical composition tuning is critical for achieving the best hybridization effect.

(5) The exfoliated 2D sheets have a large surface area that can be used to anchor a wide range of functional groups that tailor key surface characteristics, including surface structure and surface charge, hydrophilicity, and hydrophobicity. The ability to manipulate the surface characteristics improves the photocatalytic activity of the hybrid structure.

The above-mentioned extraordinary properties of layered materials provide a wealth of opportunities for the synthesis of a wide range of functional hybrid heterostructures with tunable characteristics. An intimate integration with layered sheet structures is effective in delivering promising hybridization effects. This current chapter deals with the survey of research work related to 2D nanosheets and their heterostructures for the purpose of creating exceptional hybrid photocatalysts.

8.1.2.3 Types of Interfaces in 2D–2D Heterostructures

Two 2D materials form a great combination of heterostructures as they promote separation of photoexcited charges and improve light harvesting. The 2D–2D photocatalyst systems are classified as Z-scheme and Schottky junction systems, as shown in Figure 8.5, and Type I and Type II systems, as shown in Figure 8.6.[7]

The Z-scheme, which was inspired by the natural photosynthesis performed by green plants, has proven to be an ingenious way to separate the photoinduced charges. In this scheme, the photogenerated electrons in the conduction band (CB) of photocatalyst A are transferred and combined with the holes in the valence band (VB) of photocatalyst B, as depicted in Figure 8.5. Meanwhile, the photogenerated electrons on the photocatalyst B are readily available for the reduction reaction, and holes on the photocatalyst A are available for oxidation reaction. Thus, the recombination of photoinduced charges within the semiconductor is inhibited and active charges are readily made available for redox reaction. This kind of exceptional system helps in enhancing the overall photoactivity efficiency of photocatalysts. This system consists of two photocatalysts with bands aligned in such a way that the charge transfer process gives a look of alphabet letter 'Z'. One photocatalyst displays reduction photoactivity, whereas the other displays oxidation photoactivity. They efficiently manage the charges to be consumed for specific reaction, ensuring total utilization of photoinduced charges with minimal recombination. For example, the 2D–2D interface with Schottky junction is also an effective way to inhibit recombination of photogenerated charges. In this type of junction, the two materials in contact can transfer the charges (higher Fermi level to lower) due to the aligned Fermi level. Consequently, a space charge layer is created at the interface, which bends the CB

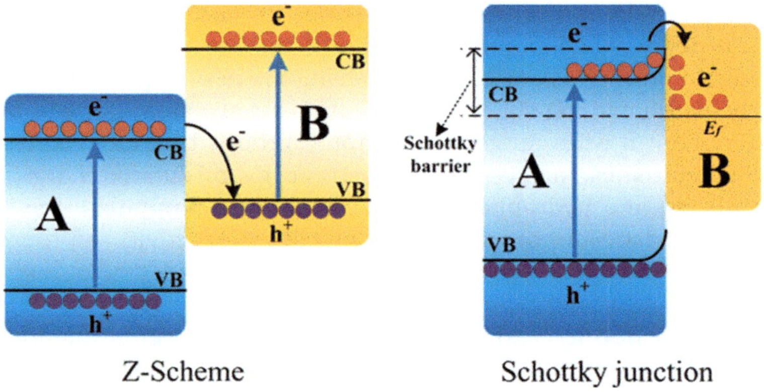

Figure 8.5 Schematic diagram for charge transfer in Z-scheme and Schottky junction systems.[7] Reproduced from ref. 7 with permission from American Chemical Society, Copyright 2018.

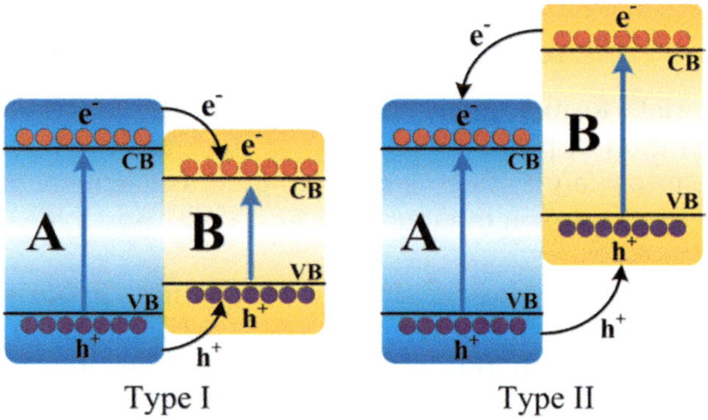

Figure 8.6 Schematic diagram for charge transfer in Type I and Type II heterostructure systems. Reproduced from ref. 7 with permission from American Chemical Society, Copyright 2018.

and VB upward at the interface. As a result, a Schottky junction is created, which serves as a trap to efficiently catch the photoinduced electrons and enhances the photocatalytic activity.

There are two different types of Schottky junction: type I and type II. In type I, electrons and holes are accumulated in the semiconductor B, as shown in Figure 8.6, as it is more positive than semiconductor A. Type II is identical to the Z-scheme but with different electron transfer directions. In type II, the CB electrons of the semiconductor B transfer to the CB of semiconductor A, while the VB holes of semiconductor A transfer to the VB of the semiconductor B. This is a kind of p–n junction where p- and n-type materials come into contact, resulting in the accumulation of photoexcited electrons at the n-type side and holes on the p-type side.[7] This results in a concentration gradient of electrons and holes at the interface, which promotes transportation of charge carriers and suppression of photoinduced charge recombination. This is an essential criterion for enhancing the efficiency of any kind of solar-energy conversion.

8.2 2D Nanomaterials-based Heterostructures for Water Splitting

8.2.1 Metal Oxides

Metal oxides are the most widely studied photocatalysts for hydrogen generation and pollutant degradation due to their low cost, abundance, non-toxicity and photostability.[12] Several metal oxides with 2D structures, such as TiO_2, ZnO, SnO, Fe_2O_3, Cu_2O, In_2O_3, Bi_2WO_6, and others, have also been explored for photocatalysis applications.[3,12,13] Among these, TiO_2 has attracted the most attention due to its suitable electronic structure and light absorption properties. Exfoliation of layered titanate results in the formation of layered $Ti_{0.91}O_2$ with a wider bandgap of ~3.8 eV than its anatase TiO_2 counterpart (3.2 eV).[12] Due to the quantum size effect of very thin sheets, optical properties can be fine-tuned by adjusting the composition of the parent layered compound. Apart from optical properties, the surface properties of lepidocrocite-type $Ti_{0.91}O_2$ also differ from those of anatase/rutile TiO_2. It has surface with unsaturated oxygen ions, which facilitated electric coupling of hybridized species *via* chemical bonding, resulting in improved visible light harvesting. It has been discovered that the hybridization of TiO_2 nanosheets with narrow bandgap semiconductors can provide a powerful tool for the exploration of new efficient visible light active photocatalysts. For example, coupling CdS,[14] or graphene[15] with TiO_2 nanosheets results in a hybrid structure with unique electronic and optical properties. The hybridization of the layered titanate with the visible light absorbing CdS resulted in a remarkable red shift in absorption from 2.9 eV to 2.5 eV, indicating the strong coupling between the components of nanohybrid. Such electronic coupling allows for an effective electron transfer from the valence band of CdS to the conduction band of titanate. This particular study demonstrates the advantages of 2D titanate nanosheets as a building block for hybrid structures.[14] A huge H_2 evolution rate of 18 mmol g^{-1} h^{-1} of was observed in a face-to-face 2D–2D heterostructure of TiO_2 nanosheet of thickness ~3 nm and graphitic carbon nitride of ultra-thickness 1.4 nm. The reason is that the formation of a heterostructure between the two types of nanosheets allows for efficient charge separation and transfer.[16] Analogously, MoS_2/TiO_2 demonstrated 36 times higher photoactivity than pristine TiO_2, with an apparent quantum yield (AQY) of 6.4% at 360 nm due to an intimate and large contact interface between the nanojunction of photocatalyst and co-catalyst.[17] Another Earth abundant layered oxide explored for photocatalytic applications is Cu_2O. Mateo *et al.*[18] developed Cu_2O nanoplates decorated on few layered graphene with higher photocatalytic H_2 production of almost 19.5 mmol g^{-1} h^{-1} from overall water splitting than Cu_2O nanoparticles due to the distinctive Cu_2O/graphene crafting. Graphene served as a co-catalyst in the 2D–2D construction, increasing not only the activity but also the photostability of Cu_2O.

Apart from the aforementioned oxides, there is another class of oxides known as inorganic perovskites, which have the general formula ABO_3, where cation A is 12-fold coordinated and cation B is 6-fold coordinated to oxygen anion. In layered perovskite due to the deficiency of oxygen, cation A site is reduced to coordination 8.[19] This large family of oxides is considered the promising photocatalysts because of their structural simplicity, flexibility, and photostability. The ability of exfoliated perovskite layers to self-assemble simplifies the construction of hybrid structure.[19,20] Restacking creates a platform for guest ions or species to disperse uniformly in the interlayer space, resulting in enhanced light absorption and charge separation. For example, CdS is introduced into the interlayer of $HCa_2Nb_3O_{10}$ by self-assembly of Cd^{2+} and $HCa_2Nb_3O_{10}$ nanosheets followed by sulfuration (Figure 8.7). The hybrid structure has type II band alignment, which results in rapid charge separation and reduced charge carrier recombination.[21] Core–shell samples with hetero-stacked by two different sheets $CaNb_3O_{10}$ and $PA_2K_2Nb_6O_{17}$ are more active than homo-staked samples, with an excellent H_2 evolution rate of 19 000 μmol g^{-1} h^{-1}, proving that heterojunction promotes photochemical charge separation.[22] Ida *et al.* demonstrated that n–p ($KCa_2Nb_3O_{10}$/NiO) junction surfaces have superior photocatalytic activity than p–n (NiO/CNO) junction surfaces due to potential gradient. Such gradient results in separated reaction sites, which inhibits recombination.[23]

Recently, 2D layered structures were produced from naturally occurring ores such as chromiteen, hemateen, and ilmenene, which is still an emerging research topic. However, high incident photon to current conversion efficiency at visible wavelengths suggests their wide range of applications in photocatalysis.[24]

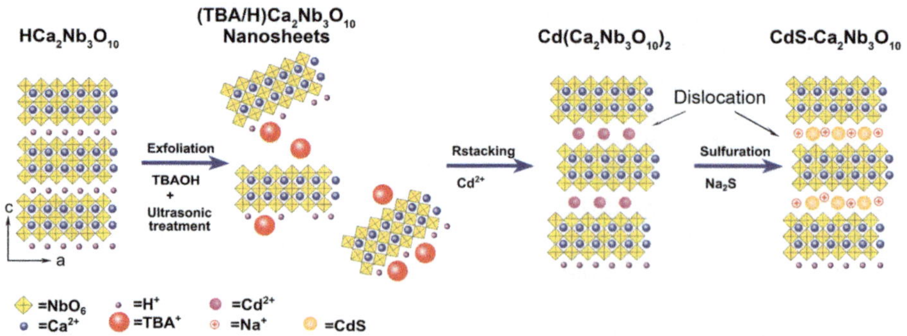

Figure 8.7 Schematic diagram for the preparation of CdS-$Ca_2Nb_3O_{10}$. Reproduced from ref. 21 with permission from Elsevier, Copyright 2018.

8.2.2 Metal Chalcogenides

Metal chalcogenides are Earth-abundant semiconductors that are explored as co-catalysts in photocatalysis due to their distinct electronic, optical, and catalytic properties.[25,26] Several studies have found that layered transition metal dichalcogenides (TMD) can improve the photoactivity of a photocatalyst for two reasons: (1) the active S atom at the edge of 2D TMDs can form strong bonds with H^+ ions, allowing it to be reduced to H_2; and (2) suitable bandgap and band positions that can be tuned depending on the number of layers.[27] Notably, TMDs have comparable catalytic activities to noble metals. For example, Chen *et al.*[28] prepared MoS_2/CdS by simple ball-milling followed by calcination method, which generated H_2 with a rate of 13 100 μmol g^{-1} h^{-1} when compared to pristine CdS nanoparticles (850 μmol g^{-1} h^{-1}). The close contact of CdS nanoparticles on the MoS_2 nanosheets is responsible for the exceptional performance. Similarly, Chang and group[29] used layered MoS_2 in conjunction with layered graphene as a co-catalyst to enhance the CdS performance (Figure 8.8). This multi-component catalyst exhibited H_2 evolution of 1.8 mmol h^{-1} in lactic acid solution with an AQY of 28% at 420 nm, which is greater than Pt/CdS. The incorporation of graphene indicated

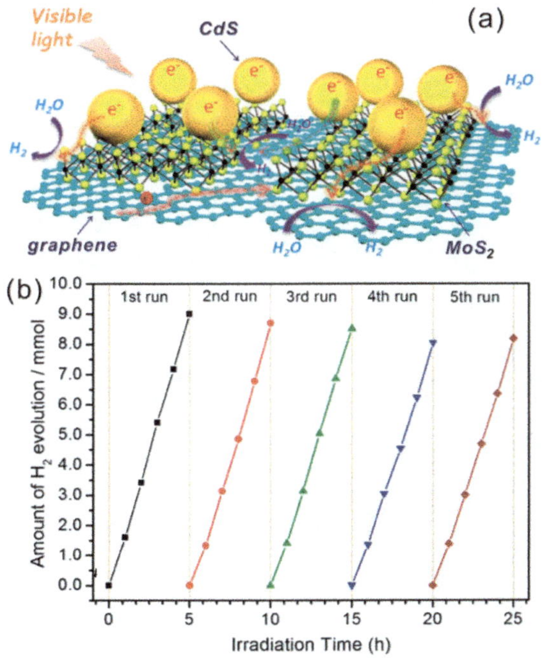

Figure 8.8 (a) Schematic diagram of charge flow in MoS_2/G-CdS photocatalyst under visible light. (b) Cyclic test of photocatalytic hydrogen evolution of MoS_2/G-CdS composite.[29] Reproduced from ref. 29 with permission from American Chemical Society, Copyright 2014.

that a carbon-based co-catalyst replacing expensive Pt could improve electron transport and prevent electron–hole pair recombination, resulting in improved photocatalytic performance.

Likewise, 2D–2D face contact of MoS_2/Cu-$ZnInS_2$ displayed commendable (5463 μmol g^{-1} h^{-1}) 72 times higher performance than $ZnInS_2$/Pt with point contact and even more than other noble metal co-catalysts due to the intimate nanojunction between the MoS_2 and $ZnInS_2$ nanosheets facilitating efficient charge separation (Figure 8.9).[30] Shen *et al.* demonstrated a MoS_2 nanosheets/TiO_2 nanowires hybrid nanostructure with remarkable performance for H_2 generation at a rate of 16.7 mmol g^{-1} h^{-1} under visible light and with Eosin Y sensitizer.[31] SnS_2 is also a unique 2D TMD that can be coupled with different photocatalysts such as TiO_2,[32] g-C_3N_4,[33] graphene[34] and others for improved photocatalytic H_2 generation and dye degradation under visible light because of its layered structure. The WS_2@Cu nanohybrid presents a stable and incredible H_2-evolution rate of 64 mmol g^{-1} h^{-1}, which is about 40 and 2.2 times higher than that of WS_2 and Cu, respectively. The surface plasmon resonance (SPR) of Cu NPs facilitates hot electron excitation and transfer to WS_2 NSs, with spontaneous charge separation across the Schottky interface, resulting in enhanced photocatalytic activity with a response extending to the NIR range beyond $\lambda > 750$ nm.[35]

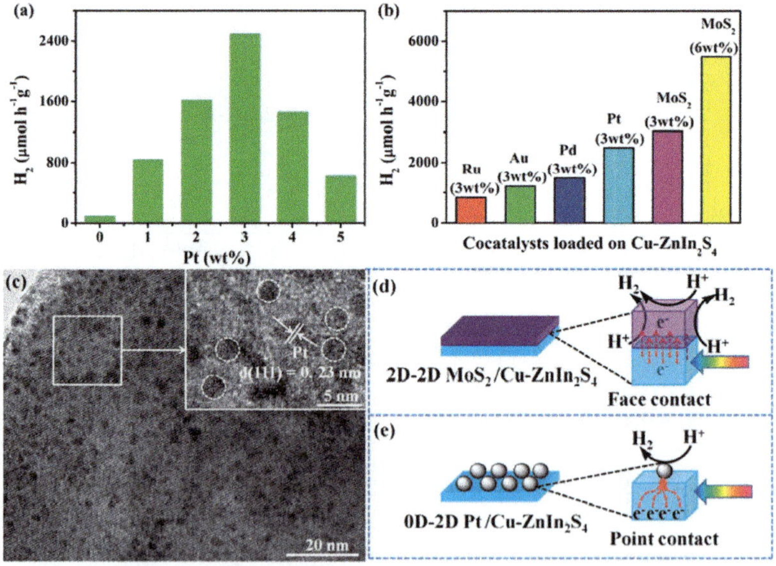

Figure 8.9 (a) H_2 evolution rate of Pt/Cu-$ZnIn_2S_2$ composite loaded with different amount of Pt under visible light. (b) H_2 evolution rate of Pt/Cu-$ZnIn_2S_2$ composite loaded with different co-catalyst under visible light. (c) HRTEM image of Pt/Cu-$ZnIn_2S_2$ photocatalyst. (d) and (e) Schematic diagram for 2D–2D MoS_2/Cu-$ZnIn_2S_2$ and 0D–2D/Cu-$ZnIn_2S_2$ for comparative study. Reproduced from ref. 30 with permission from the Royal Society of Chemistry.

8.2.3 Graphene-based 2D Nanomaterials

Graphene is a monoatomically thick sp^2 hybridized 2D carbon layer arranged in a hexagonal pattern with a π–π stacking network. Graphene has emerged as a miracle material in the world of nanoscience since its discovery by Geim and Novoselov[36] in 2004, due to its astonishing features such as high surface area (2630 $m^2\ g^{-1}$), high electron mobility (200 000 $cm^2\ V^{-1}\ s^{-1}$), high thermal conductivity (3000 $W\ m^{-1}\ K^{-1}$), high mechanical strength (1060 GPa), and exceptional conductivity (106 S cm^{-1}).[37] Such remarkable properties made graphene applicable in various energy conversion applications. Particularly, graphene has demonstrated remarkable advantages for the application of solar driven hydrogen generation from water splitting. First, it acts as a sink or electron transport bridge to expedite charge separation in photocatalysis owing to its high work function and good conductivity. Second, it has emerged as an excellent co-catalyst for the substitute of commonly used noble metals because of its reduction potential. Third, functionalized graphene can attain suitable oxidation level, making it photoactive for photocatalysis. Fourth, graphene with a tunable bandgap can act as a sensitizer in some cases to extend light absorption. Hence, numerous semiconductors/photocatalysts such as TiO_2, SnO_2, ZnO, Cu_2O, NiO, $BiVO_4$, Bi_2WO_6, Sr_2Ta_2O, CdS, CdSe, ZnS, MoS_2, SnS_2, g-C_3N_4, LDH, metals, and others were coupled with graphene to construct efficient hybrid composites that benefit from the aforementioned assets of graphene.[37–41] Graphene is produced by first oxidizing graphite to graphene oxide (GO), which is then reduced to produce reduced graphitic oxide (rGO). When GO was coupled with UV-light active TiO_2 by a simple hydrothermal method, the resulting RGO/TiO_2 nanocomposite generated about 380 μmol h^{-1} of H_2 with an apparent quantum yield (AQY) of 8.2% at 420 nm. The mechanism behind such enhanced efficiency is attributed to RGO, which served as a photosensitizer, expanding the light response of TiO_2 to visible region. The photoexcited electron from RGO is transferred to TiO_2 due to d–π interaction and transferred to active site of TiO_2 surface.[42]

However, RGO acts as an electron transporter when coupled with a visible light-responsive photocatalyst. For instance, RGO/CdS/Pt exhibited the highest photoactivity of 1.12 mmol $g^{-1}\ h^{-1}$ with AQY of 22.5% at 420 nm under visible light, which is 4.84 times superior than CdS/Pt.[43] Wang and colleagues[44] reported that when CdS was coupled with MoS_2 co-catalyst, its photocatalytic performance increased from 0.18 mmol h^{-1} to 1.34 mmol h^{-1}. However, with further addition of graphene to CdS/MoS_2, the resulting ternary system produced more H_2 with an evolution rate of 2.32 mmol h^{-1} and an exceptional AQY of 65.8% at 420 nm. The reason lies in high surface area of the ternary system, and efficient electron transfer from CdS to MoS_2 through graphene sheets, which suppresses charge recombination. The use of graphene not only boosts the photocatalytic activity but also protects the stability of semiconductors such as, ZnO, Cu_2O and CdS, which is a major concern for the practical application of any photocatalyst.[18,45] During the prolonged photocatalytic reactions, semiconductors undergo corrosion, which diminishes their photoactivity. Graphene has the ability to rapidly capture photogenerated charges and promote their migration to respective reaction centres, thereby preventing photoinduced charges from interacting with the photocatalyst for an extended period of time.[40]

Graphene or RGO has been the most well-known transporter or solid-state mediator between the two photocatalysts since the introduction of the Z-scheme in the field of photocatalysis. The presence of electron mediator is crucial for increasing the efficiency of photo-charges transfer. Because of its unique 2D structure and electrical conductivity, graphene is an effective mediator in a photocatalytic system. Amal and group[46] demonstrated the efficacy of photoreduced-graphene oxide (P-RGO) as a mediator between $BiVO_4$ and Ru/$SrTiO_3$:Rh as O_2 and H_2 photocatalysts, respectively. They found that the photocatalytic H_2 production rate of the Z-scheme system was greater than that of the system without P-RGO (Figure 8.10). This type of research opens up a new avenue for the coherence design of novel and efficient photocatalytic systems for water splitting that use graphene as a mediator. Many other Z-scheme systems that use graphene or RGO as a mediator have been successfully exploited, such as in $CoFe_2O_4$/graphene/BiOBr,[47] CdS/RGO/WO_3,[48] $ZnInS_2$/RGO/CoO_x:Bi_2MoO_6,[49] AgBr/graphene/TiO_2,[50] $(BiO)_2CO_3$/graphene/BiO_{2-x}, and Bi_2WO_6/graphene/tetra(4-carboxyphenyl)porphyrin (TCPP).[51]

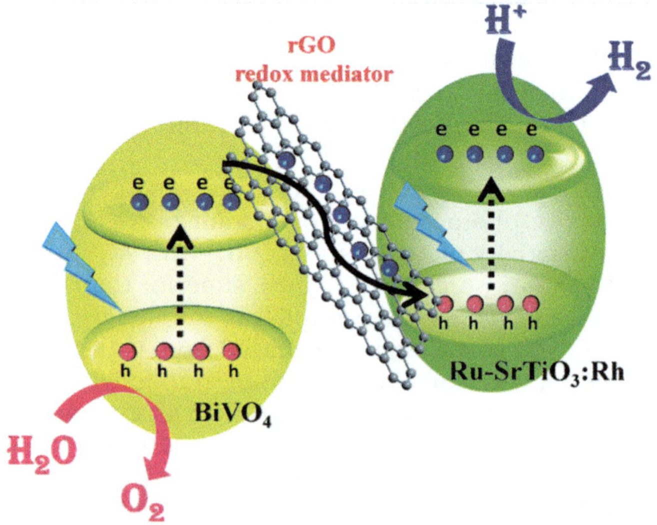

Figure 8.10 $BiVO_4$/Ru-$SrTiO_3$: Rh based Z-scheme photocatalyst with graphene sheet as a mediator. Reproduced from ref. 37 with permission from the Royal Society of Chemistry.

8.2.4 g-C_3N_4-based 2D Nanomaterials

Since the first report of Wang *et al.*[52] in 2009, g-C_3N_4 has emerged as a promising catalyst in the field of photocatalysis. It is characterized as a visible light active metal-free polymeric semiconductor with a narrow bandgap (2.4–2.7 eV) and a band structure appropriate for hydrogen generation from water splitting. Moreover, it is used to construct tailorable composite photocatalysts with adjustable composition, size, thickness, and pore structure. In view of its unique advantages, it is considered as the 'holy grail' of a next generation photocatalyst for the H_2 evolution reaction.[53] However, its practical applicability is still impeded due to shortcomings like high recombination rate of charge carriers, low electrical conductivity, and a lack of visible light absorption beyond 460 nm. Therefore, several strategies are adopted to modify g-C_3N_4 to overcome its barriers and to improve its performance. The development of g-C_3N_4-based heterojunctions proved to be the best strategy for retarding the charge recombination and improving the photocatalytic activity of the g-C_3N_4-based photocatalysts. For this reason, several metal oxides, chalcogenides, carbon materials, phosphides, oxyhalides, MOFs, polymers, metals, and other compounds are coupled with 2D g-C_3N_4.[53–56]

Krishnan *et al.*[57] demonstrated a unique 2D–2D heterojunction of N-ZnO and g-C_3N_4 in which nitrogen doping into ZnO introduces an impurity band near the valence band, allowing for visible light absorption. When N-ZnO forms a 2D–2D heterojunction with 30 wt% of g-C_3N_4, it demonstrated a remarkable H_2 evolution rate of 18 836 μmol g^{-1} h^{-1} in the presence of sacrificial agents. This high performance is a result of the synergistic effect of due to 2D–2D heterojunction and N-doping with an enlarged interfacial contact area, which retards the recombination of photoinduced charges. When another 2D metal oxide, Fe_2O_3, formed a tight 2D–2D hybrid nanostructure with g-C_3N_4 nanosheets, the resulting Z-scheme hybrid nanostructure presented an exceptional H_2 production of 30 mmol g^{-1} h^{-1} with an AQY of 44.35% at 420 nm, which was reported to be the highest H_2 production among the oxide/g-C_3N_4 composite-based photocatalysts. The enhanced performance is attributed to the tight interface between 2D Fe_2O_3 and 2D g-C_3N_4 that enable electron transfer, and the Z-scheme mechanism that suppresses electron–hole pair recombination.[58] Similarly, a rationally designed direct Z-scheme WO_3/g-C_3N_4 photocatalyst demonstrated a host–guest architecture with the WO_3 nano-cuboid embedded as a guest in a 2D g-C_3N_4 host (Figure 8.11).[59] This Z-scheme hybrid system demonstrated significantly improved H_2 generation performance as a result of quick electron transfer from WO_3 to impede charge recombination within g-C_3N_4 as well as longer charge carrier life duration. When CdS nanoparticles are decorated on g-C_3N_4 nanosheets *via* a solvothermal method, the H_2 evolution rate is enhanced (4494 μmol g^{-1} h^{-1}), which is 115-fold higher than that of pure g-C_3N_4.[60]

Most of the other metal chalcogenides with a narrow band gap act as co-catalysts for g-C_3N_4. For example, Zhao *et al.*[61] reported that NiS-coupled g-C_3N_4 demonstrated a high H_2 production rate of 16 400 μmol g^{-1} h^{-1}, which

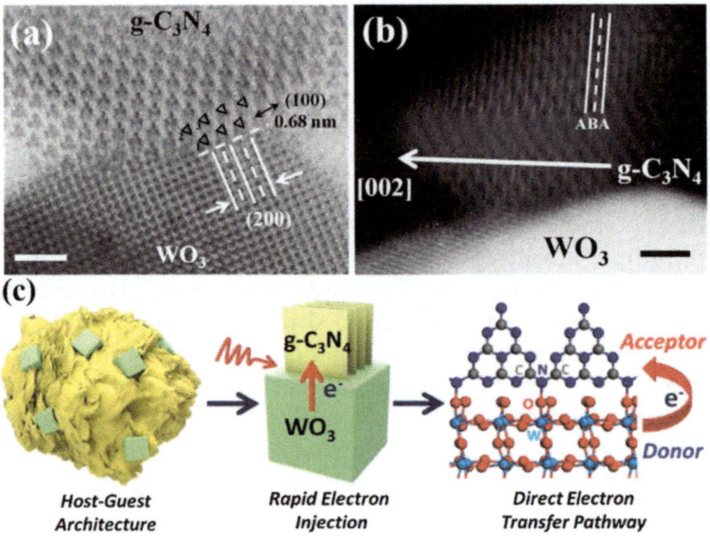

Figure 8.11 STEM image of well-defined interface between g-C_3N_4 and WO_3 (a) in-plane, (b) inter layer structure, (c) schematic diagram for the charge transfer from WO_3 nanocuboid to vertically standing g-C_3N_4 on it. Reproduced from ref. 59 with permission from Elsevier, Copyright 2017.

is almost 2500 times more than pure g-C_3N_4. Furthermore, its photocatalytic performance was the same after 40 h with no discernible decline. The exceptional performance is credited to the effective separation of photogenerated charges. Similarly, carbon-based materials such as graphene, carbon nanotubes, fullerenes, and carbon dots either act as a support for g-C_3N_4 or as a co-catalyst to boost g-C_3N_4 photoactivity.[62–64]

8.2.5 Black Phosphorous

Recently, black phosphorous (BP) has emerged as a rising star, capturing enormous attention owing to its distinctive properties such as adjustable layer dependent direct bandgap (0.3 eV–2 eV) that endows widespread absorption ranging until infrared, high carrier mobility (1000 $cm^2\ V^{-1}\ s^{-1}$), and large specific surface area.[65] Among the available strategies, formation of a heterojunction with BP has proven to be the most effective way to boost the photoactivity due to its feasibility to separate the photoinduced charges. Lu *et al.*[66] reported 2D–2D BP/Bi_2WO_6 Z-scheme heterojunction for NO removal from the air and H_2 evolution, which delivered the highest H_2 evolution (21 042 μmol g^{-1}) among the oxide photocatalysts coupled with BP due to interface synergic effect and effective charge separation. With layered Ni_2P loaded g-C_3N_4, 2D BP demonstrated a type I heterojunction with 50 times greater performance than pristine g-C_3N_4 as a result of an increase in active proton reduction sites and an efficient charge transfer mechanism (Figure

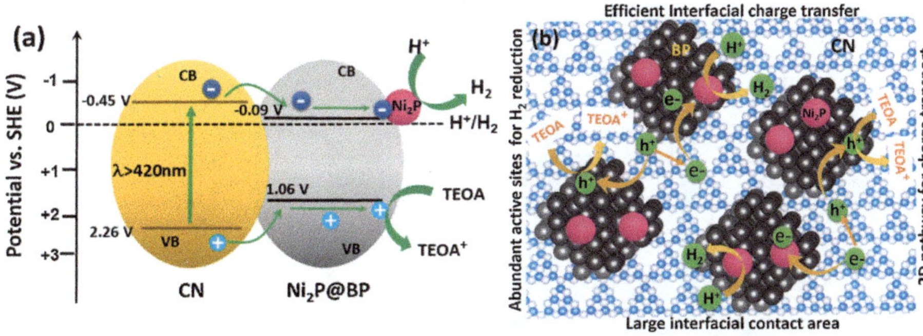

Figure 8.12 (a) Schematic presentation of Ni_2P@BP/CN along with the mechanism of photocatalytic mechanism. (b) Schematic representation of structural benefits of Ni2P@BP/CN photocatalyst. Reproduced from ref. 67 with permission from Elsevier, Copyright 2019.

8.12).[67] Lu and colleagues[68] reported that the combination of porous g-C_3N_4 (PCN) and metal–organic framework (MOF) coupled with BP displayed a H_2 evolution rate of 7380 μmol g^{-1} h^{-1}. The superior performance is attributed to the layered structure of BP, the high surface area of g-C_3N_4, and the high absorption ability of MOF, as well as the close contact between the three components. Besides, CdS coupled BP demonstrated type II heterojunction and delivered outstanding H_2 production of 11 192 μmol g^{-1} h^{-1} and AQY of 34.7% at 420 nm in lactic acid medium without the addition of co-catalyst. The high surface area and abundant active sites on the phosphorene sheet, as well as the favorable band structure and strong interfacial coupling, are attributable to such a significant performance.[69] In another study, a noble-metal-free nanohybrid: MoS_2 as a co-catalyst loaded on a CdS nanorod/BP manifested a remarkable H_2 evolution rate of 183.24 mmol g^{-1} h^{-1} and photostability 72 times higher than pristine CdS. The authors attributed it to the high catalytic sites, light harvesting, and charge migration rate.[70]

8.2.6 MXenes

MXenes are a new member of the 2D nanomaterial family with the general formula $M_{n+1}X_nT_x$, where M is a transition metal, X is nitrogen or carbon, and T is a functional group. $Ti_3C_2T_x$ was the first MXene discovered and is the most widely studied Mxene in the field of photocatalysis.[71] It mainly acts as a co-catalyst for the photocatalyst, accelerating the hydrogen evolution reaction (HER). For example, Meng *et al.*[72] reported a 1D–2D photocatalyst system of TiO_2 nanofiber/Ti_3C_2 with heterogeneous interfacial contact that delivered 6.979 mmol g^{-1} h^{-1} of H_2. With the aid of theoretical predications, Qiao and co-workers[73] rationally constructed –OH terminated Ti_3C_2 and coupled with visible light CdS photocatalyst, resulting in a Schottky junction between the two with rapid charge shuttling. Markedly, this composite produced a maximum of 14 342 μmol g^{-1} h^{-1} of H_2 due to the synergistic effect of charge

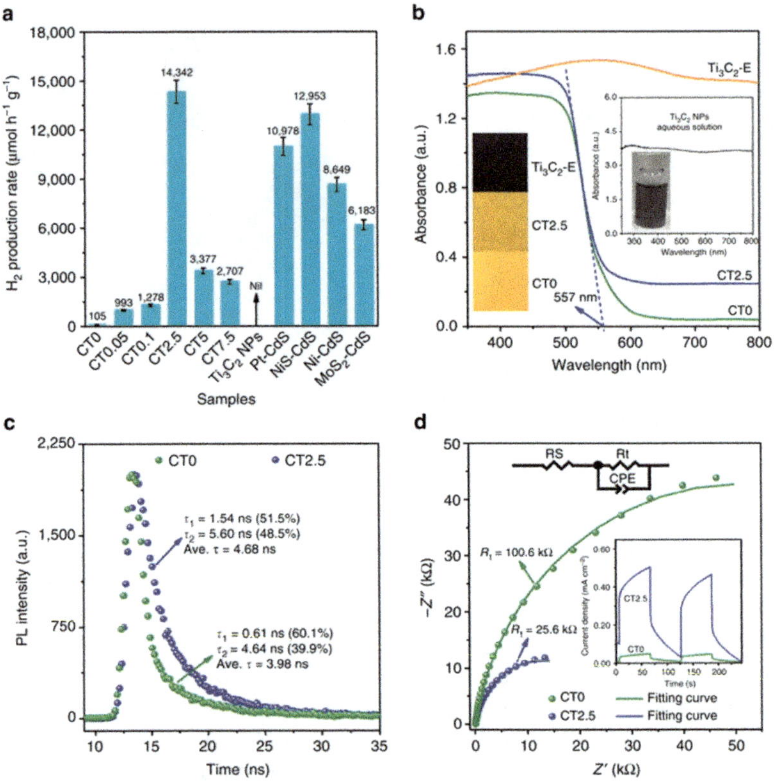

Figure 8.13 (a) Comparison of photocatalytic H_2 evolution activities of different catalyst. (b) UV-visible diffuse reflectance spectroscopy of pristine MXene, CdS and composite. (c) Time-resolved spectroscopy of pristine MXene and composite photocatalyst. (d) Nyquist plot of pristine MXene and composite photocatalyst. Reproduced from ref. 73, https://doi.org/10.1038/ncomms13907, under the terms of the CC BY 4.0 license https://creativecommons.org/licenses/by/4.0/.

separation and migration, as evidenced by the lifetime measurements and electrochemical impedance spectroscopy results in Figure 8.13. Similarly, $Zn_2In_2S_5$ with a flower-like morphology delivered 12 983.8 µmol g^{-1} of H_2 *via* water splitting due to the quasi-alloying effect and efficient electron transfer from $Zn_2In_2S_5$ to $Ti_3C_2(O,OH)_x$.[74] Tang *et al.*[75] synthesized 2D–2D heterojunction by electrostatic assembly of positively charged oxygen-doped g-C_3N_4 and negatively charged Ti_3C_2, resulting in intimate contact by Schottky junction. This resulted in a shorter migration pathway for charge transfer and thus increased charge separation. It demonstrated remarkable photoactivity, producing 25 124 µmol g^{-1} h^{-1} of H_2, which is two-fold higher than the pristine counterpart.

8.2.7 Layered Double Hydroxides (LDH)

LDH is a breakthrough finding in the fields of energy and environment due to its exceptional structural properties. It is also called as 'hydrotalcite' and has the general chemical formula $[M^{2+}_{(1-x)}M^{3+}_x(OH)_2]^{x+}(A^{n-})_{x/n}\cdot mH_2O$, where M^{2+} is a divalent cation, M^{3+} is a trivalent cation, and A^{n-} is the valence interlayer anion.[76] In short, it is composed of two positively charged hydroxide layers balanced by interlayer anions with interlayer bonding. The noteworthy fact is the positively charged MO_6 octahedra are uniformly distributed throughout the structure, which facilitates easy migration and separation of photogenerated charges and aids in the stability of the multilayer structure. The overall structure has a guest–host connection as the interlayer bonds are weak and the interlayer galleries can accommodate bulkier ions with only an anion exchange, which has an incremental effect on the photocatalytic performance.[77] With such a unique structural ability, both pristine and functionalized pristine LDH demonstrated excellent photocatalytic activity for H_2 and O_2 generation. However, particularly for H_2 generation, hybridization with the LDH revealed some remarkable cases that stand out due to their outstanding performance. For example, CeO_2 from the oxide family demonstrated 990 μmol for 2 h (16 483 μmol g^{-1} h^{-1}) of hydrogen evolution rate with MgAl-LDH. The authors recognized that the improved activity is due not only to the strong interfacial contact, but also to the uniform distribution of CeO_2 over the layered LDH, which suppressed carrier recombination over the Ce^{3+}–O trap sites.[78] However, the ternary WO_{3-x}/Ag/ZnCr-LDH heterostructure exhibited Z-scheme charge transfer, with Ag nanoparticle acting as a mediator linking the two components and significantly enhancing performance with an excellent H_2 evolution rate of 29 375 μmol g^{-1} h^{-1}.[79] Similarly, one more ternary 1D/0D/2D Z-scheme photocatalyst system $BiVO_4$/CdS/MgAl-LDH displayed 17 350 μmol g^{-1} h^{-1} and 33 000 μmol g^{-1} h^{-1} of HER and OER, respectively under UV irradiation.[80] Lee *et al.*[81] used a pulse laser ablation method to fabricate a CdS nanorod/NiFe-LDH photocatalyst that demonstrated almost 72 mmol g^{-1} h^{-1} of H_2 in lactic acid solution. This performance is the highest among the sulfide-LDH family, attributed to the reduced lifetime (0.15 ns *vs.* 0.19 s), indicating intimate contact and carrier transfer from CdS to NiFe-LDH. Qu *et al.*[82] fabricated a $Ni_xCo_{1-x}O$@C nanocube co-catalyst from NiCo-LDH and ZIF-67 MOF that, when sensitized by CdS, demonstrated significantly improved H_2 production from 0.7 mmol g^{-1} h^{-1} to 56 mmol g^{-1} h^{-1}. This 80-time enhancement is attributed to the electrostatic attraction between CdS and $Ni_xCo_{1-x}O$@C, which promotes separation of photoexcited electron–hole pair. Likewise, Kim and co-workers[83] synthesized a novel visible-light-driven hybrid composites of Zn-Cr LDH and Ti-based MOF (NH_2-MIL-125Ti) that demonstrated on an average of 127.6 mmol g^{-1} h^{-1} of H_2 without the addition of any co-catalyst (Figure 8.14). The hybrid composites are produced by solvothermal growth of NH_2-MIL-125(Ti) in the presence of exfoliated ZnCr-LDH NSs. The incredible yield is the result of strong electronic coupling between the two components, a large surface

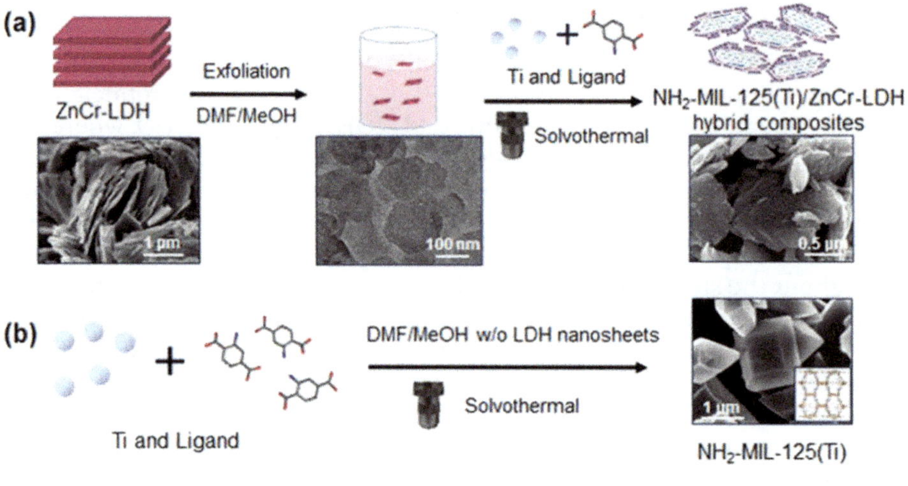

Figure 8.14 Illustration of synthesis of (a) hybrid composite of NH_2-MIL-125 (Ti) and ZnCr-LDH. (b) Pristine NH_2-MIL-125 (Ti). Reproduced from ref. 83, https://doi.org/10.1038/s41598-019-44008-6, under the terms of the CC BY 4.0 license https://creativecommons.org/licenses/by/4.0/.

area, expanded light absorption and band alignment to enhance the lifetime of photogenerated charge carriers.

8.2.8 Other Rising Stars

Bismuth oxyhalide (BiOX, where X = Cl, Br, and I) has recently emerged as a new class of photocatalyst with V–VI–VII semiconductors. Basically, the slabs [X–Bi–O–Bi–X] are combined together through van der Waals forces. A decahedral structure is formed when four oxygen atoms and four halogen atoms surround each central bismuth ion in the slab.[84,85] The excellent optical, electronic, and structural properties of these layered material generate a cooperative effect of powerful covalent bonding and van der Waals interaction among the atomic layers. The construction of heterojunction is an effective way to enhance the photocatalytic activity of Bi-based oxyhalides. Until now, metal oxide such as ZnO, TiO_2, $BiVO_4$, $CdWO_4$, $BiWO_4$ and $ZnWO_4$ have been coupled with various bismuth oxyhalides by p–n junction for effective pollutant degradation while enhancing solar light harvesting.[85] However, metal sulfides were coupled as co-catalysts for effective charge separation and migration of photoinduced charges. For example, Zhang *et al.*[86] recently reported $B_{12}O_{17}Cl_2/MoS_2$, in which MoS_2 acted as a co-catalyst for oxyhalide. They designed and crafted a Janus (Cl_2)-$(B_{12}O_{17})$-MoS_2 bilayer junction by selectively and chemically assembling the MoS_2 monolayer on to the $(B_{12}O_{17})$ end faces, where the built-in electric field separated the charges and aided in electron migration *via* a Bi–S bond (Figure 8.15). Consequently, such atomic level engineering geared up the performance to 33 mmol g^{-1} h^{-1} of hydrogen.

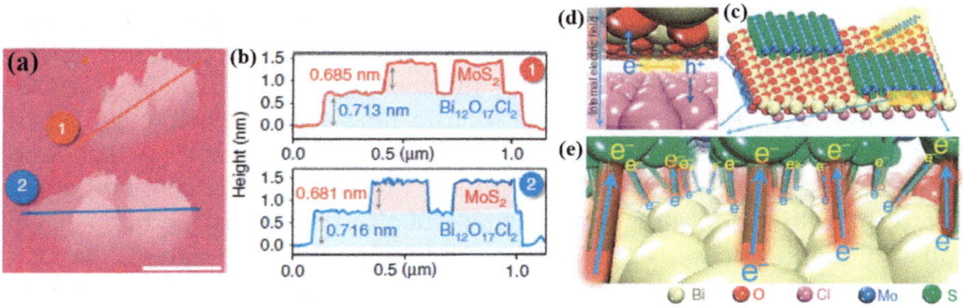

Figure 8.15 (a) AFM image and (b) corresponding height profile diagram of MoS_2/$Bi_{12}O_{17}C_{l2}$. (c) Schematic illustration of the crystal structure of MoS_2/$Bi_{12}O_{17}C_{l2}$. (d) and (e) Charge flow processes in MoS_2/$Bi_{12}O_{17}C_{l2}$. Reproduced from ref. 86, https://doi.org/10.1038/ncomms11480, under the terms of the CC BY 4.0 license https://creativecommons.org/licenses/by/4.0/.

Another emerging 2D catalyst that has become a hot spot in the field of catalysis is hexagonal boron nitride (hBN). It is also called 'white graphene' because it possesses alternatively bonded B and N with strong in-plane covalent sp^2 bonds. It has a large specific area, high thermochemical stability, and unique optical and electronic properties. However, it is rarely considered as a photocatalyst because of its large band gap of 4–6 eV.[87,88] The most effective method to minimize the bandgap of hBN and employ it as a photocatalyst is hybridization. Therefore, it has been coupled with TiO_2, CdS, C_3N_4 and graphene for dye degradation and H_2O splitting reactions to date.[87] In general, it worked as an electron acceptor and transporter in the case of TiO_2, whereas it improved the conductivity of CdS due to B and N defect vacancies. With C_3N_4, hBN formed a 2D–2D heterostructure by restacking *via* van der Waals interaction, resulting in a hybrid photocatalyst with a narrower bandgap and improved light absorption.[77] Liu *et al.*[89] constructed a hierarchical 3D core–shell structure of Co@BN that demonstrated about 13.78 mmol g^{-1} h^{-1} of Eosin Y-sensitized photocatalytic H_2 evolution. The nanocomposite exhibited enhanced photocatalytic activity by enhancing the transfer of photoexcited electrons and boosting the reduction reaction.

8.3 2D Nanomaterials-based Heterostructures for CO_2 Reduction

8.3.1 Metal Oxides

In recent years, 2D metal oxides have been investigated for photocatalytic CO_2 reduction, with TiO_2 being the widely studied metal oxide due to its remarkable stability, eco-friendliness, and cost-effectiveness.[12] For instance, Xu *et al.*[90] fabricated extremely thin TiO_2 nanosheets with a thickness of 2 nm and 95% of exposed (100) facets, which demonstrated approximately five-fold the CO_2 reduction performance of the reference TiO_2 cuboids with 53% of exposed (100) facets. The larger surface area and abundant exposed (100) facets, which enabled a greater number of more active sites for CO_2 conversion, were the main reasons for the improved performance. When TiO_2 nanosheets are combined with another semiconductor[91] or conducting material,[92] their activity can be greatly improved. Tu *et al.*[92] fabricated ultrathin multilayer hollow spheres made of alternating $Ti_{0.91}O_2$ nanosheets and CdS nanoparticles using a layer-by-layer self-assembly strategy. The CO_2 reduction activity of this $Ti_{0.91}O_2$/CdS hybrid catalyst for CH_4 generation was seven times higher when compared to bare $Ti_{0.91}O_2$ nanosheet hollow spheres. The remarkable performance of this hybrid catalyst was due primarily to its artificial Z-scheme architecture, which results in the effective separation of photoexcited charges and greatly increased lifespan of charge carriers. Besides, Di *et al.*[93] prepared Bi_2MoO_6 ultrathin nanosheets that exhibited remarkable CO_2 photoconversion performance with CO production of 3.62 μmol h^{-1} g^{-1}, 2.55-fold greater than their bulk counterparts. Of note, this activity was achieved without the use of a co-catalyst, photosensitizer, or sacrificial reagent. This impressive performance of Bi_2MoO_6 ultrathin nanosheets was attributed mainly to the improved light absorption, CO_2 adsorption and activation, and charge separation efficiency. Furthermore, Liang *et al.*[94] synthesized Bi_2WO_6 sheets with a single layer thickness that demonstrated a methanol yield of 75 μmol h^{-1} g^{-1} from CO_2 photoconversion, which is 125 folds greater than the bulk Bi_2WO_6 counterpart. The ultrathin thickness of Bi_2WO_6 enables easy migration of the photoexcited charges from the interior to the surface, thereby halting charge recombination. In addition to this, large specific surface area of the ultrathin Bi_2WO_6 facilitates stronger CO_2 adsorption and better optical absorption than the bulk Bi_2WO_6. Benefitting from these advantageous properties, the single layer Bi_2WO_6 nanosheets displayed superior CO_2 reduction performance than their bulk counterparts.

2D ternary metal oxides containing vanadium, including $BiVO_4$[95] and $InVO_4$,[96] have recently been identified as viable options for CO_2 photoconversion to produce solar fuels. For instance, one-unit-cell $BiVO_4$ layers with high vanadium vacancies showed outstanding CO_2 reduction activity, producing 398.3 μmol g^{-1} h^{-1} of methanol at a rate significantly higher than that of its counterpart with poor vacancies.[95] Notably, the catalyst with rich vacancies exhibited an apparent quantum yield of 5.96% when illuminated with 350 nm monochromatic light. The higher vanadium vacancies result in enhanced optical responsiveness and electronic conductivity, as well as improved charge separation efficiency, all of which contribute to the outstanding CO_2 reduction activity of one-unit-cell $BiVO_4$ layers. Zhu *et al.*[97] developed 2D monolayer $SnNb_2O_6$ sheets and applied them in CO_2 photoreduction to produce methane under visible light without the use of sacrificial reagents or co-catalysts. These $SnNb_2O_6$ sheets with a thickness of approximately 1 nm were prepared using a one-pot hydrothermal route without the use of any toxic organic additives. Because of the improved surface area, surface basicity, and photoelectric properties, the monolayer $SnNb_2O_6$ sheets produced a significant amount of methane (110.9 μL h^{-1} g^{-1}), almost 45 times higher than the multi-layered $SnNb_2O_6$. It has been shown that 2D gallates, such as $ZnGa_2O_4$,[98] can also be used as potential catalysts for CO_2 to hydrocarbon conversion.

8.3.2 Metal Chalcogenides

In recent years, 2D metal chalcogenides have garnered considerable attention in CO_2 photoreduction owing to their peculiar electronic and optical characteristics.[99] Among metal chalcogenides, MoS_2 has received extensive research attention owing to its stability, abundance, nontoxicity, and tunable optical characteristics.[100] Bulk MoS_2 has an indirect bandgap energy of approximately 1.3 eV.[101] Unfortunately, this bulk MoS_2 cannot be used effectively in photocatalysis because of its inadequate redox potentials. However, 2D MoS_2 sheets have a direct bandgap of approximately 1.9 eV, allowing them to absorb visible light and thus being suitable for photocatalytic applications.[102] 2D MoS_2 sheets have frequently been used as a support material for anchoring metals or semiconductors without affecting their mobility, and at a decreased risk of aggregate formation.[103] These MoS_2 sheets have also been reported to be used as an efficient co-catalysts in composite materials to improve CO_2 reduction efficiency. For instance, Zhou and group[104] developed a 2D hybrid nanojunction of MoS_2–TiO_2 for the efficient and selective reduction of CO_2 to methanol. When exposed to UV-vis light, this MoS_2–TiO_2 nanojunction demonstrated marked CO_2 reduction activity for methanol production, outperformed TiO_2 loaded with Pt, Au, or Ag noble metals in terms of activity and selectivity. As such, MoS_2 was considered to be a potential low-cost alternative to traditional noble metal-based co-catalysts. The 2D interfacial contact between the MoS_2 and TiO_2 nanosheets allows photoinduced electrons to migrate easily from TiO_2 to MoS_2, extending the lifespan of the charges. Moreover, a high d-electron density on the Mo-terminated edges of MoS_2 sheets facilitates electron transport and stabilizes CH_xO_y intermediates, resulting in high activity and selectivity for CO_2 photoreduction. In addition, TiO_2-graphene coupled with few-layered MoS_2 was found to display exceptional CO_2 conversion activity for selective CO generation, with a rate of 92.33 μmol h^{-1} g^{-1}, which is several orders of magnitude higher than bare TiO_2 and TiO_2-graphene.[105]

Besides, Jiao *et al.*[106] prepared partially oxidized SnS_2 atomic layers that exhibited significantly boosted CO_2 reduction activity for selective CO production at a rate of 12.28 μmol g^{-1} h^{-1} under visible light, which was 2.6 and 2.3 folds greater than the rates of the SnS_2 atomic layers and the poorly oxidized SnS_2 atomic layers, respectively. The boosted activity was due to the addition of local oxide domains that could serve as effective active sites and could aid the rate limiting COOH* adsorption process by stabilizing COOH* intermediates, thereby lowering their activation energy. The same group developed one-unit-cell $ZnIn_2S_4$ layers with variable defect concentrations to unveil the direct association between defect structures and charge separation characteristics at the atomic level.[107] The CO evolution of the $ZnIn_2S_4$ layers with abundant zinc vacancies was 33.2 μmol g^{-1} h^{-1}, which is about 3.6 times that of the $ZnIn_2S_4$ layers with poor zinc vacancies. The presence of zinc vacancies ensures improved charge density and accelerated charge transportation, which in turn results in improved lifespan of photoexcited charges. Moreover, the abundance of zinc vacancies improves the light absorption, CO_2 adsorption capacity, and also surface hydrophilicity. These characteristics, taken together, contributed to the outstanding CO_2 photoreduction performance of the $ZnIn_2S_4$ layers with rich zinc vacancy concentrations. In another study,[108] the same group developed a low-cost and earth abundant CuS atomic layers that afforded closely 100% selectivity for CO with a formation rate of up to 14.5 μmol g^{-1} h^{-1} under IR light illumination. This is the first study to show that IR light can induce CO_2 conversion in a metallic conductor.

8.3.3 Graphene-based 2D Nanomaterials

As previously mentioned, the successful exfoliation of 2D single layer of graphene by Geim and Novoselov[36] in 2004 sparked considerable research interest in graphene due to its fascinating characteristics, which include extraordinary optical transparency, excellent electronic conductivity, strong CO_2 adsorption, and a large surface area. The most studied forms of graphene materials are graphene oxide (GO), reduced GO (RGO), and virgin graphene, which can be synthesized using a variety of methods. The functional groups that are linked to the edges and surfaces distinguish these from one another, and each has its own set of properties.[109] For instance, Hsu *et al.*[110] developed a series of GO catalysts for solar fuel production from CO_2 photoreduction. Under simulated sunlight irradiation, the synthesized GO catalysts with bandgap energies ranging from 2.9 to 4.4 eV demonstrated impressive CO_2 conversion performance for methanol production, with an optimum yield of 0.172 μmol g^{-1} h^{-1}, which is significantly higher than P25 TiO_2. Notably, when GO was decorated with Cu nanoparticles, its CO_2 photoreduction activity was boosted even further.[111] The Cu/GO photocatalyst containing 10 wt% Cu afforded the best CO_2 conversion activity for methanol and formaldehyde generation, with a rate of 6.84 μmol g^{-1} h^{-1}, which was 60 and 240 times greater than that observed for bare GO and P25 TiO_2, respectively.

Besides, when synthesizing virgin graphene or RGO from GO, most of the oxygenated functional groups are eliminated, which allows the sp^2 hybrid carbon framework to be restored. This restoration merges all electronic energy levels of virgin graphene or RGO, resulting in a narrowing of the bandgap to zero and the loss of semiconductor properties. Research into the use of graphene for photocatalytic CO_2 reduction applications has advanced dramatically since Liang *et al.*[112] published their first study in 2011. In the first study, Liang *et al.*[112] conceptualized the role of defects in graphene in the photocatalytic CO_2 reduction performance of graphene-TiO_2 composites. In detail,

RGO (prepared *via* a conventional method) and solvent-exfoliated graphene (SEG) are coupled with TiO_2 to form RGO–TiO_2 and SEG–TiO_2, respectively. The defect density of SEG was found to be lower than that of RGO. In contrast to the conventional role of defects in catalysis, the less defective SEG–TiO_2 exhibited higher photocatalytic activity than the RGO–TiO_2 for the reduction of CO_2 to CH_4, with up to a seven-fold enhancement compared to pure TiO_2 under visible light illumination. This is because the reduced defect density of SEG sheets improved electrical mobility and enabled a greater number of photoexcited electrons to transfer efficiently to reactive sites, thus enhancing the CO_2 reduction. The same team compared the photocatalytic activities of carbon nanotube (1D)/TiO_2 nanosheet (2D) and SEG (2D)/TiO_2 nanosheet (2D) hybrid catalysts to see how the dimensionality of carbon nanomaterial affected CO_2 photoreduction.[113] In comparison to a 1D/2D carbon nanotube/TiO_2 hybrid, the face-to-face 2D/2D SEG/TiO_2 hybrid had a greater intimate interfacial contact area, which led to a better photoreduction of CO_2.

2D graphene can be used as a low-cost co-catalyst to replace expensive noble metals in CO_2 photoconversion. For example, Yu *et al.*[114] demonstrated that optimized RGO–CdS hybrid exhibited a remarkable CO_2 conversion activity to yield methane, with a methane production of 2.51 $\mu mol\ h^{-1}\ g^{-1}$, outperforming CdS and Pt–CdS catalysts under the same experimental conditions. The large surface area of RGO can increase the surface active sites on the hybrid catalyst, which is beneficial for enhancing the CO_2 photoreduction activity. The π–π conjugation interaction between RGO and CO_2 molecules can destabilize and activate CO_2 molecules, resulting in an exceptional CO_2 reduction performance of the RGO–CdS hybrid. Graphene sheets have been found to be useful as support materials for improving the CO_2 reduction performance of photocatalytically unstable semiconductors. For example, the photostability of CdS dramatically improved when it is encapsulated with a few layers of amine-functionalized graphene.[115] As a result, the graphene/CdS hybrid demonstrated stable activity even after 10 photocatalytic CO_2 runs. Tonda and his group[116] rationally developed a Bi_2WO_6/RGO/g-C_3N_4 Z-scheme system with 2D/2D/2D configuration. RGO played two prominent roles in this system. First, it was served as a redox mediator to promote Z-scheme charge transfer between g-C_3N_4 and Bi_2WO_6, resulting in a greater extent of charge separation in the Bi_2WO_6/RGO/g-C_3N_4 hybrid (Figure 8.16(a)). Second, it acts as an electron acceptor, capturing electrons from g-C_3N_4 for the subsequent CO_2 reduction reaction. As a result, the Bi_2WO_6/RGO/g-C_3N_4 Z-scheme system achieved record performance in the yields of carbonaceous products (CO + CH_4) with a selectivity of 92% against H_2 (Figure 8.16(b)).

8.3.4 g-C_3N_4-based 2D Nanomaterials

g-C_3N_4 is considered as the most stable metal free photocatalyst with appropriate band positioning for CO_2 photoreduction and this is the reason it has attracted much more attention in the scientific community.[117] However, in

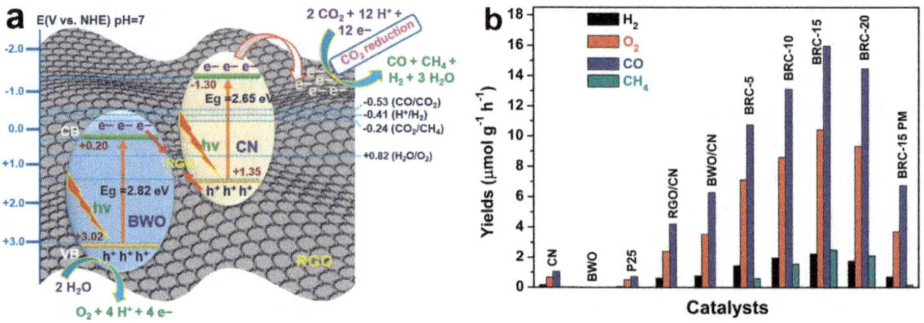

Figure 8.16 (a) Schematic illustration of proposed mechanism of CO_2 photoreduction in Bi_2WO_6/RGO/g-C_3N_4. (b) Evolution rates of different gases over Bi_2WO_6/RGO/g-C_3N_4 photocatalysts. Reproduced from ref. 116 with permission from Elsevier, Copyright 2018.

spite of the excellent properties, g-C_3N_4 suffers from poor solar conversion efficiency due to rapid recombination of charges and low surface area.[117,118] Formation of heterostructure with the other materials has capacity to address the aforementioned issue. Therefore, the literature of g-C_3N_4 in last few years has bloomed with g-C_3N_4-based composites. The g-C_3N_4 has been combined with metal oxides (TiO_2, ZnO, NiO, Cu_2O, SnO_2, MnO_2, Fe_2O_3, WO_3), metal sulfide (MoS_2, SnS, SnS_2, Cu_2S, $CuInS_2$), carbon (graphene and graphene oxide), oxyhalides (BiOBr, BiOCl, $Bi_4O_5I_2$), LDH, MXenes and metal organic framework and so on.[118,119] Generally, g-C_3N_4 based composites show either Type II heterojunction or Z-scheme heterostructure. A binary composite of g-C_3N_4/ZnO was constructed *via* single step calcination process which demonstrated direct Z-scheme heterojunction. The composite exhibited 2.5 times higher performance due to efficient electron transfer from ZnO to g-C_3N_4 and due to extended light absorption in the visible region.[120] Adekoya *et al.*[121] prepared Cu loaded g-C_3N_4/TiO_2 nanocomposite which showed CH_3OH (2574 μmol g^{-1}) and HCHO (5069 μmol g^{-1}), respectively, upon CO_2 photoreduction. The resultant enhanced performance is attributed to the location of Cu metal in the composite and creation of Schottky barrier promoting the better charge separation of photoexcited electrons. A novel 2D/2D MnO_2/g-C_3N_4 demonstrated largely enhanced photocatalytic activity due to efficient separation of charges and increased surface area of the layered heterojunction.[122] Ogale and co-workers[123] reported nature inspired Z-scheme g-C_3N_4/$FeWO_4$ system which delivered six times higher performance for CO_2 photoreduction to CO with excellent selectivity compared to pristine g-C_3N_4. Similarly, in the case of metal sulfide based composites, ternary Z-scheme photocatalyst g-C_3N_4/Au/$ZnIn_2S_4$ is developed by Lu *et al.*[124] demonstrating remarkable CO production of 243.2 μmol g^{-1} h^{-1} with 94.1% selectivity. The improved performance is ascribed to the Au which acted as mediator for efficient charge transfer promoting the better charge separation. Su *et al.*[125]

reported carbon coated hollow tubular Cu_xS derived from classical MOF HKUST-1 precursor and further loaded on g-C_3N_4 to form three component heterojunction (C-Cu_xS@g-C_3N_4). This unique photocatalyst facilitated extended light absorption till 800 nm unlike g-C_3N_4 and the carbon coating facilitated electron–hole separation. The given benefits were very well reflected on the photocatalytic activity where reactivity of 1062.6 μmol g^{-1} and selectivity of 97% is achieved. A new class of layered material—layered double hydroxide (LDH)—emerged as a promising 2D material which is also explored as a photocatalyst to couple with g-C_3N_4 through electrostatic attraction. Ogale *et al.*[126] reported NiAl LDH to construct 2D–2D heterostructure with g-C_3N_4 which delivered almost five times and nine times higher CO production compared to pristine g-C_3N_4 and pristine LDH, respectively (Figure 8.17). The originated result is ascribed to the synergetic effect of g-C_3N_4 and LDH including suppressed recombination and increased charge transfer. Tonda *et al.*[127] fabricated hybrid heterostructure with a nitrogen-doped carbon dot (NCD) coupled with CoAl LDH/g-C_3N_4, which displayed highest CH_4 production of 25 μmol g^{-1} h^{-1} with 99% selectivity. From the literature it is evident that 2D–2D structure of LDH and g-C_3N_4 foster electron–hole pair separation but in this case with the addition of NCDs the process became much more efficient. More importantly, NCDs has capability to absorb longer wavelength of visible light and emit shorter wavelength light. So, the converted shorter wavelength light could also excite g-C_3N_4 and LDH promoting more production of electrons and holes. Additionally, H_2 generation is sluggish process on NCDs which promoted CO_2 reduction over proton reduction resulting in higher selectivity of the products. Layered material from oxyhalide family is also employed to minimize the recombination of photoinduced charges in g-C_3N_4. For example, indirect Z-scheme based hybridization of g-C_3N_4 with $Bi_4O_5I_2$ is designed by Wong *et al.*[128] *via* I_3^-/I^- as a redox mediator. This system showed the remarkable yield of CO at the rate of 45.6 μmol g^{-1} h^{-1}. The reason lies in the efficient charge transfer *via* redox couple as evident from electrochemical impedance spectroscopy and radical test.

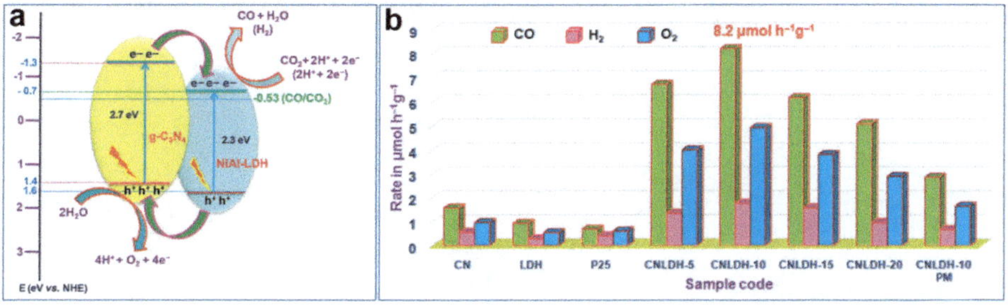

Figure 8.17 (a) Schematic illustration of proposed mechanism of CO_2 photoreduction in g-C_3N_4/NiAl-LDH. (b) Evolution rates of different gases over g-C_3N_4/NiAl-LDH photocatalysts. Reproduced from ref. 126 with permission from American Chemical Society, Copyright 2018.

8.3.5 Layered Double Hydroxide (LDH)

LDH has attracted much more attention in the scientific community due to its layered structure, tunable and narrow band gap.[129] However, it exhibits poor solar quantum efficiency due to slow charge carrier mobility and electron–hole recombination. Heterojunction formation with layered LDH is considered to be the best way to improve the abovementioned issue of LDH.[130,131] Kumar *et al.*[132] synthesized P25@CoAl LDH by one pot hydrothermal method. CoAl LDH is good photocatalyst for oxygen evolution reaction whereas TiO_2 is good electron acceptor. Therefore, the construction of heterojunction between the two facilitates effective driving force to drive the oxidation and reduction reaction without any sacrificial agent. Additionally, basicity of CoAl LDH enhances the CO_2 absorption on its surface resulting in the more yield and selectivity. Similarly, Yan *et al.*[133] reported MgAl LDH/Fe_3O_4 composite, which delivered CO and CH_4 at the rate of 442.2 μmol g^{-1} h^{-1} and 223.9 μmol g^{-1} h^{-1}, respectively. The exceptional performance is attributed to the synergetic effect of ultra-thin MgAl LDH and Fe_3O_4 in fostering the charge separation and in reduction of charge transfer resistance.

Another oxide Ga_2O_3 is coupled with MgAl LDH *via* Ag as co-catalyst that delivered 211.7 μmol h^{-1} of CO under UV-light irradiation. High surface area and high CO_2 absorption ability of MgAl LDH concentrates CO_2 near Ag co-catalyst which results in enhance selective CO_2 reduction to CO using water as reductant.[134] Recently, Shen and co-workers[135] reported a unique g-C_3N_4 porous foam substrate loaded *in situ* with NiFe LDH nanoplates (Figure 8.18), which demonstrated excellent CO_2 absorption of 1.52 mmol g^{-1} along with CO_2 reduction to CO (55 μmol g^{-1} h^{-1}). The strong synergistic effect of CN and LDH lead to improved light harvesting, CO_2 capture and strong interfacial charge transfer resulting in high CO yield.

8.3.6 Oxyhalides

In recent times, a new class of layered photocatalyst comprising of (V–VI–VII) semiconductors such as bismuth oxyhalides (BiOX, X= Cl, Br and I) has emerged as a promising candidate for CO_2 reduction.[136–138] However,

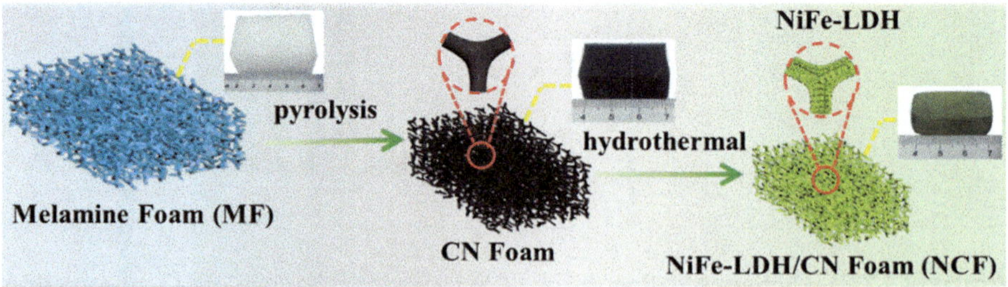

Figure 8.18 Schematic illustration of preparation of porous NiFe-LDH/CN (NCF) foam. Reproduced from ref. 135 with permission from Elsevier, Copyright 2022.

mismatch energy level positioning with respect to CO_2 reduction potentials hinders its performance for photocatalytic CO_2 reduction. Thanks to the spatial separation of photogenerated charges by hybridization or by formation of heterojunction with other materials. For example, Ye *et al.*[139] loaded Bi co-catalyst on $Bi_{14}O_5I_2$, which displayed outstanding photothermal-catalytic reduction of CO_2 to CO (40.2 μmol g^{-1} h^{-1}) and CH_4 (7.19 μmol g^{-1} h^{-1}), respectively, under UV-vis-IR irradiation. The augmented performance of $Bi/Bi_{14}O_5I_2$ is credited to the co-catalyst and photothermal effect of Bi nanoparticles which prohibited the electron–hole recombination and accelerated the light to thermal energy conversion, respectively. In another similar example, single isolated Co atom is loaded on ultrathin Bi_3O_4Br layers which minimized the CO_2 activation energy barrier and favored CO_2 absorption, activation, and charge carrier separation. This resulted in CO formation rate of 107.1 μmol g^{-1} h^{-1}, which is four times and 32 times higher than atomic layered Bi_3O_4Br and bulk Bi_3O_4Br, respectively.[140] Jain and co-workers[141] crafted Co(I) phthalocyanine embedded bismuth oxyhalide ($CoPc@Bi_{24}O_{31}Br_{10}$) for the photoreduction of CO_2 into methanol with the yield of 3485.3 μmol g^{-1} h^{-1} using trimethylamine as a sacrificial donor under visible light illumination. The highest yield is due to the higher CO_2 concentration on the surface and continuous electron supply from the complex to the CB of BOB.

Recently, 0D/2D $CsPbBr_3$ quantum dots/Bi_2WO_6 nanosheets (CPB/BWO), Z-scheme based photocatalyst demonstrated CH_4 production of 503 μmol g^{-1} which is 9.5 times higher than $CsPbBr_3$. It is stated that photoexcited holes in the VB of $CsPbBr_3$ would recombine with the holes of CB of Bi_2WO_6, leaving spatially separated electrons and holes at $CsPbr_3$ QDs and Bi_2WO_6 nanosheets, respectively, and contributing to enhanced performance (Figure 8.19).[142]

8.3.7 MXenes

MXene, a young member of the 2D family of nanomaterials holds promising features to be employed in CO_2 photoreduction.[143,144] Among the oxides, TiO_2 (P25) is coupled with the surface alkalinized Ti_3C_2 which delivered CO (11.74 μmol g^{-1} h^{-1}) and CH_4 (16.61 μmol g^{-1} h^{-1}), respectively. Here, the MXene acted as co-catalyst and the significant yield is attributed to its superior electrical conductivity. Moreover, alkalinized Ti_3C_2 provided site for enhanced CO_2 adsorption and activation.[145] On the similar line, g-C_3N_4 is also decorated on the surface alkalinized Ti_3C_2 which showed almost 5.9 times higher CO evolution rate compared to pristine g-C_3N_4.[146] Yu *et al.*[147] reported 2D–2D heterojunction of ultra-thin MXene/Bi_2WO_6 nanosheet, ensuring short migration distance and large surface area of contact. This resulted in CH_3OH and CH_4 products after CO_2 photoreduction which is 4.6 times higher than the pristine Bi_2WO_6. In another report $CsPbBr_3$ nanocrystal were grown *in situ* on the monolayer of MXene where MXene played the important role of efficient co-catalyst. Quenching of fluorescence of $CsPbBr_3$ NCs and conductivity measurement confirmed the efficient charge transfer between MXene and

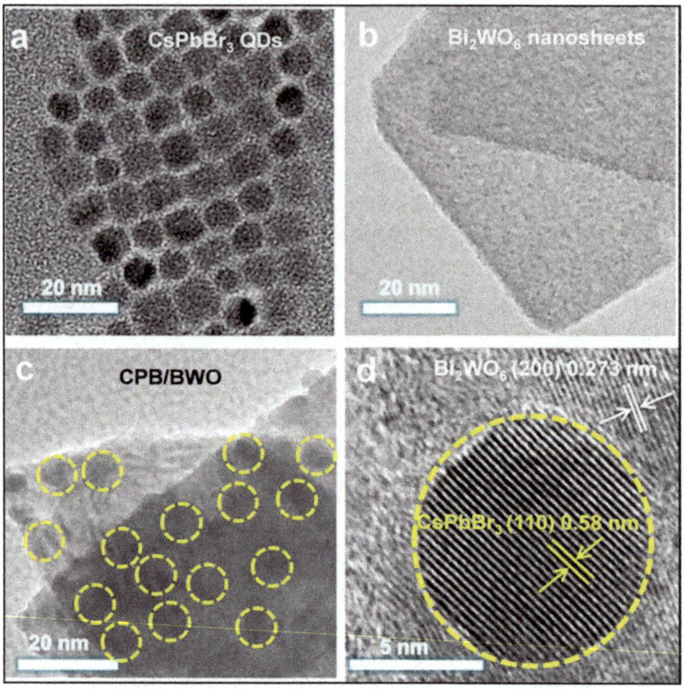

Figure 8.19 TEM image of (a) $CsPbBr_3$ quantum dots, (b) Bi_2WO_6 nanosheets, (c) and (d) HRTEM image of heterojunction of CPN/BWO. Reproduced from ref. 142 with permission from American Chemical Society, Copyright 2020.

$CsPbBr_3$.[148] Recently, Xu and co-workers[149] published 2D $Ti_3C_2T_x$ MXene activated and anchored by single Co atom loading by forming Co-oxygen–carbon bond. As a result, the ensemble delivered an extraordinary production rate of 6.06 mmol g^{-1} h^{-1} of CO, which exceeds the performance of MXene based photocatalysts reported up till now (Figure 8.20).

8.3.8 Other Rising Stars

Black phosphorous (BP) has just emerged in recent decades for photocatalytic applications. BP is made up of stacked layered structure with sp^3 hybridization leading to puckered like honeycomb structure. It has wide range of band gap ranging from 0.3 eV to 2 eV. It has become star material for CO_2 photoreduction due to its lower conductive band positioning, which is propitious for the CO_2 reduction to the value added products.[150] Ye and co-authors[151] reported noble metal free co-catalyst, single Co atom decorated BP nanosheets which presented almost 18-fold increment in the performance of CO evolution rate (88.6 μmol h^{-1}) compared to pristine BP in the presence of Ru-complex as photosensitizer. Loading of Co on BP opened a new charge transfer channel and making new sites available for CO_2 reduction. 2D g-C_3N_4 formed type-II heterojunction with 2D BP, which boosted the

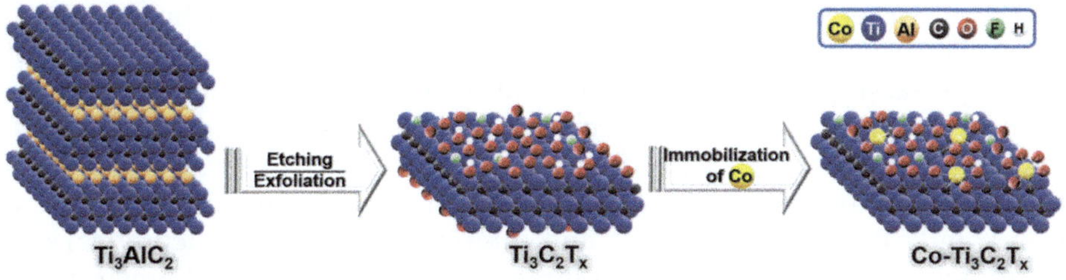

Figure 8.20 Schematic diagram for the preparation of Co-$Ti_3C_2T_x$. Reproduced from ref. 149 with permission from Elsevier, Copyright 2021.

separation of electron–hole pairs, resulting in the production rate of 187.7 μmol g^{-1} h^{-1} of CO.[152] Recently, perovskite with 2D BP demonstrated CO production rate of 76.6 μmol g^{-1} h^{-1} due to efficient charge transfer *vis* aromatic molecules as a mediator.[153]

Hexagonal boron nitride (hBN) is another emerging metal free layered catalyst. However, low optical absorption in the visible region and recombination of excited charges restricts its application for CO_2 photoreduction.[88] In 2015, Wang *et al.*[154] introduced carbon doped hBN to form ternary layered alloy of B–C–N for the CO_2 photoreduction. On the similar line Zhou *et al.*[155] reported ultra-thin oxygen modified hBN for enhance photocatalytic CO_2 performance of 12.5 μmol g^{-1} h^{-1} of CO. To tune the performance of boron carbon nitride (BCN) for visible responsive photocatalysis, CdS nanoparticles were coupled which showed remarkable CO production rate of 250 μmol g^{-1} h^{-1} with triethanolamine as reducing agent.[156]

Another 2D material, metal carbides, are explored now days for CO_2 photoreduction to value added fuels. Excellent chemical stability and appropriate bandgap makes it approved candidate for CO_2 reduction. Ultra-thin SiC nanosheets delivered improved 3.11 μmol g^{-1} h^{-1} of CH_4 and 90.6% of selectivity compared to SiC micro size (0.76 μmol g^{-1} h^{-1}).[157] Goodenough and co-authors[158] reported Mo_2C/CdS for CO_2 photoreduction which delivered 29.3 μmol h^{-1} of CO with 98.3% of selectivity.

8.4 Conclusion and Outlook

In summary, we have reviewed almost all the 2D materials, including the most recent emerging materials, for photocatalytic hydrogen generation from water splitting and CO_2 reduction. More importantly, the focus of this chapter was directed towards the formation of unique hybrid nanostructures employing 2D materials with astonishing structural and optoelectronic properties that benefit the given applications. Thus, the hybridization strategy offers a broad and diverse opportunity to create novel functional materials with unexpected properties. Such innovative materials with unprecedented features can be explored and made resourceful in other applications of the clean energy sector including batteries, supercapacitors, fuel cells, solar cells, and so on. Nevertheless, 2D materials, like any other materials, have a variety of hurdles which need to be addressed such as controlling the number of layers, preserving photostability, preventing agglomeration, and large-scale production. Therefore, much more innovative synthesis methods, strategies, and *in situ* operando characterization techniques are necessary to address the aforementioned hurdles of 2D materials. Apart from this, more attention should be given on surface and interface study using theoretical calculations. A better insight and comprehension of the unique characteristics of the hybridization procedure can be obtained from the theoretical study. From the future perspective, 2D materials will undoubtingly offer an exciting platform for other new rising stars and a wide range of applications.

References

1. A. Fujishima and K. Honda, *Nature*, 1972, **238**, 37–38.
2. Y. Wang, J. Mao, X. Meng, L. Yu, D. Deng and X. Bao, *Chem. Rev.*, 2018, **119**, 1806–1854.
3. J. Di, J. Xiong, H. Li and Z. Liu, *Adv. Mater.*, 2018, **30**, 1704548.
4. S. Z. Butler, S. M. Hollen, L. Cao, Y. Cui, J. A. Gupta, H. R. Gutiérrez, T. F. Heinz, S. S. Hong, J. Huang, A. F. Ismach, E. Johnston-Haperin, M. Kuno, V. V. Plashnitsa, R. D. Robinson, R. S. Ruoff, S. Salahuddin, J. Shan, L. Shi, M. G. Spencer, M. Terrones, W. Windl and J. E. Goldberger, *ACS Nano*, 2013, **7**, 2898–2926.
5. R. Bhosale, S. Tonda, S. Kumar and S. B. Ogale, Two-dimensional materials for photocatalytic water splitting and CO_2 reduction, *Nanostructured Photocatalysts*, Elsevier Ltd, 2020, DOI: 10.1016/b978-0-12-817836-2.00007-7.
6. D. Deng, K. S. Novoselov, Q. Fu, N. Zheng, Z. Tian and X. Bao, *Nat. Nanotechnol.*, 2016, **11**, 218–230.
7. T. Su, Q. Shao, Z. Qin, Z. Guo and Z. Wu, *ACS Catal.*, 2018, **8**, 2253–2276.
8. A. H. Castro Neto, F. Guinea, N. M. R. Peres, K. S. Novoselov and A. K. Geim, *Rev. Mod. Phys.*, 2009, **81**, 109–162.
9. J. Low, S. Cao, J. Yu and S. Wageh, *Chem. Commun.*, 2014, **50**, 10768–10777.
10. J. Azadmanjiri, C. C. Berndt, J. Wang, A. Kapoor and V. K. Srivastava, *RSC Adv.*, 2016, **6**, 109361–109385.
11. I. Y. Kim, Y. K. Jo, J. M. Lee, L. Wang and S. Hwang, *J. Phys. Chem. Lett.*, 2014, **5**, 4149–4161.
12. L. Wang and T. Sasaki, *Chem. Rev.*, 2014, **114**, 9455–9486.
13. S. Ida and T. Ishihara, *J. Phys. Chem. Lett.*, 2014, **5**, 2533–2542.
14. H. N. Kim, T. W. Kim, I. Y. Kim and S. Hwang, *Adv. Funct. Mater.*, 2011, **21**, 3111–3118.
15. K. Dai, L. Lu, Q. Liu, G. Zhu, Q. Liu and Z. Liu, *Dalton Trans.*, 2014, **43**, 2202–2210.
16. W. Gu, F. Lu, C. Wang, S. Kuga, L. Wu, Y. Huang and M. Wu, *ACS Appl. Mater. Interfaces*, 2017, **9**, 28674–28684.

17. Y. Yuan, Z. Ye, H. Lu, B. Hu, Y. Li, D. Chen, J. Zhong, Z. Yu and Z. Zou, *ACS Catal.*, 2016, **6**, 532–541.
18. D. Mateo, E. Iván, J. Albero, A. Primo and H. García, *Appl. Catal., B*, 2017, **201**, 582–590.
19. Y. Hu, L. Mao, X. Guan, K. A. Tucker, H. Xie, X. Wu and J. Shi, *Renewable Sustainable Energy Rev.*, 2020, **119**, 109527.
20. G. Zhang, G. Liu, L. Wang and J. T. S. Irvine, *Chem. Soc. Rev.*, 2016, **45**, 5951–5984.
21. Y. Hu, G. Li, S. Zong, J. Shi and L. Guo, *Catal. Today*, 2018, **315**, 117–125.
22. H. Zhou, E. M. Sabio, T. K. Townsend, T. Fan, D. Zhang and F. E. Osterloh, *Chem. Mater.*, 2010, **22**, 3362–3368.
23. S. Ida, A. Takashiba, S. Koga, H. Hagiwara and T. Ishihara, *J. Am. Chem. Soc.*, 2014, **136**, 1872–1878.
24. P. Kumbhakar, C. C. Gowda, P. L. Mahapatra, M. Mukherjee, K. D. Malviya, M. Chaker, A. Chandra, B. Lahiri, P. M. Ajayan, D. Jariwala, A. Singh and C. S. Tiwary, *Mater. Today*, 2021, **45**, 142–168.
25. D. Voiry, J. Yang and M. Chhowalla, *Adv. Mater.*, 2016, **28**, 6197–6206.
26. Q. Lu, Y. Yu, Q. Ma, B. Chen and H. Zhang, *Adv. Mater.*, 2016, **28**, 1917–1933.
27. M. Chhowalla, H. S. Shin, G. Eda, L. Li, K. P. Loh and H. Zhang, *Nat. Chem.*, 2013, **5**, 263–275.
28. G. Chen, D. Li, F. Li, Y. Fan, H. Zhao, Y. Luo, R. Yu and Q. Meng, *Appl. Catal., A*, 2012, **443**, 138–144.
29. K. Chang, Z. Mei, T. Wang, Q. kang, S. Ouyang and J. Ye, *ACS Nano*, 2014, **8**, 7078–7087.
30. Y. Yuan, D. Chen, J. Zhong, L. Yang, J. Wang, M. Liu, W. Tu, Z. Yu and Z. Zou, *J. Mater. Chem. A*, 2017, **5**, 15771–15779.
31. M. Shen, Z. Yan, L. Yang, P. Du, J. Zhang and B. Xiang, *Chem. Commun.*, 2014, **50**, 15447–15449.
32. J. Yu, C. Xu, F. Ma, S. Hu, Y. Zhang and L. Zhen, *ACS Appl. Mater. Interfaces*, 2014, **6**, 22370–22377.
33. L. Jing, Y. Xu, Z. Chen, M. He, M. Xie, J. Liu, H. Xu, S. Huang and H. Li, *ACS Sustainable Chem. Eng.*, 2018, **6**, 5132–5141.
34. H. Chauhan, K. Soni, M. Kumar and S. Deka, *ACS Omega*, 2016, **1**, 127–137.
35. X. Xu, F. Luo, W. Tang, J. Hu, H. Zeng and Y. Zhou, *Adv. Funct. Mater.*, 2018, **28**, 1804055.
36. K. S. Novoselov, A. K. Geim, S. V. Morozov, D. Jiang, Y. Zhang, S. V. Dubonos, I. V. Grigorieva and A. A. Firsov, *Science*, 2004, **306**, 666–669.
37. P. Kumar, R. Boukherroub and K. Shankar, *J. Mater. Chem. A*, 2018, **6**, 12876–12931.
38. G. Xie, K. Zhang, B. Guo, Q. Liu, L. Fang and J. R. Gong, *Adv. Mater.*, 2013, **25**, 3820–3839.
39. J. Zhang, Z. Xiong and X. S. Zhao, *J. Mater. Chem.*, 2011, **21**, 3634–3640.
40. Q. Xiang, J. Yu and M. Jaroniec, *Chem. Soc. Rev.*, 2012, **41**, 782–796.
41. C. Xu, X. Wang and J. Zhu, *J. Phys. Chem. C*, 2008, **112**, 19841–19845.

42. P. Zeng, Q. Zhang, X. Zhang and T. Peng, *J. Alloys Compd.*, 2012, **516**, 85–90.
43. Q. Li, B. Guo, J. Yu, J. Ran, B. Zhang, H. Yan and J. R. Gong, *J. Am. Chem. Soc.*, 2011, **133**, 10878–10884.
44. M. Liu, F. Li, Z. Sun, L. Ma, L. Xu and Y. Wang, *Chem. Commun.*, 2014, **50**, 11004–11007.
45. A. Kudo and Y. Miseki, *Chem. Soc. Rev.*, 2009, **38**, 253–278.
46. A. Iwase, Y. H. Ng, Y. Ishiguro, A. Kudo and R. Amal, *J. Am. Chem. Soc.*, 2011, **133**, 11054–11057.
47. M. Li, C. Song, Y. Wu, M. Wang, Z. Pan, Y. Sun, L. Meng, S. Han, L. Xu and L. Gan, *Appl. Surf. Sci.*, 2019, **478**, 744–753.
48. Y. Huang, Y. Liu, D. Zhu, Y. Xin and B. Zhang, *J. Mater. Chem. A*, 2016, **4**, 13626–13635.
49. S. Wan, M. Ou, Q. Zhong, S. Zhang and F. Song, *Chem. Eng. J.*, 2017, **325**, 690–699.
50. N. Shehzad, M. Tahir, K. Johari, T. Murugesan and M. Hussain, *Appl. Surf. Sci.*, 2019, **463**, 445–455.
51. K. Hu, C. Chen, Y. Zhu, G. Zeng, B. Huang, W. Chen, S. Liu, C. Lei, B. Li and Y. Yang, *J. Colloid Interface Sci.*, 2019, **540**, 115–125.
52. X. Wang, K. Maeda, A. Thomas, K. Takanabe, G. Xin, J. M. Carlsson, K. Domen and M. Antonietti, *Nat. Mater.*, 2009, **8**, 76–80.
53. G. Liao, Y. Gong, L. Zhang, H. Gao, G. Yang and B. Fang, *Energy Environ. Sci.*, 2019, **12**, 2080–2147.
54. S. Cao, J. Low, J. Yu and M. Jaroniec, *Adv. Mater.*, 2015, **27**, 2150–2176.
55. J. Fu, J. Yu, C. Jiang and B. Cheng, *Adv. Energy Mater.*, 2018, **8**, 1701503.
56. W. Ong, L. Tan, Y. H. Ng, S. Yong and S. Chai, *Chem. Rev.*, 2016, **116**, 7159–7329.
57. S. Kumar, N. L. Reddy, A. Kumar, M. V. Shankar and V. Krishnan, *Int. J. Hydrogen Energy*, 2018, **43**, 3988–4002.
58. X. She, J. Wu, H. Xu, J. Zhong, Y. Wang, Y. Song, K. Nie, Y. Liu, Y. Yang, M. F. Rodrigues, R. Vajtai, J. Lou, D. Du, H. Li and P. M. Ajayan, *Adv. Energy Mater.*, 2017, **7**, 1700025.
59. W. Yu, J. Chen, T. Shang, L. Chen, L. Gu and T. Peng, *Appl. Catal., B*, 2017, **219**, 693–704.
60. S. Cao, Y. Yuan, J. Fang, M. M. Shahjamali, F. Y. C. Boey, J. Barber, S. C. J. Loo and C. Xue, *Int. J. Hydrogen Energy*, 2013, **38**, 1258–1266.
61. H. Zhao, H. Zhang, G. Cui, Y. Dong, G. Wang, P. Jiang, X. Wu and N. Zhao, *Appl. Catal., B*, 2017, **225**, 284–290.
62. Y. Chen, J. Li, Z. Hong, B. Shen, B. Lin and B. Gao, *Phys. Chem. Chem. Phys.*, 2014, **16**, 8106–8113.
63. G. S. Shanker, R. Bhosale, S. Ogale and A. Nag, *Adv. Mater. Interfaces*, 2018, **5**, 1801488.
64. B. Chai, X. Liao, F. Song and H. Zhou, *Dalton Trans.*, 2014, **43**, 982–989.
65. F. Liu, C. Huang, C. Liu, R. Shi and Y. Chen, *Chem. – Eur. J.*, 2020, **26**, 4449–4460.

66. J. Hu, D. Chen, Z. Mo, N. Li, Q. Xu, H. Li, J. He, H. Xu and J. Lu, *Angew. Chem., Int. Ed.*, 2019, **58**, 2073–2077.
67. R. Boppella, W. Yang, J. Tan, H. Kwon, J. Park and J. Moon, *Appl. Catal., B*, 2019, **242**, 422–430.
68. J. Hu, Y. Ji, Z. Mo, N. Li, Q. Xu, Y. Li, H. Xu, D. Chen and J. Lu, *J. Mater. Chem. A*, 2019, **7**, 4408–4414.
69. J. Ran, B. Zhu and S. Qiao, *Angew. Chem., Int. Ed.*, 2017, **56**, 10373–10377.
70. A. Reddy, E. H. Kim, M. Gopannagari, Y. Kim, P. Kumar and T. K. Kim, *Appl. Catal., B*, 2019, **241**, 491–498.
71. X. Li, Y. Bai, X. Shi, N. Su, G. Nie, R. Zhang, H. Nie and L. Ye, *Mater. Adv.*, 2021, **2**, 1570–1594.
72. Y. Zhuang, Y. Liu and X. Meng, *Appl. Surf. Sci.*, 2019, **496**, 143647.
73. J. Ran, G. Gao, F. Li, T. Ma, A. Du and S. Qiao, *Nat. Commun.*, 2017, **8**, 13907.
74. H. Wang, Y. Sun, Y. Wu, W. Tu, S. Wu, X. Yuan, G. Zeng, Z. J. Xu, S. Li and J. W. Chew, *Appl. Catal., B*, 2019, **245**, 290–301.
75. P. Lin, J. Shen, X. Yu, Q. Liu, D. Li and H. Tang, *Ceram. Int.*, 2019, **45**, 24656–24663.
76. H. Boumeriame, E. S. D. Silva, A. S. Cherevan, T. Chafik, J. L. Faria and D. Eder, *J. Energy Chem.*, 2022, **64**, 406–431.
77. L. Mohapatra and K. Parida, *J. Mater. Chem. A*, 2016, **4**, 10744–10766.
78. S. Nayak and K. M. Parida, Nanostructured, *Int. J. Hydrogen Energy*, 2016, **41**, 21166–21180.
79. D. P. Sahoo, S. Patnaik and K. Parida, *ACS Omega*, 2019, **4**, 14721–14741.
80. D. Kandi, D. P. Sahoo, S. Martha and K. Parida, *Adv. Mater. Interfaces*, 2019, **6**, 1900370.
81. H. Lee, D. A. Reddy, Y. Kim, S. Y. Chun, R. Ma, D. P. Kumar, J. K. Song and T. K. Kim, *ACS Sustainable Chem. Eng.*, 2018, **6**, 16734–16743.
82. J. Wang, Q. Xu, X. Wei, N. Zhang, Z. Wang and P. Qu, *Int. J. Hydrogen Energy*, 2020, **45**, 28437–28446.
83. M. Sohail, H. Kim and T. W. Kim, *Sci. Rep.*, 2019, **9**, 7584.
84. J. Di, J. Xia, H. Li, S. Guo and S. Dai, *Nano Energy*, 2017, **41**, 172–192.
85. X. Wei, M. U. Akbar, A. Raza and G. Li, *Nanoscale Adv.*, 2021, **3**, 3353–3372.
86. J. Li, G. Zhan, Y. Yu and L. Zhang, *Nat. Commun.*, 2016, **7**, 11480.
87. X. Li, J. Zhgang, S. Zhang, S. Xu, X. Wu, J. Chang and Z. He, *J. Alloys Compd.*, 2021, **864**, 158153.
88. N. Wang, G. Yang, H. Wang, R. Sun and C. Wong, *Front. Chem.*, 2018, **6**, 440.
89. Q. Liu, Z. Li, Q. Liu, C. Cheng, M. Song and C. Huang, *Appl. Surf. Sci.*, 2020, **514**, 146096.
90. H. Xu, S. Ouyang, P. Li, T. Kako and J. Ye, *ACS Appl. Mater. Interfaces*, 2013, **5**, 1348–1354.
91. W. Tu, Y. Zhou, S. Feng, Q. Xu, P. Li, X. Wang, M. Xiao and Z. Zou, *Chem. Commun.*, 2015, **51**, 13354–13357.
92. W. Tu, Y. Zhou, Q. Liu, Z. Tian, J. Gao, X. Chen, H. Zhang, J. Liu and Z. Zou, *Adv. Funct. Mater.*, 2012, **22**, 1215–1221.

93. J. Di, X. Zhao, C. Lian, M. Ji, J. Xia, J. Xiong, W. Zhou, X. Cao, Y. She, H. Liu, K. P. Loh, S. J. Pennycook, H. Li and Z. Liu, *Nano Energy*, 2019, **61**, 54–59.
94. L. Liang, F. Lei, S. Gao, Y. Sun, X. Jiao, J. Wu, S. Qamar and Y. Xie, *Angew. Chem., Int. Ed.*, 2015, **54**, 13971–13974.
95. S. Gao, B. Gu, X. Jiao, Y. Sun, X. Zu, F. Yang, W. Zhu, C. Wang, Z. Feng, B. Ye and Y. Xie, *J. Am. Chem. Soc.*, 2017, **139**, 3438–3445.
96. Q. Han, X. Bai, Z. Man, H. He, L. Li, J. Hu, A. Alsaedi, T. Hayat, Z. Yu, W. Zhang, J. Wang, Y. Zhou and Z. Zou, *J. Am. Chem. Soc.*, 2019, **141**, 4209–4213.
97. S. Zhu, S. Liang, J. Bi, M. Liu, L. Zhou, L. Wu and X. Wang, *Green Chem.*, 2016, **18**, 1355–1363.
98. Q. Liu, D. Wu, Y. Zhou, H. Su, R. Wang, C. Zhang, S. Yan, M. Xiao and Z. Zou, *ACS Appl. Mater. Interfaces*, 2014, **6**, 2356–2361.
99. B. Luo, G. Liu and L. Wang, *Nanoscale*, 2016, **8**, 6904–6920.
100. M. Asadi, B. Kumar, A. Behranginia, B. A. Rosen, A. Baskin, N. Repnin, D. Pisasale, P. Phillips, W. Zhu, R. Haasch, R. F. Klie, P. Král, J. Abiade and A. Salehi-Khojin, *Nat. Commun.*, 2014, **5**, 4470.
101. Z. Wang and B. Mi, *Environ. Sci. Technol.*, 2017, **51**, 8229–8244.
102. D. Wang, Z. Wang, L. Wang, L. Hu and J. Jin, *Nanoscale*, 2015, **7**, 17649–17652.
103. S. Singh, A. Modak, K. K. Pant, A. Sinhamahapatra and P. Biswas, *ACS Appl. Nano Mater.*, 2021, **4**, 8644–8667.
104. W. Tu, Y. Li, L. Kuai, Y. Zhou, Q. Xu, H. Li, X. Wang, M. Xiao and Z. Zou, *Nanoscale*, 2017, **9**, 9065–9070.
105. H. Jung, K. M. Cho, K. H. Kim, H. Yoo, A. Al-Saggaf, I. Gereige and H. Jung, *ACS Sustainable Chem. Eng.*, 2018, **6**, 5718–5724.
106. X. Jiao, X. Li, X. Jin, Y. Sun, J. Xu, L. Liang, H. Ju, J. Zhu, Y. Pan, W. Yan, Y. Lin and Y. Xie, *J. Am. Chem. Soc.*, 2017, **139**, 18044–18051.
107. X. Jiao, Z. Chen, X. Li, Y. Sun, S. Gao, W. Yan, C. Wang, Q. Zhang, Y. Lin and Y. Xie, *J. Am. Chem. Soc.*, 2017, **139**, 7586–7594.
108. X. Li, L. Liang, Y. Sun, J. Xu, X. Jiao, X. Xu, H. Ju, Y. Pan, J. Zhu and Y. Xie, *J. Am. Chem. Soc.*, 2019, **141**, 423–430.
109. N. Vu, S. Kaliaguine and T. Do, *Adv. Funct. Mater.*, 2019, **29**, 1901825.
110. H. Hsu, I. Shown, H. Wei, Y. Chang, H. Du, Y. Lin, C. Tseng, C. Wang, L. Chen, Y. Lin and K. Chen, *Nanoscale*, 2013, **5**, 262–268.
111. I. Shown, H. Hsu, Y. Chang, C. Lin, P. Kumar Roy, A. Ganguly, C. Wang, J. Chang, C. Wu, L. Chen and K. Chen, Highly efficient visible light photocatalytic reduction of CO_2 to hydrocarbon fuels by cu-nanoparticle decorated graphene oxide, *Nano Lett.*, 2014, **14**, 6097–6103.
112. Y. T. Liang, B. K. Vijayan, K. A. Gray and M. C. Hersam, *Nano Lett.*, 2011, **11**, 2865–2870.
113. Y. T. Liang, B. K. Vijayan, O. Lyandres, K. A. Gray and M. C. Hersam, *J. Phys. Chem. Lett.*, 2012, **3**, 1760–1765.
114. J. Yu, J. Jin, B. Cheng and M. Jaroniec, *J. Mater. Chem. A*, 2014, **2**, 3407–3416.

115. K. M. Cho, K. H. Kim, K. Park, C. Kim, S. Kim, A. Al-Saggaf, I. Gereige and H. Jung, *ACS Catal.*, 2017, **7**, 7064–7069.
116. W. Jo, S. Kumar, S. Eslava and S. Tonda, *Appl. Catal., B*, 2018, **239**, 586–598.
117. K. Li, B. Peng and T. Peng, *ACS Catal.*, 2016, **6**, 7485–7527.
118. R. Liu, Z. Chen, Y. Yao, Y. Li, W. A. Cheema, D. Wang and S. Zhu, *RSC Adv.*, 2020, **10**, 29408–29418.
119. J. Wen, J. Xie, X. Chen and X. Li, *Appl. Surf. Sci.*, 2017, **391**, 72–123.
120. W. Yu, D. Xu and T. Peng, *J. Mater. Chem. A*, 2015, **3**, 19936–19947.
121. D. O. Adekoya, M. Tahir and N. A. S. Amin, *J. CO2 Util.*, 2017, **18**, 261–274.
122. M. Wang, M. Shen, L. Zhang, J. Tian, X. Jin, Y. Zhou and J. Shi, *Carbon*, 2017, **120**, 23–31.
123. R. Bhosale, S. Jain, C. P. Vinod, S. Kumar and S. Ogale, *ACS Appl. Mater. Interfaces*, 2019, **11**, 6174–6183.
124. G. Zhang, X. Zhu, D. Chen, N. Li, Q. Xu, H. Li, J. He, H. Xu and J. Lu, *Environ. Sci.: Nano*, 2020, **7**, 676–687.
125. C. Hu, J. Zhou, C. Sun, M. Chen, X. Wang and Z. Su, *Chem. – Eur. J.*, 2019, **25**, 379–385.
126. S. Tonda, S. Kumar, M. Bhardwaj, P. Yadav and S. Ogale, *ACS Appl. Mater. Interfaces*, 2018, **10**, 2667–2678.
127. W. Jo, S. Kumar and S. Tonda, *Composites, Part B*, 2019, **176**, 107212.
128. Y. Bai, L. Ye, L. Wang, X. Shi, P. Wang, W. Bai and P. K. Wong, *Appl. Catal., B*, 2016, **194**, 98–104.
129. X. Bian, S. Zhang, Y. Zhao, R. Shi and T. Zhang, *InfoMat*, 2021, **3**, 719–738.
130. N. Dewangan, W. M. Hui, S. Jayaprakash, A. Bawah, A. J. Poerjoto, T. Jie, A. Jangam, K. Hidajat and S. Kawi, *Catal. Today*, 2020, **356**, 490–513.
131. S. Ng, M. Y. L. Lau and W. Ong, *Sol. RRL*, 2021, **5**, 2000535.
132. S. Kumar, M. A. Isaacs, R. Trofimovaite, L. Durndell, C. M. A. Parlett, R. E. Douthwaite, B. Coulson, M. C. R. Cockett, K. Wilson and A. F. Lee, *Appl. Catal., B*, 2017, **209**, 394–404.
133. G. Gao, Z. Zhu, J. Zheng, Z. Liu, Q. Wang and Y. Yan, *J. Colloid Interface Sci.*, 2019, **555**, 1–10.
134. S. Iguchi, Y. Hasegawa, K. Teramura, S. Kidera, S. Kikkawa, S. Hosokawa, H. Asakura and T. Tanaka, *Sustainable Energy Fuels*, 2017, **1**, 1740–1747.
135. B. Zhu, Q. Xu, X. Bao, H. Yin, Y. Qin and X. Shen, *Chem. Eng. J.*, 2022, **429**, 132284.
136. X. Wei, M. U. Akbar, A. Razav and G. Li, *Nanoscale Adv.*, 2021, **3**, 3353–3372.
137. X. Liu, J. Xiao, S. Ma, C. Shi, L. Pan and J. Zou, *ChemNanoMat*, 2021, **7**, 684–698.
138. J. Xiong, P. Song, J. Di and H. Li, *J. Mater. Chem. A*, 2020, **8**, 21434–21454.
139. Y. Bai, P. Yang, P. Wang, H. Xie, H. Dang and L. Ye, *J. CO_2 Util.*, 2018, **23**, 51–60.
140. J. Di, C. Chen, S. Yang, S. Chen, M. Duan, J. Xiong, C. Zhu, R. Long, W. Hao, Z. Chi, H. Chen, Y. Weng, J. Xia, L. Song, S. Li, H. Li and Z. Liu, *Nat. Commun.*, 2019, **10**, 2840.

141. P. K. Prajapati and S. L. Jain, *Dalton Trans.*, 2019, **48**, 4941–4948.
142. J. Wang, J. Wang, N. Li, X. Du, J. Ma, C. He and Z. Li, *ACS Appl. Mater. Interfaces*, 2020, **12**, 31477–31485.
143. J. Shen, Z. Wu, C. Li, C. Zhang, A. Genest, G. Rupprechiter and L. He, *FlatChem*, 2021, **28**, 100252.
144. Y. Zhao, M. Que, J. Chen and C. Yang, *J. Mater. Chem. C*, 2020, **8**, 16258–16281.
145. M. Ye, X. Wang, E. Liu, J. Ye and D. Wang, *ChemSusChem*, 2018, **11**, 1606–1611.
146. Q. Tang, Z. Sun, S. Deng, H. Wang and Z. Wu, *J. Colloid Interface Sci.*, 2020, **564**, 406–417.
147. S. Cao, B. Shen, T. Tong, J. Fu and J. Yu, *Adv. Funct. Mater.*, 2018, **28**, 1800136.
148. A. Pan, X. Ma, S. Huang, Y. Wu, M. Jia, Y. Shi, Y. Liu, P. Wangyang, L. He and Y. Liu, *J. Phys. Chem. Lett.*, 2019, **10**, 6590–6597.
149. Y. Chen, M. Qi, Y. Li, Z. Tang, T. Wang, J. Gong and Y. Xu, *Cell Rep. Phys. Sci.*, 2021, **2**, 100371.
150. G. Zhao, J. Hu, X. Long, J. Zou, J. Yu and F. Jiao, *Small*, 2021, **17**, 2102155.
151. X. Ren, L. Shi, Y. Li, S. Song, Q. Wang, S. Luo, L. Ren, H. Zhang, Y. Izumi, X. Peng, D. Philo, F. Ichiharam and J. Ye, *ChemCatChem*, 2020, **12**, 3870–3879.
152. G. Zhou, J. Yang, X. Zhu, Q. Li, Q. Yu, W. El-alami, C. Wang, Y. She, J. Qian, H. Xu and H. Li, *J. Energy Chem.*, 2020, **49**, 89–95.
153. Y. Gong, J. Shen, Y. Zhu, W. Yan, J. Zhu, L. Hou, D. Xie and C. Li, *Appl. Surf. Sci.*, 2021, **545**, 149012.
154. C. Huang, C. Chen, M. Zhang, L. Lin, X. Ye, S. Lin, M. Antonietti and X. Wang, *Nat. Commun.*, 2015, **6**, 7698.
155. Y. Cao, R. Zhang, T. Zhou, S. Jin, J. Haung, L. Ye, Z. Haung, F. Wang and Y. Zhou, *ACS Appl. Mater. Interfaces*, 2020, **12**, 9935–9943.
156. M. Zhou, S. Wang, P. Yang, C. Huang and X. Wang, *ACS Catal.*, 2018, **8**, 4928–4936.
157. C. Han, B. Wang, C. Wu, S. Shen, X. Zhang, L. Sun, Q. Tian, Y. Lei and Y. Wang, *ChemistrySelect*, 2019, **4**, 2211–2217.
158. Y. Men, Y. You, Y. Pan, H. Gao, Y. Xia, D. Cheng, J. Song, D. Cui, N. Wu, Y. Li, S. Xin and J. B. Goodenough, *J. Am. Chem. Soc.*, 2018, **140**, 13071–13077.

CHAPTER 9

Role of Co-catalysts for Photocatalytic H_2O Splitting and CO_2 Reduction

CHRISTIAN M. PICHLER*[a,b]

[a]Institute of Applied Physics, Vienna University of Technology, Wiedner Hauptstraße 8-10, 1040 Vienna, Austria; [b]Centre of Electrochemical and Surface Technology, Viktor Kaplan Straße 2, 2700 Wiener Neustadt, Austria
*E-mail: christian.pichler@cest.at

9.1 Background

Global primary energy consumption has been rising over recent decades and this trend is expected to continue, reaching 18 billion toe (ton of oil equivalent, approx. 209 000 TWh) by 2040.[1] Today, most of the energy demand is met by the consumption of fossil fuels, which is raising major environmental concerns, due to the massive emission of CO_2. Utilizing solar energy would be a sustainable and viable alternative. The fluctuating nature of solar energy requires a pathway to store this energy to ensure a constant supply of it, also during phases of no solar energy input (night). Utilizing solar energy to drive chemical reactions and use the reaction products for energy storage is a viable way forward. Photovoltaic devices can produce electricity, which can then drive the electrolysis of water to hydrogen or of CO_2 to hydrocarbons or alcohols.[2] All products (hydrogen, hydrocarbons, and alcohols) are well suited

Sustainable Energy No. 1
Recent Developments in Functional Materials for Artificial Photosynthesis
Edited by Srabanti Ghosh and Qian Wang

Published by the Royal Society of Chemistry, www.rsc.org

as energy storage compounds as they can be combusted later to release the stored energy. In the case of hydrogen, fuel cells can be utilized for electricity production.[3] Instead of combining photovoltaics with electrolysis, photocatalysts can be applied to directly drive chemical reactions. Photocatalysts are materials or compounds that absorb light and catalyze a variety of reactions using the absorbed light energy. Photocatalysis can be also applied to catalyze organic reactions for the synthesis of valuable fine chemicals.[4] However, here the focus will be on reactions that can potentially store solar energy on a larger scale, which are the water splitting reaction (splitting of H_2O into H_2 and O_2, where H_2 serves as an energy storage compound) and the CO_2 reduction into hydrocarbons and alcohols.[5,6]

9.2 General Mechanism

When light is absorbed by the photocatalytic material, an electron from the catalyst's valence band is excited into the conduction band, forming an excited electron–hole pair, with the excited electron being localized in the conduction band and the hole remaining in the valence band. In theory, the excited electron can engage in reduction reactions and the hole takes part in oxidation reactions, closing the charge balance, as illustrated in Figure 9.1. Decisive parameters for photocatalysts are the bandgap and the band levels of valence and conduction band.[7] The bandgap determines the light energy that is necessary to excite an electron from the valence into the conduction band. For example, a large bandgap, *e.g.*, 3.3 eV in TiO_2, requires high-energetic UV-light, to ensure the electron excitation. Reducing the bandgap enables also larger wavelengths (lower energies) to excite electrons.[8] To enable a reaction, the band levels of the photocatalyst must be adequate as well. The conduction band must lie at a more negative potential than the desired reductive half reaction, while the valence band requires a more positive potential than the anticipated oxidative half reaction (Figure 9.1). To drive the water splitting reaction, the conduction band of an anticipated photocatalyst must lie lower than 0 V (*vs.* SHE) and the valence band should lie at a potential greater than 1.23 V (*vs.* SHE). For a different combination of half reactions at different reaction potentials, the band levels have to be adjusted. Without those very elementary preconditions, no photocatalytic reaction is possible.[9] A selection of various photocatalysts with their bandgaps and band levels is depicted in Figure 9.2.

These photocatalytic materials alone are however not sufficient to drive reactions such as water splitting or CO_2 reduction. To enable those reactions, several requirements must be fulfilled:

- The excited charge carriers (electrons or holes) have to reach the material surface.
- The material should possess active sites where the desired reaction can proceed.
- The charge carriers must have a sufficient lifetime.

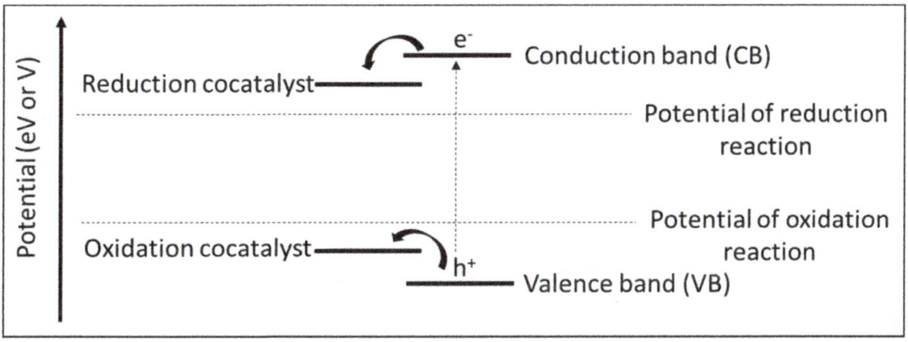

Figure 9.1 General scheme of photocatalyst with co-catalyst.

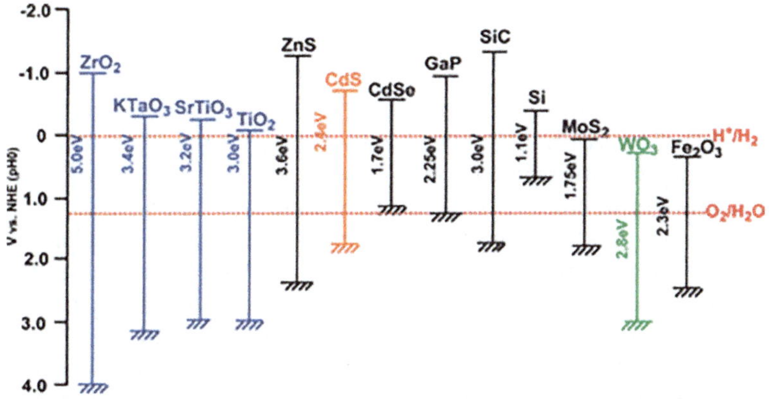

Figure 9.2 Bandgaps and levels of several photocatalysts. Adapted from ref. 10 with permission from the Royal Society of Chemistry.

In unmodified, pure photocatalytic materials, the lifetime of the excited charge carriers is mostly too short, so the electrons and holes recombine before they can reach the material surface and catalyze the reaction. Furthermore, not all photoactive materials possess suitable catalytic sites for the desired reactions. To overcome those challenges, photoactive materials are combined with co-catalysts, that are playing a pivotal role for enabling the photocatalytic reaction. In the following, it will be shown that there exists a great variety of co-catalysts in terms of materials and chemical composition. All those co-catalysts serve the following purposes.[7]

- Co-catalysts act as a sink for the excited charge carriers, and thereby significantly reduce the charge recombination by facilitating the separation of the electron–hole pair. When a photoactive material is in contact with a co-catalyst, a so-called heterojunction is created. To enable an efficient charge extraction the energy levels at this heterojunction

are crucial. This means that the co-catalyst must possess the right energy levels, so the excited charge carriers can be efficiently extracted (Figure 9.1).
- Co-catalysts provide active sites for the desired photoreaction and lower the overpotential or activation energy for the reaction. Especially for challenging half reactions such as the oxygen evolution reaction (OER) in water splitting, this is a crucial factor. The OER is a complex four electron process and by using a suitable co-catalyst with a low overpotential for this reaction, the efficiency of the overall water splitting can be significantly improved.
- Co-catalysts can improve the stability of the photocatalytic material, by reducing photocorrosion. Sulfide or nitride based photocatalysts can be oxidized and deactivated by the photogenerated holes. To avoid this, suitable co-catalysts are applied that extract the holes, before they can react with the photocatalyst, thereby increasing the stability of the material.

In this chapter, the different varieties of co-catalyst for the water splitting and CO_2 reduction will be discussed. Depending on the chemical nature of the co-catalyst, different phenomena have to be considered, when the co-catalyst is brought in contact with the photocatalytic material.

Widely used co-catalysts for various purposes are (noble) metals. If a metal is brought in close contact to a semiconductor (photocatalytic material), a Schottky barrier is formed at the interface. The nature of this barrier is determined by the work function of the metal co-catalyst and the electron affinity of the semiconductor material (energy difference between the lower limit of the conduction band and vacuum energy).[11] A larger work function of the metal co-catalyst (describes how much energy is needed to remove an electron from the metal), accounts for a lower Fermi level and metal co-catalysts with lower Fermi levels are expected to trap electrons more efficiently. Hence, platinum (and other noble metals) with its large work function is especially well suited as co-catalyst and therefore also widely used. The band structure of the semiconductor will be distorted by an "upwards bending" of the conduction band and the formation of an electric field between the metal and the semiconductor. This also contributes to the efficient electron–hole separation.[12]

Instead of a metal, semiconductors can also be used as co-catalysts. In this case, other effects must be considered, depending on the respective band energy levels of both semiconductors. If one semiconductor has a valence band with higher and a conduction band with lower energy, a type I heterojunction is formed (see Figure 9.3). This means that electrons and holes would accumulate at only one semiconductor, fostering the undesired charge recombination. Therefore, such a heterojunction rarely results in suitable photocatalyst–co-catalyst combinations. If the conduction band as well as the valence band energies are lower for one of the semiconductors, a type II heterojunction is present (see Figure 9.3). In this case, effective charge

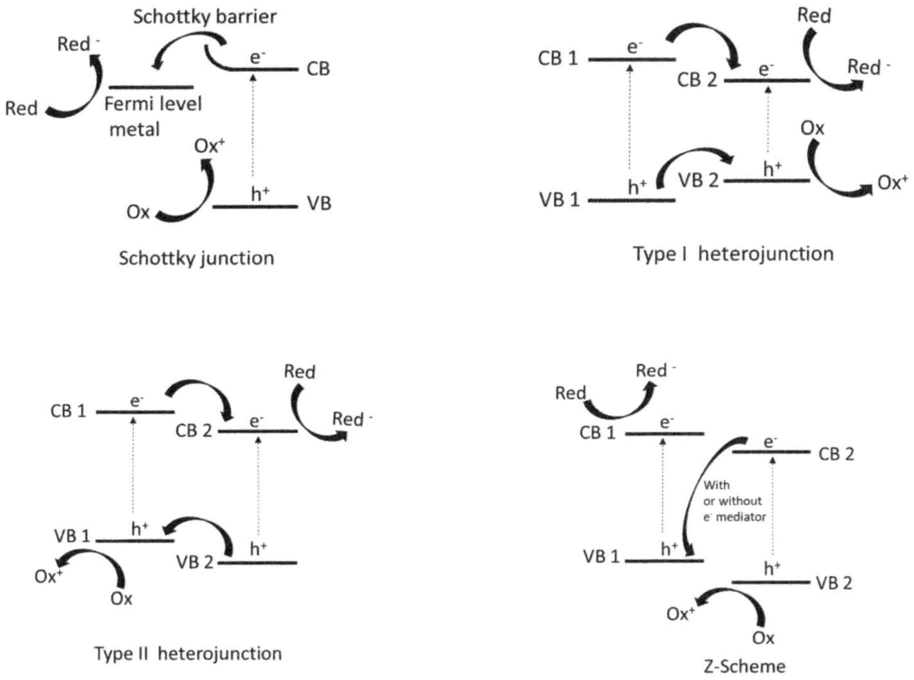

Figure 9.3 Band level schemes for Schottky barrier, Type I, Type II heterojunction and Z-scheme.

separation can be achieved, as the electrons will flow towards the conduction band with the lowest energy and the holes will accumulate in the valence band with the higher energy. Such heterojunctions can result in effective photocatalytic systems. There also exists a type III heterojunction, where the conduction band of one semiconductor lies below the valence band of the other one. In this case, the difference in the respective band levels is too vast and no charge transfer between the semiconductors is happening. Hence, no benefit of improved charge separation is expected.[13,14]

Recently, advanced combinations of multiple photocatalytic materials and co-catalysts have been developed and improved. One of those assemblies is the Z-scheme system (see Figure 9.3). In a Z-scheme two photocatalysts are combined with each other. Like in a type II heterojunction the conduction band as well as the valence band energies are lower for one of the semiconductors. The striking difference between a Z-scheme and a type II photocatalyst is, that on a Z-scheme assembly the reduction reaction is taking place on the photocatalyst with the higher conduction band and the oxidation is proceeding at the material with the lower valence band. In a type II heterojunction, the reaction sequence is taking place the other way round. To enable a working Z-scheme, efficient electron transfer between the two photoactive materials has to be ensured, which can be done with a liquid phase or solid-state electron mediator. Also, direct Z-schemes without any mediator have been reported. The Z-scheme photocatalyst system can then be loaded with co-catalysts to enhance the reactions on the specific semiconductor where they are expected to take place.[5,13]

9.3 Co-catalysts for the Hydrogen Evolution Reaction (HER)

Hydrogen can play a crucial role in a renewable energy system. It can be used directly as an energy vector and be utilized in fuel cells for the generation of electricity. Alternatively, it can be used in chemical processes to produce synthetic carbon-based fuels or base chemicals such as ammonia. The direct photocatalytic production of hydrogen is hence a very appealing goal, since Fujishima and Honda made their seminal report on the utilization of TiO_2 as photocatalytic material for water splitting.[15] Co-catalysts play a decisive role in this process and in the following several of the most important classes of co-catalysts for this reaction are mentioned. When comparing H_2 evolution activities (or any other photocatalytic reactions) obtained in different studies, care has to be taken, as there is a great variety of parameters that can influence the results and impede direct comparison. Some of these factors that should be considered are:

- complete water splitting system reported or just the HER by using a sacrificial agent
- type and concentration of sacrificial agent
- catalyst loading
- reaction time
- catalyst stability
- light intensity
- photoreactor setup.

9.3.1 Noble Metals

9.3.1.1 Platinum (Pt)

As it is already known from electrocatalytic applications, noble metals possess excellent properties for the HER. Therefore, they are widely used as co-catalysts in photocatalytic systems. The most prominent metal is platinum, which is the most widely used co-catalyst for the HER. It has been used in combination with a great multitude of different photocatalysts and is often the first choice to assess the activity of a new photocatalytic material, as the HER proceeds smoothly and reliably on the Pt co-catalyst.[9] Nevertheless, there are several factors that influence the performance of the Pt co-catalyst and thereby the catalytic activity of the whole photocatalytic system. The first parameter is the deposition method of Pt onto the photocatalyst. A very common method is photo deposition, where a Pt precursor such as

H_2PtCl_6 is added to a suspension of the photocatalyst together with a sacrificial agent such as triethanolamine (TEOA). By shining light onto the suspension, the excited electrons will reduce the Pt precursor to metallic Pt, which will be deposited as nanoparticles on the photocatalyst surface.[16] By changing the concentration of the Pt precursor, the final loading of metallic Pt on the photocatalyst can be tuned. Furthermore, the type of sacrificial agent has to be matched with the photocatalyst as not every catalyst exhibits the same oxidation activity for every sacrificial agent. Finally, even the sequence of the addition of the sacrificial agent and the Pt precursor influences the final catalytic activity. A four-fold increase in H_2 evolution activity was achieved by just changing this addition sequence for a Pt/carbon nitride photocatalyst system, because the sacrificial agent can interact with the Pt precursor and hinder the deposition of the metallic Pt nanoparticles.[16]

Other deposition methods include the chemical reduction method, where the Pt precursor is reduced chemically (*e.g.*, with $NaBH_4$) to form metallic nanoparticles on the photocatalyst. In this case, the type and concentration of the reduction agent influence the Pt deposition.[17,18] Complex methods such as atomic layer deposition (ALD) can as well be utilized to deposit metallic Pt nanoparticles with a controlled particle size. The particle size of Pt is an important factor for the catalytic activity. For Pt on TiO_2, it has been determined that a Pt particle size of approx. 1 nm is ideal for catalyzing the HER. Too small clusters or even single atom catalysts exhibited a lower activity.[19] These differences are caused by a complex interplay between electron transfer between TiO_2 and Pt and the surface catalytic activity. For carbon nitride, however, it could be shown that also single atom Pt catalysts exhibit excellent activity for the HER, as carbon nitride contains surface sites that can stabilize the Pt single atoms.[20] This underlines, that for each photocatalytic system, the co-catalyst has to be specifically optimized (this is valid for all co-catalysts not only noble metals). If the co-catalyst loading is too low, the full potential of the photocatalytic system cannot be achieved. However, if the loading is too high, larger co-catalyst particles will form, that are often less active and cause an inefficient utilization of the co-catalyst. Furthermore, excessive loading of the photocatalyst can block the incident light and thereby reduce the activity.[3] This is also valid for all types of co-catalysts.

9.3.1.2 Palladium (Pd), Rhenium (Rh) and Ruthenium (Ru)

Besides Pt, other noble metals can also be used as co-catalysts. Metals like Pd, Rh or Ru can catalyze the HER and with their large work function, they are also suitable electron sinks.[9] The investigation of those other metals is often fueled by economic considerations as the prices of platinum group metals undergo major fluctuations. Especially, if this technology shall be implied at a large scale, the price of the catalyst is a crucial factor. Generally, Ru and Pd have a lower price when compared to Pt and are therefore attractive alternatives. In some studies, that directly compared Pt and Pd co-catalyst loaded photocatalysts, the Pd loaded material exhibited higher activity for the HER. When Pd or Pt on P25 TiO_2 catalysts are compared, the Pd/P25 material gives an approx. 30% higher H_2 yield, utilizing glucose or cellulose as sacrificial agent (approx. 0.18 *vs.* 0.13 mmol H_2 per g per h).[21] Another comprehensive study tested seven different alcoholic sacrificial agents for the photo conversion and for all of them Pd/TiO_2 was more active than Pt/TiO_2. The reason for this could be the lower electron affinity of Pd, facilitating the electron transfer from the metal for the HER. Additionally, Pd possesses a higher density of states in the vicinity of the Fermi level, which can also contribute to the enhanced catalytic activity.[22] Also, Ru has been successfully applied as a co-catalyst. For example, Ru can be employed as single atoms, where its activity can be significantly enhanced, compared to nanoparticles. A multi-edged TiO_2 photocatalyst was shown to stabilize Ru single atoms and give a catalyst system with high activity and stability for the HER.[23] Another example was shown in a multicomponent Z-scheme where Ru was utilized as HER catalyst on a $SrTiO_3$:La,Rh photoactive material. The other part of the Z-scheme system was comprised of a $BiVO_4$:Mo material, connected with Au contacts to enable charge transfer. In this case, the Ru co-catalyst was further modified by deposition of a Cr_2O_3 layer, which reduced the back reaction and thereby increased H_2 yield and catalyst stability.[24] In the mentioned $SrTiO_3$ based system, another noble metal is utilized with Rh. However, Rh is a dopant in the semiconductor and does not act as co-catalyst. Even though Rh has a large work function and can potentially catalyze the HER, it is rarely utilized as HER co-catalyst in the metallic state. Nevertheless, there are some examples where Rh is used in combination with transition metal oxides such as CrO_x and MoO_x, as HER co-catalyst, with $SrTiO_3$ as the photocatalytic material.[25] Another example employed Rh single atoms as HER co-catalyst, by integrating them into a TiO_2 photocatalyst.[26] This approach also delivered high yields of hydrogen.

9.3.1.3 Gold (Au)

The next noble metal that is intensively investigated for the HER is Au. The reason for this, is not only its activity for the HER, but the capability to engage in localized surface plasmon resonance (LSPR). The LSPR occurs when the free electrons in a metal nanoparticle are engaging in oscillation, stimulated by incoming photons. Thereby, the metal nanoparticles strongly absorb photon energy, resulting in an additional light absorption peak.[9] Au nanoparticles on a TiO_2 photocatalyst, for example, showed an additional absorption band around 570 nm. Au/TiO_2 has been shown to be active for HER and the additional light absorption of the Au improves the overall catalytic activity. To maximize the contributions of the LSPR effect, the shape and size of the Au particles must be optimized. It could be shown that 18 nm sized spherical Au particles give the highest HER activity on a carbon nitride photocatalyst when compared with rod shaped or differently sized particles, because of the contribution of the LSPR effect.[27]

9.3.1.4 *Silver (Ag)*

Ag is rarely used as a co-catalyst for the HER, as it is not particularly well suited for that reaction. Nevertheless, several studies can be found where Ag has been applied as co-catalyst, mostly to compare it with other co-catalysts. In one exceptional study, however, Ag single atoms were combined with carbon nitride as photocatalyst and demonstrated a higher HER activity when compared with a Pt nanoparticle co-catalyst.[28]

9.4 Non-noble Metals

Although noble metals are effective co-catalysts for the HER, their high price often impedes their utilization on a larger scale. Hence, alternatives have been sought and found among non-noble metals. Nickel (Ni) is a promising candidate as it has a high work function and is known to possess active sites for the HER. Metallic nickel as a co-catalyst has been used with a wide variety of photocatalysts. With carbon nitride, it was shown that a comparably high metal loading of 7.4 wt% was necessary to obtain the highest activity of 4.3 mmol H_2 per g per h.[29] For the widely used P25 TiO_2 material an activity of 2.5 mmol H_2 per g per h was reported but at a much lower Ni loading of approximately 0.25 wt%.[30] This underlines again the difficulty of comparing catalytic results from different studies with each other and shows furthermore, that every photocatalyst requires optimization studies regarding its co-catalyst as the optimal co-catalyst loading can vary significantly. But also, for other photocatalysts such as graphene oxide or CdS composites, Ni was utilized as a co-catalyst.[31,32] Besides the loading, the size and type of metal particle are crucial, as it was shown that smaller nanoparticles result in higher activity for the HER reaction and Ni nanoparticles with a hexagonally closed packed (hcp) crystal structure exhibited higher activity than face centered cubic (fcc) particles, as the hcp crystal structure allows an energetically more favorable accommodation of the hydrogen intermediates.[33]

Similar results could be achieved with metallic cobalt (Co) nanoparticles as co-catalyst, that were used in combination with TiO_2 or CdS as photocatalysts, giving HER activities of 1.2 and 1.3 mmol H_2 per g per h, respectively.[34,35] With these values, the activity of Co and Ni is around a factor of three lower when compared to Pt co-catalysts. However, the price of the non-noble metals is around three orders of magnitude lower. When comparing Ni with Co, it has to be considered that Ni is a more abundant element and Co is also widely used as electrode material for Li-ion batteries, which poses some competition and might influence the metal price, especially with the emergence of electric vehicles.

Another suitable metal as co-catalyst is copper (Cu), which is also widely abundant at a comparably moderate price. Metallic Cu nanoparticles were shown to be effective co-catalysts in combination with TiO_2, graphene oxide and carbon nanodots as photoactive materials.[36,37] For a mesoporous TiO_2 photocatalyst, Cu has been shown to be an especially active co-catalyst and exhibited higher activity than a comparable Ni/meso-TiO_2 material, because Cu has stronger electronic interactions with TiO_2.[38] However, it must be noted that this may change under different reaction conditions and other photocatalysts. Nevertheless, it shows that the interaction between the co-catalyst and the photocatalytic material is a crucial factor that must be considered. Cu also possesses plasmonic properties such as Au and therefore can improve the light absorption properties of photocatalysts, especially for higher wavelengths. Thereby, the overall activity of the whole photocatalytic system can be increased.

9.5 Metal Alloys

Instead of using only one metal at a time, alloying can prove to be beneficial for the catalytic activity. By alloying, the electronic structure of the metals can be modified, which can have positive and negative effects, as has been shown for alloyed AuPt and AuPd clusters, respectively.[39] Another benefit of alloys is that reaction intermediates can be stabilized by a certain surface composition, which can also lead to improved catalytic activity. This is the reason why there is often a defined mixing ratio for many two metal alloys, where the stabilization of the intermediates is optimal, which was shown for PtPd and PtAu nanoparticles.[40,41] For those effects, the surface composition is decisive (and not the overall composition), hence a core–shell particle can have completely different catalytic behavior, when compared to an alloyed one, as has been demonstrated for AuCu particles, where the alloyed co-catalyst, showed increased catalytic activity.[42] For a PtCo alloy, an increased number of surface defects proved to be helpful and was increasing catalytic activity, while at the same time the changed Fermi level was beneficial as well.[43] In this case the combination of surface and electronic effects was the key to success. For alloys with Au or Ag, the plasmonic effects can also play a role. For PdAu and PdAg, the high catalytic activity of Pd was improved by the presence of the plasmonic metals Au and Ag, that enabled more efficient light absorption.[44,45] The benefit of alloyed co-catalysts can sometimes be also connected to a specific photocatalyst. For CdS as photocatalyst, it was shown that a PdNi alloy co-catalyst exhibited higher activity than pure Pd, as the electronic interactions of the alloy with the photocatalyst are stronger (54 *vs.* 32 mmol H_2 per g per h).[46] Alloying a noble metal with a cheaper metal, has of course also the advantage of reducing the costs of the co-catalyst. Nevertheless, in many cases alloying brings additional benefits, due to the alteration of the electronic and surface structure of the co-catalyst, hence it is often a worthwhile option to try out in the course of optimizing a photocatalytic system.

9.6 Oxides and Sulfides

Besides metallic co-catalysts, also other compounds are suited for this purpose, such as transition metal oxides. Again, the cheaper price, when compared to noble metals, is a driver of new development of those co-catalysts. Already in the late 1980s, NiO had been reported as a co-catalyst for the HER with TiO_2 photocatalysts.[47] In contrast to pure metal co-catalyst, the metal oxides require different synthesis methods. Often wet, or incipient wetness, impregnation of metal precursors is followed by calcination steps, to convert the precursor (*e.g.*, nitrates or other compounds that have a favorable thermal decomposition profile) into oxides. The synthesis method of the oxide co-catalyst has a significant influence on the catalytic activity, as it could be shown that a sol–gel synthesis resulted in a more active NiO/TiO_2 photocatalytic system, when compared to an impregnation method.[48] As for all co-catalysts, also in this case the surface structure is a relevant factor. For a NiO on carbon nitride (CN_x) system, it could be shown that amorphous NiO gives higher catalytic than its crystalline modification due to a combination of improvements, such as increased visible light absorption, a higher number of active sites for the HER, better charge separation over the NiO–CN_x heterojunction, which is furthermore improved by the formation of C–O–Ni bonds between the photocatalyst and the NiO particles.[49] However, when a metallic Ni co-catalyst, was compared with NiO with graphene oxide as photocatalyst, it was shown, that the NiO material exhibited lower catalytic activity, as NiO had a lower electron trapping efficiency. It should be remembered that on the semiconductor metal interface (Ni/GO) a Schottky contact will be formed, while the combination of two semiconductors (NiO as a bandgap of 3.5 eV) results in the formation of a heterojunction, as shown in Figure 9.3.[32] These differences can lead to different electron trapping efficiencies. However, these differences might change between different photocatalyst–co-catalyst combinations. A viable strategy is to combine the strengths of both co-catalyst types, by synthesizing Ni–NiO core–shell particles. Ni–NiO core–shell co-catalysts have higher catalytic activities when compared to Ni or NiO alone, in combination with a TiO_2 photocatalyst.[50] The Ni core enables an efficient transfer of the excited electron and the NiO shell provides active sites for hydrogen evolution that are well stabilized in the NiO matrix. Nevertheless, also stability issues have been reported for those catalysts, as it was shown that the Ni core will suffer from photocorrosion, leading to catalyst deactivation.[51] Excellent results were achieved with a related co-catalyst type, namely $Ni(OH)_2$, that gave hydrogen yields, comparable to Pt loaded systems, in combination with CdS and carbon nitride photocatalysts (5.1 *vs.* 3.9 mmol H_2 per g per h).[52,53] Although the Ni loading (23 mol%) was very high for the $Ni(OH)_2$/CdS system. $Ni(OH)_2$ did not only provide the active centers for the HER, it also increased light absorption in the visible range and thereby improving the overall catalytic activity. Even though it is proposed that $Ni(OH)_2$ forms metallic Ni intermediates that serve as active sites for the

HER, it was demonstrated that $Ni(OH)_2$ is superior to metallic Ni co-catalysts. One reason for this can be the better dispersion and stabilization of the Ni^0 active sites for HER, in the $Ni(OH)_2$ matrix.

Another widely investigated metal oxide co-catalyst is copper oxide (CuO). It can be loaded onto P25 TiO_2 with a facile impregnation, calcination synthesis and this CuO/TiO_2 system can reach hydrogen yields comparable with noble metal co-catalysts (while using methanol as sacrificial agent).[54,55] The optimal Cu loading was around 9 mol%. Deeper investigations showed, that the CuO synthesis procedure and the resulting copper oxide species are crucial for the final HER activity. Among the comparison of sol–gel synthesis, photodeposition, wet impregnation and chemical reduction, the sol–gel synthesized material showed the most favorable results, as this material delivered the highest H_2 yield over a longer period of time.[56] Even though the chemical reduction material, exhibited high initial activity, its deactivation was also more rapid. The active species in the chemical reduction material was Cu_2O, in contrast to CuO for the sol–gel approach. As the Cu_2O material is not stable under the reaction conditions over longer time periods, the overall H_2 yield is also lower for that photocatalytic system. CuO can be also utilized for different types of photocatalysts such as mesoporous TiO_2, where Cu precursor is introduced during the photocatalyst synthesis step, enabling strong $CuO–TiO_2$ interactions.[57] Also, with CdS as the light absorber, good HER activities could be achieved, making CuO a versatile co-catalyst that can be combined with various photocatalysts and produced *via* different synthesis procedures.[58] Finally, the hydroxide $Cu(OH)_2$ can also be used as a co-catalyst, that forms metallic Cu clusters embedded in $Cu(OH)_2$ in operando, which are the actual active species, similar to the $Ni(OH)_2$ co-catalysts mentioned previously.[59]

As mentioned previously, many of these metal oxides are semiconductors themselves and are often used as photoactive materials as well and not solely as HER co-catalyst. For example, cobalt oxides and zinc oxide, can both be used as HER co-catalysts in combination with other photoactive materials such as carbon nitride, ZnS, TiO_2 or CdS. But they can themselves act as light absorbers alone or with other co-catalysts to boost the HER.[60–63]

Another prominent class of HER co-catalysts is metal sulfides and among them, MoS_2 is one of the most intensively investigated compounds, with more than 160 published studies in 2019.[64] MoS_2 has been investigated in combination with a very broad variety of light absorbers including other sulfidic compounds, the "classic" TiO_2 and carbon nitride materials and many more (4.2 mmol H_2 per g per h).[65,66] For the MoS_2 loading step itself various procedures have been reported, including many hydrothermal or solvothermal methods, that require comparably high reaction temperatures and elevated pressures. Also, impregnation and sulfidization procedures are reported, where the MoS_2 compound is formed by reaction with H_2S (toxic).[64] The array of synthesis also includes exfoliation (chemical or ultrasonic) and more complex approaches such as chemical vapor deposition.[67–69] Even photodeposition methods have been reported.[70] With such a wide array of available methods, the synthesis can be specifically adjusted to the desired parameters (price, handling, material quality, toxicity, light absorber, reaction environment, *etc.*). MoS_2 preferably forms layered, sheet-like particles. This can be of advantage when MoS_2 is combined with sheet-like photocatalysts such as carbon nitride, as the interaction area between the two components can be maximized, increasing stability, and facilitating charge transfer. The active sites for HER in MoS_2 co-catalysts are not fully coordinated Mo sites, which can be often found at the edges of the sheets.[64] Depending on the synthesis method those undercoordinated or edge sites can be tuned, to improve HER activity.

Other metal sulfides such as NiS are comparable to MoS_2, regarding their HER activity and their synthesis pathway, *e.g.*, hydrothermal methods are widely used for the preparation of NiS. A combined NiS/CdS photocatalytic system showed excellent activity for the HER, surpassing the Pt/CdS system, as electron transfer between NiS and CdS are accelerated as well as the elementary steps of the HER.[71,72] In combination with carbon nitride, a photodeposition method was utilized to form the NiS.[73] Further metal sulfides include WS, CoS and CuS, which can be also used in a combination with various photocatalysts.[73] Nevertheless, MoS_2 seems to be the most promising candidate among the sulfide catalysts, due to its high catalytic activity and as it is the best understood material, because of intensive research.

9.7 Phosphides

The next notable class of HER co-catalysts is metal phosphides, with Ni_2P as the most prominent example. Ni_2P has been used with a variety of photocatalytic materials such as metal sulfides, TiO_2 or carbon nitride.[74–77] The Ni_2P/CdS system yield with approx. 33 mmol g^{-1} h^{-1} nearly twice as much H_2, as the comparable Pt/CdS system, which underlines the high catalytic activity of Ni_2P.[76] Nevertheless, a common problem with Ni_2P is, that Ni^{2+} will leach from it, especially in neutral or acidic conditions, diminishing its longtime stability. Alkaline media are better suited, as the nickel leaching can be reduced under those conditions. Regarding the synthesis methods, Ni_2P can be synthesized as free nanoparticles *via* high temperature phosphidation synthesis of metallic Ni particles.[78] Another method is the hydrothermal synthesis combing a nickel precursor with elemental phosphorous.[79] The free Ni_2P nanoparticles can then be mixed (*e.g.*, ultrasonication) with a light absorber. Nevertheless, the contact between light absorber and Ni_2P nanoparticles might be of lower quality in this case. For the synthesis of free Ni_2P, often capping agents are required, that can have a detrimental impact on the catalytic activity, as they can block surface sites. When the Ni_2P particles are directly formed on the photocatalyst, better contact and interaction between those two components can be ensured. This was shown with an impregnation/reduction method, where a nickel salt is mixed with sodium hypophosphite and impregnated on a carbon nitride material. Thermal treatment of those mixtures leads to formation of Ni_2P particles directly on the photocatalyst and yields a Ni_2P/CN_x material with excellent HER activity.[75] Furthermore, this synthesis method is more convenient compared to high temperature and hydrothermal approaches. Apart from that, also other synthesis methods, *e.g.*, involving mechanochemistry were reported. The reason for the high HER activity of Ni_2P is its platinum group metal-like electronic structure. Hence, Ni_2P can form Schottky contacts with semiconductors such as carbon nitride, which improves charge separation. The performance of Ni_2P can be even further improved by doping with other metals such as Co, Mo, or Fe. By changing the electronic structure (*e.g.*, reduction of electron density on Ni centers) of the material and influencing the reaction intermediates (number of adsorbed H species), the HER activity is improved.[80] Other metal phosphide phases $Ni_{12}P_5$ or Ni_3P were not as active for HER as Ni_2P.[81] These results indicated that the Ni–P ratio is important, as the P sites (and not only the Ni sites) are also involved in the reaction mechanism as they are acting as acceptor sites for protons and adsorbed H atoms.

Other metal phosphides such as CoP and FeP, were also shown to be suitable co-catalysts, although they are not as intensively investigated as Ni_2P. Excellent, results were achieved upon combination with carbon nitride as photocatalyst, as both CoP and FeP could form N–metal–P bonds, that led to high catalyst stability and enhanced charge transfer.[82,83] But also, in combination with other photoactive materials, those other metal phosphides could show good HER activity.

9.8 Carbides and Borides

Material classes that have very similar properties to phosphides, are metal carbides and borides. Metal carbides have been investigated for several applications in heterogeneous catalysis, as they have a similar electronic structure (and therefore similar catalytic reactivity as platinum group metals). Tungsten carbide (WC) is an exemplary compound as has been used as co-catalyst and WC has delivered excellent results in combination with several photocatalysts such as CdS and carbon nitride.[84,85] A disadvantage is the extremely high synthesis temperatures that are required to form WC (approx. 900 °C) which makes preparation of nanosized carbide particles challenging. Nevertheless, it could be also shown that micrometer-sized carbide particles, in combination with a nanosized ZnSe light absorber, can be an active material for the HER (approx. 0.6 mmol H_2 per g per h).[86] Alternatively, also other carbide materials such as Co_2C or VC are suitable co-catalysts that can be combined with various sulfide and selenide based photocatalysts.[87,88] Also, Ni can form carbides such as Ni_3C, which requires lower synthesis temperatures compared to the tungsten carbides but is also a suitable co-catalyst. Although, it shows signs of degradation under acidic conditions.[89] Most of those carbide materials have metallic characteristics, hence they form a Schottky contact with semiconductors. A comparably recent development is the application of so-called MXene materials as co-catalysts. MXenes are thin layered, two-dimensional carbides or carbonitrides with the general formula $M_{n+1}X_nT_x$ (M = transition metal, often Ti or Mo, X = carbon and/or partially nitrogen, T = termination groups such as OH, O or F). They show excellent activity for the HER and have been combined with photocatalysts such as CdS, TiO_2 or carbon nitride.[90,91] The catalytic nature of the MXenes can be significantly changed by altering its composition. Especially, the termination groups "T" are important, as it could be shown that hydrogen preferentially adsorbs on those groups and by exchanging oxygen containing groups with F, the HER activity decreases.[92] The electronic character of the MXenes can as well be altered from a metallic to a semiconductor characteristic. Another peculiarity of those materials is, that they form larger micrometer-sized sheets, while the nanosized photocatalysts are then deposited on those MXene sheets. Just in contrast to many other materials, where the nanosized co-catalysts are deposited on the light absorbers. There is still a variety of different carbide phases and compositions that has not been investigated, so further advancement for this material class is likely.

Regarding metal borides, MoB and Ni_2B have been reported as suitable co-catalysts in combination with carbon nitride[93,94] As the metal phosphides, also the borides can form N–metal–B bonds with carbon nitride, which facilitates the charge transfer between the carbon nitride and the metal boride. But NiB can be also used with other photocatalysts such as ZnCdS.[95,96] When compared to phosphides, sulfides and carbides, the synthesis methods for borides require less harsh and low temperature conditions, but it is also difficult to control the crystallinity and phase composition of the resulting borides. Therefore, they are less intensively investigated compared to the other mentioned compounds.

9.9 Metal Complexes

The next important class of co-catalysts is metal complexes. In metal complexes, a defined reaction center (metal) with a controlled environment (ligand) is present, which makes the investigation of the active catalytic center easier compared with more complex heterogeneous systems, where it is not always clear, what the actual active site is and how it can be modified. One major advantage of metal complexes is their tunability. There are various possible combinations of metal centers and ligands and the metal center as well as the ligand can be modified and optimized to improve several parameters such as electronic structure, steric environment, interaction with the photocatalyst (anchoring group), hydrophilicity/hydrophobicity, *etc.* For the HER the majority of the investigated metal complexes are based on Ni, Co, and Fe. Those complexes can either be applied in completely homogeneous systems, where the HER metal complex is combined with a soluble photosensitizer (often an organic dye or another metal complex); or they are immobilized on a heterogeneous light absorber (in this case also soluble quantum dots would be counted as heterogeneous material).[97] In this chapter only metal complexes in combination with heterogeneous materials are discussed. There are extensive reviews about the design and application of those metal complexes, such as ref. 97. For Ni^{2+}, many of the most active complexes are based on the DuBois ligand type which is a bis(1,5-R′-diphospha-3,7-R″-diazacyclooctane) structure. The metal-containing complex would be abbreviated as $[Ni\text{-}(P_2{}^{R'}N_2{}^{R''})_2]^{2+}$.[98] To improve the interaction of the metal complex with heterogeneous light absorbers, modifying the ligand with anchoring groups such as phosphonic acid groups, has proven to be very effective. Such a phosphonated Ni complex has been used in combination with TiO_2, ZrO_2, carbon nitride or $CuCrO_2$ and delivered excellent activity for the HER.[98–100] The phosphonic acid group enables a strong interaction of the complex with the heterogeneous light absorber, thereby improving charge transfer and stability of the overall system. Even though the DuBois type Ni^{2+}-complexes are the most prominent examples, there are also Ni complexes based on other ligands such as phthalocyanines.[101] Much more frequently, porphyrin and phthalocyanine complexes are used as light absorbers instead of HER co-catalysts.

Among the Co catalysts, the cobaloximes are the most prominent and active HER catalysts. In those complexes, a Co^{3+} cation is bound by two glyoximate ligands in the equatorial plane, while the two axial ligands can theoretically be varied.[102] Often a pyridine and a halogenide ligand can be found in these positions. By structural modification of the equatorial glyoximates and the axial pyridine, the activity of the complex and especially its compatibility with light absorbers can be greatly influenced. The ligands can be modified with anchoring groups such as phosphonic acids, to improve interactions with light absorbers such as TiO_2. It could be shown that the distance between the TiO_2 surface and the Co reaction centre, significantly influences the overall HER activity.[103] A greater distance between TiO_2 and Co (9 instead of 7 Å) reduces the electron–hole recombination by a factor of up to 10^4. The anchoring site on the ligand can be matched with specific photoactive materials, *e.g.*, phosphonate groups have been shown to interact well with ZnS–CdS particles, while replacing the O with a S in the glyoximate, was beneficial for combination with CdTe materials, as the S interacted strongly with the CdTe surface.[104,105] As for all co-catalysts, a strong interaction with the photoactive material is crucial. This is further emphasized by an example where a cobaloxime co-catalyst was modified with a pyrene group, to improve its interaction with a graphitic carbon nitride light absorber.[99] This shows that also π–π interactions or van der Waals forces are suited for this purpose and not only covalent bonds.

Finally, Fe complexes are also active for the HER. These Fe complexes are generally inspired by Fe-containing hydrogenase enzymes that catalyze the HER in nature. Hence, mainly binuclear complexes with sulfide and carbonyl ligands are found. Those binuclear complexes (or clusters) have been used in combination with ZnS and CdSe as light absorbers.[106–110] Also, for this catalyst type the significance of the anchoring groups was shown, as the introduction of additional aniline groups, improved the catalytic activity for a Fe-complex ZnS system.[108]

Although the metal complexes have several advantages such as flexibility, tunability and high HER activity, there are also some factors that should be considered. Often the stability of those complexes is an issue and many of those complexes will be deactivated after several hours or a few days. Furthermore, many complexes can neither work in the presence of oxygen nor in pure water (very often mixtures of organic solvents and water are used in the studies).

9.10 Enzymes

The inspiration for several metal complexes is drawn from the active sites of enzymes. But those enzymes can be also used themselves as co-catalysts. In nature hydrogenase enzymes are found to catalyze the HER and there are several types of those enzymes. Although most of these active sites are based on metal centers and clusters, different types of metal clusters are found, dividing the enzymes into [Fe–Fe], [Fe–Ni] or [NiFeSe] hydrogenases. Those enzymes have been combined with various photocatalysts such as TiO_2, carbon nitride, CdS and CdTe.[111–114] The [NiFeSe] hydrogenase shows even good oxygen tolerance and high HER activities, with turnover numbers (TON) of up to 50 000.[99] A crucial factor is the interaction of the enzyme with the photocatalyst. Electrostatic interactions have been shown to have a tremendous influence on the HER results. This was proven by combining a hydrogenase enzyme with carbon nanodots as photocatalyst. With negatively charged surface groups, no interaction between the enzymes and the CNDs was possible (due to electrostatic repulsion), hence no H_2 was formed. By changing the CND surface groups to positive functional groups and excellent HER was achieved.[115] Even the sacrificial agent (used for the oxidative side reaction) can influence these interactions. Switching the sacrificial agent from EDTA to TEOA resulted in an HER boost, as the TEOA favorably altered the electronic interactions between the enzyme and the CND photocatalyst.[116] Nevertheless, a common problem for enzymes is their laborious and costly production and handling, as well as their sensitivity.

9.11 Organic Compounds—Carbons

So far, all the mentioned co-catalysts contained metal species of some kind. However, there are also purely carbon-based materials. The exceptional electronic properties of carbon nanostructures such as carbon nanodots (CND), carbon nanotubes (CNT) and graphene enable the utilization of those materials as HER co-catalysts. CNDs can be synthesized from different precursors (citric acid, aspartic acid, *etc.*) and with different electronic properties, as many of those CND materials have semiconductor characteristics and the bandgap can be partially influenced by size and composition of those

CNDs.[115,116] CNTs and graphene, are derived from the graphite structure, and share its excellent electric conductivity and metal-like electronic properties. Previously, carbon dots were mentioned as photoactive materials, however, they can be also used as co-catalysts. In combination with materials such as TiO_2 and carbon nitride, the CND (sometimes also named carbon quantum dots CQD) acted as electron sink and even as active site for the HER.[117,118] The combined carbon-dot TiO_2 or CN_x materials exhibited good HER activity (H_2 formation of 3.5 mmol g^{-1} h^{-1} was reported for the CND–CN_x system).[119] To obtain the optimal activity, the right synthesis protocol must be selected among the great variability of available procedures for the synthesis of carbon dots and combined CND-light absorber materials. For the previously mentioned CND–CN_x system, the CNDs contained amino groups and they were added directly to the reaction mixture for CN_x synthesis. This allowed an efficient incorporation of the CND into the CN_x matrix, and the amino groups of the CND, facilitated this incorporation and enabled an effective and strong interaction between those two components.

The metal-like electronic properties of carbon nanotubes (CNT), cause the formation of metal–semiconductor junction, when in contact with photocatalyst. This junction has already been mentioned to facilitate the charge separation and CNTs can act as electron sinks and again also as active sites for the HER. This was shown by combining CNT with photocatalysts such as TiO_2, CN_x or ZnCdS.[120–122] A good physical contact between the photocatalyst and the CNTs is essential, which can be achieved by combining electrospinning and sol–gel synthesis methods to produce a CNT–TiO_2 composite with good activity for the HER.[123] Another example for this was a combination of a TiO_2 photocatalyst with CNT + C_{60} fullerene materials. A large contact area between TiO_2 and CNT facilitated the charge separation and in a second step the extracted electrons migrate from the CNT to the C_{60} sites, where they engage in the HER. The incorporation of C_{60} fullerenes increases the HER activity compared to the sole CNT modification.[124] Especially, when working with CNTs, the synthesis method of the CNT should be considered. Most synthesis methods involve the presence of transition metal catalysts. Even though the CNT materials undergo after-treatments to remove those metals, it should be always investigated (*e.g.*, by elemental analysis) if still metal residues are present. Because even very low concentrations of metals (ppm range) can significantly influence the HER activity and thereby lead to an overestimation of the HER activity for CNTs. Nevertheless, very few studies consider this issue.

The last carbon material that should be considered is graphene, with its excellent electronic conductivity. Graphene was used in combination with a MoS_2–TiO_2 system, where it mainly acted as electron sink and not as active site for the HER. Graphene facilitated the charge separation and electron extraction and "transported" the electrons to MoS_2, where the HER evolution took place.[125] A similar study investigated a Cu–graphene–TiO_2 system. Also in this case, it could be shown that the graphene significantly improved charge separation and electron extraction.[126] Furthermore, this study demonstrated that a combination of only graphene and TiO_2, was also active for HER, indicating that the HER could also proceed on graphene.

9.12 Oxygen Evolution

The previously mentioned studies were mainly investigating solely the HER reaction. To balance the charges, organic sacrificial agents were used for the oxidative half reaction. If hydrogen shall be produced directly from water splitting, the sacrificial agents have to be replaced and the catalytic system must be capable of performing the oxygen evolution reaction (OER). The OER is a challenging reaction as it involves the transfer of four electrons. This fosters the occurrence of side reactions, which are responsible for the generally sluggish kinetics of the OER (eqn (9.1)).

$$2H_2O + 4h^+ \rightarrow 4H^+ + O_2 \qquad (9.1)$$

Therefore, co-catalysts are required that enable the OER in a smooth manner and facilitate this four-electron transfer reaction. As their pendants in the HER reaction, the OER-co-catalysts shall facilitate the charge separation and provide active sites for the OER. In contrast to the HER materials, the OER co-catalyst shall extract the hole, that is generated by the photoexcitation in the photoactive material. The required positions of the band levels of photocatalyst and co-catalyst are shown in Figure 9.3.

9.12.1 Noble Metal Oxides

Noble metals were among the first investigated catalyst types, as they show excellent results in the electrocatalytic OER. In photocatalytic applications, it seems to be surprising, that noble metals can act as co-catalysts for the HER as well as the OER. However, in the vast majority of OER examples, the noble metals are not present in their metallic state but rather as oxides or at least the surface of the noble metals contains oxygen groups. Au seems to be an exception as there are several reports of metallic Au as co-catalyst for the OER. However, in these cases, Au is acting as light absorbing material, due to its plasmonic properties. This causes the injection of excited electrons from the Au metal into the actual photocatalyst. Thereby, holes are generated in the Au which can engage in the OER. In these studies, the size of the Au particles was found to be crucial, as the plasmonic properties of Au are determined by the particle size. Larger particles of 11 nm yielded better OER activity compared to 2 nm Au particles, on Fe_2O_3 as support.[127] For a photoelectrochemical Au/TiO_2 system, the optimal Au size was found to be 5 nm. On larger particles it was claimed that the electron injection from Au to TiO_2 was hindered, resulting in reduced OER activity.[128] These two examples show that always the complete photocatalytic system has to be considered and that detrimental material characteristics in one system can be beneficial in another one.

Other noble metals such as Pt, are rarely used as co-catalysts for the OER, if they are utilized in OER studies they rather act as electron sinks (to catalyze to cathodic reaction), which was shown in a Pt/WO_3 system, where the reduction of IO_3^- to I^- was the "sacrificial" reduction reaction. For the OER itself mainly noble metal oxides such as RuO_2 are active. The mentioned study showed that already low loadings of 0.001–0.01 wt% of RuO_2 on the Pt/WO_3 system increased the OER activity.[129] RuO_2 was also applied on a TaON photoactive material where it was shown to simultaneously catalyze the anodic OER reaction and the cathodic IO_3^- reduction.[130] It is uncommon to have one co-catalyst material that can catalyze both half reactions. In this case, the RuO_2/TaON system was combined with a Pt/ZrO_2/TaON material to generate a Z-scheme. In this manner complete water splitting was achieved, with the OER taking place at the RuO_2/TaON material and the HER at the Pt/ZrO_2/TaON catalyst, while the IO_3^-/I^- couple was acting as electron shuttle between the two materials.

For the TaON photocatalyst, RhO_x was also shown to be an active OER catalyst, with even higher activity compared to the RuO_2 species. For the RhO_x/TaON system it could be shown how the synthesis procedure influences the final OER activity.[131] Firstly, the Rh-precursor was decisive as $Na_3[RhCl_6]{\cdot}nH_2O$ gave better results for the OER than $Rh(NO_3)_3{\cdot}nH_2O$, after impregnation and calcination. The XPS analysis showed differences for the formed RhO_x species, depending on the precursor type. For the $Na_3[RhCl_6]{\cdot}nH_2O$ derived material, oxidic Rh species with partly metallic Rh were found, while for the $Rh(NO_3)_3{\cdot}nH_2O$ derived material, no metallic contributions were detected. No precise Rh compound could be determined. Hence, the formed Rh species are denoted as RhO_x. Furthermore, the calcination temperature for the $Na_3[RhCl_6]{\cdot}nH_2O$ derived compound was important for the OER reactivity. Higher calcination temperatures of 500 °C led to improved OER activity, as at lower calcination temperatures the decomposition of the $Na_3[RhCl_6]{\cdot}nH_2O$ precursor may be incomplete, as indicated by higher amounts of residual Cl^-.

Another frequently used compound is IrO_2, which is among the best materials for electrochemical OER, in water electrolysis, but it can be also applied in photocatalysis. Laser assisted transition absorption spectroscopy clearly showed for a IrO_2/TiO_2 system, that IrO_2 is effective in extracting the holes from the photocatalyst.[132] From electrocatalysis it is anyway already known, that IrO_2 provides suitable catalytic sites for the OER. IrO_2 is often used as a reference OER co-catalyst to benchmark novel photocatalytic materials or to compare new co-catalysts. For photocatalysts such as $Sm_2Ti_2S_2O_5$ or GaN:ZnO, IrO_2 exhibited decent OER activity.[133,134] Especially for (oxy) nitridic or sulfidic photocatalysts, the utilization of OER co-catalysts is crucial, as they are especially sensitive to oxidative deactivation, caused by the photogenerated holes (photocorrosion). The importance of the co-catalyst loading was demonstrated by a study that compared a classical impregnation method with an *in situ* hydrothermal method, to synthesize a IrO_2/TiO_2 material.[135] The hydrothermal method yielded smaller and better dispersed

IrO_2 nanoparticles and clusters (2 nm size), that showed higher OER activity (around double) compared to the impregnated material.

9.12.2 Non-noble Metal Compounds

When comparing the noble metal oxide catalysts directly with non-noble metal, frequently the non-noble metal ones, especially Co oxide species exhibit better results and demonstrate excellent activity for the OER. Studies that compared several noble and non-noble metal oxides as co-catalysts, with different photoactive materials such as TiO_2, $LaTiO_2N$ or $NaTaO_3$:La, showed that oxidic Co species such as CoO and CoO_x showed the highest activity for OER.[135,136] This excellent catalytic activity is enabled by the combination of two factors, extraction efficiency of holes and inherent activity for OER. CoO_x was shown to combine both characteristics. Other metal oxides such as Cu_2O are still capable to extract the holes from the photocatalyst, however the localized holes on the Cu_2O do not readily engage in the OER. For CoO_x, the band level is just right for the OER and it possesses catalytic sites for this reaction. The effectivity of CoO_x as co-catalyst, is also underlined by the extremely long lifetime for the excited charge carriers, that was determined to be one second, for the $CoO_x/LaTiO_2N$ system. The morphology of the CoO_x and its interaction with the photocatalyst are also crucial for its activity. For a $CoO_x/BaTaO_2N$ increasing the calcination temperature (773 K to 1123 K) as well as a subsequent reductive H_2 treatment have shown to double the OER activity for this catalytic system.[137] The higher calcination temperature causes more intimate contact between the two materials, while the H_2 treatment causes aggregation of the CoO_x particles, freeing active sites for the OER but also for the reductive half-reaction on the $BaTaO_2N$ material. When loading co-catalysts onto a photoactive material, it always has to be considered how this loading procedure affects the photocatalyst. For a $Sm_2Ti_2S_2O_5$ photocatalyst, a sulfurization assisted loading process of CoO_x, was greatly beneficial, as the presence of sulfur during the CoO_x loading, reduces defect formation and enhances stability of the photocatalyst.[133] Thereby, the final $CoO_x/Sm_2Ti_2S_2O_5$ exhibits a significantly higher activity for OER. The activity of CoO_x species can be improved further, as it was demonstrated that doping of small amounts (0.05 wt%) of Rh or Ru into CoO_x (2 wt% loading on Ta_3N_5) increases the OER activity.[138] Doping of those metals improves the hole transfer from the Ta_3N_5 into the co-catalyst, and the presence of low amount of dopant modifies the active sites for the OER in an energetically favorable way, resulting in greater OER activity. Higher doping amounts would reduce the activity again as the dopants will not be that well dispersed in the CoO_x anymore, reducing its beneficial effect on the active sites.

Other metal oxides such as MnO_x, Cr_2O_3 or mixed oxides such as $NiCo_2O_4$ or $CaFe_2O_4$ can as well give good OER activities over an array of photocatalysts such as TiO_2, $Ba_5Ta_4O_{15}$, carbon nitride or Fe_2O_3.[134,139–141] With such a great number of different potential co-catalysts, the desired co-catalyst–photocatalyst combination shall be kept in mind, sometimes the loading procedures for a specific co-catalyst can be well combined with a matching photocatalyst. Furthermore, factors such as stability, costs and reaction environment shall be considered as well to choose the optimal catalytic system.

Another prominent OER catalyst is cobalt oxide phosphate, often abbreviates as Co-Pi. Co-Pi can be applied in combination with different photocatalysts such as ZnO, carbon nitride or Fe_2O_3, for all of which respectable catalytic activities are reported.[142–144] A common method for Co-Pi preparation is photodeposition using $Co(NO_3)_2$ and K_3PO_4 as precursors. Although the nature of Co-Pi is mostly amorphous and it is a mixture of oxidic and phosphate species, the Co^{3+} centers have been identified as important active sites. Co-Pi can also be prepared in a two-step process, combining the deposition of metallic Co nanoparticles on carbon nitride, followed by an *in situ* photocatalytic oxidation in the presence of a phosphate buffer, which will form Co_2O_3 nanoparticles with a phosphate modified (Co-Pi) surface layer. This approach facilitates the preparation of bigger amounts of this material, compared with the one-step photodeposition. Another phosphate-based material is Ag_3PO_4, that can be utilized as an OER co-catalyst.[145,146] Ag_3PO_4 is a semiconductor photocatalyst itself with limited stability for overall water splitting. In a Z-scheme with carbon nitride as the second semiconductor and metallic Ag to facilitate the charge transfer between those two photocatalysts, Ag_3PO_4, showed good activity for the OER.

A common characteristic of many OER co-catalysts and photocatalysts is their oxidic nature. This means, that the semiconductor photocatalysts, can in some cases efficiently engage in the OER, without an additional co-catalyst. However, modification or post-treatment of the photocatalysts can significantly increase the inherent catalytic activity of those materials. The activity of plain TiO_2 for the OER can be greatly improved by increasing the number of surface titanium vacancy sites. The titanium vacancies serve as active sites for the OER, as the water will adsorb *via* its two H-atoms onto the two-vicinal surface oxygen atoms, generated by the formation of at Ti-vacancy.[147] On non-defective TiO_2, the water molecule would adsorb *via* the oxygen atom on the surface Ti site. The H-coordination on the oxygen sites lowers the required activation energy for the OER, which is the reason for increased OER activity, for Ti-vacancy rich materials. At the same time, the water adsorption is energetically more favorable on those vacancy sites. Ti-Vacancy rich TiO_2 can be prepared by a two-step treatment of single crystalline TiO_2, with acid (to protonate the surface oxygen sites) followed by an extensive treatment with tetrabutylammonium hydroxide. The defective TiO_2 can be combined with a WO_3 photocatalyst and form a heterojunction. Thereby the charge separation is improved and the OER will take place on the vacancy rich TiO_2. A similar concept was also reported for pure WO_3 materials, where vacancies also improved the OER activity of pure WO_3 materials. Surface modification is also beneficial for other photocatalysts such as $BiVO_4$

where surface modification with Cl^- has proven to increase the activity for OER.[148] Simple addition of NaCl during the $BiVO_4$ synthesis, is sufficient to load enough Cl^- onto the $BiVO_4$ surface. The Cl^- on the surface enhances the water adsorption, but facilitates the O_2 desorption at the same time, which combined increase the OER activity. In addition, also the charge carrier separation is improved by the presence of Cl^-. $BiVO_4$ shows a facet-dependent activity for photocatalysis; after light absorption, the electrons accumulate at the (010) crystal facet, while the hole preferentially migrates to the (110) facet, as those two facets have a slight difference in their conduction/valence band level.

Besides purely oxidic materials and oxide clusters also single metal atoms can exhibit activity for OER. For this purpose, the stabilization of the single metal atoms in a matrix, such as carbon nitride, is necessary. Carbon nitride is especially well suited for stabilizing metal atoms, with its variety of nitrogen functional groups. This approach has proven to be successful for Mn-carbon nitride or Co-carbon nitride materials.[149,150] In both cases the metal atoms are well dispersed and stabilized over the whole carbon nitride surface where they form surface junctions, that are responsible for the hole extraction and are working as active sites for the OER. A similar approach is the utilization of Co-porphyrins as co-catalyst.[151] Also, in this case, a single Co^{3+} ion is stabilized by a nitrogen containing framework, that is in this case provided by the porphyrin ligand.

9.12.3 Combined Systems Without Sacrificial Agents

In the catalytic systems mentioned so far, the focus was mostly on either the HER or OER, so only one of these reactions was investigated at a time. In those cases, mostly sacrificial agents are employed to balance the reaction. For example, the oxidation of molecular substrates like ethylenediaminetetraacetic acid (EDTA) or TEOA is applied for HER studies or the reduction of Ag^+ or NaS_2O_8 for OER studies. For real world applications, the utilization of sacrificial agents is not viable and an overall water splitting system is required, meaning a catalytic system on which the HER and OER proceed in parallel. Overall water splitting systems can work with only one co-catalyst (either HER or OER) if the complementary reaction can proceed at the photocatalyst itself. Nevertheless, it has been proven to be very efficient to apply two co-catalysts in one system, even though such systems are conceptually more challenging. This was demonstrated with a system based on a GaN:ZnO photocatalyst.[152] When loaded with a combination of Rh/Cr_2O_3 as HER co-catalyst, decent activity for overall water splitting was observed, by constant formation of H_2 and O_2 gas. In addition, a Mn_2O_3 co-catalyst can be loaded as well onto the photocatalyst substrate, which doubles the catalytic activity (and the $H_2 + O_2$ yield). The Mn_2O_3 provides the active sites for the OER. The simultaneous presence of a HER and an OER co-catalyst improves the charge separation, as electrons migrate to the HER co-catalyst and holes to the OER one; and the presence of the active sites for the particular reactions is also beneficial for the overall activity. The Mn_2O_3 co-catalyst was loaded onto the GaN:ZnO material by deposition of previously synthesized nanoparticles. In contrast, the Rh/Cr_2O_3 material was formed by an *in situ* photodeposition method. The Rh/Cr_2O_3 co-catalyst is especially suitable for overall water splitting applications, as it mitigates the reverse reaction of the just evolved H_2 with O_2, back to water. Cr_2O_3 forms a shell on the Rh nanoparticles that is permeable for H^+ and H_2 species, but not for O_2. Therefore, the H_2 oxidation (reverse reaction) cannot proceed on the Rh surface. Without this Cr_2O_3 the back reaction can proceed and the catalytic efficiency for water splitting decreases.

The catalytic performance of such systems can be even improved, by spatial separation of the respective co-catalysts. Hollow carbon nitride structures were loaded with Pt as HER and Co_3O_4 as OER co-catalyst.[153] The hollow structure of the carbon nitride material and a suitable synthesis protocol enabled the selective deposition of Pt on the inside of the hollow structure, while the Co_3O_4 was attached to the outside of the hollow carbon nitride. This architecture enabled a directed migration of the charge carriers (electrons inside towards the Pt and holes outside towards the Co_3O_4), which significantly reduced the charge recombination. A random distribution of both co-catalysts showed a lower water splitting activity, as the charge recombination was increased again.

Separation of co-catalysts can be also conducted on photocatalysts with distinct facets. A $SrTiO_3$:Al photocatalyst was loaded *via* photodeposition methods with a Rh/Cr_2O_3 HER and a CoOOH OER co-catalyst.[154] As the excited charge carriers in $SrTiO_3$:Al preferentially migrate towards different facets of the material, also the co-catalysts will be deposited on separate facets of the photocatalyst, (100) facet for the Rh/Cr_2O_3 and (110) for the CoOOH. This enables good charge separation and high catalytic efficiency, an external quantum efficiency of 96% was achieved for wavelengths between 350–360 nm.

By separating not only the co-catalysts but providing two distinct photocatalytic systems, even more impressive results could be achieved for water splitting. For the HER a Ru/Cr_2O_3–$SrTiO_3$:La,Rh system was designed.[24] The Ru/Cr_2O_3 part acted as HER co-catalyst and formed nanoparticles on the $SrTiO_3$:La,Rh substrate, which was the photocatalyst. Doping of La and Rh into the $SrTiO_3$ improved the photocatalytic properties of this material. For the OER a $RuO_x/BiVO_4$:Mo material was utilized, while RuO_x nanoparticles acted as co-catalysts. To improve the contact between the HER and OER components they were immobilized on glass sheets and a Au layer was applied below the photocatalyst particles, which facilitated the charge transfer. Thereby a photocatalytic Z-scheme was established, with excellent charge separation properties and high catalytic activity. The overall solar to hydrogen conversion efficiency was higher than 1% (at a quantum yield of 30% for the wavelength of 419 nm), which is an exceptionally high value for an overall water splitting system.

9.12.4 Replacement of O_2 Evolution—Photoreforming

The product of interest for the water splitting reaction is hydrogen. Oxygen is merely a byproduct of low economic value. In addition, the OER is challenging and sluggish. Replacing the OER with another oxidation reaction is therefore a viable concept, provided the substrate for the alternative oxidation reaction is cheap and widely available (in contrast to the common sacrificial agents). Suitable oxidizable substrates would be biomass or waste derived compounds (lignocellulose, plastic waste, or derivatives thereof). The photocatalytic oxidation of such substrates for the generation of hydrogen, is also called photoreforming.[155] To further improve the feasibility of photoreforming, it is aimed to obtain useful chemical products from the oxidation reaction. Thereby, a second product, in addition to hydrogen, could be generated. Possibilities are the oxidation of 5-hydroxymethylfurfural (HMF) to the corresponding furandicarboxylic acid (FDCA).[156] The oxidation of those alternative substrates depends as well on the applied photocatalysts or co-catalyst. For the photoreforming of cellulose and sugars, it could be shown that the type of TiO_2 modification, influences the yield of organic products.[18] Rutile based photocatalysts give higher yields (of erythrose, arabinose and formic acid) compared to anatase based materials. Rutile engages in a more selective direct hole transfer mechanism with the sugar substrate, while anatase (or P25) forms OH radicals, that tend to over oxidize the substrates, yielding lower yields of organics.

9.13 CO_2 Reduction

The photocatalytic CO_2 reduction has attracted great interest during the last years, as the conversion of CO_2 into chemicals or other compounds is expected to be a crucial part in mitigating the CO_2 emissions.[157] The photocatalytic reaction is mimicking the natural photosynthesis, as CO_2 is converted into organic products, utilizing light as sole energy input. When compared to the HER, the CO_2 reduction is more complex as CO_2 can react towards various products *via* different reaction pathways. A selection of possible reactions with their equilibrium potential are shown in eqn (9.2)–(9.5).[158]

$$CO_2 + 2e^- + 2H^+ \rightarrow HCOOH_{(aq)} \quad (-0.12\ \text{V}\ \textit{vs.}\ \text{RHE};\ \Delta H^\circ = -31.2\ \text{kJ mol}^{-1}) \tag{9.2}$$

$$CO_2 + 2e^- + 2H^+ \rightarrow CO_{(g)} + H_2O \quad (-0.10\ \text{V}\ \textit{vs.}\ \text{RHE};\ \Delta H^\circ = -41.2\ \text{kJ mol}^{-1}) \tag{9.3}$$

$$CO_2 + 6e^- + 6H^+ \rightarrow CH_3OH_{(aq)} + H_2O \quad (0.03\ \text{V}\ \textit{vs.}\ \text{RHE};\ \Delta H^\circ = -131.0\ \text{kJ mol}^{-1}) \tag{9.4}$$

$$CO_2 + 8e^- + 8H^+ \rightarrow CH_{4\,(g)} + 2H_2O \quad (0.17\ \text{V}\ \textit{vs.}\ \text{RHE};\ \Delta H^\circ = -253.0\ \text{kJ mol}^{-1}) \tag{9.5}$$

The actual product distribution of the CO_2 reduction depends on several factors, such as the band levels of the photocatalyst and the co-catalyst, as well as the reaction environment (gas or liquid phase, pH value, *etc.*). The catalyst selection and potential surface modification are other crucial factors as the related properties (adsorption and stabilization of starting materials, intermediates, and products) highly influence the reaction outcome. Under certain circumstances also C_{2+} products can be formed, that is especially attractive, as they generally have higher economic values compared to the previously listed C_1 compounds.[158] Nevertheless, more electrons and protons are transferred in those reactions forming C_2 products. This makes their selective formation significantly more challenging.

$$2CO_2 + 8e^- + 8H^+ \rightarrow CH_3COOH_{\,(aq)} + 2H_2O \quad (0.11\ \text{V}\ \textit{vs.}\ \text{RHE};\ \Delta H^\circ = 122.1\ \text{kJ mol}^{-1}) \tag{9.6}$$

$$2CO_2 + 10e^- + 10H^+ \rightarrow CH_3CHO_{\,(aq)} + 3H_2O \quad (0.06\ \text{V}\ \textit{vs.}\ \text{RHE};\ \Delta H^\circ = 130.1\ \text{kJ mol}^{-1}) \tag{9.7}$$

$$2CO_2 + 12e^- + 12H^+ \rightarrow CH_3CH_2OH_{\,(aq)} + 3H_2O \quad (0.09\ \text{V}\ \textit{vs.}\ \text{RHE};\ \Delta H^\circ = 216.1\ \text{kJ mol}^{-1}) \tag{9.8}$$

$$CO_2 + 12e^- + 12H^+ \rightarrow C_2H_{4(g)} + 4H_2O \quad (0.08\ \text{V}\ \textit{vs.}\ \text{RHE};\ \Delta H^\circ = 127.8\ \text{kJ mol}^{-1}) \tag{9.9}$$

$$CO_2 + 14e^- + 14H^+ \rightarrow C_2H_{6(g)} + 4H_2O \quad (0.14\ \text{V}\ \textit{vs.}\ \text{RHE};\ \Delta H^\circ = 264.3\ \text{kJ mol}^{-1}) \tag{9.10}$$

9.13.1 Noble Metals

Although noble metals are often excellent co-catalysts for the HER, they can be as well utilized for the CO_2 reduction reaction. Pt loading was shown to improve the CO_2 photoreduction to ethanol on a TaON photocatalyst.[159] Even though TaON alone also exhibited activity towards this reaction, an additional loading of 0.5 wt% Pt increased the ethanol formation rate by approximately 20% (2 µmol g^{-1} h^{-1}). Simultaneously, the yield of the acetaldehyde side product is decreased in presence of Pt, as the presence of Pt improves the hydrogen transfer to the adsorbed CO_2 intermediate. Additionally, Pt improves the charge separation, by acting as electron sink, which makes the overall reaction more efficient. The optimal size for the Pt nanoparticles was determined to be 1 nm on a TiO_2 photocatalyst.[160] As the Fermi level of the 1 nm Pt clusters matches optimally with the TiO_2 band level, to enable the formation of a Schottky barrier, with smaller or larger particle sizes, the Fermi level of the Pt particle shifts, making the electron transfer process less favorable.

Another widely used noble metal is Pd, which is frequently applied as co-catalyst for the CO_2 reduction. When Pd is loaded on CN_x and applied as photocatalyst the main product is methanol with CH_4 as a side product, with a gas phase approach (CO_2 and H_2O being present in the gas phase).[161] By comparing tetrahedral and cubic Pd particles it was shown that, the catalytic activity of Pd is facet dependent, as the (111) facet in the tetrahedrons gave higher yields of methanol than the (100) facet in the cubes. This is because CO_2 can adsorb better on the (111) facet, while at the same time the formed methanol can easier desorb from the same surface. Both factors together explain the improved catalytic activity. The surface structure of Pd has proven to be decisive also in other studies. For a Pd/TiO_2 system it could be shown that, a thin amorphous layer on the Pd particles would increase the selectivity of the CO_2 reduction to CO and CH_4, while decreasing the H_2 side product formation (from water splitting).[162] The optimal results were achieved with crystalline core and amorphous shell Pd particles, as the crystallinity of the core facilitated the electron transfer from the photocatalyst and the amorphous layer modified the adsorption properties of CO_2 and H_2O, which led to a preferred reduction of CO_2. Another possibility to improve the CO_2 reduction activity, is to increase the number of Pd edge sites. For another Pd/TiO_2 system, Pd hexagonal nanosheets were immobilized on TiO_2 support, while decreasing the sheet size, increased the density of edge sites, which improved the CO_2 reduction activity to CO and CH_4.[163] In the mentioned studies the main product was mostly CO and CH_4 was only the side product, but this selectivity can be shifted, by generating Pd-hydrides. When Pd is converted to hydrides (Pd–H) *via* a hydrothermal procedure, the resulting Pd–H/TiO_2 photocatalyst has a CH_4 selectivity of 63% for cubic and 73% for tetrahedral Pd particles, compared to 3% and 16% for the respective metallic Pd particles.[164] In the Pd hydrides significantly more H atoms are available on the surface, which results in the preferred formation of CH_4.

A significantly cheaper co-catalyst is Ag, that is frequently used with several photocatalytic materials. Ag has been shown to be selective for the CO_2 to CO reduction, with comparably low HER activity as side reaction.[157] Among different layered perovskites, $BaLa_4Ti_4O_{15}$ showed the highest activity for CO_2 to CO reduction.[165] As side product small amounts of formic acid were formed. Liquid phase Ag deposition was shown to be more effective compared to photodeposition, as the liquid phase deposition resulted in smaller Ag particles. During the CO_2 reduction, the Ag on the planes of the $BaLa_4Ti_4O_{15}$ layers will dissolve and redeposit on the layer edges. This is due to preferred migration of the charge carriers, as the holes migrate to the perovskite planes and the electrons to the edges. It is common for CO_2 reduction tests, that the oxidative counter reaction is OER. Hence, addition of another OER co-catalyst benefits the CO_2 reduction. For an Ag/$SrTiO_3$:Al system, the inclusion of a

FeOOH OER co-catalyst significantly increased the CO_2 to CO reduction rate (76 μmol CO per g per h).[166] Also, in this case, a facet dependent activity was observed, with a selective deposition of the Ag particles on the (100) facet of the $SrTiO_3$:Al particles (where the CO_2 reduction is proceeding) and FeOOH is only formed on the (110) facet. But also, with other photocatalysts such as CdS or $CaZrO_3$, Ag is a suitable co-catalyst with good selectivity towards CO, although with the Ag/CdS system a high share of H_2 was produced, as the target product was syngas.[167,168]

Also, Au can be used as co-catalyst for the CO_2 reduction. In combination with a brookite TiO_2 photocatalyst, Au has been shown to increase the yield of methanol as the main reduction production.[169] Au facilitates the charge separation and the electron transfer to the CO_2 substrate. For brookite it was shown that there was a facet dependent selectivity, as the oxidation half reaction was proceeding at the (212) facet and the reduction with the help of the Au particles at the (210) facet. The reaction was carried out in liquid phase, in a CO_2 saturated aqueous solution. For several porous titanate photocatalysts (such as $SrTiO_3$), it was shown that the chemical reduction deposition of Au, resulted in a higher activity for CO_2 reduction, compared with photodeposited Au.[170] Most likely, the interaction of the Au particles with the photocatalyst support is improved for the chemical reduction method. In this study, a special porous architecture of the photocatalyst was utilized to enhance the interaction area with the gas phase. This gas phase reduction approach mainly yielded CO and CH_4 as products (in contrast to the methanol, in the previously mentioned liquid phase approach). Another study proved that the yield of CH_4 is increased in the presence of Au, on a porous TiO_2 photocatalyst.[171] Without Au, a lower CO_2 conversion rate, but a higher selectivity towards multi-carbon products is observed.

Gold is frequently used in bimetallic alloy co-catalysts, that can further increase the catalytic activity. Bimetallic Au–Cu particles on P25, gave a high activity CO_2 to CH_4 reduction of 2 mmol g^{-1} h^{-1}, with a selectivity of 97%.[172] The benefit of the Au–Cu alloy particles is the presence of isolated Cu sites that can bind the CO intermediate and facilitate its further conversion to CH_4. For gas phase approaches like this example, it is mostly the case, that also H_2O is present in the gas phase to act as H donor (unless CO is the only product). The efficiency of this bimetallic system was also used for another application where defined Au_3Cu nanoparticles were deposited on a mixed $SrTiO_3$–TiO_2 nanotube array.[173] In this case hydrazine was present at the catalyst surface to facilitate the CO_2 reduction. The main product was CO (3.8 mmol g^{-1} h^{-1}), but also mixed hydrocarbons, mainly CH_4, were formed with an activity of 0.7 mmol g^{-1} h^{-1}. The proximity between the catalytic sites of the Au_3Cu particles and the reductive hydrazine functionalities proved to be beneficial in this case. Alloyed CuPt particles on TiO_2, did not reach such a high activity, as the CH_4 formation rate was only at 0.01 mmol g^{-1} h^{-1}, but it was found that smaller nanoparticles (approx. 1 nm) gave the highest activity, as the stability on the support was increased and those particles possessed the optimal ratio of Cu and Pt edge atoms, where Cu could stabilize CO intermediates and Pt was responsible for the H transfer intermediate steps.[174]

9.13.2 Non-noble Metals

The suitability of Cu as a co-catalyst for the CO_2 reduction, is widely known from electrocatalytic CO_2 reduction where Cu is widely used as a catalyst. Also, in the previous alloy examples, the beneficial effect of Cu is demonstrated. Hence, it is not surprising that pure Cu is as well a suitable co-catalyst for the photocatalytic CO_2 reduction. For a Cu/TiO_2 system, a narrow window for the optimal Cu loading was determined (0.4 wt% Cu).[175] Loading of co-catalysts is always a balance, as a low loading might not provide sufficient active sites. But too high loading can lead to particle agglomeration and shadowing of the underlying photocatalyst, thereby reducing the possible light absorption. The main product in this example is CH_4 with a formation rate of 8 μmol g^{-1} h^{-1}. For a Cu on graphene oxide (GO) system, the optimal Cu loading was with 10 wt% significantly higher, demonstrating that the co-catalyst loading has always to be matched with the respective photocatalyst.[176] In this case a liquid phase reaction was conducted, yielding methanol and acetaldehyde as main products (formation rate, 6.8 μmol g^{-1} h^{-1}). Ethanol was not formed as the band level of the catalyst system was not suitable to catalyze the CO_2 to ethanol reduction, hence acetaldehyde and methanol were formed. In the electrocatalytic CO_2 reduction, Cu is known for its tendency to form multi-carbon products. Such multi-carbon products are also detected in photocatalytic approaches, *e.g.*, in a combined CdS–$Na_xH_{2-x}Ti_3O_7$–Cu system, where the CdS acts as main photocatalysts, and the excited electrons are transferred *via* the trititanate substrate to the Cu particles, where the CO_2 reduction is taking place.[177] Gaseous hydrocarbons such as CH_4, C_2H_6, C_3H_8, C_2H_4 and C_3H_6 are detected. The proclivity of this system towards the formation of multi-carbon products was investigated with electron spin resonance (EPR) spectroscopy and it was found that copper could stabilize the CH_3-radical species. This radical is a central intermediate for the formation of those products. For a Cu nanoparticle on carbon nitride (CN_x) system, the main product for the CO_2 reduction was found to be CO even though H_2O was present.[178] The reasons for the finding of CO as sole product can either be that the analysis of liquid phase product was not conducted, or that the Cu/CN_x system does not possess active sites that can partake in H transfer reaction steps. For a single Cu atom/CN_x also CO was the main product, but in this case, CH_4 and CH_3OH were detected as minor side products.[179]

Other transition metals such as Ni, are rarely used for the CO_2 photoreduction, as they mostly favor the HER. Nevertheless, with suitable photocatalysts also metallic nickel clusters show activity for CO_2 reduction. In combination with black TiO_2, which contains a high concentration of surface oxygen vacancies, Ni can facilitate the CO_2 reduction. In this example, CO_2 adsorbs *via* its O atom at the TiO_2 oxygen vacancy and the C atom is adsorbed at the Ni cluster.[180] This enables selective CO_2 reduction to acetaldehyde. It has to be mentioned that the oxygen vacancies do not only modify the CO_2 adsorption but shift the band level energies of TiO_2. Apart from that, Ni can be active as single atom species in covalent organic frameworks (COF), where it showed excellent CO_2 to CO selectivity and high activity (4 mmol g^{-1} h^{-1}).[181]

9.13.3 Sulfides, Oxides, Nitrides, Carbides and Phosphides

Transition metals can be more frequently found in their oxidic state, when being applied as CO_2 reduction co-catalysts. Copper oxide is widely investigated in various studies, with different photocatalysts. Often copper oxide is found to have a high selectivity towards the formation of methanol. Because the surface of copper oxides will be hydroxylated under reaction conditions, which facilitates hydrogen transfer reactions. Furthermore, copper oxide can stabilize the formate intermediate, so formic acid will not be released but further reduced to methanol.[157] If the copper oxide is reduced to metallic copper, the methanol selectivity and catalyst activity will decrease. As the reduction leads to particle agglomeration and to loss of the active sites. Mostly, TiO_2 is utilized as photocatalyst in combination with copper oxide, but also SiC has been reported.[182,183] If other photocatalysts such as $SrTiO_3$ or $Nb_3O_8^-$ nanoclusters are used, the product selectivity shifts to CO as main product.[184,185] This illustrates that the co-catalyst has always to be considered in a complete catalytic system, as there exist many cooperative mechanisms in which the photocatalyst is involved as well. Preparing mixed metal oxides shifts the product selectivity. If mixed Cu_2O–Fe_2O_3 co-catalysts are deposited on TiO_2 methane and ethylene are formed as main products. The presence of Fe_2O_3 was shown to suppress the methane route and favor ethylene, by altering the dispersion of Cu_2O, which influences its selectivity.[186]

Other oxides such as NiO_x or CoO_x are suitable co-catalysts as well for the CO_2 reduction. For NiO_x, core–shell Ni–NiO_x particles have shown to be especially suited for the reduction of CO_2 to methanol, as the Ni core improves the electron transport from the photocatalyst to the co-catalyst surface.[187] This core–shell catalyst has been utilized with several photocatalysts such as $InTaO_4$ or $InVO_4$.[188,189] If $Ni(OH)_2$ is utilized as co-catalyst, methane will be the main product. In this case, Ni-clusters will be generated *in situ* by photoreduction and the basic nature of $Ni(OH)_2$, supports the adsorption of the acid CO_2.[190] CoO_x gives different product selectivities, depending on which photocatalyst it is immobilized. On carbon nitride, CO is the main product, while for a core–shell CoO_x–ZnO material CH_4 is formed.[191,192] This is due to the different band energy levels of the photocatalysts but also their different capability to engage in hydrogen transfer reactions.

Sulfidic co-catalysts, that have been showing promising activity for the HER, have also been tested for the photocatalytic CO_2 reduction. NiS and CuS have both shown promising activity for the CO_2 to CH_4 reduction with TiO_2 as photocatalyst.[193,194] Also, MoS_2 has shown interesting results for the CO_2 reduction and was utilized together with a Bi_2WO_6 photocatalyst to form methanol and ethanol (in an aqueous solution).[195] With TiO_2 and a gas phase reaction, the main product was CO.[196] Here it should be emphasized again that the product selectivity of the photocatalytic CO_2 reduction is highly dependent on the reaction set up (phase and reactor type), besides the catalytic system. It can be seen that co-catalysts for the HER can frequently be used for the CO_2 reduction as well, because these materials support the charge separation and often their active sites are suitable for other reactions as well. Hence, materials such as transition metal phosphides and carbides, with their metal-like electronic structure, can be used as co-catalysts for the CO_2 reduction. MoP was used on a carbon nitride photocatalyst for the conversion of CO_2 to CO.[197] In this case, the HER was however even more dominant and H_2 was the main product. Nevertheless, the CO/H_2 (syngas) mixture is useful for several applications in chemical industry. CoP also prefers the water splitting reaction and produces large amounts of H_2, even in presence of CO_2. This can be avoided by coating the CoP surface with a thin layer of N-doped carbon, that increases the hydrophobicity and enhances CO_2 adsorption.[198] Thereby, a high selectivity for the CO_2 to CO reaction could be achieved. In this study, a Ru-complex was used as homogeneous light-absorber. The influence of the N-doped carbon exemplarily demonstrates the significance of the surface properties and composition for catalysis. Another intermetallic species is Co_2N that showed good activity (0.07 mmol g^{-1} h^{-1}) for the CO_2 to CO reduction with a BiOBr photocatalyst.[199] Co_2N has a strong interaction with BiOBr and facilitates the charge separation and provides active sites for the CO_2 reduction.

Transition metal carbides often have similar electronic properties as their corresponding phosphides and nitrides. A special class of carbides (or oxy-carbides) that have recently been widely utilized for the photocatalytic CO_2 reduction are MXenes, two-dimensional (oxy-carbide) materials, with special electronic and catalytic properties. Ti_3C_2 is a good example and showed activity for the CO_2 reduction in combination with a TiO_2 photocatalyst.[200] The surface termination of the Ti_3C_2 has proven crucial for the overall activity and by replacing the F-termination with OH-groups, the catalytic activity could be enhanced by a factor of 277 for the formation of CH_4. Also, the formation of the CO secondary product was improved by a factor of three. The presence of OH groups improved the charge transfer and favorably modified the band layers. Finally, the hydroxyl groups served as suitable sites for the adsorption of CO_2 molecules.

9.13.4 Metal Complexes

Besides metal nanoparticles also metal complexes can be applied as co-catalysts for the CO_2 reduction. Frequently, these complexes are applied in a homogeneous fashion. Either they act as light absorbers and catalysts simultaneously or they are combined with a homogeneous light absorber (Ru-complexes, all organic dyes, *etc.*). Those complexes are mainly based on transition metals such as Co, Ni, Cu or Zn, that can accommodate several oxidation states. Mostly, nitrogen containing ligands are utilized, such as pyridines, bipyridines or terpyridines, porphyrins, phthalocyanines or tetraamines. But especially for Ni and Co also phosphide ligands can be found.[97] Although such a completely homogeneous system can achieve good catalytic activity and are crucial for the mechanistic understanding of the CO_2 reduction, in this case, the focus will be put on systems where those catalysts are combined with particulate or heterogeneous photoactive materials. A good example of such a system is a Fe-porphyrin complex that is combined with carbon nitride photocatalyst.[201] This system exhibited good selectivity for the CO_2 to CO reaction but had a comparably low turnover number (TON) of 5.7. It was demonstrated that the presence of carboxylic acid groups on the porphyrin ligand is highly beneficial, as this enhances the interaction between the metal complex and the carbon nitride. This system was however operating in a mainly organic solvent (95% acetonitrile, 5% water). This is also the case for a Co based catalyst (with a chlorine ligand), that is immobilized on carbon nanotubes and combined with an organic light absorber (giving a very high TON of 501 for CO formation).[202] To enable more widespread application of those systems, they should be able to operate in purely aqueous conditions. This could be achieved with Ni-terpyridine complexes in combination with CdS quantum dots as light absorbers.[203] In this case, it could be clearly shown that suitable anchoring groups on the terpy-ligand could significantly increase the catalytic activity. In this system, thiol groups proved to be a better choice compared to carboxylic of phosphonic acid groups, as they showed the most stable interaction with the CdS surface. Thereby, the Ni complexes would be in closer proximity to the surface (improving charge transfer) and stay there for an extended amount of time (improving stability). The importance of strong interactions between metal complex and photocatalyst was again elucidated in another study, in which the porphyrin ring of a Fe complex was modified with positively charged trimethylamine groups. The photocatalyst was $CuInS_2/ZnS$ core–shell quantum dots, which were modified with mercaptopropionic acid.[204] The Coulombic interactions between the negatively charged carboxylic acid and the positively charges side groups of the porphyrin ligand, enabled stable adsorption of the metal complex on the modified quantum dot surface, resulting in good catalytic activity with a TON of 450. A drawback of those metal complex-based systems is often the complicated synthesis of the metal complex and limited stability of those compounds. Furthermore, in most cases, expensive sacrificial agents have to be used for the balancing oxidation reaction. This issue can be solved by replacing the sacrificial agents with cheap and available substrates such as biomass. This approach has been successfully demonstrated with a Co-bipyridine catalyst combined with TiO_2.[205] While the Co catalyst was selectively reducing CO_2 to CO, the oxidation reaction was the conversion of enzymatically pretreated cellulose (to convert it into soluble sugars) into formate. With this approach two useful products (CO and formate) can be generated, which increases the economic feasibility of those processes.

9.13.5 Enzymes

Some of those metal complexes try to mimic the functionality of enzymes such as CO_2-reductases, that can be also used as co-catalysts. These reductases contain a Ni-active center have been immobilized on TiO_2 and deliver excellent activity and selectivity towards CO, when combined with an organic light absorber or even Ag-clusters (as light absorber).[206,207] The Ag-clusters have better light absorbing properties, resulting in TONs for CO formation of 250 000. Alternatively, CdS nanorods can also be used as photocatalyst, as they have a higher driving force for the reaction and absorb visible light, making the utilization of additional light absorbers as for TiO_2, obsolete.[208]

9.13.6 Carbons

There are, however, co-catalysts that work without any active metal center. Nanostructured carbon materials have been widely used as co-catalysts, in combination with various photocatalysts. Graphene oxide (GO) and especially reduced graphene oxide (RGO) have been demonstrated to be a suitable co-catalyst with various photocatalysts such as TiO_2, carbon nitride, CdS, ZnO, Cu_2O, $CsPbBr_3$ and others.[209–213] In the gaseous phase the main reaction product is CH_4 or CO, in liquid phase approaches methanol is formed. Those material combinations profit from the sheet-like structure of RGO, as it enables facile and stable deposition of nanoparticulate photocatalysts. In contrast to most other approaches where "small" co-catalyst particles are deposited on "larger" photocatalyst sheets. The typical catalytic activities for CO and/or CH_4 formation of such materials lies in the range of 1–5 $\mu mol\ g^{-1}\ h^{-1}$. Those graphene-derived structures can be further modified by doping or functionalization with heteroatoms. This can improve the electronic properties (electron extraction efficiency or electric conductivity) and the interaction with the photocatalyst material, both leading to higher catalytic activity. Lowering the number of defect sites (*e.g.*, by better synthesis methods) in the graphene materials, has also beneficial effects.[214]

Carbon nanotubes have been utilized as well as co-catalysts, especially in combination with TiO_2 photocatalysts. Depending on the TiO_2 crystal structure the main reaction product is either ethanol (for anatase) or formic acid (for rutile) in liquid phase.[215,216] For gas phase approaches CO and CH_4 were detected. Carbon nanotubes were also combined with Ag–AgBr core–shell nanoparticles for the reduction of CO_2 to CH_4 and methanol.[217] It was proven that the longer, better ordered carbon tubes gave better catalytic results, as they exhibited better electron transfer properties and lesser charge recombination. Sometimes the carbon species does not necessarily have to possess a specific nanostructure. Even thin-layered deposits of amorphous carbon on ZnO, Cu_2O or carbon nitride improve the catalytic activity towards the formation of methanol, CH_4 or CO by increasing the number of adsorption sites for CO_2.[218–220] Furthermore, the carbon layer reduces photo corrosion, by extracting charge carriers.

9.14 Conclusion and Outlook

Co-catalysts are an essential part for photocatalytic systems as they can influence all aspects of the catalytic process such as activity, selectivity, and stability. It was shown that a broad range of compounds can be applied for that purpose including not only metals but also enzymes and carbon materials. To further develop photocatalytic approaches and bring them closer to real world applications, co-catalysts will play a decisive role. A real-world system must function without the addition of sacrificial agents. As target reactions are either the HER or the CO_2 reduction, there is intensive research for co-catalysts for these reactions. However, the oxidation reactions must not be neglected, as it was shown for overall photocatalytic water splitting systems, where the OER co-catalysts are crucial. In this context also alternative oxidation reactions (photoreforming) should be considered where cheap and widely available substrates (biomass, plastic) are consumed in the oxidation process. In the optimal case, these substrates are not completely oxidized to CO_2 but selectively to defined products. Especially for those selective oxidation reactions, novel co-catalysts have to be developed and better understanding of the oxidation mechanisms is required, as currently the oxidation reactions are proceeding at the photocatalyst surface. The goal should be, as for water splitting systems, photocatalyst systems with two co-catalysts engaging in the reduction reaction (HER or CO_2 reduction) and a selective oxidation reaction of cheap substrates. But also, the co-catalysts for the reduction reactions shall be further developed to improve stability and activity.

Acknowledgements

C.M.P. gratefully acknowledges financial support from the Austrian Research Fund (FWF) (Schrödinger Fellowship J-4381) and the Austrian Research Promotion Agency (FFG) (COMET-Program 2018–2022). Prof. Qian Wang is gratefully acknowledged for her editorial work.

References

1. BP Energy Outlook, *IOP Conf. Ser.: Mater. Sci. Eng.*, 2018, ISSN 1757-899X, 342, 9 p.
2. Z. J. Schiffer and K. Manthiram, *Joule*, 2017, **1**, 10–14.
3. Z. Wang, C. Li and K. Domen, *Chem. Soc. Rev.*, 2019, **48**, 2109–2125.
4. L. Marzo, S. K. Pagire, O. Reiser and B. König, *Angew. Chem., Int. Ed.*, 2018, **57**, 10034–10072.
5. W. Zhang, A. R. Mohamed and W. J. Ong, *Angew. Chem.*, 2020, **59**, 22894–22915.
6. S. Chen, T. Takata and K. Domen, *Nat. Rev. Mater.*, 2017, **2**, 17050.
7. J. Ran, J. Zhang, J. Yu, M. Jaroniec and S. Z. Qiao, *Chem. Soc. Rev.*, 2014, **43**, 7787–7812.

8. T. Su, Q. Shao, Z. Qin, Z. Guo and Z. Wu, *ACS Catal.*, 2018, **8**, 2253–2276.
9. N. Xiao, S. Li, X. Li, L. Ge, Y. Gao and N. Li, *Chin. J. Catal.*, 2020, **41**, 642–671.
10. A. Kudo and Y. Miseki, *Chem. Soc. Rev.*, 2009, **38**, 253–278.
11. M. R. Khan, T. W. Chuan, A. Yousuf, M. N. K. Chowdhury and C. K. Cheng, *Catal. Sci. Technol.*, 2015, **5**, 2522–2531.
12. X. She, H. Xu, Y. Yu, L. Li, X. Zhu, Z. Mo, Y. Song, J. Wu, S. Yuan and H. Li, *Small*, 2019, **15**, 1804613.
13. J. Low, J. Yu, M. Jaroniec, S. Wageh and A. A. Al-Ghamdi, *Adv. Mater.*, 2017, **29**, 1601694.
14. Y. Wang, Q. Wang, X. Zhan, F. Wang, M. Safdar and J. He, *Nanoscale*, 2013, **5**, 8326–8339.
15. A. Fujishima and K. Honda, *Nature*, 1972, **238**, 38–40.
16. M. Liu, P. Xia, L. Zhang, B. Cheng and J. Yu, *ACS Sustainable Chem. Eng.*, 2018, **6**, 10472–10480.
17. H. Xing, S. Teng, Z. Xing, L. Bi, Q. Bu, T. Xie and W. Yang, *Appl. Surf. Sci.*, 2020, **532**, 147000.
18. C. M. Pichler, T. Uekert and E. Reisner, *Chem. Commun.*, 2020, **56**, 5743–5746.
19. C. Dessal, L. Martínez, C. Maheu, T. Len, F. Morfin, J. L. Rousset, E. Puzenat, P. Afanasiev, M. Aouine, L. Soler, J. Llorca and L. Piccolo, *J. Catal.*, 2019, **375**, 155–163.
20. Y. Zhu, T. Wang, T. Xu, Y. Li and C. Wang, *Appl. Surf. Sci.*, 2019, **464**, 36–42.
21. A. Caravaca, W. Jones, C. Hardacre and M. Bowker, *Proc. R. Soc. A*, 2016, **472**, 20160054.
22. Z. H. N. Al-Azri, W. T. Chen, A. Chan, V. Jovic, T. Ina, H. Idriss and G. I. N. Waterhouse, *J. Catal.*, 2015, **329**, 355–367.
23. H. Zhang, S. Zuo, M. Qiu, S. Wang, Y. Zhang, J. Zhang and X. W. D. Lou, *Sci. Adv.*, 2020, **6**, 2–10.
24. Q. Wang, T. Hisatomi, Q. Jia, H. Tokudome, M. Zhong, C. Wang, Z. Pan, T. Takata, M. Nakabayashi, N. Shibata, Y. Li, I. D. Sharp, A. Kudo, T. Yamada and K. Domen, *Nat. Mater.*, 2016, **15**, 611–615.
25. T. H. Chiang, H. Lyu, T. Hisatomi, Y. Goto, T. Takata, M. Katayama, T. Minegishi and K. Domen, *ACS Catal.*, 2018, **8**, 2782–2788.
26. C. Jin, Y. Dai, W. Wei, X. Ma, M. Li and B. Huang, *Appl. Surf. Sci.*, 2017, **426**, 639–646.
27. Y. Guo, H. Jia, J. Yang, H. Yin, Z. Yang, J. Wang and B. Yang, *Phys. Chem. Chem. Phys.*, 2018, **20**, 22296–22307.
28. X. H. Jiang, L. S. Zhang, H. Y. Liu, D. S. Wu, F. Y. Wu, L. Tian, L. L. Liu, J. P. Zou, S. L. Luo and B. B. Chen, *Angew. Chem.*, 2020, **59**, 23112–23116.
29. L. Kong, Y. Dong, P. Jiang, G. Wang, H. Zhang and N. Zhao, *J. Mater. Chem. A*, 2016, **4**, 9998–10007.
30. W. Wang, S. Liu, L. Nie, B. Cheng and J. Yu, *Phys. Chem. Chem. Phys.*, 2013, **15**, 12033–12039.

31. C. T. Dinh, M. H. Pham, F. Kleitz and T. O. Do, *J. Mater. Chem. A*, 2013, **1**, 13308–13313.
32. A. K. Agegnehu, C. J. Pan, J. Rick, J. F. Lee, W. N. Su and B. J. Hwang, *J. Mater. Chem.*, 2012, **22**, 13849–13854.
33. S. Fukuzumi and Y. Yamada, *ChemSusChem*, 2013, **6**, 1834–1847.
34. P. D. Tran, L. Xi, S. K. Batabyal, L. H. Wong, J. Barber and J. S. Chye Loo, *Phys. Chem. Chem. Phys.*, 2012, **14**, 11596–11599.
35. W. Chen, Y. Wang, M. Liu, L. Gao, L. Mao, Z. Fan and W. Shangguan, *Appl. Surf. Sci.*, 2018, **444**, 485–490.
36. P. Zhang, T. Song, T. Wang and H. Zeng, *Appl. Catal., B*, 2017, **206**, 328–335.
37. P. Zhang, T. Song, T. Wang and H. Zeng, *Appl. Catal., B*, 2018, **225**, 172–179.
38. A. V. Korzhak, N. I. Ermokhina, A. L. Stroyuk, V. K. Bukhtiyarov, A. E. Raevskaya, V. I. Litvin, S. Y. Kuchmiy, V. G. Ilyin and P. A. Manorik, *J. Photochem. Photobiol., A*, 2008, **198**, 126–134.
39. W. Kurashige, R. Hayashi, K. Wakamatsu, Y. Kataoka, S. Hossain, A. Iwase, A. Kudo, S. Yamazoe and Y. Negishi, *ACS Appl. Energy Mater.*, 2019, **2**, 4175–4187.
40. N. Xiao, S. Li, S. Liu, B. Xu, Y. Li, Y. Gao, L. Ge and G. Lu, *Chin. J. Catal.*, 2019, **40**, 352–361.
41. K. Bhunia, M. Chandra, S. Khilari and D. Pradhan, *ACS Appl. Mater. Interfaces*, 2019, **11**, 478–488.
42. J. Ding, X. Li, L. Chen, X. Zhang and X. Tian, *Appl. Catal., B*, 2018, **224**, 322–329.
43. C. Han, Y. Lu, J. Zhang, L. Ge, Y. Li, C. Chen, Y. Xin, L. Wu and S. Fang, *J. Mater. Chem. A*, 2015, **3**, 23274–23282.
44. I. Majeed, U. Manzoor, F. K. Kanodarwala, M. A. Nadeem, M. A. Nadeem, E. Hussain, H. Ali, A. Badshah and J. A. Stride, *Catal. Sci. Technol.*, 2018, **8**, 1183–1193.
45. C. Han, L. Wu, L. Ge, Y. Li and Z. Zhao, *Carbon*, 2015, **92**, 31–40.
46. Q. Ba, X. Jia, L. Huang, X. Li, W. Chen and L. Mao, *Int. J. Hydrogen Energy*, 2019, **44**, 5872–5880.
47. A. Kudo, K. Domen, K. ichi Maruya and T. Onishi, *Chem. Phys. Lett.*, 1987, **133**, 517–519.
48. T. Sreethawong, Y. Suzuki and S. Yoshikawa, *Int. J. Hydrogen Energy*, 2005, **30**, 1053–1062.
49. J. Liu, Q. Jia, J. Long, X. Wang, Z. Gao and Q. Gu, *Appl. Catal., B*, 2018, **222**, 35–43.
50. E. Cui and G. Lu, *Int. J. Hydrogen Energy*, 2014, **39**, 8959–8968.
51. L. Zhang, Q. Liu, T. Aoki and P. A. Crozier, *J. Phys. Chem. C*, 2015, **119**, 7207–7214.
52. J. Ran, J. Yu and M. Jaroniec, *Green Chem.*, 2011, **13**, 2708–2713.
53. J. Yu, S. Wang, B. Cheng, Z. Lin and F. Huang, *Catal. Sci. Technol.*, 2013, **3**, 1782–1789.
54. S. Xu and D. D. Sun, *Int. J. Hydrogen Energy*, 2009, **34**, 6096–6104.

55. Y. H. Li, L. R. Zheng and H. G. Yang, *Int. J. Hydrogen Energy*, 2015, **40**, 15454–15459.
56. S. Xu, J. Ng, X. Zhang, H. Bai and D. D. Sun, *Int. J. Hydrogen Energy*, 2010, **35**, 5254–5261.
57. Y. Te Liao, Y. Y. Huang, H. M. Chen, K. Komaguchi, C. H. Hou, J. Henzie, Y. Yamauchi, Y. Ide and K. C. W. Wu, *ACS Appl. Mater. Interfaces*, 2017, **9**, 42425–42429.
58. J. Wang, Q. Zhu, Y. Liao, H. Fu, J. Chang, Y. Zhang, T. Kan, H. Gao and W. Huang, *J. Alloys Compd.*, 2021, **883**, 160832.
59. J. Yu and J. Ran, *Energy Environ. Sci.*, 2011, **4**, 1364–1371.
60. J. Chu, G. Sun, X. Han, X. Chen, J. Wang, W. Hu, I. Waluyo, A. Hunt, Y. Du, B. Song and P. Xu, *Nanoscale*, 2019, **11**, 15633–15640.
61. Z. Yan, H. Wu, A. Han, X. Yu and P. Du, *Int. J. Hydrogen Energy*, 2014, **39**, 13353–13360.
62. A. Kheradmand, Y. Zhu, W. Zhang, A. Marianov and Y. Jiang, *Int. J. Hydrogen Energy*, 2019, **44**, 17930–17942.
63. A. Wu, L. Jing, J. Wang, Y. Qu, Y. Xie, B. Jiang, C. Tian and H. Fu, *Sci. Rep.*, 2015, **5**, 1–9.
64. Z. Liang, R. Shen, Y. H. Ng, P. Zhang, Q. Xiang and X. Li, *J. Mater. Sci. Technol.*, 2020, **56**, 89–121.
65. H. He, J. Lin, W. Fu, X. Wang, H. Wang, Q. Zeng, Q. Gu, Y. Li, C. Yan, B. K. Tay, C. Xue, X. Hu, S. T. Pantelides, W. Zhou and Z. Liu, *Adv. Energy Mater.*, 2016, **6**, 1600464.
66. H. Xu, J. Yi, X. She, Q. Liu, L. Song, S. Chen, Y. Yang, Y. Song, R. Vajtai, J. Lou, H. Li, S. Yuan, J. Wu and P. M. Ajayan, *Appl. Catal., B*, 2018, **220**, 379–385.
67. T. Huang, W. Chen, T. Y. Liu, Q. L. Hao and X. H. Liu, *Int. J. Hydrogen Energy*, 2017, **42**, 12254–12261.
68. X. Hao, Z. Jin, H. Yang, G. Lu and Y. Bi, *Appl. Catal., B*, 2017, **210**, 45–56.
69. L. Guo, Z. Yang, K. Marcus, Z. Li, B. Luo, L. Zhou, X. Wang, Y. Du and Y. Yang, *Energy Environ. Sci.*, 2018, **11**, 106–114.
70. H. Zhao, Y. Dong, P. Jiang, H. Miao, G. Wang and J. Zhang, *J. Mater. Chem. A*, 2015, **3**, 7375–7381.
71. W. Zhang, Y. Wang, Z. Wang, Z. Zhong and R. Xu, *Chem. Commun.*, 2010, **46**, 7631–7633.
72. J. Hong, Y. Wang, Y. Wang, W. Zhang and R. Xu, *ChemSusChem*, 2013, **6**, 2263–2268.
73. H. Zhao, H. Zhang, G. Cui, Y. Dong, G. Wang, P. Jiang, X. Wu and N. Zhao, *Appl. Catal., B*, 2018, **225**, 284–290.
74. R. Song, B. Luo, J. Geng, D. Song and D. Jing, *Ind. Eng. Chem. Res.*, 2018, **57**, 7846–7854.
75. A. Indra, A. Acharjya, P. W. Menezes, C. Merschjann, D. Hollmann, M. Schwarze, M. Aktas, A. Friedrich, S. Lochbrunner, A. Thomas and M. Driess, *Angew. Chem., Int. Ed.*, 2017, **56**, 1653–1657.
76. D. P. Kumar, J. Choi, S. Hong, D. A. Reddy, S. Lee and T. K. Kim, *ACS Sustainable Chem. Eng.*, 2016, **4**, 7158–7166.

77. X. Wang, X. Li, C. Liu, F. Li, Y. Li, J. Zhao, R. Liu and G. Li, *Int. J. Hydrogen Energy*, 2018, **43**, 219–228.
78. K. Cho, H. Seo and Y. Lee, *Catal. Commun.*, 2011, **12**, 470–474.
79. Z. Chen, Y. Yu, X. She, K. Xia, Z. Mo, H. Chen, Y. Song, J. Huang, H. Li and H. Xu, *Appl. Surf. Sci.*, 2019, **495**, 143528.
80. H. W. Man, C. S. Tsang, M. M. J. Li, J. Mo, B. Huang, L. Y. S. Lee, Y. chung Leung, K. Y. Wong and S. C. E. Tsang, *Appl. Catal., B*, 2019, **242**, 186–193.
81. Z. Sun, M. Zhu, M. Fujitsuka, A. Wang, C. Shi and T. Majima, *ACS Appl. Mater. Interfaces*, 2017, **9**, 30583–30590.
82. X. Yue, S. Yi, R. Wang, Z. Zhang and S. Qiu, *Small*, 2017, **13**, 1–9.
83. P. Wang, S. Zhan, H. Wang, Y. Xia, Q. Hou, Q. Zhou, Y. Li and R. R. Kumar, *Appl. Catal., B*, 2018, **230**, 210–219.
84. J. S. Jang, D. J. Ham, N. Lakshminarasimhan, W. yong Choi and J. S. Lee, *Appl. Catal., A*, 2008, **346**, 149–154.
85. A. T. Garcia-Esparza, D. Cha, Y. Ou, J. Kubota, K. Domen and K. Takanabe, *ChemSusChem*, 2013, **6**, 168–181.
86. C. Ren, W. Li, S. Gu, X. Liu, X. Li, H. Fan, K. Han and X. Ma, *Appl. Mater. Today*, 2020, **20**, 100731.
87. L. Tian, S. Min, Y. Lei, S. Chen and F. Wang, *Chem. Commun.*, 2019, **55**, 6870–6873.
88. Q. Guo, F. Liang, X. Y. Gao, Q. C. Gan, X. B. Li, J. Li, Z. S. Lin, C. H. Tung and L. Z. Wu, *ACS Catal.*, 2018, **8**, 5890–5895.
89. Z. Luo, X. Zhao, H. Zhang and Y. Jiang, *Appl. Catal., A*, 2019, **582**, 117115.
90. N. Zhao, Y. Hu, J. Du, G. Liu, B. Dong, Y. Yang, J. Peng, J. Li and M. Zhai, *Appl. Surf. Sci.*, 2020, **530**, 147247.
91. L. Cheng, X. Li, H. Zhang and Q. Xiang, *J. Phys. Chem. Lett.*, 2019, **10**, 3488–3494.
92. A. D. Handoko, K. D. Fredrickson, B. Anasori, K. W. Convey, L. R. Johnson, Y. Gogotsi, A. Vojvodic and Z. W. Seh, *ACS Appl. Energy Mater.*, 2018, **1**, 173–180.
93. Z. Zhuang, Y. Li, Z. Li, F. Lv, Z. Lang, K. Zhao, L. Zhou, L. Moskaleva, S. Guo and L. Mai, *Angew. Chem.*, 2018, **130**, 505–509.
94. M. Cao, X. Zhang, J. Qin and R. Liu, *ACS Sustainable Chem. Eng.*, 2018, **6**, 16198–16204.
95. L. Song, S. Zhang, D. Liu, S. Sun and J. Wei, *Int. J. Hydrogen Energy*, 2020, **45**, 8234–8242.
96. L. Li, Z. Deng, L. Yu, Z. Lin, W. Wang and G. Yang, *Nano Energy*, 2016, **27**, 103–113.
97. K. E. Dalle, J. Warnan, J. J. Leung, B. Reuillard, I. S. Karmel and E. Reisner, *Chem. Rev.*, 2019, **119**, 2752–2875.
98. M. A. Gross, A. Reynal, J. R. Durrant and E. Reisner, *J. Am. Chem. Soc.*, 2014, **136**, 356–366.
99. C. A. Caputo, M. A. Gross, V. W. Lau, C. Cavazza, B. V Lotsch and E. Reisner, *Angew. Chem., Int. Ed.*, 2014, **53**, 11538–11542.
100. C. E. Creissen, J. Warnan and E. Reisner, *Chem. Sci.*, 2018, **9**, 1395–1704.

101. J. Feng, J. Bian, L. Bai, S. Xi, Y. Wang, C. Chen and L. Jing, *Appl. Catal., B*, 2021, **295**, 120260.
102. V. Artero, M. Chavarot-Kerlidou and M. Fontecave, *Angew. Chem., Int. Ed.*, 2011, **50**, 7238–7266.
103. A. Reynal, J. Willkomm, N. M. Muresan, F. Lakadamyali, M. Planells, E. Reisner and J. R. Durrant, *Chem. Commun.*, 2014, **50**, 12768–12771.
104. J. Huang, K. L. Mulfort, P. Du and L. X. Chen, *J. Am. Chem. Soc.*, 2012, **134**, 16472–16475.
105. K. Han, M. Wang, S. Zhang, S. Wu, Y. Yang and L. Sun, *Chem. Commun.*, 2015, **51**, 7008–7011.
106. Y. Wang, L. Shi, B. Qian, X. Huang, X. Sun, B. Wang, X. Song, H. Huang, Y. Zhang and T. Ma, *Chem. Eng. J.*, 2021, **413**, 127551.
107. F. Wen, J. Yang, X. Zong, B. Ma, D. Wang and C. Li, *J. Catal.*, 2011, **281**, 318–324.
108. F. Wen, X. Wang, L. Huang, G. Ma, J. Yang and C. Li, *ChemSusChem*, 2012, **2**, 849–853.
109. X. Song, H. Wen, C. Ma, M. Hu, H. Chen, H. Cui and C. Chen, *Appl. Organomet. Chem.*, 2014, **28**, 267–273.
110. M. Wen, X. Li, J. Jian, X. Wang, H. Wu, B. Chen and C. Tung, *Sci. Rep.*, 2016, **6**, 29851.
111. K. A. Brown, S. Dayal, X. Ai, G. Rumbles and P. W. King, *J. Am. Chem. Soc.*, 2010, **132**, 9672–9680.
112. K. A. Brown, M. B. Wilker, M. Boehm, G. Dukovic and P. W. King, *J. Am. Chem. Soc.*, 2012, **134**, 5627–5636.
113. E. Reisner, D. J. Powell, C. Cavazza, J. C. Fontecilla-camps and F. A. Armstrong, *J. Am. Chem. Soc.*, 2009, 18457–18466.
114. C. A. Caputo, L. Wang, R. Beranek and E. Reisner, *Chem. Sci.*, 2015, **6**, 5690–5694.
115. G. A. M. Hutton, B. Reuillard, B. C. M. Martindale, C. A. Caputo, C. W. J. Lockwood, J. N. Butt and E. Reisner, *J. Am. Chem. Soc.*, 2016, **138**, 16722–16730.
116. K. Hola, M. V. Pavliuk, B. Nemeth, P. Huang, L. Zrazil, H. Land, G. Berggren and H. Tian, *ACS Catal.*, 2020, **10**, 9943–9952.
117. X. Xia, N. Deng, G. Cui, J. Xie, X. Shi, Y. Zhao, Q. Wang, W. Wang and B. Tang, *Chem. Commun.*, 2015, **51**, 10899–10902.
118. Y. Sui, L. Wu, S. Zhong and Q. Liu, *Appl. Surf. Sci.*, 2019, **480**, 810–816.
119. Y. Wang, X. Liu, J. Liu, B. Han, X. Hu, F. Yang, Z. Xu, Y. Li, S. Jia, Z. Li and Y. Zhao, *Angew. Chem., Int. Ed.*, 2018, **430074**, 5867–5873.
120. Y. Chen, J. Li, Z. Hong, B. Shen, B. Lin and B. Gao, *Phys. Chem. Chem. Phys.*, 2014, **16**, 8106–8113.
121. N. R. Reddy, M. M. Kumari, K. K. Cheralathan and M. V. Shankar, *Int. J. Hydrogen Energy*, 2018, **43**, 4036–4043.
122. L. Wang, Z. Yao, F. Jia, B. Chen and Z. Jiang, *Dalton Trans.*, 2013, **42**, 9976–9981.
123. A. Moya, A. Cherevan, S. Marchesan, P. Gebhardt, M. Prato, D. Eder and J. J. Vilatela, *Appl. Catal., B*, 2015, **179**, 574–582.

124. B. Chai, T. Peng, X. Zhang, J. Mao, K. Li and X. Zhang, *Dalton Trans.*, 2013, **42**, 3402–3409.
125. Q. Xiang, J. Yu and M. Jaroniec, *J. Am. Chem. Soc.*, 2012, **134**, 23–26.
126. X.-J. LV, S.-X. Zhou, C. Zhang, H.-X. Chang, Y. Chen and W.-F. Fu, *J. Mater. Chem.*, 2012, **22**, 18542–18549.
127. S. W. Cao, J. Fang, M. M. Shahjamali, Z. Wang, Z. Yin, Y. Yang, F. Y. C. Boey, J. Barber, S. C. J. Loo and C. Xue, *CrystEngComm*, 2012, **14**, 7229–7235.
128. S. Y. Moon, H. C. Song, E. H. Gwag, I. I. Nedrygailov, C. Lee, J. J. Kim, W. H. Doh and J. Y. Park, *Nanoscale*, 2018, **10**, 22180–22188.
129. S. S. K. Ma, K. Maeda, R. Abe and K. Domen, *Energy Environ. Sci.*, 2012, **5**, 8390–8397.
130. K. Maeda, R. Abe and K. Domen, *J. Phys. Chem. C*, 2011, **115**, 3057–3064.
131. Y. Iwase, O. Tomita, M. Higashi, A. Nakada and R. Abe, *Sustainable Energy Fuels*, 2019, **3**, 1501–1508.
132. B. H. Meekins and P. V. Kamat, *J. Phys. Chem. Lett.*, 2011, **2**, 2304–2310.
133. R. Li, Z. Chen, W. Zhao, F. Zhang, K. Maeda, B. Huang, S. Shen, K. Domen and C. Li, *J. Phys. Chem. C*, 2013, **117**, 376–382.
134. A. Xiong, T. Yoshinaga, T. Ikeda, M. Takashima, T. Hisatomi, K. Maeda, T. Setoyama, T. Teranishi and K. Domen, *Eur. J. Inorg. Chem.*, 2014, 767–772.
135. L. Liu, Z. Ji, W. Zou, X. Gu, Y. Deng, F. Gao, C. Tang and L. Dong, *ACS Catal.*, 2013, **3**, 2052–2061.
136. F. Zhang, A. Yamakata, K. Maeda, Y. Moriya, T. Takata, J. Kubota, K. Teshima, S. Oishi and K. Domen, *J. Am. Chem. Soc.*, 2012, **134**, 8348–8351.
137. H. Okamoto, M. Kodera, T. Hisatomi, M. Katayama, T. Minegishi and K. Domen, *Catal. Today*, 2020, **354**, 204–210.
138. E. Nurlaela, H. Wang, T. Shinagawa, S. Flanagan, S. Ould-Chikh, M. Qureshi, Z. Mics, P. Sautet, T. Le Bahers, E. Cánovas, M. Bonn and K. Takanabe, *ACS Catal.*, 2016, **6**, 4117–4126.
139. E. S. Kim, N. Nishimura, G. Magesh, J. Y. Kim, J. W. Jang, H. Jun, J. Kubota, K. Domen and J. S. Lee, *J. Am. Chem. Soc.*, 2013, **135**, 5375–5383.
140. J. Soldat, G. W. Busser, M. Muhler and M. Wark, *ChemCatChem*, 2016, **8**, 153–156.
141. C. Cheng, L. Mao, J. Shi, F. Xue, S. Zong, B. Zheng and L. Guo, *J. Mater. Chem. A*, 2021, **9**, 12299–12306.
142. Y. Wang, Y. Wang, R. Jiang and R. Xu, *Ind. Eng. Chem. Res.*, 2012, **51**, 9945–9951.
143. R. L. Lee, P. D. Tran, S. S. Pramana, S. Y. Chiam, Y. Ren, S. Meng, L. H. Wong and J. Barber, *Catal. Sci. Technol.*, 2013, **3**, 1694–1698.
144. M. M. Maitani, T. Yamada, H. Mashiko, K. Yoshimatsu, T. Oshima, A. Ohtomo and Y. Wada, *ACS Appl. Mater. Interfaces*, 2017, **9**, 10349–10354.
145. X. Yang, H. Tang, J. Xu, M. Antonietti and M. Shalom, *ChemSusChem*, 2015, **8**, 1350–1358.

146. L. Tian, X. Xian, X. Cui, H. Tang and X. Yang, *Appl. Surf. Sci.*, 2018, **430**, 301–308.
147. T. Wang, L. Liu, G. Ge, M. Liu, W. Zhou, K. Chang, F. Yang, D. Wang and J. Ye, *J. Catal.*, 2018, **367**, 296–305.
148. Q. Zhang, M. Liu, W. Zhou, Y. Zhang, W. Hao, Y. Kuang, H. Liu, D. Wang, L. Liu and J. Ye, *Nano Energy*, 2021, **81**, 105651.
149. G. Zhang, C. Huang and X. Wang, *Small*, 2015, **11**, 1215–1221.
150. S. Sun, G. Shen, J. Jiang, W. Mi, X. Liu, L. Pan, X. Zhang and J.-J. Zou, *Adv. Energy Mater.*, 2019, **9**, 1901505.
151. Z. Zhong, J. Liu, X. Xu, A. Cao, Z. Tao, W. You and L. Kang, *J. Mater. Chem. A*, 2021, **9**, 2404–2413.
152. K. Maeda, A. Xiong, T. Yoshinaga, T. Ikeda, N. Sakamoto, T. Hisatomi, M. Takashima, D. Lu, M. Kanehara, T. Setoyama, T. Teranishi and K. Domen, *Angew. Chem.*, 2010, **122**, 4190–4193.
153. D. Zheng, X. N. Cao and X. Wang, *Angew. Chem., Int. Ed.*, 2016, **55**, 11512–11516.
154. T. Takata, J. Jiang, Y. Sakata, M. Nakabayashi, N. Shibata, V. Nandal, K. Seki, T. Hisatomi and K. Domen, *Nature*, 2020, **581**, 411–414.
155. T. Uekert, C. M. Pichler, T. Schubert and E. Reisner, *Nat. Sustainability*, 2021, **4**, 383–391.
156. C. Li, J. Li, L. Qin, P. Yang and D. G. Vlachos, *ACS Catal.*, 2021, **11**, 11336–11359.
157. X. Li, J. Yu, M. Jaroniec and X. Chen, *Chem. Rev.*, 2019, **119**, 3962–4179.
158. J. Albero, Y. Peng and H. García, *ACS Catal.*, 2020, **10**, 5734–5749.
159. Q. Han, Y. Zhou, L. Tang, P. Li, W. Tu, L. Li, H. Li and Z. Zou, *RSC Adv.*, 2016, **6**, 90792–90796.
160. W. N. Wang, W. J. An, B. Ramalingam, S. Mukherjee, D. M. Niedzwiedzki, S. Gangopadhyay and P. Biswas, *J. Am. Chem. Soc.*, 2012, **134**, 11276–11281.
161. S. Cao, Y. Li, B. Zhu, M. Jaroniec and J. Yu, *J. Catal.*, 2017, **349**, 208–217.
162. Q. Chen, W. Mo, G. Yang, S. Zhong, H. Lin, J. Chen and S. Bai, *Small*, 2021, 2102105.
163. Y. Zhu, Z. Xu, W. Jiang, S. Zhong, L. Zhao and S. Bai, *J. Mater. Chem. A*, 2017, **5**, 2619–2628.
164. Y. Zhu, C. Gao, S. Bai, S. Chen, R. Long, L. Song, Z. Li and Y. Xiong, *Nano Res.*, 2017, **10**, 3396–3406.
165. K. Iizuka, T. Wato, Y. Miseki, K. Saito and A. Kudo, *J. Am. Chem. Soc.*, 2011, **133**, 20863–20868.
166. S. Wang, K. Teramura, T. Hisatomi, K. Domen, H. Asakura, S. Hosokawa and T. Tanaka, *ACS Sustainable Chem. Eng.*, 2021, **9**, 9327–9335.
167. Z. Zhu, J. Qin, M. Jiang, Z. Ding and Y. Hou, *Appl. Surf. Sci.*, 2017, **391**, 572–579.
168. T. Ishii, A. Anzai, A. Yamamoto and H. Yoshida, *Appl. Catal., B*, 2020, **277**, 119192.

169. T. Ohno, T. Higo, N. Murakami, H. Saito, Q. Zhang, Y. Yang and T. Tsubota, *Appl. Catal., B*, 2014, **152–153**, 309–316.
170. H. Zhou, J. Guo, P. Li, T. Fan, D. Zhang and J. Ye, *Sci. Rep.*, 2013, **3**, 1–9.
171. B. Mei, A. Pougin and J. Strunk, *J. Catal.*, 2013, **306**, 184–189.
172. ˛tefan Neatu, J. A. Maciá-Agullò, P. Concepciòn and H. Garcia, *J. Am. Chem. Soc.*, 2014, **136**, 15969–15976.
173. Q. Kang, T. Wang, P. Li, L. Liu, K. Chang, M. Li and J. Ye, *Angew. Chem.*, 2015, **127**, 855–859.
174. S. Lee, S. Jeong, W. D. Kim, S. Lee, K. Lee, W. K. Bae, J. H. Moon, S. Lee and D. C. Lee, *Nanoscale*, 2016, **8**, 10043–10048.
175. L. Liu, F. Gao, H. Zhao and Y. Li, *Appl. Catal., B*, 2013, **134–135**, 349–358.
176. I. Shown, H. C. Hsu, Y. C. Chang, C. H. Lin, P. K. Roy, A. Ganguly, C. H. Wang, J. K. Chang, C. I. Wu, L. C. Chen and K. H. Chen, *Nano Lett.*, 2014, **14**, 6097–6103.
177. H. Park, H. H. Ou, A. J. Colussi and M. R. Hoffmann, *J. Phys. Chem. A*, 2015, **119**, 4658–4666.
178. G. Shi, L. Yang, Z. Liu, X. Chen, J. Zhou and Y. Yu, *Appl. Surf. Sci.*, 2018, **427**, 1165–1173.
179. J. Wang, T. Heil, B. Zhu, C. W. Tung, J. Yu, H. M. Chen, M. Antonietti and S. Cao, *ACS Nano*, 2020, **14**, 8584–8593.
180. T. Billo, F. Y. Fu, P. Raghunath, I. Shown, W. F. Chen, H. T. Lien, T. H. Shen, J. F. Lee, T. S. Chan, K. Y. Huang, C. I. Wu, M. C. Lin, J. S. Hwang, C. H. Lee, L. C. Chen and K. H. Chen, *Small*, 2018, **14**, 1–11.
181. W. Zhong, R. Sa, L. Li, Y. He, L. Li, J. Bi, Z. Zhuang, Y. Yu and Z. Zou, *J. Am. Chem. Soc.*, 2019, **141**, 7615–7621.
182. J. C. S. Wu, H. M. Lin and C. L. Lai, *Appl. Catal., A*, 2005, **296**, 194–200.
183. H. Li, Y. Lei, Y. Huang, Y. Fang, Y. Xu, L. Zhu and X. Li, *J. Nat. Gas Chem.*, 2011, **20**, 145–150.
184. S. Shoji, G. Yin, M. Nishikawa, D. Atarashi, E. Sakai and M. Miyauchi, *Chem. Phys. Lett.*, 2016, **658**, 309–314.
185. G. Yin, M. Nishikawa, Y. Nosaka, N. Srinivasan, D. Atarashi, E. Sakai and M. Miyauchi, *ACS Nano*, 2015, **9**, 2111–2119.
186. T. V. Nguyen and J. C. S. Wu, *Appl. Catal., A*, 2008, **335**, 112–120.
187. C. Han, R. Zhang, Y. Ye, L. Wang, Z. Ma, F. Su, H. Xie, Y. Zhou, P. K. Wong and L. Ye, *J. Mater. Chem. A*, 2019, **7**, 9726–9735.
188. C. Tsai, H. M. Chen, R. Liu, K. Asakura and T. Chan, *J. Phys. Chem. C*, 2011, **115**, 10180–10186.
189. H. Y. Lin, Y. F. Chen and Y. W. Chen, *Int. J. Hydrogen Energy*, 2007, **32**, 86–92.
190. A. Meng, S. Wu, B. Cheng, J. Yu and J. Xu, *J. Mater. Chem. A*, 2018, **6**, 4729–4736.
191. T. Wang, L. Shi, J. Tang, V. Malgras, S. Asahina, G. Liu, H. Zhang, X. Meng, K. Chang, J. He, O. Terasaki, Y. Yamauchi and J. Ye, *Nanoscale*, 2016, **8**, 6712–6720.

192. X. Zhu, H. Ji, J. Yi, J. Yang, X. She, P. Ding, L. Li, J. Deng, J. Qian, H. Xu and H. Li, *Ind. Eng. Chem. Res.*, 2018, **57**, 17394–17400.
193. H. Lee, B. S. Kwak, N. K. Park, J. I. Baek, H. J. Ryu and M. Kang, *Appl. Surf. Sci.*, 2017, **393**, 385–396.
194. J. H. Lee, S. Il Kim, S. M. Park and M. Kang, *Ceram. Int.*, 2017, **43**, 1768–1774.
195. W. Dai, J. Yu, Y. Deng, X. Hu, T. Wang and X. Luo, *Appl. Surf. Sci.*, 2017, **403**, 230–239.
196. H. Jung, K. M. Cho, K. H. Kim, H. W. Yoo, A. Al-Saggaf, I. Gereige and H. T. Jung, *ACS Sustainable Chem. Eng.*, 2018, **6**, 5718–5724.
197. J. Tang, D. Yang, W. Zhou, R. Guo, W. Pan and C. Huang, *J. Catal.*, 2019, **370**, 79–87.
198. P. Niu, Z. Pan, S. Wang and X. Wang, *ChemCatChem*, 2021, **13**, 3581–3587.
199. J. Di, C. Chen, C. Zhu, P. Song, M. Duan, J. Xiong, R. Long, M. Xu, L. Kang, S. Guo, S. Chen, H. Chen, Z. Chi, Y. X. Weng, H. Li, L. Song, M. Wu, Q. Yan, S. Li and Z. Liu, *Nano Energy*, 2021, **79**, 105429.
200. M. Ye, X. Wang, E. Liu, J. Ye and D. Wang, *ChemSusChem*, 2018, **11**, 1577.
201. L. Lin, C. Hou, X. Zhang, Y. Wang, Y. Chen and T. He, *Appl. Catal., B*, 2018, **221**, 312–319.
202. S. Aoi, K. Mase, K. Ohkubo and S. Fukuzumi, *Catal. Sci. Technol.*, 2016, **6**, 4077–4080.
203. M. F. Kuehnel, K. L. Orchard, K. E. Dalle and E. Reisner, *J. Am. Chem. Soc.*, 2017, **139**, 7217–7223.
204. S. Lian, M. S. Kodaimati and E. A. Weiss, *ACS Nano*, 2018, **12**, 568–575.
205. E. Lam and E. Reisner, *Angew. Chem.*, 2021, **60**, 23306–23312.
206. L. Zhang, M. Can, S. W. Ragsdale and F. A. Armstrong, *ACS Catal.*, 2018, **8**, 2789–2795.
207. T. W. Woolerton, S. Sheard, E. Pierce, S. W. Ragsdale and F. A. Armstrong, *Energy Environ. Sci.*, 2011, **4**, 2393–2399.
208. Y. S. Chaudhary, T. W. Woolerton, C. S. Allen, J. H. Warner, E. Pierce, S. W. Ragsdale and F. A. Armstrong, *Chem. Commun.*, 2012, **48**, 58–60.
209. L. L. Tan, W. J. Ong, S. P. Chai and A. R. Mohamed, *Nanoscale Res. Lett.*, 2013, **8**, 1–9.
210. W. J. Ong, L. L. Tan, S. P. Chai and S. T. Yong, *Chem. Commun.*, 2015, **51**, 858–861.
211. J. Yu, J. Jin, B. Cheng and M. Jaroniec, *J. Mater. Chem. A*, 2014, **2**, 3407–3416.
212. X. Li, Q. Wang, Y. Zhao, W. Wu, J. Chen and H. Meng, *J. Colloid Interface Sci.*, 2013, **411**, 69–75.
213. X. An, K. Li and J. Tang, *ChemSusChem*, 2014, **7**, 1086–1093.
214. Y. T. Liang, B. K. Vijayan, K. a Gray and M. C. Hersam, *Nano Lett.*, 2011, 2865–2870.
215. X. H. Xia, Z. J. Jia, Y. Yu, Y. Liang, Z. Wang and L. L. Ma, *Carbon*, 2007, **45**, 717–721.

216. M. M. Gui, S. P. Chai, B. Q. Xu and A. R. Mohamed, *Sol. Energy Mater. Sol. Cells*, 2014, **122**, 183–189.
217. M. Abou Asi, L. Zhu, C. He, V. K. Sharma, D. Shu, S. Li, J. Yang and Y. Xiong, *Catal. Today*, 2013, **216**, 268–275.
218. L. Yu, G. Li, X. Zhang, X. Ba, G. Shi, Y. Li, P. K. Wong, J. C. Yu and Y. Yu, *ACS Catal.*, 2016, **6**, 6444–6454.
219. L. Y. Lin, S. Kavadiya, B. B. Karakocak, Y. Nie, R. Raliya, S. T. Wang, M. Y. Berezin and P. Biswas, *Appl. Catal., B*, 2018, **230**, 36–48.
220. Y. Wang, X. Bai, H. Qin, F. Wang, Y. Li, X. Li, S. Kang, Y. Zuo and L. Cui, *ACS Appl. Mater. Interfaces*, 2016, **8**, 17212–17219.

CHAPTER 10

Semiconductor Quantum Dots for Water Splitting and CO_2 Photoreduction

N. LAKSHMANA REDDY[a], N. RAMESH REDDY[b], SANTOSH S. PATIL[c], M. MAMATHA KUMARI[d] AND M. V. SHANKAR*[d]

[a]Division of Advanced Materials Engineering, Jeonbuk National University, Jeonju 54896, South Korea; [b]School of Mechanical Engineering, Yeungnam University, Gyeongsan 38541, Republic of Korea; [c]Department of Chemistry and Chemical Engineering, Inha University, 100 Inha-ro, Michuhol-gu, 22212, Incheon, South Korea; [d]Nanocatalysis and Solar Fuels Research Laboratory, Department of Materials Science & Nanotechnology, Yogi Vemana University, Kadapa-516005, Andhra Pradesh, India
*E-mail: shankar@yogivemanauniversity.ac.in

10.1 Introduction

The huge amounts of CO_2 released into the atmosphere is causing severe environmental damage to the world. Energy demands have been increasing due to the increased population and increased utilization of various electrical and electronic devices. To tackle these problems many researchers and scientists have researched many techniques and efficient nanomaterials. Photocatalysis (PC) and/or photoelectrochemical (PEC) processes are considered to be the best techniques to overcome these problems through water splitting

Sustainable Energy No. 1
Recent Developments in Functional Materials for Artificial Photosynthesis
Edited by Srabanti Ghosh and Qian Wang

Published by the Royal Society of Chemistry, www.rsc.org

and photocatalytic CO_2 reduction mechanisms.[1–4] Hydrogen (H_2) is considered to be one of the best alternative clean energies for the world's future energy demands. H_2 generation from PC and PEC processes *via* water splitting works under ambient conditions. Moreover, it utilizes renewable energies like natural solar light and seawater for the water-splitting process.[5,6] Similarly, CO_2 reduction through PC and PEC processes is considered an emerging technique by using various types of semiconductor photocatalysts. A wide range of nanomaterials including 0D nanoparticles, quantum dots, 1D nanostructures including nanotubes, nanorods, nanofibers, 2D nanosheets and 3D nanostructures have been used for energy and environmental applications. Among them, the 0D quantum dots (QDs) based photocatalysts have been considered as promising materials for PC and PEC water splitting and CO_2 reduction applications due to their unique properties. These properties include high surface area, a huge number of active sites, discrete number of energy levels for charge transportation, *etc.*[7–11] In this chapter we discuss the various types of QDs based photocatalysts including metal oxide based QDs, metal sulfides based QDs, metal selenides based QDs, graphene-based QDs for water splitting and CO_2 reduction applications.

10.2 Metal Oxide Based QDs

QDs are gaining a lot of attention in current research due to their wide exploitation in many applications and superior optoelectronic properties. The QDs photoluminescence emission is adjustable with light irradiation from UV to IR region and they have "size-tunable" fluorescence properties. QDs can be prepared by using different chemical synthesis routes that regulate the shape, size, constituent materials, and surface chemistry. For instance, for the synthesis of QDs, colloidal chemistry is a commonly used synthesis route due to the surface of the nanocrystals functionalization through the construction process. This procedure permits the nanocrystals to interrelate with narrow size delivery, high luminescence efficiency as well as selected species. Moreover, to prevent agglomeration, QDs should be dispersed in solvent to stabilize them. As a result, it is crucial for surface modification of QDs and the selection of appropriate ligand materials. Because of their crucial involvement in many applications, several QDs such as carbon QDs, metal sulfide QDs, and metal oxide QDs have been reported. Metal oxide QDs are widely used due to their non-toxicity, abundant availability, and easy synthesis at ambient conditions.

Because of their superior properties, QDs are used for many applications such as biological, biomedical, coatings, drug delivery systems, cosmetics, printing, computer electronics, and environmental and energy storage applications.[12–14] QDs are showing their remarkable performance in energy conversion and environmental applications such as water splitting and CO_2 photoreduction. Especially, metal oxide QDs with their semiconductor nature offer suitable bandgaps, oxidation and reduction potentials, visible absorptions, high surface areas and reactive sites. They are also able to act as co-catalysts when combined with other semiconductors. For example, the Co_3O_4 QDs decorated TiO_2 nanobelts reported for the overall water splitting by Jie *et al.*[15] In this work, Co_3O_4 QDs/TiO_2 nanobelts showed an enhanced (1735.1 μmol g^{-1} h^{-1}) H_2 production efficiency when compared with pristine TiO_2 (1363.1 μmol g^{-1} h^{-1}). This is due to the QDs size of Co_3O_4 (below 10 nm), which facilitates electron trapping and suppressing the recombination of the photo-excited electron–hole pair. Therefore, QD sizes not only enhance the light sensitivity but also change the work function of semiconductors by electron trapping and participation in redox reactions on the surface for H_2 generation. The CuO QDs-decorated TiO_2 nanosheets have also been reported for photocatalytic water splitting.[16]

10.2.1 Water Splitting

The ever increasing demand for sustainable and renewable power sources has led to the expansion of novel materials for photocatalytic water separation. However, increasing photocatalytic capacity remains a major problem. Usually, metal oxide nanomaterials are used for water splitting due to their remarkable surface area, chemical, thermal stability, and mechanical flexibility. However, the major limitation is most of the metal oxides are absorbed in the UV region. Few absorb in the visible region but suffer from quick electron–hole recombination. Hence, to develop strong materials for water splitting with visible absorption and reduce the recombination rate, bandgap modulation is essential. In terms of theoretical analysis, unoccupied orbital molecules of metal cations from the conduction bands (CB) of oxide semiconductors and O2p orbits form the valence band (VB). Therefore, to narrow the bandgap in a metal oxide based system, a new VB or electron donor level of elements excluding O2p orbits is needed so that the CB level is never reduced.[17] To develop a visible absorption semiconductor, several strategies have been taken, *e.g.*, doping of non-metal,[18–21] co-doping of metal and non-metal,[22,23] single-phase material,[24] carbon materials,[25–27] and quantum dots sensitization.[28,29]

Metal oxide sensitization with QDs has shown significant enhancement in light absorption, thereby improving the overall water splitting efficiency. For example, the SnO_2 QDs-interspersed multiphase TiO_2 nanorod heterojunction for photocatalytic water splitting reported by Sun *et al.*[30] In this work, TiO_2 nanorods (with anatase and rutile phase) were developed and then combined with SnO_2 QDs with various ratios of Ti. The developed materials were used and the evolution of H_2 and O_2 was evaluated. The various Ti-doped Sn time-dependent photocurrent densities and their water splitting mechanism with TiO_2@SnO_2 QDs is shown in Figure 10.1(a)–(g). The Sn–Ti = 1:3 (Figure 10.1(c)) catalyst displayed a stable photocurrent density of ~2.25 mA cm^{-2} with continuous on/off cycles. The quantum confinement nature of the SnO_2 QDs and uniform nature of the TiO_2 nanorods are favorable to a stable

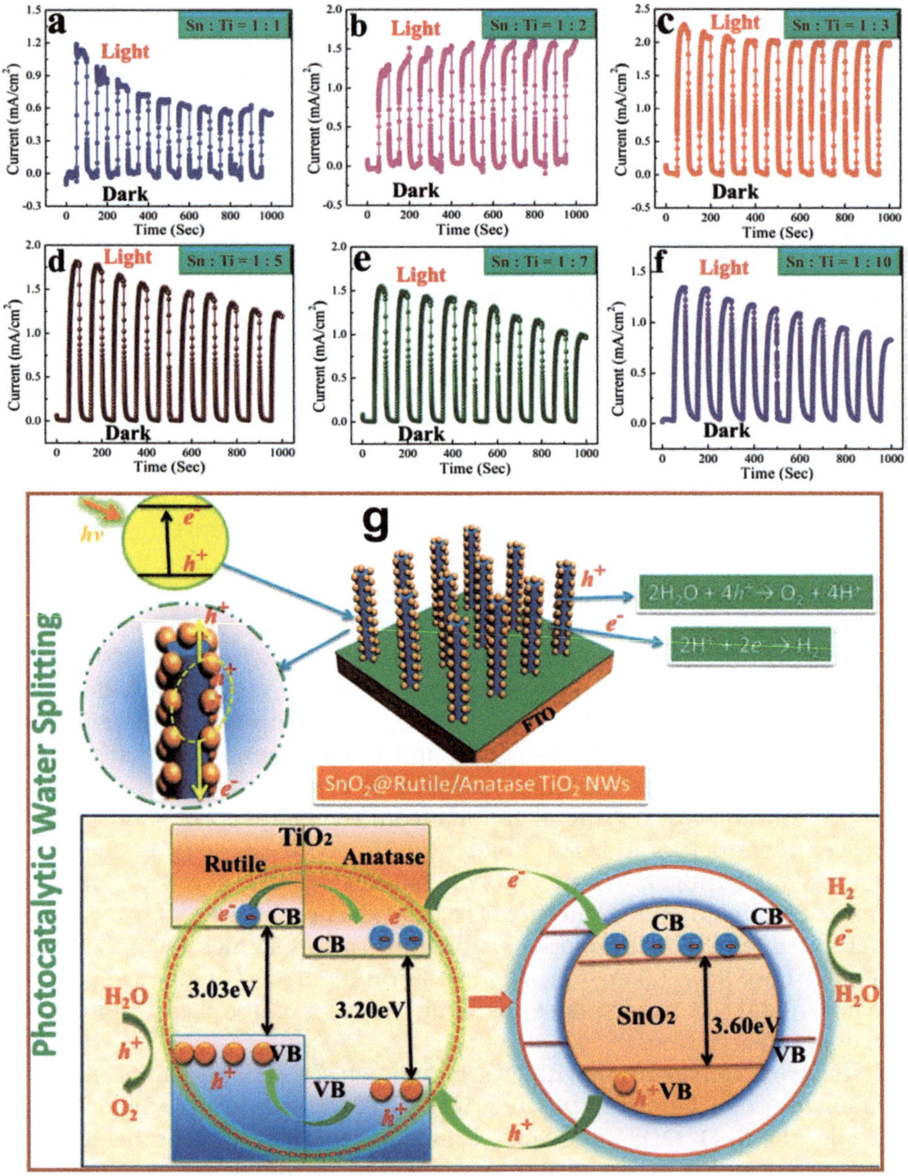

Figure 10.1 Time-dependent photocurrent density of the various SnO_2/RA@TiO_2 NR photoanodes with light/dark (on/off) cycles with simulated sunlight irradiation. (a) Sn–Ti = 1 : 1, (b) Sn–Ti = 1 : 2, (c) Sn–Ti = 1 : 3, (d) Sn–Ti = 1 : 5, (e) Sn–Ti = 1 : 7, (f) Sn–Ti = 1 : 10, and (g) the photocatalytic water splitting process with heterojunction electrode showing the electron–hole transfer processes in their VB and CB edges. Reproduced from ref. 30 with permission from American Chemical Society, Copyright 2019.

structure and electronic properties thereby improving the light absorption capability as well as reducing the recombination of photogenerated charge carriers. From the water splitting mechanism (Figure 10.1(g)), the QDs with strong electronic coupling and high surface area facilitates the electron–hole pair separation and migration to redox reactions. Songmei *et al.* reported quantum-sized $BiVO_4$ for pure water splitting without any co-catalyst.[31] In this work, they achieved a higher hydrogen value with quantum sized $BiVO_4$ (66.8 μmol) when compared to the nano-sized $BiVO_4$ (almost null). They used 15% aqueous methanol solution for the water splitting. This is due to the durable reduction capacity of the photogenerated electrons in the quantum sized $BiVO_4$ when compared with nanoscale $BiVO_4$. Furthermore, the CdS and CdSe QDs sensitized ZnO nanowire photoanode has been fabricated for water splitting by Wang *et al.*[32] These combinations displayed durable absorption photoanodes with remarkable higher IPCE value *i.e.* ~45% at 0 V *vs.* Ag/AgCl. This is due to the appropriate band arrangement of the QDs and the ZnO NWs, which enables proficient migration of photogenerated electrons from the QDs to the NWs. The Fermi levels of ZnO NWs, CdSe QDs, and CdS are associated and CB positions of the QDs are closely present, thereby photoexcited electrons formed in CdSe QDs are sent to the ZnO NWs over an intermediate layer of CdS. Along with that, ZnO NWs electrons travel to the cathode where hydrogen is produced in the PEC device and the holes are left at the anode–electrolyte interface for the process of oxidation. The IPCE value is significantly enlarged due to the electron transfer and migration from CdSe QDs to ZnO NWs and thereby band alignment between CdS and CdSe.

Metal oxide QDs are not only useful as direct semiconductors for water splitting. They can also be used for shifting the conduction band bottom level (CBB) of another semiconductor. For example, Xie *et al.* reported the $BiVO_4$ QDs decorated screw-like SnO_2 nanostructure for hydrogen evolution.[33] In this work, the authors investigated the function of bulk materials and $BiVO_4$ photoanode roles for hydrogen evolution. Interestingly, QDs $BiVO_4$ showed a higher bandgap (2.6 eV) than the bulk $BiVO_4$ (2.4 eV). A Pt co-catalyst added to the $BiVO_4$@SnO_2 catalyst also showed higher hydrogen production. The morphology and hydrogen production, as well as the water splitting mechanism, is shown in Figure 10.2(a)–(e). The QDs sized $BiVO_4$ widens their bandgap by 0.2 eV, which shows that the CBB of $BiVO_4$ could be higher than the bulk size. Hence, it can be because of the higher reduction potential. The Fermi level of $BiVO_4$ is higher than that of SnO_2 thereby the CBB of SnO_2 could be shifted when coupling with $BiVO_4$ due to their realignment of the energy levels occurring at the interface of the metal oxide with various fermi levels.[33–35] Therefore, as shown in Figure 10.2(e), the electron generated in $BiVO_4$ QDs under light illumination transfers to the CB of SnO_2 and then taken to the Pt. In this, holes are trapped by methanol. This reduces the recombination of generated electrons and holes, which enhances the water splitting efficiency.[35] Various metal oxides QDs for water splitting and their conditions are summarized in Table 10.1.

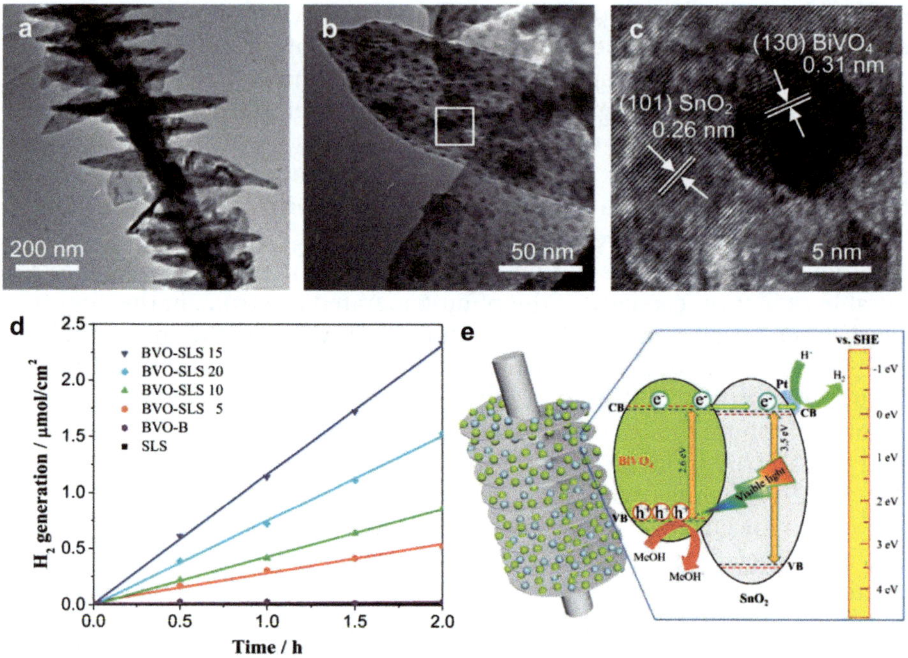

Figure 10.2 (a)–(c) TEM and HR-TEM images of screw-like SnO_2 and $BiVO_4$ QDs and their *d* spacing values with lattice fringes. (d) Resultant hydrogen evolution with various catalysts. (e) Charge carrier separation and migration with Pt@$BiVO_4$/SnO_2 catalyst. Reproduced from ref. 33 with permission from the Royal Society of Chemistry.

In summary, metal oxide QDs are highly attractive materials for enhancing the efficiency of water splitting. They show an improvement in light absorption, the trapping of electrons from the generated charge carriers, reduce the recombination of charge carriers, and change the energy levels of other semiconductors due to their Fermi levels. QDs also show high surface areas, reactive sites, and fast electron generation. These all efficiently enhance the overall water splitting efficiency. However, QDs synthesis is a little difficult when compared to other metal oxide nanostructures. This is due to their smaller size and that it is difficult to collect them after the chemical synthesis process.

10.2.2 CO_2 Photoreduction

Solar energy-driven carbon dioxide (CO_2) reduction to respected solar fuels/chemicals using particulate photocatalysts is considered to be one of the most favorable and effective ways to combat energy scarcity and global warming.[48] The evolution of nanotechnology plays a leading role in improving CO_2 reduction capabilities, which are the decisive factors for high CO_2

Table 10.1 Various metal oxide QDs for water splitting, their reaction conditions and resultant hydrogen and/or oxygen evolutions.

S.No	Metal oxide QDs based composite	Morphology	Synthesis method	Light source/reaction solution	Photocurrent density, H_2 (or) O_2 evolution	Reference
1	TiO_2& CdS QDs/SnO_2	Porous nanosheet	Atomic layer deposition	1.5G simulated sunlight/ methanol	4.7 mA cm^{-2}	36
2	$BiVO_4$ QDs	QDS	Hydrothermal	500 W Xe lamp/methanol	66.6 μmol	31
3	$BiVO_4$ QDs QDs/SnO_2	Screw	Chemical vapor deposition	500 W Xe lamp/methanol	2.2 μmol cm^{-2}	33
4	ZnO QDs/Fe_2O_3	Thin films	Electrodeposition	500 W Xe lamp/1 M NaOH	2.84 mA cm^{-2}	37
5	ZnO QDs/HS	QDS	Sol–Gel	100 mW cm^{-2} of white light illumination from Xe lamp.	260 μmol h^{-1}	38
6	ZnO QDs/RGO	Film	Modified Hummers	300 W Xe lamp/0.1 M Na_2SO_4	0.35 mA cm^{-2}	39
7	$BiVO_4$ QDs/WO_3	Nanosheet	Ionic layer adsorption and reaction	1.5G simulated sun-light/0.5 M Na_2SO_4	0.75 μmol h^{-1}	40
8	TiO_2 QDs/MXene	Nanosheet	Hydrothermal	300 W Xe lamp/0.5 M Na_2SO_4	62.5 μmol h^{-1}	41
9	CuO QDs/TiO_2	Nanosheet	Hydrothermal	300 W Xe lamp/0.5 M Na_2SO_4	0.04 mmol h^{-1}	16
10	Co_3O_4 QDs	QDs	Reverse micelle	UV and Visible/ethanol	66.6 μmol	42
11	$BiVO_4$ QDs/Carbon dot	Dots	Hydrothermal	300 W Solar light	0.92 μmol h^{-1}	43
12	$BiVO_4$ QDs/SiC	Facets	Hydrothermal	300 W Xe lamp	2.5 μA cm^{-2}	44
13	ZnO QDs/poly nanosheets	Nanosheets	Ionothermal	Visible light/TEOA	22 μmol h^{-1}	45
14	Ta^{4+}@Ta_2O_5 QDS/g-C_3N_4	Cotton	Hydrolysis	300 W Xe lamp/0.1 M Na_2SO_4	624.9 μmol g^{-1} h^{-1}	46
15	Fe_2O_3 QDs/$Cd_xZn_{1-x}S$@ RGO	Nanorods	Hydrothermal	Xe lamp/0.35 M Na_2SO_3	26.8 mmol g^{-1} h^{-1}	47

reduction by providing opportunities to achieve better surface reaction, solar light absorption, and charge separation by modifying the morphology of photocatalysts in nanoscale governance.[4] Metal oxide QDs with sizes below 20 nm are gaining a lot of attention due to their higher surface atoms, size-dependent energy band positions, very short charge migration lengths, and unique optical properties. Metal oxide semiconductor QDs such as ZnO, TiO_2, MgO, ZrO_2, WO_3, Fe_2O_3, CuO, and Co_3O_4 have been widely studied for the reduction of CO_2 as co-catalyst and catalyst.[49–51] To improve the photocatalytic reduction of CO_2, zero-dimensional (0D) metal oxide QDs (MOQDs) have been studied along with their bulk counterparts, because QDs are offering a lot of benefits like high surface area, good dispersion, economical, eco-friendly, and well-capable light absorption towards to visible region. For instance, Dai *et al.* developed the Bi_2MoO_6 QDs on RGO layers grown *in situ* for efficient photoreduction of CO_2.[11] In this work, the authors achieved the output alcohol up to 142.3 μmol g^{-1} by the benefit of improved transport, electron extraction from abundant electrons. These results are 4.4 and 2.2 times higher than the flower Bi_2MoO_6 and Bi_2MoO_6 QDs, respectively. Therefore, the individual results of Bi_2MoO_6 indicated that the QDs have a higher reduction efficiency than the nanoscale and micro-scale nanostructures. The resultant yields of CH_3OH and C_2H_5OH alcohols from CO_2 photocatalytic reduction with various ratios of RGO catalyst and unit cell model of Bi_2MoO_6 and electron transfer process over with Bi_2MoO_6 QDs/RGO are shown in Figure 10.3(a)–(c). In Figure 10.3(a) and (b) the BM/10 rGO catalyst showed a higher CH_3OH and C_2H_5OH reduction than the other catalysts. Under irradiation, the Bi_2MoO_6 electrons in VB migrate to the CB of Bi_2MoO_6, which is trapped in the orbital of Mo 4d. Figure 10.3(d) represents the XRD pattern after stability, which represents no changes with respect to the original crystallinity of the catalyst. In the unit cell model (Figure 10.3(c)), the electron-rich Mo atoms have higher active sites for attracting CO_2 molecules due to high electron density on the surface. After that, the generated electron is directly trapped by the rGO due to its conductive nature, thereby improving the electron-rich interface. However, the electron excitation process and CO_2 reduction follow eqn (10.1)–(10.5).

$$Bi_2MoO_6 + light \rightarrow e^- + h^+ \tag{10.1}$$

$$CO_2 + 6H^+ + 6e^- \rightarrow CH_3OH + H_2O \tag{10.2}$$

$$2CO_2 + 12H^+ + 6e^- \rightarrow C_2H_5OH + 3H_2O \tag{10.3}$$

$$CO_2 + H_2O \Leftrightarrow H_2CO_3 \Leftrightarrow H^+ + HCO_3^- \Leftrightarrow H^+ + CO_3^{2-} \tag{10.4}$$

$$H_2CO_3 + 6H^+ + 6e^- \rightarrow CH_3OH + H_2O \tag{10.5}$$

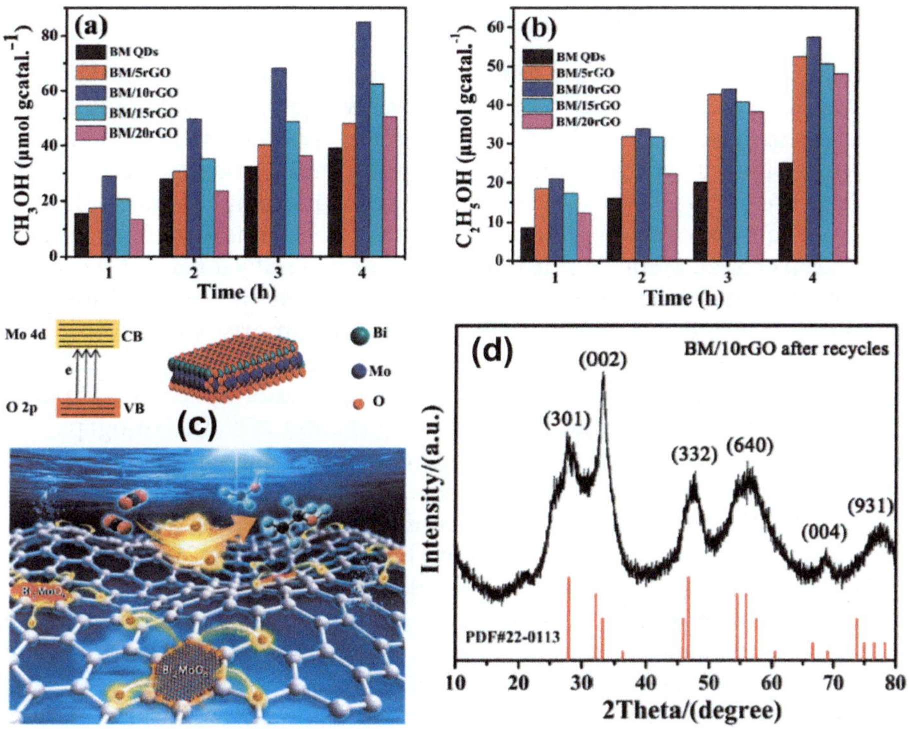

Figure 10.3 Yields of (a) CH_3OH and (b) C_2H_5OH from CO_2 photocatalytic reduction over various Bi_2MoO_6/rGO catalysts. (c) Unit cell model of Bi_2MoO_6, and electron transfer performance over Bi_2MoO_6 QDs/rGO. (d) XRD pattern after recycling. Reproduced from ref. 11 with permission from American Chemical Society, Copyright 2020.

Chen *et al.* also reported the $BiVO_4$ QDs/rGO composite for CO_2 reduction.[52] In this work, they showed 3.5 times higher photocatalytic CO_2 reduction with the composite, and $BiVO_4$ QDs synthesized using a modified hydrothermal synthesis. Furthermore, Zhongkai *et al.* constructed a CuO QDs/WO_3 nanosheets heterojunction catalyst for CO_2 reduction.[53] In this work, they concluded that the CO_2 reduction mechanism is based on 0D–2D structural benefits and Z-scheme charge transfer. Therefore, the constructed 0D–2D structural composite represents the higher CO_2 reduction under visible light irradiation.

In summary, metal oxides QDs show remarkable efficiency in CO_2 reduction due to their higher surface atoms, size-dependent energy band positions, very short charge migration lengths, and unique optical properties. The photocatalytic CO_2 reduction includes three primary steps: (1) under light illumination the semiconductor photocatalyst generates charge carriers; (2) the photogenerated charge carriers are separated and transported to the surface of the semiconductor; and (3) redox reactions facilitated by electrons and holes take place at the active sites on the surface.

10.3 Metal Sulfides Based QDs

Metal sulfide QDs, which have low cost, exotic electronic properties[54,55] and stability, have gained considerable attention owing to potential applications in various fields including electronics,[56] optics,[54] photocatalysis,[57] magnetic storage,[58] biomedical[59] and pharmaceutical.[60] Unlike metal oxides, metal sulfides have narrow bandgaps, wider light absorption ranges and higher VB positions because of the higher position of the S 3p orbital when compared to the O 2p orbital. Metal sulfide photocatalysts generally form metal cations with d^0, d^5 and d^{10} electron configurations. The CBs of metal sulfides are composed of d and sp orbitals, while the VBs are comprised of S 3p orbitals exhibiting a more negative conduction band, and probably higher reduction capability. In addition, sulfide moieties have high stability when compared to other chalcogenide compounds, featuring their QDs static bandgaps and optoelectronic and photocatalysis implications.[61] Many transition metal sulfide QDs have been explored in recent years such as PbS, ZnS, CdS and Ni_2S_3, MoS_2 QDs, *etc.*, where quantum confinement effects, size tunable light absorption and emission properties have been documented.

Top-down approaches that rely on lithography or milling of particulate materials to quantum sizes have been employed to produce metal sulfide (MS) QDs. For instance, Yang *et al.* used a top-down approach to prepare well-dispersed colloidal PbS QDs with controllable sizes by adopting two-step laser ablation.[62] The reagent 1-dodecanethiol (DDT, $C_{12}H_{26}S$) was used, which not only acts as surfactant but also serves to be sulfur source. In this synthesis, first raw PbS nanocrystals (polydisperse) were obtained by means of laser ablation of a lead plate in DDT. The authors examined the change in solution to black indicating the formation of raw PbS QDs. With second laser ablation (20 min), the monodispersed PbS QDs were generated with a significant reduction in size. The universality of this process was tested by extending the synthesis to different chalcogenides such as CdSe, PbSe, PbTe, *etc.*, demonstrating prominent quantum confinement effects. Dieleman *et al.* demonstrated direct patterning of colloidal PbS QDs down to tens of nanometers by means of an electron beam lithography technique.[63] The photoluminescence efficiency was preserved even after pattern making, which suggests implications in a wide range of optical device applications. Additionally, bottom-up strategies that rely on the assembly of atomic building units and are determined by built-in strain growth style have been explored to develop various MSQDs.[55] The hot injection technique is a commonly used technique to obtain monodisperse colloidal QDs with narrow size distributions. For example, Yarema *et al.* used the under pressure-assisted hot-injection method to produce multigram-scale high-quality PbS QDs with narrow size distributions, interesting optical properties, and reproducible compositions. This work highlights implementing a large injection volume does not necessarily limit the upscaling process in terms of quality and size of colloidal QDs.[64] In another study, water-soluble CuS QDs were accomplished *via* adopting L-cysteine (NAC) as a surfactant and injecting Na_2S solution to a mixture of $CuCl_2$ and NaOH *via* a room temperature one-step synthesis. Subsequently, these CuS QDs were applied as efficient luminescent probes for the detection of heavy metals such as Hg^{2+}, Ag^{+} and Au^{3+}. Lin *et al.* demonstrated the synthesis of 2D WS_2 quantum dots (8–15 nm) *via* solution (EtOH + H_2O) processed exfoliation of WS_2 flakes with ultrasonication. Benefitting from the higher bandgap (3.16 eV), these 2D WS_2 QDs provided an excellent photoluminescence quantum yield (4.0%).[65] In another study, Zhang *et al.* reported the synthesis of gradient-alloyed $CdSe_xS_{1-x}$ QDs from cadmium acetate ($Cd(OA)_2$), tri-*n*-octylphosphine selenide (TOPSe), tri-*n*-octylphosphine sulfide (TOPS), and diphenyl phosphine ($HPPh_2$) in 1-octadecene (ODE) at 180 °C *via* adopting a non-hot-injection method, which displays a high photoluminescence quantum yield (79%).[66]

10.3.1 Water Splitting

The selection of photocatalyst is critical in this process, which needs to satisfy some basic requirements such as: (1) effective absorption of light energy and generation of electron–hole pairs; (2) the CB of the photocatalyst should be more negative in relation to the standard hydrogen reduction potential (0.0 V *vs.* NHE); and (3) the bandgap of the catalyst should be in the range of 1.8–3.0 eV to allows solar light and relatively low overpotential to drive the hydrogen evolution reaction (HER) at an impressive rate.[67,68] Metal sulfide QDs have been rigorously investigated for H_2 production from the water splitting reaction.[69] CdS QDs are one of the most promising photocatalysts for the HER due to their low cost, well-suited bandgap (2.4 eV) and CB and VB edge potentials.[70–72] Jiang *et al.* prepared CdS and ZnS QDs by employing a modified gel crystal growth method in which hydrogel (HR) serves as the substrate to effectively confine the CdS and ZnS nanoparticles. In this case, the chemical adsorption effect can inhibit the agglomeration of nanoparticles.[72] This study further reports the use of hydrogel (HR) as a miniature hydrogen production reactor, which boosts the H_2 evolution reaction when compared with conventional bulk metal sulfides.[72] Colloidal Se-doped CdS quantum dots were synthesized by adopting a simple solvothermal method (Figure 10.4(A)(i)).[73] In this colloidal synthesis, first Cd $(NO_3)_2{\cdot}4H_2O$ was dissolved in oleylamine (OAm) to form a Cd–oleyl amine complex. Following this, a OAm solution with a certain molar ratio of Se to S was injected into the above solution. As the temperature increased, the CdS_xSe_{1-x} nanocrystals were grown with uniform particle size (~2.5 nm) and good dispersibility (Figure 10.4(A)(ii) and (iii)). The DDT served as ligands that can control the growth and prevent particle agglomeration. They found that Se doping can elevate the Fermi level of CdS to a higher position, which enables efficient capture of photoinduced electrons through the trapping sites and suppresses the unwanted recombination of electron–hole pairs. This presents a prolonged lifetime of charge carriers, strong light absorption, and charge separation in the resultant $CdS_{0.9}Se_{0.1}$ QDs with a remarkably high H_2 production rate of 29.12 mmol h^{-1} g^{-1}, which is ~23 times higher than pristine

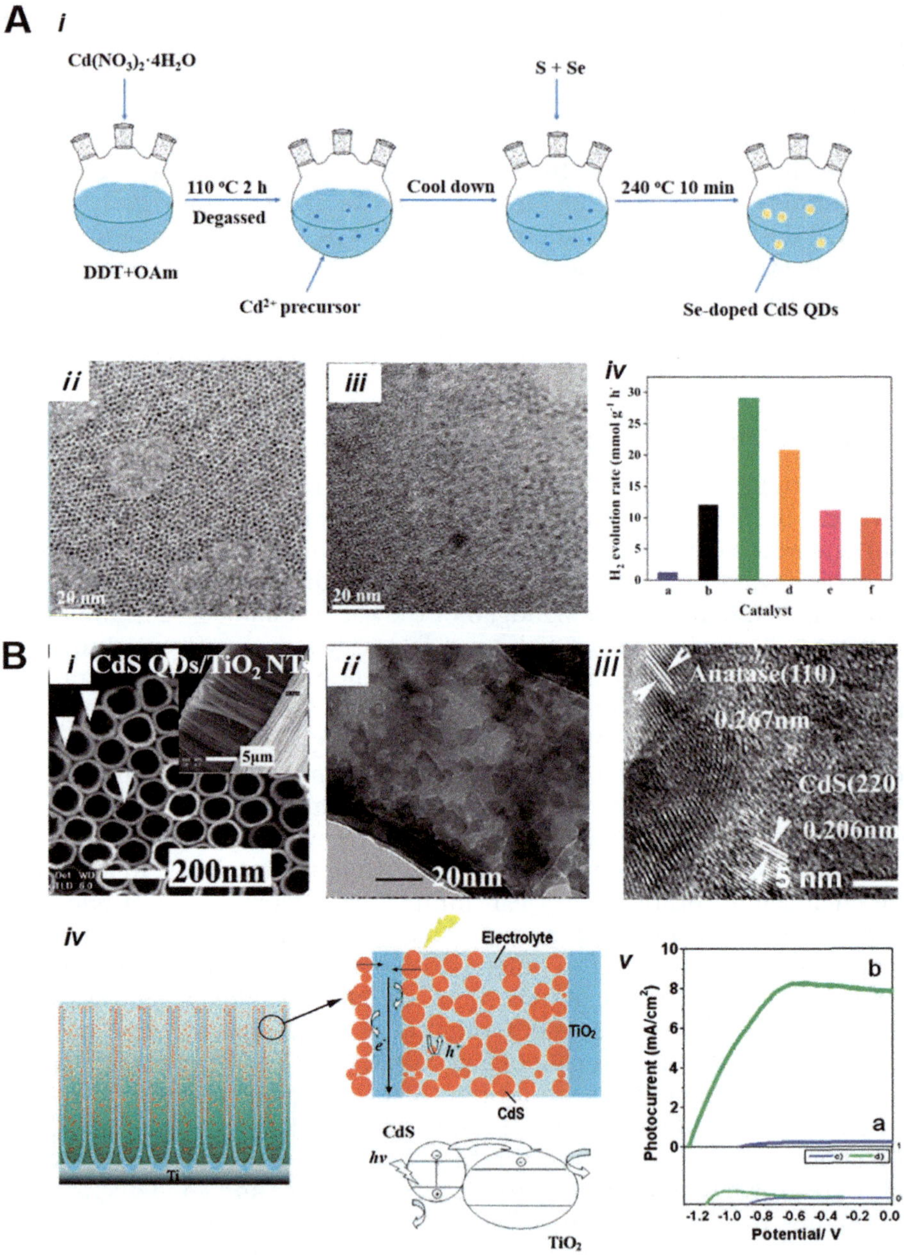

Figure 10.4 (A) Se-doped CdS quantum dots: (i) schematic representation of the synthesis strategy, (ii) TEM image of pure CdS QDs, (iii) TEM image of $CdS_{0.9}Se_{0.1}$ QDs, (iv) photocatalytic H_2 evolution rates over CdS (a), $CdS_{0.947}Se_{0.053}$ (b), $CdS_{0.9}Se_{0.1}$ (c), $CdS_{0.75}Se_{0.25}$ (d) $CdS_{0.5}Se_{0.5}$ (e) and CdSe QDs (f), respectively. Reproduced from ref. 73 with permission from Elsevier, Copyright 2019. (B) CdS quantum dots sensitized TiO_2 nanotube arrays: (i) SEM image of TiO_2 nanotubes. Inset shows a

CdS QDs (Figure 10.4(A)(iv)). While CdS has a good visible light absorption range and good reducing capability, the photocatalytic efficiency of CdS is greatly limited by fast recombination of charge carriers and photo corrosion in which surface S atoms are easily oxidized by photogenerated holes as shown in eqn (10.6).[69]

$$CdS + 2h^{+} \rightarrow Cd^{2+} + S \qquad (10.6)$$

To overcome these issues, various strategies have been employed such as, elemental doping, coupling of CdS with other semiconductors to form heterostructures (Z-scheme),[70,74] use as a photosensitizer[32] and CdS-based graphene composites,[75] *etc*. $Zn_xCd_{1-x}S$ has shown promising photoactivity by virtue of the tunable band structure *via* fine-tuning the Cd–Zn atomic ratio.[72] Moreover, reports reveal that $Zn_xCd_{1-x}S$ has lower toxicity and better stability when compared to pure CdS.[76] Further, Xie *et al.* reported S vacancy modulated $Zn_xCd_{1-x}S$/CoP QDs heterojunctions by combined combustion and wet chemical reactions. The vacancy engineering together with quantum effect had a positive effect on strong visible light absorption, and transfer of photoinduced electron–hole carriers dramatically showing high photocatalytic activity towards HER (17.53 mmol g^{-1} h^{-1}).[76] Ma *et al.* rationally designed a CdS/ZnO heterojunction composed of 0D CdS QDs and 2D ZnO nanosheets.[71] First, 2D ZnO NSs were fabricated through an ultrasonic-assisted hydrothermal reaction between zinc nitrate ($ZnNO_3{\cdot}5H_2O$) and NaOH solution (0.06 M). Next, CdS QDs were deposited on ZnO NS through *in situ* successive ionic-layer absorption and reaction (SILAR) method. The as-synthesized CdS/ZnO-12 heterojunction showed an excellent hydrogen evolution rate of 22.12 mmol h^{-1} g^{-1}, which was 13- and 138-fold higher than pristine CdS (1.68 mmol h^{-1} g^{-1}) and ZnO (0.16 mmol h^{-1} g^{-1}), respectively. The remarkable improvement in photoactivity was attributed to several key factors including Z-scheme charge transfer, quantum size effect of 0D CdS QDs and special 2D surface morphological features of ZnO NSs offering large active surface area. In addition, CdS QDs have been used as photosensitizers together with TiO_2 nanotube arrays photoelectrodes. The CdS QDs of size (2–10 nm) were uniformly deposited onto anodic TiO_2 arrays by simple chemical bath deposition (CBD). In this case, a unique TiO_2 nanotube array structure is reported to have superior electron transfer channels while CdS QDs further assist to

cross-sectional view of TiO_2nanotubes, (ii) SEM image of CdS quantum dot loaded TiO_2 nanotubes, (iii) and (iv) TEM and HR-TEM images of CdS deposited TiO_2 nanotubes, (v) schematic illustration of CdS QDs deposited TiO_2 nanotube electrode and associated electron transfer mechanism, (vi) linear sweep voltammograms (LSVs) in 1 M Na_2S under AM 1.5G at 138.4 mW cm^{-2} irradiation: (a) pristine TiO_2, (b) CdS QDs deposited TiO_2, (c) and (d) represents corresponding LSV curves under darkness. Reproduced from ref. 70 with permission from American Chemical Society, Copyright 2008.

extend the photo-response of TiO_2. As a result, a significantly high PEC cell efficiency (4.15%) and photocurrent density (7.82 mA cm^{-2}) were obtained due to the combined effect of strong light absorption and efficient charge transfer between TiO_2 and CdS QDs under light visible irradiation (Figure 10.4(B)(i)–(v)).[70] Further, Lee *et al.* reported the CdS/CdSe co-sensitized TiO_2 to form cascade photoelectrode by adopting a combined spin coating and CBD method.[77] HRTEM analysis displayed highly crystalline counterparts and clear lattice interfaces between them as confirmed by corresponding lattice fringes (Figure 10.5(A)(i)). They propose the Fermi-level alignment between CdS and CdSe to form a stepwise band structure in the TiO_2/CdS/CdSe photoelectrode (Figure 10.5(A)(ii)), which likely contributes to improve the PEC performance of the electrode. By modifying photoelectrode with ZnS layer, which apparently ZnS acted as a passivation layer to offer higher stability. The resulting cascade TiO_2/CdS/CdSe/ZnS showed the highest reported photoelectrochemical H_2 evolution rate (220 μmol cm^2 h^{-1}) as shown in Figure 10.5(A)(iii). Although the charge transfer during the PEC water splitting reaction seems reasonable, there was no straight evidence in this system. Because of its ecofriendly nature and wide range of applications, nowadays Cd, Hg, and Pb-free QDs have gained huge research interests. Recently, a narrow bandgap semiconductor (Ag_2S QDs) was prepared by heating a mixture of silver nitrate $AgNO_3$, ethylene glycol (EG), and 3-mercaptopropionic acid (MPA).[78] A highly crystalline and well dispersed Ag_2S QDs with an average size of 7.2 nm and 8.5 nm, respectively, were obtained as confirmed by TEM analysis (Figure 10.5(B)(i)–(iii)). The Ag_2S QDs showed two different excitonic absorbance peaks at 800 (Ag_2S QD800) and 1000 nm (Ag_2S QD1000), respectively, (Figure 10.5(B)(iv)). Owing to excellent physicochemical properties, the Ag_2S QDs were examined as both photocatalysts and electrocatalysts. Ag_2S QDS with an absorption peak at 800 nm (Ag_2S QD800) displayed the highest electrocatalytic HER activity with a Tafel slope of 89 mV/dec and a low overpotential of 0.32 V (Figure 10.5(B)(v)). The Ag_2S QDs 800 also show the utmost photocatalytic H_2 evolution rate of 858 μmol h^{-1} g^{-1} and was found to be dependent on a size-tunable quantum confinement effect (Figure 10.5(B)(vi)). Additionally, enormous efforts have been dedicated to the synthesis of alloyed MSQDs *via* changing the surface atomic ratio and ligand exchange reactions.[68] Recently, Tan *et al.* fabricated a series of alloyed Cu–In–Zn–S QDs (size: 3 to 5 nm) to tailor the bandgaps *via* systematic control over the Cu content and adopting a simple hydrothermal method. Upon adjusting the Cu content, they found that a color change in the material (from yellow to dark red) indicated substantial changes in electronic structures. The bandgap energy was tuned from 2.90 eV to 1.98 eV increasing the light absorption capability due to lifting of the VB through the Cu 3d orbital as this is relatively higher than that of the S 3p orbital. Therefore, this enhances the photocatalytic activity for the H_2 production reaction. A composite (0D/2D) Ag–In–Zn–S QDs/NiFe-LDH with increased active surface area and accelerated interfacial charge transfer was realized by Zhang *et al.*[79] by using a simple hydrothermal method. This resulted in doubled electrochemical or photoelectrochemical performances. Similarly, a range of different Cd-free metal sulfide QDs

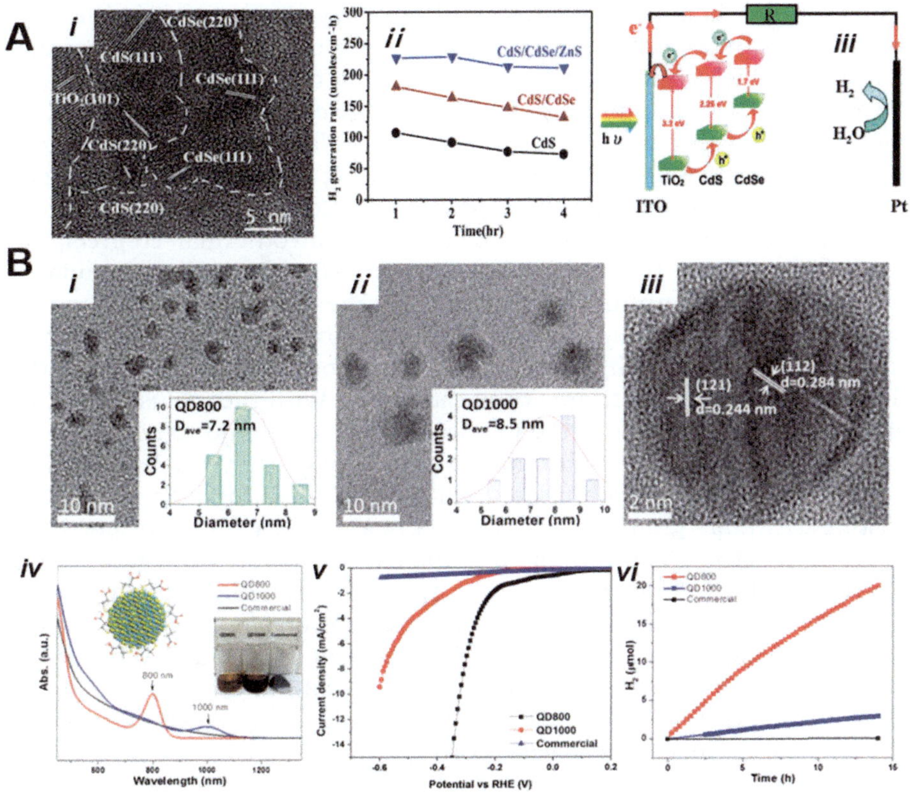

Figure 10.5 (A) CdS/CdSe QDs co-sensitized TiO_2 photoelectrode for *water* splitting: (i) HRTEM image of TiO_2/CdS/CdSe electrode, (ii) a comparison of H_2 production rates with different photoelectrodes, (iii) schematic for the electron transfer mechanism. Reproduced from ref. 77 with permission from American Chemical Society, Copyright 2010. (B) Ag_2S QDs as an electrocatalyst and photocatalyst. TEM images of Ag_2S: (i) QD800 and (ii) QD1000 samples (insets show the distribution and average diameters of Ag_2S QDs), (iii) HR-TEM image of an Ag_2S nanocrystal, which exhibits the hexagonal crystalline structure, (iv) UV-vis absorbance spectra of Ag_2S samples, (v) polarization curves, and (vi) photocatalytic H_2 evolution over Ag_2S QD800, Ag_2S QD1000, and commercial Ag_2S. Reproduced from ref. 78 with permission from American Chemical Society, Copyright 2019.

have been successfully developed such as Mo-doped Ni_3S_2,[80] 0D/2D/2D NbS_2 QDs,[81] MoS_2,[82] alloyed-Ag–In–Zn–S,[79] Zn-doped $AgIn_5S_8$.[83] These have been used for the H_2 production reaction *via* water splitting.

10.3.2 CO_2 Photoreduction

Metal sulfide QDs have been extensively studied for their use in CO_2 photoreduction to produce chemical fuels including CO, HCOOH, CH_3OH, CH_4, *etc.*[84] For instance, Kuehnel *et al.* integrated CdS QDs with nickel bis(terpyridine) complexes ($[Ni(terpy)_2]^{2+}$) to carry out photocatalytic reduction of CO_2 with water. A selectivity of >90% was realized for the CO production reaction, which was attributed to intimate contact between CdS and $[Ni(terpyS)_2]^{2+}$. Effective migration of the charge carrier can trigger the CO_2 photoreduction reaction. However, this molecular catalyst showed a loss in CO_2 photoreduction performance after a long run time (8 h) and needs to be further improved. Yuan *et al.* fabricated a $Cd_{0.5}Zn_{0.5}S$ (CZS) QDs modified $CdIn_2S_4$ (CIS) heterojunction by a simple solvothermal–co-precipitation method (Figure 10.6(A)(i)).[85] The TEM and HRTEM analysis elucidated that CZS QDs were uniformly dispersed on the surface of the CIS nano-octahedron (Figure 10.6(A)(ii) and (iii)), and forms a 0D/3D heterostructure. The type II band alignment was found to be effective to separate the photoinduced charge carriers at the interfaces. With the inclusion of CZS QDs in CZS/CIS composites the CO production was found to be increased and the highest CO production rate (14.07 μmol g^{-1} h^{-1}) was realized with an optimal 10% CIS combination, which was much higher than that of pristine CZS (5.51 μmol g^{-1} h^{-1}) and CIS (4.17 μmol g^{-1} h^{-1}), respectively, (Figure 10.6(A)(iv)). The stability of photocatalyst is very critical to yield a high CO production reaction rate, the optimal 10% CZS/CIS photocatalyst did not show significant loss in CO production even after repeated use (18 h) (Figure 10.6(A)(v)). Thus, several factors likely contribute to boost the CO_2 photoreduction reaction: (1) the intricate 0D/3D interface between CZS and CIS inhabits the recombination of the photogenerated charge carriers (Figure 10.6(A)(vi)); (2) the special morphologies of 0D CZS QDs and 3D CIS nano-octahedron were prone to provide more electron transfer channels; and (3) the CZS/CIS heterojunction resulted in remarkable photocatalytic CO_2 reduction due to more accessible specific active sites and charge migration. In another case, binary semiconductor II-VI QDs (*e.g.*, PbS, Bi_2S_3, and ZnS), ternary I–III–VI QDs (*e.g.*, Ag–In–S, $CuInS_2$, $Cd_{0.5}Zn_{0.5}S$, and $CuAlS_2$), and perovskite-based QDs (*e.g.*, $CsPbBr_3$, $CH_3NH_3PbBr_3$, and $Cs_2AgBiBr_6$) and their composites have been investigated for CO_2 photoreduction.[84] Hu *et al.* combined cubic perovskite (CsPbBr) QDs with PbS nanoplates to form 0D–2D mixed composites *via* epitaxy-hot injection-mechanical stirring methods and employed them to be highly active photocatalysts for CO_2 photoreduction. The size of CsPbBr QDs was found to be remarkably reduced when prepared on the 2D-PbS nanoplates except in solvent and showed low bandgap and strong reduction potential for CO_2 photoreduction reaction. The perovskite QDs cubes were uniformly attached to the 2D-PbS, and it was proved that HBr pre-treatment is required for anchoring of perovskite QDs. The Br^- ions tend to form the dangling bonds with PbS QDs resulting in mixed dimensional photocatalyst without severe aggregation. Thus, excellent CO_2 photoreduction activity with formation of CO (81.6 μmol g^{-1}) and CH_4 (52.7 μmol g^{-1}) and selectivity (>90%) was realized which was nine times higher when compared to that of pristine CsPbBr QDs (Figure 10.6(B)(i)–(iii)).[86]

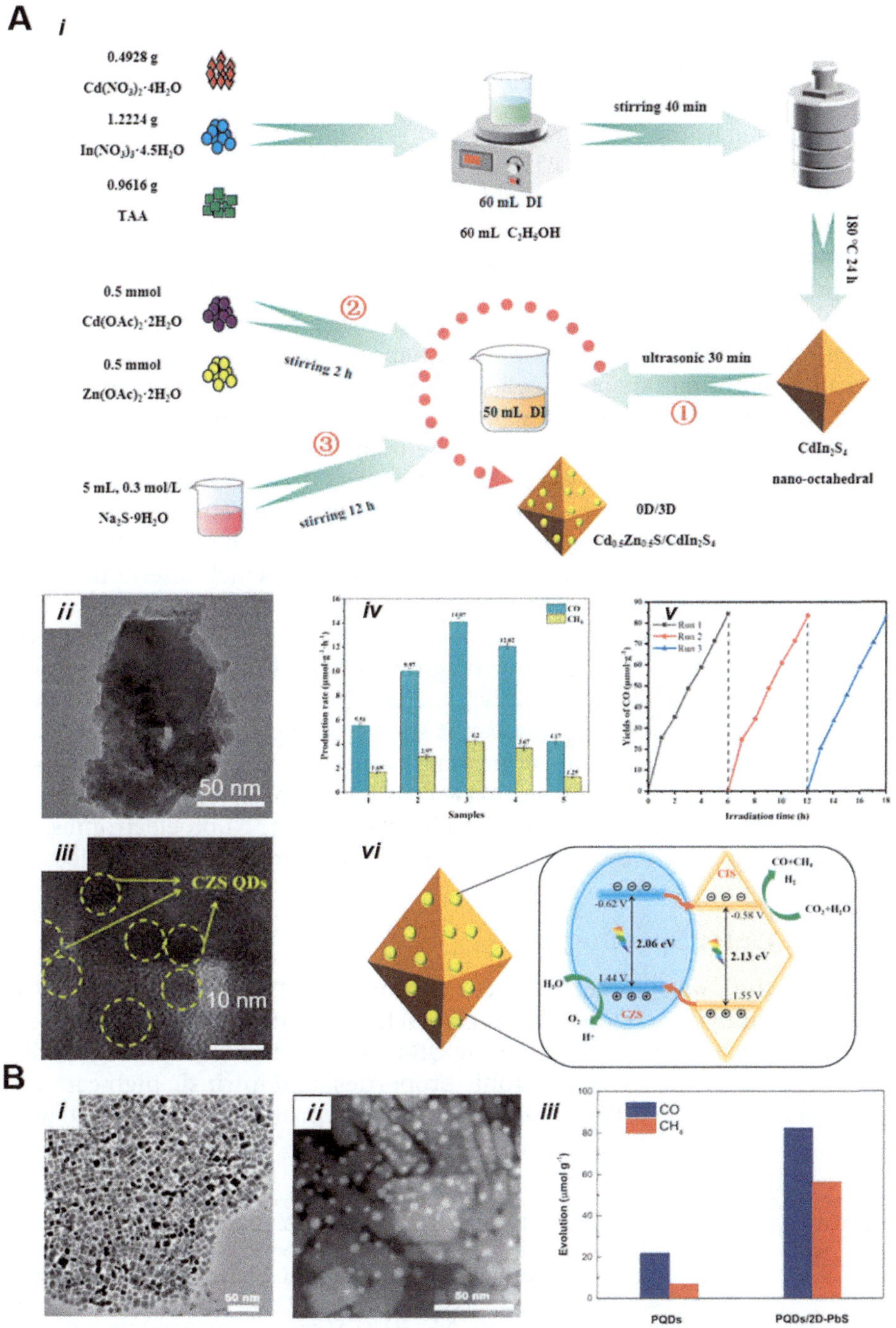

Figure 10.6 (A): (i) Synthetic protocol for $Cd_{0.5}Zn_{0.5}S$ QDs-modified $CdIn_2S_4$ nano-octahedron (CZS/CIS), (ii) and (iii) TEM image of the 10% CZS/CIS composite, (iv) photoconversion yields of CO and CH_4 over *(continued)*

10.4 Metal Selenides Based QDs

Metal selenides (MSes) refer to a unique class of semiconductor nanomaterials formed by the combination of a metal cation (M^{2+}) and a selenium anion (Se^{2-}). Obviously, the chemistry of metal selenides is quite similar to metal sulfides. Nonetheless, MSe QDs take an advantage of a high refractive index of Se enabling it to be most suitable for photoconductive applications such as photodetectors and sensors, *etc*. The high reactivity of Se is beneficial to be used as a template in the colloidal synthesis of various nanostructures.[87] Recently, Park *et al.* reported the reduction of $SeCl_4$ by $NaBH_4$ in the presence of $AgNO_3$, which leads to the generation of Ag_2Se nanoparticles. By executing cation exchange reactions on Ag_2Se nanoparticles it can be transformed into a range of monodisperse colloidal M–Se nanoparticles (M = Cd, Zn, Pb) exclusively at a low temperature (60 °C).[88] In another study, Yang *et al.* performed cation exchange reactions on pre-made Ag_2Se QDs with a Au precursor to form alloyed silver–gold selenide AgAuSe QDs. Meanwhile, numerous strategies such as solvothermal,[89] sonochemical,[90] electrochemical[91] and biochemical have been extensively studied for the synthesis of MSe QDs.[92] However, these syntheses have their pros and cons, which often limit their modern device system use. As electrosynthesis seems to be a convenient method because of its simple synthesis procedure and room temperature operations, it severely suffers from aggregation or stacking of MSe QDs.[93] From an environmental perspective, the biochemical methods are considered to be straightforward. However, it shows poor product yields and even low-quality QDs.[93] Enormous efforts have been directed to tailor the spectral properties of MSe QDs. These MSe QDs have subsequently been applied as photovoltaics,[94] photocatalysts,[32] electrocatalysts,[94] optoelectronics,[95,96] thermoelectrics,[97] phototherapy[98] and as electrochemical storage devices.[99]

10.4.1 Water Splitting

In this section, we summarize recent advances in MSe QDs catalyst innovations and catalytic activities for H_2 production *via* water splitting reactions. Particularly, transition metal selenide QDs are attractive electrocatalysts due to their low-cost, excellent electronic properties, and intrinsic high catalytic activity (Table 10.2).[100] The 3d orbital of Se was reported to have bonded with

Figure 10.6 *(continued)* CZS, 5% CZS/CIS, 10% CZS/CIS, 15% CZS/CIS, and CIS photocatalysts and (v) recyclability of optimum 10% CZS/CIS photocatalyst for CO_2 photoreduction. Reproduced from ref. 85 with permission from American Chemical Society, Copyright 2021. (B) Synthesis of 0D–2D mixed dimensional materials: (i) TEM image of perovskite quantum dots (PQDs), (ii) HAADF image of PQDs/2D-PbS composite materials, (iii) comparison of CO_2 photoreduction performance over PQDs and PQDs/2D-PbS. Reproduced from ref. 86 with permission from Elsevier, Copyright 2020.

Table 10.2 Comparison of electrocatalytic activities of metal chalcogenide QDs, overpotential and Tafel slope.

Catalyst	Electrolyte	Type of reaction	Tafel slope (mV dec^{-1})	Overpotential (@ 10 mA cm^{-2})	Reference
CdS QDs/Ni_3S_3	1 M KOH	HER	110	121	106
MoS_2	0.5 M H_2SO_4	HER	53	187	107
MoS_2	0.5 M H_2SO_4	HER	210	60	108
Ag_2S	0.5 M H_2SO_4	HER	89	320	78
MoS_2/GQDs	0.5 M H_2SO_4	HER	40	120	109
Mo–Ni_3S_2 QDs	1 M KOH	HER	260	115	80
Co–$Ni_{0.85}$Se/RGO	0.5 M H_2SO_4	HER	49	148	110

metal atoms by virtue of comparable energy levels with the 3s and 3p orbitals and show more metallicity, which is beneficial to boost the electron transport accelerating the rate of electrochemical reactions. In this regard, CdSe QDs have been extensively studied as photocatalysts because of their suitable bandgap energy and high efficiency. Yu *et al.* demonstrated that incorporating CdSe QDs into TiO_2 nanostructures led to a substantial improvement in photocatalytic H_2 production (up to 10.1 mmol g^{-1} h^{-1}).[101] In this study, CdSe QDs function as light-harvesting antenna and TiO_2 serves as an electron mediator to eliminate the unwanted charge recombination and thus facilitates the electron transfer to yield quantum yield up to 27.4%. In another study, Zhong *et al.* coupled CdSeQDs with g-C_3N_4 to form a CdSeQDs/g-C_3N_4 heterostructure *via* a wet chemical method and indicated significant enhancement in photocatalytic H_2 production rate (615 μmol g^{-1} h^{-1}), which is ~76 times higher than pure phase g-C_3N_4.[102] The authors indicated that type II band alignment can promote the charge separation and enhance photostability under light irradiation. Similarly, Liu *et al.* reported biocompatible synthesis of core/shell/shell CdSe/CdS/ZnS QDs using mercaptopropionic acid (MPA) as the ligand (Figure 10.7(A)(i)).[103] As per HRTEM analysis (Figure 10.7(A)(ii)), monodisperse CdSe/CdS/ZnS QDs with an average size of 7.2 nm were formed. Their XRD analysis displayed a soft shift in diffraction peaks and no prominent peaks related to individual CdS and ZnS was noticed, indicating that the shift in the diffraction peaks was related to pitaxial growth of the CdS and ZnS shell on the CdSe core indicating formation of a core–shell structure (Figure 10.7(A)(iii)). The resulting CdSe/CdS/ZnS QDs showed excellent photocatalytic H_2 production (1362.70 μmol), which was ~1.85 and ~6.32 times higher than CdSe/CdS QDs (736.05 μmol) and single-phase CdSe QDs (215.54 μmol). Su *et al.*[104] synthesized a Co–CdSeTe–NiO photocathode by immobilizing CdSeTe QDs onto NiO films prepared through a hydrothermal method with involvement of L-cysteine and cobaloxime complex (proton reduction catalyst). These L-cys CdSeTe QDs were highly crystallinity with an average particle size (3 nm) distribution (Figure 10.7(B)(i)). The intimate interfacial contact and formation of Co–CdSeTe–NiO composite was confirmed by a high magnification image (Figure 10.7(B)(ii)) with two different lattices spacing in agreement with (400) and (200) crystal planes of bulk zinc

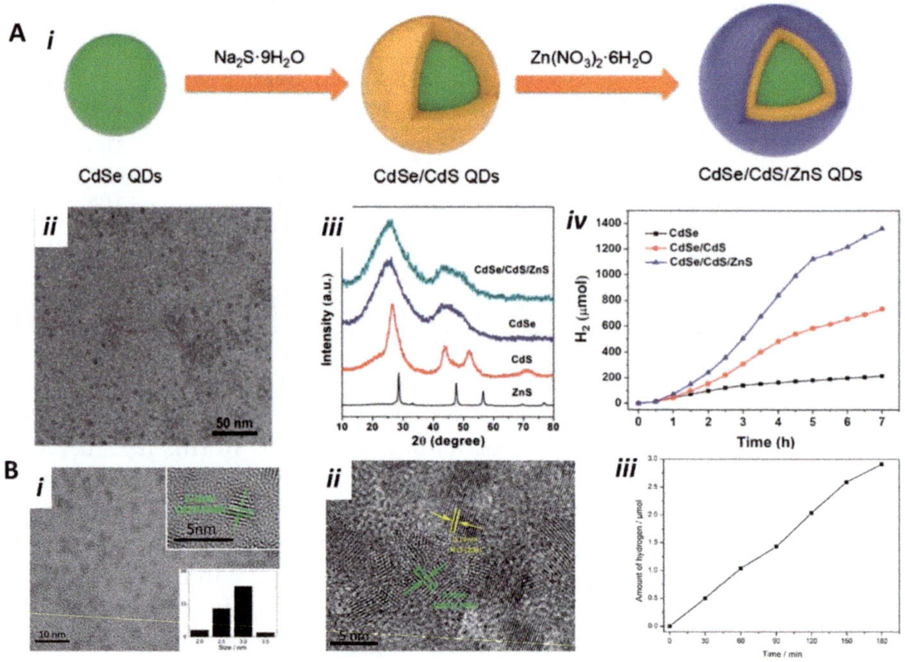

Figure 10.7 (A) Colloidal synthesis of core/shell/shell CdSe/CdS/ZnS quantum dots: (i) schematic illustration of synthesis process, (ii) TEM image, (iii) XRD pattern, (iv) photocatalytic H_2 production activity. Reproduced from ref. 103 with permission from Springer Nature, Copyright 2019. (B) CdSeTe QDs as photosensitizers on NiO thin films: (i) TEM of CdSeTe QDs (inset above shows the HRTEM image of individual QDs and inset below shows size distribution), (ii) HRTEM image of CdSeTe QDs sensitized NiO film, and (iii) H_2 production time-dependent curve. Reproduced from ref. 104 with permission from American Chemical Society, Copyright 2019.

blende CdSeTe and NiO particles. The composite Co–CdSeTe–NiO photocathode displayed a substantially high photocurrent density ($-130\ \mu A cm^{-2}$) at -0.1 V *vs.* NHE in neutral electrolyte and excellent H_2 production of 2.9 μmol, (Figure 10.7(B)(iii)) with 93% Faradaic efficiency at -0.3 V *vs.* NHE over 3 h irradiation. Similarly, Lv *et al.* prepared CdSe QDs/NiO/ITO composited *via* a simple spin-coating method and indicated that a rainbow photocathode which subsequently produces superior H_2 gas (8.4 μmol) generation from water splitting with high Faradaic efficiency (99.8%).[105]

10.4.2 CO_2 Photoreduction

Semiconductor QDs of II–VI group elements (*e.g.*, CdSe, ZnSe) were reported to have unique light-harvesting properties and are superior candidates in the CO_2 photoreduction reaction.[10] Recently, Xia and co-workers fabricated ligand-removed and Cd-rich CdSe QDs and tested them for CO_2 conversion into CO in a mixture of a triethylamine–dimethylformamide (TEA–DMF) under visible light irradiation (Figure 10.8(A)).[111] TEM analysis suggested that ligand MPA–CdSe-2 shows formation of monodispersed particles

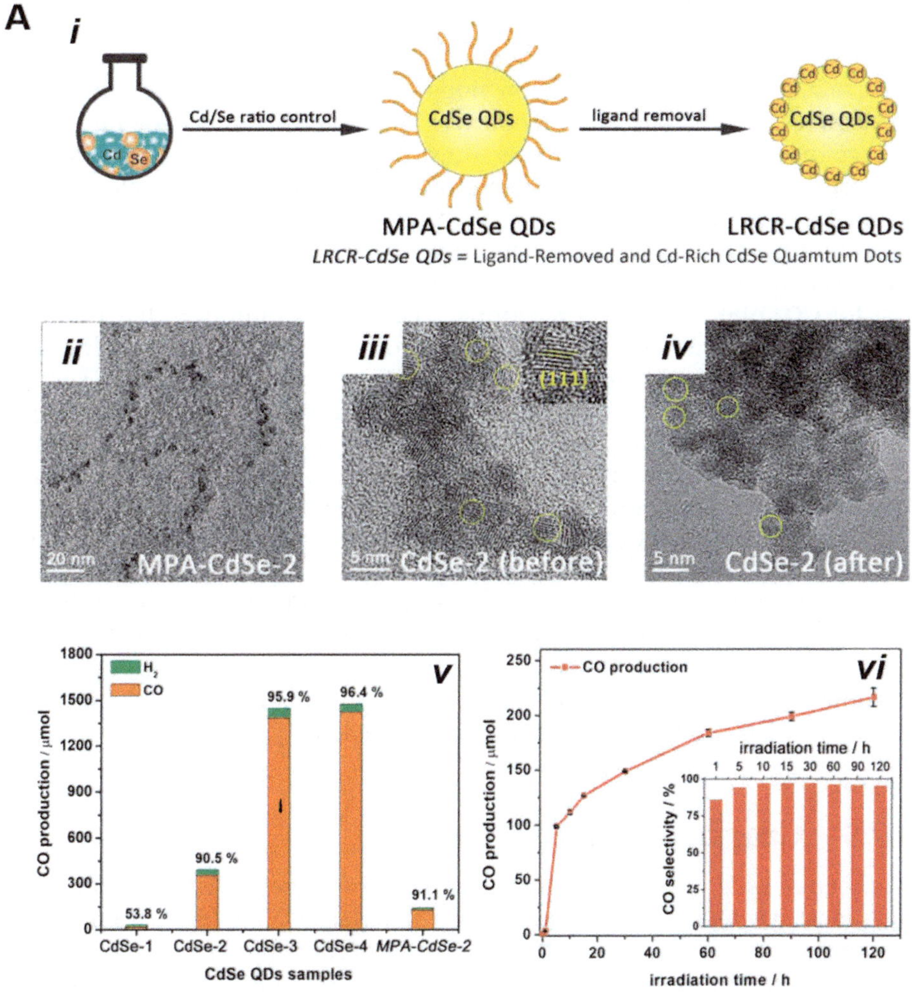

Figure 10.8 (A) Cd-rich CdSe quantum dots as photocatalysts: (i) schematic for preparation of LRCR-CdSe QDs, (ii) TEM image of MPA-CdSe-2 and HRTEM image (inset) of CdSe-2 QDs before (iii) and after (iv) illumination, (v) CO_2 photoreduction for CO and H_2 production with CdSe QDs (0.2 mg mL^{-1}) with different Cd–Se ratios and MPA-CdSe-2; [QDs] = 0.2 mg mL^{-1}, (vi) long-time CO_2 photoreduction activity by using CdSe-4 (0.005 mg mL^{-1}) (inset shows the CO production selectivity during long-time photocatalysis). Reproduced from ref. 111 with permission from John Wiley & Sons, Copyright © 2019 Wiley-VCH Verlag GmbH & Co. KGaA, Weinheim.

(Figure 10.8(A)(ii)) whereas CdSe-2 (Figure 10.8(A)(iii)) shows severe aggregation due to the high surface energy of the QDs after ligand stripping. HR-TEM analysis (Figure 10.8(A)(iv)) confirmed the well-retained core structure of CdSe nanoparticles as admitted from lattice fringes. By tuning the Cd–Se stoichiometric ratio of CdSe QDs and ligand stripping study authors suggested that more active exposed Cd atoms are crucial to the CO production activity, reaching up to 789 mmol g^{-1} h^{-1}. Overall, the surface environment and stoichiometry are critical to determine the photo reactivity (Figure 10.8(A)(v) and (vi)). Additionally, previous studies have shown that combining MSe QDs with co-catalysts is an effective approach to hamper the unwanted charge-recombination process and, thus, improves the CO_2 photoreduction performance. Especially, ternary heterostructures photocatalysts formed by MSe QDs–M–TiO_2 (QDs = CdSe; M = Pt, Au, Pd) counterparts explored to have outstanding CO_2 photoreduction activity.[112,113] Wang *et al.* demonstrated that CdSe sensitized commercial P25 TiO_2 exhibited high-efficiency photocatalyst for CO_2 photoreduction reaction.[112] The authors inferred that Pt metal incorporation *via* a wet impregnation method can deliberately provide active sites and accelerates the rate of reaction towards CH_4 (48 ppm g^{-1} h^{-1}) as the major product and several secondary products such as CH_3OH (3.3 ppm g^{-1} h^{-1}), H_2 and CO (trace amounts). In another study by Kuehnel *et al.*,[8] a Cd-free photocatalyst system was fabricated with inexpensive ZnSe QDs (photosensitizer) and Ni(cyclam) electrocatalyst, which can function under visible light. With transient absorption spectroscopy, the authors sought to understand the rationale behind the electron transfer from the photosensitizer to immobilized Ni(cyclam) at the ps time scale, explaining CO_2 reduction to CO with a high turnover number TON_{CO} > 280 and apparent quantum efficiency of 3.4%. This study further suggests that surface modification of ZnSe QDs with 2-(dimethylamino) ethanethiol (MEDA) is vital to partially suppress the H_2 production reaction and boost CO production activity.

10.5 Graphene-based QDs

Graphene-based QDs such as carbon QDs, graphene oxide (GO) QDs, rGO QDs, and g-C_3N_4 QDs are a moderately new class of low-cost carbon-based fluorescent materials. The GQDs have been predictable as excessive potential applicants for various research areas in photonics and biochemistry and applications for energy conversion/storage, sensing, and catalysis owing to their unique electrochemical, biological, chemical, and optical properties. However, GQDs should be steadily employed, as the structures are thoroughly associated with their catalytic behaviour, reactivity/selectivity, biocompatibility, chemical, and optical responses.

GQDs substituted metal oxide semiconductors, LDHs, MOFs, and perovskites are showing remarkable efficiency in water splitting and CO_2 reduction. GQDS acts as an electron trapper for initiating the redox reduction and also reduced the recombination rate. Along with that, the graphene atom doped to the semiconductor that bandgap will be reduced thereby enhancing the light absorption capability towards to the region of visible. Hence, there have been widely reported the GQDs attached to metal oxide semiconductors such as Co_3O_4, CuO, ZnO, TiO_2, Fe_2O_3 as well as LDHs to improve the overall efficiency of water splitting and CO_2 reduction. For instance, sulfur and nitrogen co-doped graphene QDs role as a photosensitizer with Ag-doped ZnO rods were reported for electrochemical conversion by Majumder *et al.*[114] They achieved the ~1.1 photoconversion efficiency with a photosensitizer-doped catalyst, which was 3.4 and 2.4 times higher than the ZnO rods and Ag-doped ZnO nanorods, respectively. Therefore, the graphene QDs sensitizer effectively reduced the recombination rate by trapping the electron thereby enhancing the photoconversion efficiency. Along with that, Ding *et al.* reported the graphene QDs sensitized Zn-MOFs for CO_2 reduction under visible light.[115] Therefore, according to this, the graphene QDs sensitized metal oxides, LDHs and MOFs showed their remarkable performance in CO_2 reduction and water splitting.

10.5.1 Water Splitting

2D graphene has a zero-bandgap due to their CB and VB being met at the Dirac point. However, graphene converted into the 0D nanostructure, especially quantum dots are opening up the quantum confinement effect. Therefore, the size-dependent graphene QDs has offered a wide spectrum of applications as mentioned previously.

Yan *et al.* reported the systematic bandgap engineering of graphene QDs and their application in CO_2 reduction and water splitting.[116] In this, they demonstrated various functionalization of GQDs and their resultant bandgap structure tailored GQDs, as shown in Figure 10.9. As of Figure 10.9(a) and (b), conjugation of polyaromatic rings lowers CBM (π^*) in a ring-size-dependent manner, while VBM remains relatively stable, whereby leading to bandgap reduction. Among various developed GQDs, the GQD-BNPTL showed the higher H_2 evolution (130 μmol h^{-1}) and methanol yield rate (0.695 μmol g^{-1} h^{-1}) of CO_2 reduction, these are due to the widest range of absorption and strongest light absorption ability. Bayat *et al.* synthesized vertically aligned rutile TiO_2 nanorods with S and N co-doped GQDs for water splitting. In this work, hydrothermal and spin coating processes were used for the fabrication of TiO_2 nanorods and, GQDs coated TiO_2 nanorods, respectively. Water splitting has been done under a solar simulator with 1 M Na_2SO_4, the N- and S-doped GQDs/TiO_2 nanorod showed a higher current density with chopping conditions.[117] Furthermore, various metal oxides QDs for water splitting and their condition are summarized in Table 10.3.

In summary, the GQDs are showing their remarkable performance in water splitting as a photosensitizer, co-catalyst, and direct photocatalyst materials. The 2D-graphene nanosheet CB and VB are overlapped but 0D GQDs are showing the semiconductor nature with visible absorption nature.

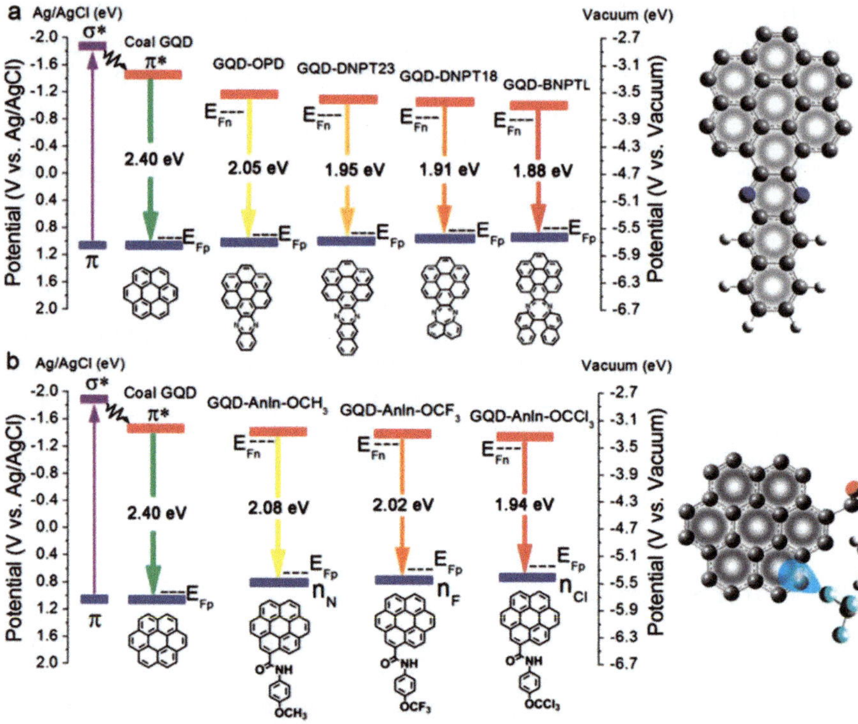

Figure 10.9 Illustration of energy level diagrams and simplified structures of (a) coal GQD, GQD-OPD, GQD-DNPT23, GQD-DNPT18, and GQD-BNPTL and (b) coal GQD, GQD-Anln-OCH3, GQD-Anln-OCF3, and GQD-Anln-OCCl3. The Fermi levels for p-type conductivity (E_{Fp}) and n-type conductivity (E_{Fn}) are indicated in the energy diagram. Reproduced from ref. 116 with permission from American Chemical Society, Copyright 2018.

Therefore, the GQDs involving in multiple works for water splitting as well as functionalization with some other semiconductor metal oxides offering the heterojunction, and Z-scheme electron transfer process. Especially, the non-metal doped such S- and N-doped GQDs are widely reported for water splitting due to the higher surface-active sites and surface area.

10.5.2 CO_2 Photoreduction

The GQDs also showed their remarkable performance in the photoreduction of CO_2 with functionalization of other metal oxide semiconductors, MOFs and LDHs. For instance, Zhipeng *et al.* fabricated N-doped GQDs on Bi_2O_3 nanosheets for high-performance CO_2 reduction. In this, they adopted the pyrolysis reflux and solvothermal process to make N-doped GQDs @Bi_2O_3 sheets.[7] The synthesis process schematic representation and resultant

Table 10.3 Various GQDs for *water* splitting, their reaction conditions and resulting hydrogen and oxygen evolutions.

S. No	Metal oxide QDs based composite	Morphology	Synthesis method	Light source/ reaction solution	Photocurrent density (or) H_2 (or) O_2 evolution	Reference
1	GQDs/TiO_2@ Pt	Cuboid	Hydrothermal	Simulated sunlight	0.535 mA cm^{-2}	118
2	S and N doped GQDs/Ag@ ZnO	Nanorods	Hydrothermal	Xe light/0.1 M Na_2SO_4	0.96 mA cm^{-2}	114
3	N-doped GQDs	QDs	Calcination	Visible light	5.5 μmol	119
4	N-doped GQDs/ ZNO @PEI	Nanowires	Wet impregnation and calcination	300 W Xe light/0.5 M Na_2SO_4	0.6 mA cm^{-2}	120
5	N-doped GQDs $BiVO_4$	Lump	Hydrothermal	300 W Xe light	3.2 mA cm^{-2}	121
6	GQDs/TiO_2	Nanotubes	Anodization	300 W Xe light/0.1 M Na_2S and 0.1 M Na_2SO_3	~59 μmol	122
7	GQDs/TiO_2	Nanorods	Sol–Gel	300 W Xe light/1 M KOH	2.4 mA cm^{-2}	123
8	C QDs/TiO_2	Nanotubes	Hydrothermal	300 W Xe light/0.1 M Na_2SO_4	30.12 μmol h^{-1}	124
9	C QDs/TiO_2	Nanorods	Hydrothermal	300 W Xe light	2.76 mA cm^{-2}	125
10	C QQs/TiO_2	Nanobelts	Hydrothermal	300 W Xe lamp/ methanol	7.42 mmol h^{-1} g^{-1}	126

morphology represented using TEM and HR-TEM are shown in Figure 10.10(a)–(d). The required elements such as Bi, O, C, and N elements were confirmed from the elemental mapping (Figure 10.10(e)). The high-resolution N1s spectra confirmed the pydridinic N, pyrrolic N and N–H groups (Figure 10.10(f) and (g)). The resultant CO_2 absorption, DFT calculation, free energy diagram, and adsorption energies are shown in Figure 10.10(h)–(k). Form these, it was confirmed that the Bi_2O_3-NGQDs showed the highest CO_2 absorption (0.192 mmol g^{-1}) when compared with the Bi_2O_3 (0.118 mmol g^{-1}), which is due to trapping of GQDs electrons from Bi_2O_3. Along with that, the DFT calculations also confirmed the enhancement of adsorption energy

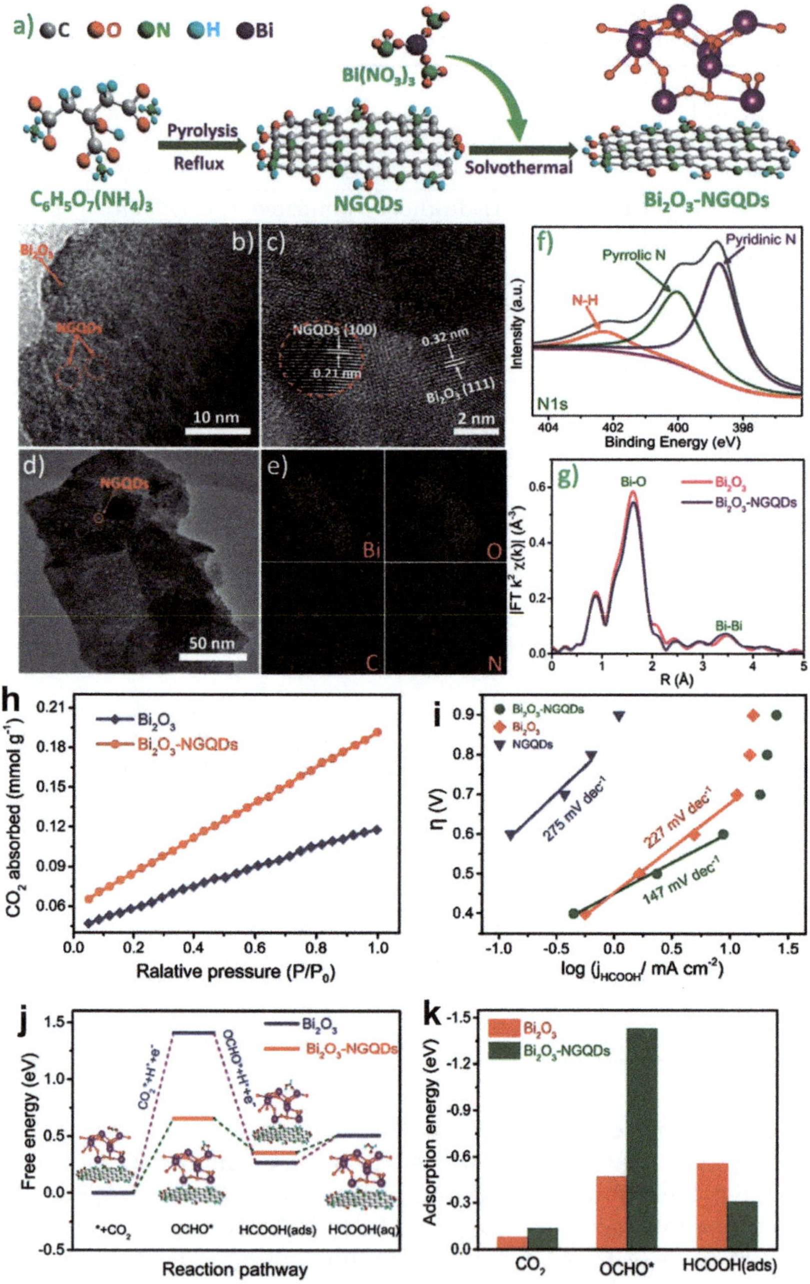

Figure 10.10 (a) Synthesis schematic process of Bi_2O_3-NGQDs, (b) and (d) TEM, (c) HR-TEM image of Bi_2O_3-NGQDS, (d) corresponding elements of Bi_2O_3-NGQDs, (e) N 1s spectrum of Bi_2O_3-NGQDs, (f) C1s high XPS spectrum, (g) the EXAFS spectra, (h) CO_2 adsorption with Bi_2O_3 and Bi_2O_3-NGQDs, (i) Tafel plots of Bi_2O_3, NGQDS, and Bi_2O_3-NGQDs, (j) calculated free energy diagram, and (k) adsorption energy from the DFT calculation. Reproduced from ref. 7 with permission from John Wiley & Sons, Copyright © 2018 Wiley-VCH Verlag GmbH & Co. KGaA, Weinheim.

of CO_2 and NGQDs react intermediates.[7] Furthermore, Dine *et al.* reported the GQDs sensitized Zn-MOFs for CO_2 reduction under visible light irradiation.[11,115] The optimal catalyst showed scalable CO_2 conversion rates of 20.9 and 3.7 μmol g^{-1} h^{-1} for CH_4 and CO, respectively. These enchantments are due to the formation of strong interactions between GQDs and Zn-MOFs and rich active sites on the MOFs defective surface.

10.6 Prospects and Outlook

Comprehensive details of various types of QDs-based photocatalysts are discussed for photocatalytic water splitting and CO_2 reduction.

- Metal oxide based QDs: CO_3O_4, CuO, TiO_2@SnO_2 QDs, and CdSe QDs sensitized ZnO nanowire photoanodes have been thoroughly studied by many researchers for energy (H_2 generation) and environmental remediation (CO_2 reduction) applications. Various synthesis techniques have been explored.
- Metal sulfide based QDs have also been studied and we have summarized the important combination of QDs and their morphological-dependent properties. Top-down approaches that rely on lithography or milling of particulate materials to quantum sizes have been employed to produce metal sulfide QDs. CdS and ZnS QDs by employing modified gel crystal growth method in which hydrogel (HR) serves as a substrate to effectively confine the CdS and ZnS nanoparticles. Mo-doped Ni_3S_2, 0D/2D/2D NbS_2 QDs, MoS_2, alloyed-Ag–In–Zn–S, and Zn-doped $AgIn_5S_8$ have been implemented for the H_2 production reaction *via* water splitting.
- Metal selenides (MSes) refer to a unique class of semiconductor nanomaterials formed by the combination of a metal cation (M^{2+}) and a selenium anion (Se^{2-}). CdSe QDs, CdSe/CdS/ZnS QDs, and CdSeTe QDs onto NiO films prepared through hydrothermal methods have been studied and explored with respect to their unique properties associated with photocatalytic properties towards H_2 production and CO_2 reduction.
- Graphene-based QDs such as carbon QDs, graphene oxide (GO) QDs, rGO QDs, and g-C_3N_4 QDs are a moderately new class of low-cost carbon-based fluorescent materials. These photocatalysts have been studied and reviewed with respects to their better combinations of composites associated with graphene QDs.
- Even though we have highlighted these special classes of photocatalysts in this chapter, there is still plenty of room available to further improve the efficiencies of these photocatalysts by advancing (and using) the latest technologies and innovative synthesis techniques. Scientists should continue to explore new synthesis methods and new combinations of materials for the application of water splitting and CO_2 reduction.

References

1. X. Li, J. Yu, J. Low, Y. Fang, J. Xiao and X. Chen, *J. Mater. Chem. A*, 2015, **3**, 2485–2534.
2. N. Lakshmana Reddy, V. Navakoteswara Rao, M. Mamatha Kumari, R. R. Kakarla, P. Ravi, M. Sathish, M. Karthik, S. Muthukonda Venkatakrishnan and Inamuddin, *Environ. Chem. Lett.*, 2018, **16**, 765–796.
3. R. Shwetharani, H. R. Chandan, M. Sakar, G. R. Balakrishna, K. R. Reddy and A. V. Raghu, *Int. J. Hydrogen Energy*, 2020, **45**, 18289–18308.
4. I. Sreedhar, U. Upadhyay, P. Roy, S. M. Thodur and C. M. Patel, *J. Cleaner Prod.*, 2021, **284**, 124712.
5. S. K. Saraswat, D. D. Rodene and R. B. Gupta, *Renewable Sustainable Energy Rev.*, 2018, **89**, 228–248.
6. N. L. Reddy, V. N. Rao, M. Vijayakumar, R. Santhosh, S. Anandan, M. Karthik, M. V. Shankar, K. R. Reddy, N. P. Shetti, M. N. Nadagouda and T. M. Aminabhavi, *Int. J. Hydrogen Energy*, 2019, **44**, 10453–10472.
7. Z. Chen, K. Mou, X. Wang and L. Liu, *Angew. Chem. Int. Ed.*, 2018, **57**, 12790–12794.
8. M. F. Kuehnel, C. D. Sahm, G. Neri, J. R. Lee, K. L. Orchard, A. J. Cowan and E. Reisner, *Chem. Sci.*, 2018, **9**, 2501–2509.
9. V. N. Rao, N. L. Reddy, M. M. Kumari, K. K. Cheralathan, P. Ravi, M. Sathish, B. Neppolian, K. R. Reddy, N. P. Shetti, P. Prathap, T. M. Aminabhavi and M. V. Shankar, *J. Environ. Manage.*, 2019, **248**, 109246.
10. H.-L. Wu, X.-B. Li, C.-H. Tung and L.-Z. Wu, *Adv. Mater.*, 2019, **31**, 1900709.
11. W. Dai, W. Xiong, J. Yu, S. Zhang, B. Li, L. Yang, T. Wang, X. Luo, J. Zou and S. Luo, *ACS Appl. Mater. Interfaces*, 2020, **12**, 25861–25874.
12. J. Yu, J. Zhang and M. Jaroniec, *Green Chem.*, 2010, **12**, 1611–1614.
13. T. Wang, X. Liu, C. Ma, Y. Liu, H. Dong, W. Ma, Z. Liu, M. Wei, C. Li and Y. Yan, *J. Taiwan Inst. Chem. Eng.*, 2018, **93**, 298–305.
14. N. Ramesh Reddy, U. Bhargav, M. Mamatha Kumari, K. K. Cheralathan and M. Sakar, *Int. J. Hydrogen Energy*, 2020, **45**, 7584–7615.
15. J. Liu, J. Ke, Y. Li, B. Liu, L. Wang, H. Xiao and S. Wang, *Appl. Catal., B*, 2018, **236**, 396–403.
16. Y. Wang, M. Zhou, Y. He, Z. Zhou and Z. Sun, *J. Alloys Compd.*, 2020, **813**, 152184.
17. S. Martha, P. Chandra Sahoo and K. M. Parida, *RSC Adv.*, 2015, **5**, 61535–61553.
18. A. Zaleska, *Recent Pat. Eng.*, 2008, **2**, 157–164.
19. L. Zhou, H. Zhang, H. Sun, S. Liu, M. O. Tade, S. Wang and W. Jin, *Catal. Sci. Technol.*, 2016, **6**, 7002–7023.
20. R. D. Tentu and S. Basu, *Curr. Opin. Electrochem.*, 2017, **5**, 56–62.
21. M. Arif, G. Yasin, M. Shakeel, M. A. Mushtaq, W. Ye, X. Fang, S. Ji and D. Yan, *Mater. Chem. Front.*, 2019, **3**, 520–531.
22. R. Sasikala, A. R. Shirole, V. Sudarsan, Jagannath, C. Sudakar, R. Naik, R. Rao and S. R. Bharadwaj, *Appl. Catal. A*, 2010, **377**, 47–54.

23. K. Lv, W. Suo, M. Shao, Y. Zhu, X. Wang, J. Feng, M. Fang and Y. Zhu, *Nano Energy*, 2019, **63**, 103834.
24. N. Lakshmana Reddy, G. Krishna Reddy, K. Mahaboob Basha, P. Krishna Mounika and M. V. Shankar, *Mater. Today: Proc.*, 2016, **3**, 1351–1358.
25. G. Xie, K. Zhang, B. Guo, Q. Liu, L. Fang and J. R. Gong, *Adv. Mater.*, 2013, **25**, 3820–3839.
26. S. G. Babu, R. Vinoth, D. Praveen Kumar, M. V Shankar, H.-L. Chou, K. Vinodgopal and B. Neppolian, *Nanoscale*, 2015, **7**, 7849–7857.
27. N. Ramesh Reddy, M. Mamatha Kumari, M. V. Shankar, K. Raghava Reddy, S. Woo Joo and T. M. Aminabhavi, *J. Environ. Manage.*, 2021, **277**, 111433.
28. H. Bai, Z. Liu and D. D. Sun, *Int. J. Hydrogen Energy*, 2012, **37**, 13998–14008.
29. T. Wang, X. Liu, D. Han, C. Ma, Y. Liu, P. Huo and Y. Yan, *J. Alloys Compd.*, 2019, **785**, 460–467.
30. B. Sun, Y. Chen, L. Tao, H. Zhao, G. Zhou, Y. Xia, H. Wang and Y. Zhao, *ACS Appl. Mater. Interfaces*, 2019, **11**, 2071–2081.
31. S. Sun, W. Wang, D. Li, L. Zhang and D. Jiang, *ACS Catal.*, 2014, **4**, 3498–3503.
32. G. Wang, X. Yang, F. Qian, J. Z. Zhang and Y. Li, *Nano Lett.*, 2010, **10**, 1088–1092.
33. M. Xie, Z. Zhang, W. Han, X. Cheng, X. Li and E. Xie, *J. Mater. Chem. A*, 2017, **5**, 10338–10346.
34. R. Saito, Y. Miseki and K. Sayama, *Chem. Commun.*, 2012, **48**, 3833–3835.
35. K. Chen, L. Ma, J. H. Wang, Z. Q. Cheng, D. J. Yang, Y. Y. Li, S. J. Ding, L. Zhou and Q. Q. Wang, *RSC Adv.*, 2017, 7, 26097–26103.
36. Z. Wang, X. Li, C. K. Tan, C. Qian, A. C. Grimsdale and A. I. Y. Tok, *Appl. Surf. Sci.*, 2019, **470**, 800–806.
37. A. Ikram, S. Sahai, S. Rai, S. Dass, R. Shrivastav and V. R. Satsangi, *Int. J. Hydrogen Energy*, 2015, **40**, 5583–5592.
38. X. Xu, G. Zhou, B. Feng, Z. Bao and J. Hu, *Mater. Lett.*, 2016, **165**, 196–199.
39. Z. Jiao, X. Guan, M. Wang, Q. Wang, B. Xu, Y. Bi and X. S. Zhao, *Chem. Eng. J.*, 2019, **356**, 781–790.
40. B. Liu, X. Jiang, X. Jiang, Y. Ma, Z. Zhang and W. Han, *Inorg. Chem.*, 2021, **60**, 3057–3064.
41. W. Li, C. Zhuang, Y. Li, C. Gao, W. Jiang, Z. Sun and K. Qi, *Ceram. Int.*, 2021, **47**, 21769–21776.
42. N. Zhang, J. Shi, S. S. Mao and L. Guo, *Chem. Commun.*, 2014, **50**, 2002–2004.
43. X. Wu, J. Zhao, S. Guo, L. Wang, W. Shi, H. Huang, Y. Liu and Z. Kang, *Nanoscale*, 2016, **8**, 17314–17321.
44. D. Wang, L. Huang, Z. Guo, S. Jin, C. Liu, W. Wang and W. Yuan, *ACS Appl. Nano Mater.*, 2018, **1**, 4594–4601.
45. H. Zhang, F. Liu, Z. Mou, X. Liu, J. Sun and W. Lei, *Chem. Commun.*, 2016, **52**, 13020–13023.
46. X. Yu, Z. Li, J. Liu and P. Hu, *Appl. Catal., B*, 2017, **205**, 271–280.

47. Q. Liu, J. Cao, Y. Ji, Y. Liu, C. Liu, G. Che, D. Wang, J. Cao, W. Li and X. Liu, *Appl. Surf. Sci.*, 2021, **570**, 151085.
48. C. Ray and T. Pal, *J. Mater. Chem. A*, 2017, **5**, 9465–9487.
49. N. Sagara, S. Kamimura, T. Tsubota and T. Ohno, *Appl. Catal. B*, 2016, **192**, 193–198.
50. B. Tahir, M. Tahir and N. A. S. Amin, *Appl. Surf. Sci.*, 2017, **419**, 875–885.
51. C. Prasad, H. Tang, Q. Q. Liu, S. Zulfiqar, S. Shah and I. Bahadur, *J. Mol. Liq.*, 2019, **289**, 111114.
52. L. Chen, M. Zhang, J. Yang, Y. Li, Y. Sivalingam, Q. Shi, M. Xie and W. Han, *Mater. Sci. Semicond. Process.*, 2019, **102**, 104578.
53. Z. Xie, Y. Xu, D. Li, L. Chen, S. Meng, D. Jiang and M. Chen, *J. Alloys Compd.*, 2021, **858**, 157668.
54. S. Shen and Q. Wang, *Chem. Mater.*, 2013, **25**, 1166–1178.
55. D. W. Su, J. Ran, Z. W. Zhuang, C. Chen, S. Z. Qiao, Y. D. Li and G. X. Wang, *Sci. Adv.*, 2021, **6**, eaaz8447.
56. C. Justin Raj, B. C. Kim, W.-J. Cho, W.-G. Lee, Y. Seo and K.-H. Yu, *J. Alloys Compd.*, 2014, **586**, 191–196.
57. C.-M. Fan, M. D. Regulacio, C. Ye, S. H. Lim, Y. Zheng, Q.-H. Xu, A.-W. Xu and M.-Y. Han, *Chem. Commun.*, 2014, **50**, 7128–7131.
58. F. L. Faita, K. Ersching, C. M. Poffo, L. C. Benetti, D. M. Trichês, S. M. Souza, A. D. C. Viegas and J. C. de Lima, *J. Alloys Compd.*, 2014, **590**, 176–183.
59. L. Carlini and J. L. Nadeau, *Chem. Commun.*, 2013, **49**, 1714–1716.
60. K. Yang, L. Zhu, L. Nie, X. Sun, L. Cheng, C. Wu, G. Niu, X. Chen and Z. Liu, *Theranostics*, 2014, **4**, 134–141.
61. M. Jouyandeh, S. S. Mousavi Khadem, S. Habibzadeh, A. Esmaeili, O. Abida, V. Vatanpour, N. Rabiee, M. Bagherzadeh, S. Iravani, M. Reza Saeb and R. S. Varma, *Green Chem.*, 2021, **23**, 4931–4954.
62. J. Yang, T. Ling, W.-T. Wu, H. Liu, M.-R. Gao, C. Ling, L. Li and X.-W. Du, *Nat. Commun.*, 2013, **4**, 1695.
63. C. D. Dieleman, W. Ding, L. Wu, N. Thakur, I. Bespalov, B. Daiber, Y. Ekinci, S. Castellanos and B. Ehrler, *Nanoscale*, 2020, **12**, 11306–11316.
64. M. Yarema, O. Yarema, W. M. M. Lin, S. Volk, N. Yazdani, D. Bozyigit and V. Wood, *Chem. Mater.*, 2017, **29**, 796–803.
65. L. Lin, Y. Xu, S. Zhang, I. M. Ross, A. C. M. Ong and D. A. Allwood, *ACS Nano*, 2013, **7**, 8214–8223.
66. J. Zhang, Q. Yang, H. Cao, C. I. Ratcliffe, D. Kingston, Q. Y. Chen, J. Ouyang, X. Wu, D. M. Leek, F. S. Riehle and K. Yu, *Chem. Mater.*, 2016, **28**, 618–625.
67. L. Paramanik, K. H. Reddy and K. M. Parida, *Nanoscale*, 2019, **11**, 22328–22342.
68. D. Prusty, L. Paramanik and K. Parida, *Energy Fuels*, 2021, **35**, 4670–4686.
69. T. Di, Q. Xu, W. Ho, H. Tang, Q. Xiang and J. Yu, *ChemCatChem*, 2019, **11**, 1394–1411.
70. W.-T. Sun, Y. Yu, H.-Y. Pan, X.-F. Gao, Q. Chen and L.-M. Peng, *J. Am. Chem. Soc.*, 2008, **130**, 1124–1125.

71. D. Ma, J. Shi, Y. Zou, Z. Fan, X. Ji and C. Niu, *ACS Appl. Mater. Interfaces*, 2017, **9**(30), 25377–25386.
72. Z. Jiang, X. Zhang, G. Yang, Z. Yuan, X. Ji, F. Kong, B. Huang, D. D. Dionysiou and J. Chen, *Chem. Eng. J.*, 2019, **373**, 814–820.
73. J.-W. Shi, D. Sun, Y. Zou, D. Ma, C. He, X. Ji and C. Niu, *Chem. Eng. J.*, 2019, **364**, 11–19.
74. L. Lei, D. Huang, S. Chen, C. Zhang, Y. Chen and R. Deng, *Coord. Chem. Rev.*, 2021, **429**, 213715.
75. M. Wang, X. Shang, X. Yu, R. Liu, Y. Xie, H. Zhao, H. Cao and G. Zhang, *Phys. Chem. Chem. Phys.*, 2014, **16**, 26016–26023.
76. Q. Xie, M. Wang, Y. Xu, X. Li, X. Zhou, L. Hong, L. Jiang and W. Lin, *J. Energy Chem.*, 2021, **61**, 210–218.
77. Y.-L. Lee, C.-F. Chi and S.-Y. Liau, *Chem. Mater.*, 2010, **22**, 922–927.
78. W. Yu, J. Yin, Y. Li, B. Lai, T. Jiang, Y. Li, H. Liu, J. Liu, C. Zhao, S. C. Singh, J. Chen, B. Lin, H. Idriss and C. Guo, *ACS Appl. Energy Mater.*, 2019, **2**, 2751–2759.
79. D. Zhang, W. Dong, Y. Liu, X. Gu, T. Yang, Q. Hong, D. Li, D. Zhang, H. Zhou, H. Huang, B. Mao, Z. Kang and W. Shi, *ACS Appl. Mater. Interfaces*, 2021, **13**, 42125–42137.
80. H. Chen, Z. Yu, R. Jiang, J. Huang, Y. Hou, Y. Zhang, H. Zhu, B. Wang, M. Wang and W. Tang, *Nanoscale*, 2021, **13**, 6644–6653.
81. B. Lin, Z. Chen, P. Song, H. Liu, L. Kang, J. Di, X. Luo, L. Chen, C. Xue, B. Ma, G. Yang, J. Tang, J. Zhou, Z. Liu and F. Liu, *Small*, 2020, **16**, 2003302.
82. G. Zhan, J. Zhang, Y. Wang, C. Yu, J. Wu, J. Cui and X. Shu, *J. Colloid Interface Sci.*, 2020, **566**, 411–418.
83. Y. Yang, B. Mao, G. Gong, D. Li, Y. Liu, W. Cao, L. Xing, J. Zeng, W. Shi and S. Yuan, *Int. J. Hydrogen Energy*, 2019, **44**, 15882–15891.
84. J. Wang, S. Lin, N. Tian, T. Ma, Y. Zhang and H. Huang, *Adv. Funct. Mater.*, 2021, **31**, 2008008.
85. Y. Yuan, R. Guo, Z. Zhang, L. Hong, X. Ji, Z. Lin and W. Pan, *Energy Fuels*, 2021, **35**, 13291–13303.
86. J. Hu, M. Yang, X. Ke, S. Yang, K. Wang, H. Huang, W. Wang, D. Luo, Z. Zheng, L. Huang, P. Xiao, C. Tu, Y. Min, N. Huo and M. Zhang, *J. Power Sources*, 2021, **481**, 228838.
87. B. Gates, Y. Wu, Y. Yin, P. Yang and Y. Xia, *J. Am. Chem. Soc.*, 2001, **123**, 11500–11501.
88. S. H. Park, J. Y. Choi, Y. H. Lee, J. T. Park and H. Song, *Chem. – Asian J.*, 2015, **10**, 1452–1456.
89. X. Ren, L. Pang, Y. Zhang, X. Ren, H. Fan and S. (Frank) Liu, *J. Mater. Chem. A*, 2015, **3**, 10693–10697.
90. H. Li and C. Han, *Chem. Mater.*, 2008, **20**, 6053–6059.
91. D. V Freitas, J. M. M. Dias, S. G. B. Passos, G. C. S. de Souza, E. T. Neto and M. Navarro, *Green Chem.*, 2014, **16**, 3247–3254.
92. Q.-Y. Luo, Y. Lin, Y. Li, L.-H. Xiong, R. Cui, Z.-X. Xie and D.-W. Pang, *Small*, 2014, **10**, 699–704.
93. D. Yao, Y. Liu, J. Li and H. Zhang, *Chin. Chem. Lett.*, 2019, **30**, 277–284.

94. H. M. Choi, I. A. Ji and J. H. Bang, *ACS Appl. Mater. Interfaces*, 2014, **6**, 2335–2343.
95. D. Thomas, H. Lee, K. C. Santiago, M. Pelzer, A. Kuti, E. Jenrette and M. Bahoura, *J. Nanomater.*, 2020, **2020**, 5056875.
96. G. S. Tito, A. S. Abolanle, A. T. Kuvarega, A. O. Idris, B. B. Mamba and U. Feleni, *Electroanalysis*, 2020, **32**, 2603–2614.
97. A. Sobhani and M. Salavati-Niasari, *Adv. Colloid Interface Sci.*, 2021, **287**, 102321.
98. C. M. Hessel, V. P. Pattani, M. Rasch, M. G. Panthani, B. Koo, J. W. Tunnell and B. A. Korgel, *Nano Lett.*, 2011, **11**, 2560–2566.
99. J. Li, W. Liu, C. Chen, X. Zhao, Z. Qiu, H. Xu, F. Sheng, Q. Hu, Y. Zheng, M. Lin, S. J. Pennycook, C. Su and J. Lu, *J. Mater. Chem. A*, 2019, **7**, 23958–23963.
100. X. Xia, L. Wang, N. Sui, V. L. Colvin and W. W. Yu, *Nanoscale*, 2020, **12**, 12249–12262.
101. S. Yu, Z.-J. Li, X.-B. Fan, J.-X. Li, F. Zhan, X.-B. Li, Y. Tao, C.-H. Tung and L.-Z. Wu, *ChemSusChem*, 2015, **8**, 642–649.
102. Y. Zhong, W. Chen, S. Yu, Z. Xie, S. Wei and Y. Zhou, *ACS Omega*, 2018, **3**, 17762–17769.
103. Y. Liu, F. Dai, R. Zhao, X. Huai, J. Han and L. Wang, *J. Mater. Sci.*, 2019, **54**, 8571–8580.
104. X. Su, Y. Chen, L. Ren, Y. He, X. Yin, Y. Liu and W. Yang, *ACS Sustainable Chem. Eng.*, 2019, **7**, 11166–11174.
105. H. Lv, C. Wang, G. Li, R. Burke, T. D. Krauss, Y. Gao and R. Eisenberg, *Proc. Natl. Acad. Sci. U. S. A.*, 2017, **114**, 11297–11302.
106. S. Qu, J. Huang, J. Yu, G. Chen, W. Hu, M. Yin, R. Zhang, S. Chu and C. Li, *ACS Appl. Mater. Interfaces*, 2017, **9**, 29660–29668.
107. G. Ou, P. Fan, X. Ke, Y. Xu, K. Huang, H. Wei, W. Yu, H. Zhang, M. Zhong, H. Wu and Y. Li, *Nano Res.*, 2018, **11**, 751–761.
108. D. Gopalakrishnan, D. Damien, B. Li, H. Gullappalli, V. K. Pillai, P. M. Ajayan and M. M. Shaijumon, *Chem. Commun.*, 2015, **51**, 6293–6296.
109. B. Guo, K. Yu, H. Li, R. Qi, Y. Zhang, H. Song, Z. Tang, Z. Zhu and M. Chen, *ACS Appl. Mater. Interfaces*, 2017, **9**, 3653–3660.
110. P. Xu, J. Zhang, Z. Ye, Y. Liu, T. Cen and D. Yuan, *Appl. Surf. Sci.*, 2019, **494**, 749–755.
111. W. Xia, J. Wu, J. C. Hu, S. Sun, M. D. Li, H. Liu, M. Lan and F. Wang, *ChemSusChem*, 2019, **12**, 4617–4622.
112. C. Wang, R. L. Thompson, J. Baltrus and C. Matranga, *J. Phys. Chem. Lett.*, 2010, **1**, 48–53.
113. A. Sarkar, E. Gracia-Espino, T. Wågberg, A. Shchukarev, M. Mohl, A.-R. Rautio, O. Pitkänen, T. Sharifi, K. Kordas and J.-P. Mikkola, *Nano Res.*, 2016, **9**, 1956–1968.
114. T. Majumder, S. Dhar, K. Debnath and S. P. Mondal, *Mater. Res. Bull.*, 2017, **93**, 214–222.
115. D. Wei, W. Tang, Y. Gan and X. Xu, *Catal. Sci. Technol.*, 2020, **10**, 5666–5676.

116. Y. Yan, J. Chen, N. Li, J. Tian, K. Li, J. Jiang, J. Liu, Q. Tian and P. Chen, *ACS Nano*, 2018, **12**, 3523–3532.
117. A. Bayat and E. Saievar-Iranizad, *J. Alloys Compd.*, 2018, **755**, 192–198.
118. W. Sangkhun, S. Wanwong, J. Wootthikanokkhan, K. Sinthiptharakoon and P. Kumnorkaew, *ChemistrySelect*, 2021, **6**, 8664–8671.
119. T. F. Yeh, C. Y. Teng, S. J. Chen and H. Teng, *Adv. Mater.*, 2014, **26**, 3297–3303.
120. Z. Zeng, F. X. Xiao, X. Gui, R. Wang, B. Liu and T. T. Yang Tan, *J. Mater. Chem. A*, 2016, **4**, 16383–16393.
121. C. Wu, R. Chen, C. Ma, R. Cheng, X. Gao, T. Wang, Y. Liu, P. Huo and Y. Yan, *Ceram. Int.*, 2019, **45**, 2088–2096.
122. Y. Yu, J. Ren and M. Meng, *Int. J. Hydrogen Energy*, 2013, **38**, 12266–12272.
123. Z. Xu, M. Yin, J. Sun, G. Ding, L. Lu, P. Chang, X. Chen and D. Li, *Nanotechnology*, 2016, **27**, 115401.
124. Q. Wang, J. Cai, G. V. Biesold-McGee, J. Huang, Y. H. Ng, H. Sun, J. Wang, Y. Lai and Z. Lin, *Nano Energy*, 2020, **78**, 105313.
125. T. Zhou, S. Chen, L. Li, J. Wang, Y. Zhang, J. Li, J. Bai, L. Xia, Q. Xu, M. Rahim and B. Zhou, *Appl. Catal. B*, 2020, **269**, 118776.
126. J. Tian, Y. Leng, Z. Zhao, Y. Xia, Y. Sang, P. Hao, J. Zhan, M. Li and H. Liu, *Nano Energy*, 2015, **11**, 419–427.

CHAPTER 11

Computational Approaches to Materials Design for Enhanced Photocatalytic Activity

TINGBO ZHANG[†a], ZHAOBO ZHOU[†a], QIAN CHEN[a] AND JINLAN WANG*[a]

[a]School of Physics, Southeast University, Nanjing 211189, China
*E-mail: jlwang@seu.edu.cn

11.1 Introduction

With the increase of environmental pollution and global warming, the concept of "carbon neutrality" has attracted widespread attention.[1,2] More precisely, the carbon dioxide (CO_2) emission and absorption within a certain period to achieve the "break-even" point. To the best of our knowledge, photocatalysis, which can reduce CO_2 emissions (generating hydrogen or NH_3) and fixing CO_2 under sunlight, is one of the most important ways to reach "carbon neutrality".[3–5] Hydrogen energy and NH_3 can be obtained by photocatalytic water splitting and nitrogen reduction reaction (NRR). Specifically, under the irradiation of sunlight, H_2O undergoes a redox reaction on the surface of photocatalysts to generate H_2 and O_2. Furthermore, using solar energy as the driving force to synthesize N_2 and H_2O into NH_3. These green and environmentally friendly processes can not only convert solar energy into

[†]These authors contributed equally.

Sustainable Energy No. 1
Recent Developments in Functional Materials for Artificial Photosynthesis
Edited by Srabanti Ghosh and Qian Wang

Published by the Royal Society of Chemistry, www.rsc.org

chemical energy to achieve energy storage,[6–10] but also avoid the participation of C-based reactants, which effectively reduces CO_2 emissions.[11–15] Converting CO_2 in the air into single-carbon or multi-carbon products through CO_2 reduction reaction (CO_2RR), such as CH_4, CH_3OH or CH_3CH_2OH, *etc.*, can effectively achieve the CO_2 fixation.[16–20]

In general, water splitting, NRR, and CO_2RR involve three key steps (Figure 11.1): (1) light absorption described by light adsorption coefficient (η_a); (2) separation and recombination of photogenerated carriers determined by the charge transfer time from photocatalysts to reactants (η_c), and (3) redox reaction at the surface of photocatalysts evaluated by energy barriers that need to be overcome in redox reaction (η_m).[21–25] η_a and η_c are mainly related to the bandgap of photocatalysts, and η_m is related to the Gibbs free energy (ΔG) between reactants and photocatalysts. Therefore, the photocatalytic efficiency η can be evaluated as follows: $\eta = \eta_a \times \eta_c \times \eta_m$.[26] In other words, to improve the photocatalytic efficiency, expanding the light absorption range, weakening photogenerated carrier recombination inside photocatalysts, and accelerating the transfer of carriers from the photocatalyst surface to reactant are inevitable. In general, introducing defects, crystal face engineering, atom doping, and building heterostructures are common

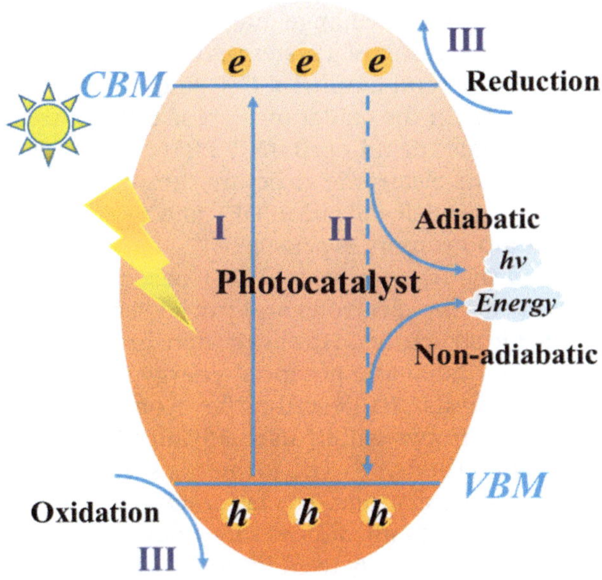

Figure 11.1 Three processes involved in photocatalysis: (I) electrons are excited from valence band (VB) to conduction band (CB) under illumination, which is mainly related to light absorption; (II) adiabatic and non-adiabatic recombination of photogenerated carriers, which is evaluated by photogenerated carrier lifetime; and (III) electrons participate in the redox reaction, which is related to active sites of photocatalysts.

engineering strategies.[27–33] However, the traditional empirical experimental method cannot rapidly obtain high-efficiency catalysts and a large number of materials may be wasted during the trial process. Moreover, it is difficult to reveal the mechanism for materials to have high-efficiency photocatalytic performance due to the complexity of experimental conditions.

With the development of computational material science, theoretical calculations have attracted wide attention from scientists in predicting high-performance photocatalysts due to the advantages of high efficiency and accuracy. Density functional theory (DFT) based on the Kohn–Sham equation is widely used,[34,35] which needs to select wisely the exchange-correlation functional, local density approximation (LDA) and generalized gradient approximation (GGA).[36,37] LDA proposes that the electron density is related to the spatial point, which cannot accurately describe the nature of transition metals with d orbital or f orbital electrons. GGA considers the impact of the uneven distribution of electrons on the exchange interaction potential to correct the total energy and energy barrier. Nevertheless, the electron self-interaction cannot be eliminated whether for LDA or GGA, which leads to inaccurate descriptions of the electronic structure. In especial, GGA seriously underestimates the bandgap of semiconductors (30–100%). To solve this dilemma, more accurate approaches including the DFT+U method or hybrid functionals (HF) have been developed.

DFT+U corrects the electron exchange interaction of the d orbital and the f orbital.[38–40] The calculated result of a system can be comparable to the experimental data by adjusting semi-empirical U (Coulomb) and J (exchange) values. However, the change of the oxidation state is difficult to describe using DFT+U. Other functionals including B3LYP, PBE0, and HSE are employed to accurately describe the electronic structure through considering the different proportions of over-delocalization DFT and under-delocalization HF, which makes computations more expensive.[41–43]

It should be noted that the traditional DFT based method on adiabatic approximation (considering electrons and nuclei separately) can only calculate static electronic structures. However, there are non-adiabatic transitions between different adiabatic potential energy surfaces in the excited state during photocatalysis. To describe the dynamics behavior of electrons and holes, there are two widely used advanced methods: mean-field and surface hopping methods.[44,45] The former is suitable for systems with weak electron–nucleus correlations. The latter not only considers electron–nucleus correlations, but also introduces thermodynamic equilibrium with long time limits, which are widely used by codes of non-adiabatic molecular dynamics (NAMD), such as Hefei-NAMD and PYXAID code. Therefore, we can use the traditional DFT to calculate band edge positions of photocatalysts to determine the light absorption, and calculate the reaction barrier to evaluate the difficulty of photocatalytic reaction. We use NAMD to assess the lifetime of electrons and holes to show the transfer and separation of photocarriers.

In this chapter, we will focus primarily on theoretical advances in photocatalysis in terms of reaction mechanism and design of high-efficiency photocatalysts. We will mainly introduce the reaction principle of water splitting, NRR, and CO_2RR in the second section. The third section describes theoretical studies related to energy band and electric field of DFT calculations, excited state dynamics, and catalytic reaction process. The final section is the conclusion and perspective for photocatalysis theory research.

11.2 The Reaction Principle of Water Splitting, NRR, and CO_2RR

Photocatalytic water splitting can perfectly convert solar energy into hydrogen energy, which has aroused widespread interest of researchers. The complete reaction of photocatalytic water splitting is $2H_2O \rightarrow 2H_2 + O_2$, which requires photocatalysts to provide an internal driving force to promote the reaction. Specifically, the bandgap of the photocatalyst is greater than 1.23 eV to meet the thermodynamic conditions for water splitting.[46,47] This means that near-infrared light, visible light, and ultraviolet light below 1008 nm can be used for water splitting. In addition, the recombination of photogenerated electrons and holes is also one of key issues restricting photocatalytic water splitting. As discussed above, when light irradiates on the semiconductor, electrons in the valence band (VB) are excited to the conduction band (CB), holes left in the VB are rapidly transferred to the valence band maximum (VBM), and excited electrons in the CB are rapidly transferred to the conduction band minimum (CBM). Electrons in the CBM and holes in the VBM may rapidly recombine, reducing the electron transfer from photocatalysts to reactants (*i.e.*, weakening the $H^+ + e^-$ process). Therefore, improving the lifetime of photogenerated carriers is of great significance to increase the photocatalytic efficiency. In this regard, nonadiabatic molecular dynamics (NAMD) can quantify the lifetime and transfer time of photogenerated carriers, which has become an important theoretical approach to design efficient photocatalysis.

The complete reaction of NRR is $N_2 + 3H_2O \rightarrow 2NH_3 + 3/2O_2$. During the process, sunlight does not provide enough energy to dissociate the N≡N triple bond.[48] Therefore, the hydrogenation reaction generally occurs first and then the NH_3 is released. There are two possible processes for hydrogenation reaction: the distal and alternative pathway.[49] The distal pathway means that the end N of the N_2 molecule adsorbs two H first, and then adsorbs one H to generate NH_3. The remaining N in the N_2 molecule is then gradually hydrogenated to generate NH_3. The alternative pathway means that two N atoms in the N_2 molecule adsorb one H to form NH*, respectively. Two NH* are added with H twice to form two NH_3 simultaneously. Due to the strong N≡N triple bond interaction, the efficiency of NRR has always been restricted. In addition, HER is a competitive reaction in NRR process. Therefore, it is necessary to find photocatalysts with easier activation of N≡N triple bonds, and preferential binding of N_2 rather than H^+ through theoretical calculations.

When compared with water splitting and NRR, the process of CO_2RR is much more complicated. According to different reaction conditions and photocatalysts, products are also different. CO, CH_4, CH_3OH, CH_3CH_2OH, and C_2H_4 are commonly reported during CO_2RR.[50–54] The reaction processes are as follows. The process of CO_2 reduction to CO: $CO_2 + 2H^+ + 2e^- \rightarrow CO + H_2O$. The process of CO_2 reduction to CH_4: $CO_2 + 8H^+ + 8e^- \rightarrow CH_4 + 2H_2O$. The process of CO_2 reduction to CH_3OH: $CO_2 + 6H^+ + 6e^- \rightarrow CH_3OH + H_2O$. The process of CO_2 reduction to CH_3CH_2OH: $2CO_2 + 12H^+ + 12e^- \rightarrow CH_3CH_2OH + 3H_2O$. The process of CO_2 reduction to C_2H_4: $2CO_2 + 8H^+ + 8e^- \rightarrow C_2H_4 + 2H_2O$.

The CO_2RR reaction products are mainly determined by the adsorption mode of CO_2 on the surface of photocatalysts and the subsequent combination of H and CO_2, which has close reduction potentials, making it difficult to obtain a single reaction product.[55] Therefore, understanding the CO_2RR reaction pathway is of great significance for obtaining high yields of single-carbon or multi-carbon products. Due to the complexity of the experimental environment, it is difficult to determine the pathway of CO_2RR reaction. DFT calculations studying the performance of photocatalysts at the atomic level can solve the problem and facilitate the development of CO_2RR.

11.3 Theoretical Methods of Designing High Efficiency Photocatalysts

11.3.1 Energy Band and Electric Field of DFT Calculations

Different from experimental trial-and-error methods, computational investigation methods exhibit advantages of high efficiency and low cost in the study of photocatalysis. Great achievements of photocatalysis utilizing computational methods have emerged in recent decades.[56–61] In particular, first-principles calculations based on DFT is available for predicting the structural, electronic, optical, and catalytic properties of photocatalysts, which not only has high computational accuracy that bandgap and lattice constants are comparable to those of experiments,[62,63] but also can reveal the catalytic reaction mechanism by analyzing the molecular orbitals.[64,65] Therefore, DFT calculations are widely used in engineering the band structure of photocatalysts with an atomic-scale perspective and three important issues will be discussed in this section.

11.3.1.1 Band Edge Positions

From a thermodynamic viewpoint, appropriate band edge positions are a prerequisite for the photocatalytic reaction to take place in whatever single semiconductor or heterostructure. Generally, the CBM of photocatalysts should be higher than the reduction reaction potential, while the VBM should be lower than the oxidation reaction potential. For example, to achieve overall water splitting, the CBM potential of the semiconductor must be more negative than that of H_2 generation (−0.42 V *vs.* NHE at pH 7) and the VBM potential must be more positive than that required for O_2 generation (0.81 V *vs.* NHE at pH 7).[66–68] Otherwise, the corresponding catalytic reaction will not occur. This means that the bandgap of selected photocatalysts for overall water splitting should be larger than 1.23 eV at least. However, the situation is more complex for CO_2 reduction. A series of CO_2 reduction potentials are adjoined.[55] The main difference between them lies in products. Therefore, if the CBM of photocatalysts is higher than the specific CO_2 reaction potential, some CO_2 reactions with lower potential, even H_2 reaction, can also occur and compete with each other. In this case, band edge positions are not the decisive factor and further calculation, *e.g.*, Gibbs free energy, should be carried out.

The formation of heterostructures is one way to control band edge positions, except for traditional heterostructures, novel Z-scheme heterostructures have also attracted the researchers' attention. For overall water splitting on Z-scheme heterostructure photocatalysts, every monolayer does not need to meet the above requirement. For example, a PtS_2/arsenene heterostructure has been reported as an efficient Z-scheme photocatalyst for overall water splitting.[69] The CBM of PtS_2 locates below the H_2 evolution potential and the VBM of arsenene locates above the O_2 evolution potential. As a result, H_2 evolution and O_2 evolution reactions are inhibited in PtS_2 and arsenene, respectively, and only half-reaction occurs on two semiconductor layers. That is to say, to design a reasonable Z-scheme heterostructure photocatalyst for overall water splitting, we just ensure that the lower VBM (higher CBM) position locates below (above) the oxidation (reduction) potential. Apart from the above-mentioned, an excellent photocatalyst also needs enough energy to overcome the photocatalytic activation barrier. That is, a sufficient energy difference between CBM (VBM) and reduction (oxidation) potential is necessary to achieve high reaction rates. For example, for O_2 evolution reaction, the overpotential δ can be expressed as follows (eqn (11.1)):[70]

$$\delta = \Delta G_{\text{max}} - 1.23\ \text{eV} \tag{11.1}$$

here the ΔG_{max} represents the Gibbs free energy difference of the rate-limiting step, and 1.23 eV is the equilibrium potential of O_2 evolution. Only when the energy difference between VBM and oxidation potential is larger than the δ, photocatalysts can facilitate the O_2 evolution from water. Therefore, a good photocatalyst requires not only the more negative CBM position (positive VBM position) of semiconductor as compared with reduction (oxidation) reaction potential, but also a considerable energy difference between CBM (VBM) and reduction (oxidation) potential to overcome the energy barrier derived from ΔG_{max} of the rate-limiting step.

11.3.1.2 Band Alignments

Beyond the single semiconductor photocatalysts, heterostructures formed by stacking semiconductors can not only enhance the light absorption but also improve the photogenerated charge separation, thus, attracting widespread attention for the design of photocatalysts. Different from single semiconductor materials, which has stringent requirement for band edge positions, heterostructure catalysts only require one of materials with CBM higher the reduction potential and the other with VBM below the oxidation potential. This means that band alignments are particularly important for designing photocatalytic heterostructures. As shown in Figure 11.2, there are three types of typical band alignments named type-I, type-II and type-III.[71] For the type-I heterostructures, the band edge position of CBM (VBM) for one layer is lower (higher) than that of another layer, as a result, it generally accumulates photogenerated carriers at the layer with a narrower bandgap, which shows high emission efficiency and is suitable for light-emitting diodes.[72] However, this band alignment cannot effectively separate the photogenerated electrons and holes, thus limiting the quantum efficiency of the carrier and reducing the catalytic capacity. The type-III heterostructure with broken-gap band alignment shows no photogenerated e–h separation at the interface. Therefore, it has the potential to be used as a tunneling field-effect transistor instead of a photocatalyst. Different from the type-I and type-III band alignments, the CBM and VBM of one layer are all higher than that of another layer for type-II band alignment. In this case, the photogenerated electrons and holes can effectively separate into two layers, respectively, and the redox reaction would take place on the corresponding layer.[73] Overall, constructing the type-II band alignment heterostructure is conducive to the theoretical design of high-efficiency photocatalysts.

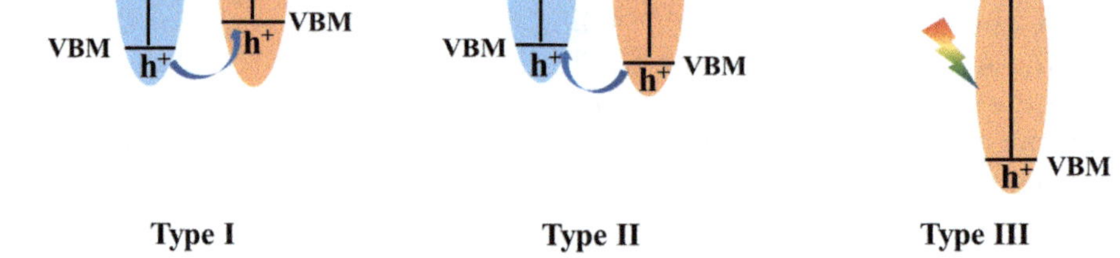

Figure 11.2 Three different band alignments of traditional semiconductor heterostructure photocatalysts: (a) type-I, (b) type-II, and (c) type-III.

11.3.1.3 Built-in Electric Field

Apart from the band alignments mentioned above, determining the direction of the built-in electric field is also important for the design of highly efficient photocatalysts. As we know, the built-in electric field provides the driving force that affects photogenerated carriers transfer path for heterostructure photocatalysts with type-II band alignment, and the magnitude of the built-in electric field depends on the work function difference between two semiconductor layers.[74,75] As shown in Figure 11.3, the work function (WF_1) of semiconductor I (SC I) is smaller than that (WF_2) of semiconductor II (SC II), which means the Fermi level (E_F) of the former is higher than that of the latter. As a result, when they are in contact with each other, free electrons will flow from SC I to SC II until the Fermi level is equilibrated. In this case, the negative charge will accumulate at the SC II layer while the positive charge will accumulate at SC I layer and a built-in electric field is formed pointing from SC I layer to the SC II layer (right panel in Figure 11.2(b) and 11.3(a)). Under light irradiation, photogenerated electrons (holes) will spontaneously transfer from CBM (VBM) of SC II (SC I) to CBM (VBM) of SC I (SC II) by means of the driving force of the built-in electric field. Meanwhile, the interlayer e–h recombination will be impeded, namely traditional type-II photocatalysts (Figure 11.2(b)). However, the situation is not necessarily so. As shown in Figure 11.3(a), if the CBM and VBM of SC I is higher than that of SC II, the formed built-in electric field will accelerate the interlayer e–h recombination and inhibits the corresponding photogenerated carriers' transfer due to the repulsive effect of the built-in electric field between two semiconductor layers. Such a type of photocatalyst is called the Z-scheme photocatalyst. When compared with the traditional type-II photocatalysts, photogenerated carriers reside in the higher CBM or lower VBM, which significantly enhances the redox reaction capacity and has attracted extensive attention recently.[76–80] Therefore, the direction of built-in electric field should be utilized in the design of highly efficient photocatalysts because it plays an

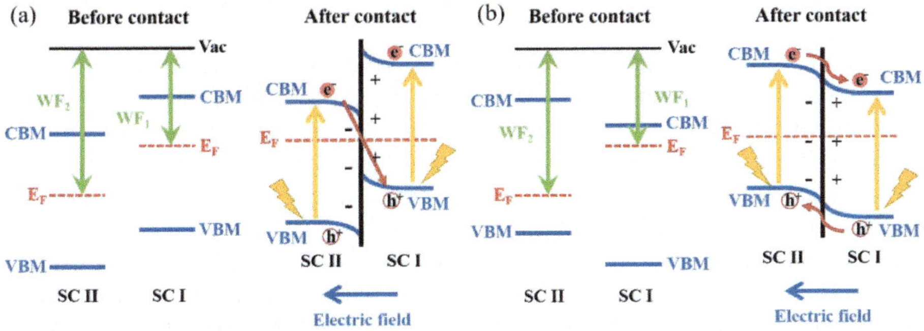

Figure 11.3 (a) and (b) Schematic diagram of semiconductor heterostructures with two different type-II band alignments before contact (left panel) and after contact (right panel). The Vac represents the vacuum energy level.

important role in judging whether designed photocatalysts are traditional type-II or Z-scheme.

11.3.2 Excited State Dynamics

Finding reasonable band edge positions and alignments by DFT calculations plays an important role in designing high-performance photocatalysis. However, the separation of photogenerated electrons and holes in the photocatalytic process is also an important factor that determines photocatalytic performance. Limited by the complexity of experimental conditions and defects of the material, it is difficult to quantitatively obtain the separation time of photogenerated carriers during the reaction. Recently, NAMD combined with DFT calculations can quantify the transfer time and lifetime of photogenerated carriers, and has been widely used in the fields of optoelectronics and photocatalysis.[81–85] This section discusses three applications of NAMD in photocatalysis.

11.3.2.1 Designing Z-scheme Photocatalysts

Z-scheme photocatalysts have attracted widespread attention due to their wide light absorption ranges and strong redox abilities.[86–88] When compared with the Type II heterostructures, the e–h recombination at the interface of Z-scheme heterostructures is faster than the electron and hole transfer at the interface.[89] Therefore, electrons accumulate on lower CBM and holes accumulate on higher VBM, which increases the bandgap range of photocatalysts to extend light absorption, and raises the potential difference of the redox reaction to promote the reaction. However, it is difficult to directly characterize Z-scheme photocatalysts in the experiment due to complex experimental conditions. Moreover, traditional DFT calculations can only give band edge positions and alignments, and are powerless for charge transfer and recombination at the interface. NAMD simulations can give direct evidence for Z-scheme heterostructures by calculating the lifetime of photogenerated carriers.

Zhou *et al.* pointed out that the introduction of VI A group vacancies of the MoSSe/WSeTe heterostructure turned the type II heterostructure into a Z-scheme heterostructure.[90] Moreover, vacancies can provide reactive sites to promote the hydrogen evolution reaction (HER). The ΔG of the HER is only −0.18 eV as shown in Figure 11.4. Unfortunately, the ΔG of oxygen evolution reaction (OER) is as high as 3.22 eV to hinder the generation of O_2. To achieve high catalytic efficiency of water splitting on Z-scheme photocatalysts and reduce the reaction barrier of OER, Fu *et al.* used transition metal carbides (MXenes) to design the $MoSe_2/Ti_2CO_2$ Z-scheme heterostructure.[91] NAMD calculations revealed that the fast e–h recombination at the heterostructure interface makes electrons and holes remain on $MoSe_2$ and Ti_2CO_2, respectively. DFT calculations showed that the OER can proceed spontaneously in

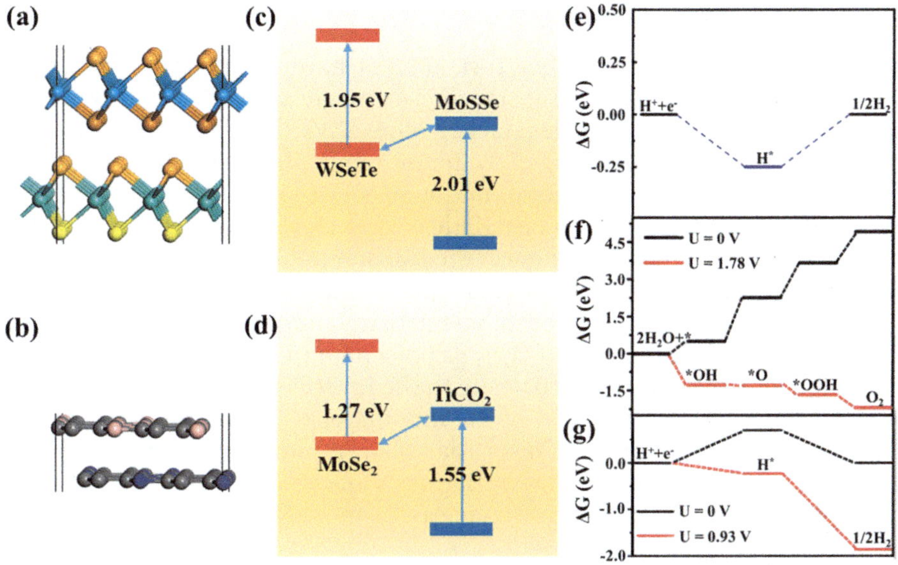

Figure 11.4 Atomic structure of MoSSe/WSeTe (a) and C_3B/C_3N (b) heterostructures. The band alignments and charge recombination of MoSSe/WSeTe (c) and C_3B/C_3N (d) heterostructures, respectively. (e)–(g) Free energy diagrams for HER of MoSSe/WSeTe, OER and HER of C_3B/C_3N, respectively.[90,92] (e) Reproduced from ref. 90 with permission from the Royal Society of Chemistry. (f) and (g) Reproduced from ref. 92 with permission from American Chemical Society, Copyright 2020.

Ti_2CO_2 under the drive of the external potential (2.10 V). More importantly, holes remained on Ti_2CO_2 have a high energy barrier to the oxidation of lattice O, which means that the heterostructure can avoid the photo-corrosion of $MoSe_2$ and Ti_2CO_2. Furthermore, the overall water splitting has been pursued in experiments, that is, photocatalysts are required to have both high HER and high OER performance. Niu *et al.* used NAMD combined with DFT calculations to design the C_3N/C_3B Z-scheme photocatalyst as shown in Figure 11.4.[92] NAMD calculations showed that electrons and holes accumulate on C_3N and C_3B, respectively, due to the built-in electric field. Under the drive of the external potential, HER and OER can proceed spontaneously to exhibit high photocatalytic efficiency.

Based on the above discussion, DFT calculations can give the bandgap of photocatalysts and the energy barrier of redox reaction. The NAMD calculations give the e–h transfer and recombination time and provide the direct evidence whether photocatalysts form the Z-scheme heterostructures. Generally speaking, the bandgap at the interface of photocatalysts for the formation of Z-scheme heterostructures should be small enough, such as OH modified phosphorene nanoribbons/OCN modified phosphorene nanoribbons (0.257 eV), BCN/C_2N (0.2 eV).[93,94] In these case, the recombination time of e–h at the interface can be faster than the electron and hole transfer time.

11.3.2.2 Layered Stacking

Most two-dimensional (2D) heterostructures have an interlayer coupling effect, which can be altered by stacking or magic angle.[95–98] For example, Zhou *et al.* found that four-layer stacked MoSSe/WSeTe heterostructures can be type I, type II, and Z-scheme.[90] Different band alignments have different effects on the transfer and recombination of photogenerated carriers, which eventually show different photocatalytic efficiency. Next, we explain how the change of interlayer stacking affects the transfer and recombination of photogenerated carriers to further regulate photocatalytic performance.

Niu *et al.* studied the photocatalytic activity and photogenerated carrier lifetime of one to four layers of C_3N_4 as presented in Figure 11.5.[99] The electron transfer time from the active site to the inactive site of the bilayer C_3N_4 is three times longer than that of the monolayer C_3N_4. Moreover, for fewer layer C_3N_4, the ΔG of HER is lower than that of monolayer C_3N_4 about 0.2 eV due to the interaction between layers. Thus, fewer layer C_3N_4 exhibits high photocatalytic HER performance than monolayer C_3N_4. In addition,

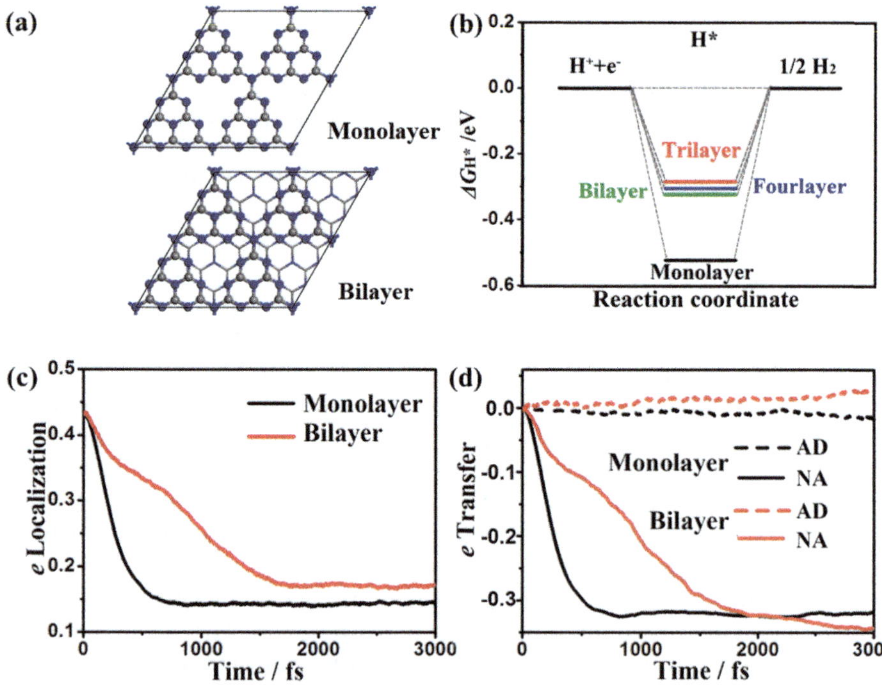

Figure 11.5 (a) Atomic structure of C_3N_4 and bilayer C_3N_4. (b) Free energy diagram for HER of C_3N_4.[99] (c) The occupied proportion of photogenerated electrons at active sites of N_2 type in C_3N_4. (d) Adiabatic and no-adiabatic contributions of photogenerated electron transfer from active site to inactive site. (b)–(d) Reproduced from ref. 99 with permission from the Royal Society of Chemistry.

the introduction of magic angle in the double-layer C_3N_4 can break original band alignment to exhibit novel photocatalytic performance. In this regard, Lu *et al.* designed a twisted bilayer g-C_3N_4 with highly efficient HER and OER, which show type II band alignment.[100] The interface charge separation time of twisted bilayer g-C_3N_4 is 120 fs, which is conducive to prolonging the lifetime of photogenerated carriers to facilitate the transfer of charge from C_3N_4 to reactants. In addition, the overpotential of OER is 0.476 V, which is much lower than that of MoS_2 (2.53 V) and $TiCO_2$ (1.99 V).[101,102] Except for twist, the type II band alignment of the homostructure can also be obtained by ferroelectric modulation. Specifically, Fan *et al.* found that the bilayer $CuInP_2S_6$ could adjust the direction of ferroelectric polarization to achieve type II band alignment.[103] Interestingly, driven by the built-in electric field, the lifetime of photogenerated carriers is as high as 1.03 ns from NAMD simulation, while the interface electron and hole transfer time is only a few ps. Therefore, there are enough photogenerated carriers inside the system to participate in redox reaction. Meanwhile, DFT calculations found that only 0.62 V overpotential is needed to convert CO_2 to CH_4 spontaneously.

11.3.2.3 Vacancies and Doping

As we all know, defects are almost unavoidable in any material preparation process. For example, during the preparation of the MoS_2 monolayer, mechanical exfoliation, chemical vapor deposition (CVD) and physical vapor deposition (PVD) all inevitably introduce S vacancies.[104–106] Vacancies have two sides to the photocatalytic performance of materials. On one hand, vacancies can provide reactive sites to reduce the energy barrier of photocatalytic, which is conductive to the reaction. On the other hand, vacancies introduce carrier trapped states, which mostly accelerates the recombination of photogenerated carriers,[107] thereby adversely affecting photocatalysis. Traditional DFT calculations can only show the effect of vacancies on the redox reaction barrier, while NAMD calculations can characterize the effect of vacancies on photogenerated carriers to have a more comprehensive understanding of the role of vacancies in the field of photocatalysis.

Specifically, Pei *et al.* considered the influence of different defects black phosphorus (BP) on the NRR and found that the rate-limiting step of DV-(5|8|5) defect ($*NH_2{\cdot}*NH_2 + *H \rightarrow *NH_2{\cdot}*NH_3$) has the lowest ΔG (0.28 eV), which is much lower than DV-(4|10|4)@BP (0.84 eV).[108] Meanwhile, NAMD calculations showed that the non-adiabatic e–h recombination time is about 2.3 ps due to the impact of defects, which is much lower than that of perfect BP (hundreds of ps).[109–111] However, the photogenerated carrier lifetime is still longer than the time of the transfer of protons (~100 fs). It means that although vacancies reduce the lifetime of photogenerated carriers, the time is sufficient for photogenerated carriers to participate in the redox reaction. Similarly, Xue *et al.* studied the effect of N vacancies on HER of polymeric carbon nitride (PCN).[112] DFT calculations showed that N vacancies are photocatalytic active sites, which is conducive to the occurrence of the HER.

More precisely, compared to perfect PCN (−1.8 eV), the ΔG (−0.13 eV) of PCN with N vacancies has a ten-fold reduction in HER, which is beneficial to the photocatalytic H_2 production of PCN. Moreover, NAMD calculations unravelled that electrons tend to accumulate in trap states of N vacancies. This means that photogenerated electrons can be quickly transferred from N vacancies to the proton H.

In addition, doping is the most common way to design high-performance photocatalysts.[113–118] Similar to defects, although doping alters photocatalyst chemical activity to provide reactive sites, it may introduce defect trapped states that adversely affect photogenerated carrier lifetime. Combining DFT calculations and NAMD simulations, the effects of doping on photocatalysts can be given at the atomic scale.

Tian *et al.* studied the effect of S-doped C_2N and C_3N_3 on the OER as displayed in Figure 11.6.[119] S-doping not only captures photogenerated holes to achieve effective carrier separation, but also serves as the OER active site to achieve low overpotential (~0.38 V). Further, Zhao *et al.* showed that in Mo-doped BiOCl, although the trapped state induced by Mo doping can capture electrons within hundreds of fs, the time of holes on the VBM transferred to the trapped state is at the ns order.[120] This means that the e–h lifetime of Mo-doped BiOCl is at the ns level. The long photogenerated carrier lifetime is beneficial to the electron transfer from the trapped state to N_2 to participate in the NRR. Moreover, the ΔG of NRR for Mo-doped BiOCl in the distal path is 0.06–1.21 eV, which is very easy to occur in an acidic environment to show high photocatalytic efficiency. However, Xie *et al.* studied Co-doped graphitic carbon nitride (g-CN) with a different conclusion.[121] Although electrons on the CBM are transferred to Co atoms within 500 fs, Co atoms are not active centers. Co doping caused the redistribution of CN charge, which make adjacent cavities have the best HER (ΔG = 0.02 eV), much lower than other sites (1.45 eV). The above studies show the introduction of trapped states by doping can rapidly capture photogenerated carriers. A doping atom can provide reaction sites, or activate the photocatalyst matrix to achieve high photocatalytic efficiency. In addition, Fan *et al.* studied B doping metal-free triphenylene–graphdiyne monolayer (B@TP–GDY) and found that doping can also increase photocarrier lifetime of photocatalyst.[122] NAMD calculations showed that although B doping introduces trapped state, the photogenerated carrier lifetime increases from 110 ps to 198 ps compared with pristine TP–GDY. Moreover, the maximum ΔG for NRR is only 0.29 eV, indicating that it can be a candidate photocatalyst.

11.3.2.4 Quantum Dot or Cluster Modification

Quantum dots or clusters with small diameter have a wide range of applications in the field of photocatalysis due to tunable bandgap and more active sites.[123–127] Wang *et al.* designed graphite quantum dots (CQDs) doping on C_3N,[123] and found that electrons and holes are rapidly transferred to CQDs and C_3N within a few ps, respectively. It implies rapid photogenerated carrier

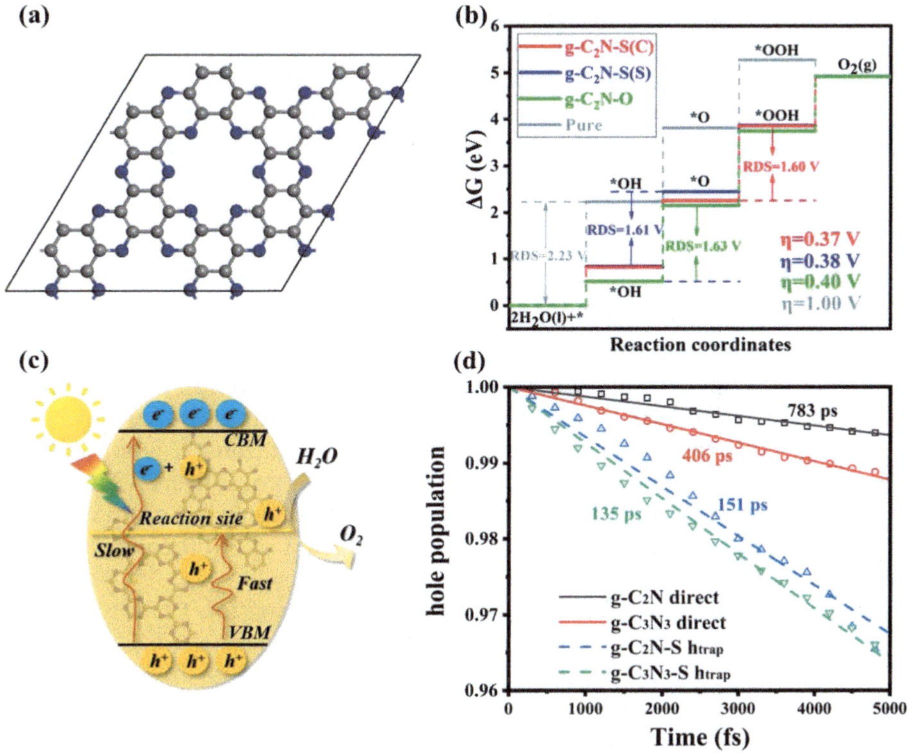

Figure 11.6 (a) Atomic structure of S-doped C_2N. (b) Free energy diagram for OER of O/S-doped g-C_2N and pristine g-C_2N.[119] (c) The electron transfer mechanism of S-doped C_2N during photocatalytic OER. (d) The population of the hole trapping and e–h recombination dynamics for S-doped g-C_2N and S-doped g-C_3N_3. (b)–(d) Reproduced from ref. 119 with permission from the Royal Society of Chemistry.

separation to weaken e–h recombination. Moreover, DFT calculations indicated that CQD doping lowers ΔG for HER by about 0.2 eV. In addition, Liu *et al.* found that g-C_3N_4 modified $(MgO)_n$ cluster forms type II band alignment.[124] The separation of photogenerated carriers occurs in hundreds of fs, corresponding to a long photogenerated carrier lifetime, which facilitates the transfer of photogenerated electrons from the photocatalyst to the reactant. DFT calculations showed that the ΔG of HER is only −0.04 eV, which is even lower than that of Pt (111) (ΔG = 0.10 eV).[125]

11.3.3 Redox Reaction

Photongenerated carriers that migrate to the surface of photocatalysts participate in the redox reaction, which is the last step of the photocatalytic process. Although the favorable band edge positions thermodynamically promotes the transfer of photogenerated electrons and holes to the reactant, it may require an external electric field to drive the surface reaction in kinetics due to high reaction barrier. Therefore, how to reduce the reaction barrier has become one of the key steps in determining the photocatalytic performance. Defect, doping and heterostructures are effective ways, and related theoretical works are discussed as follows.

11.3.3.1 Defect

It is generally known that defects play an important role in photocatalysis, which can change the electronic structure of the photocatalyst surface. The reactant can be rapidly adsorbed and desorbed completing the electron transfer around the defect, thereby making the photocatalytic process. However, due to the complexity of experimental conditions, it is difficult to directly characterize the role of certain defects in photocatalysis. DFT calculations enable the study of photocatalyst performance at the atomic scale, which has great implications for insight and design of photocatalysts with high efficiency.

Specifically, Zhang *et al.* found that the ΔG of formation COOH* on the O-Co_3O_4 system is 0.123 eV lower than that on the perfect Co_3O_4 system, implying that the CO_2 reduction on O-Co_3O_4 surface is more favorable than that on perfect Co_3O_4 surface.[127] Similarly, Liu *et al.* also found that O-vacancy Bi_2WO_6 can significantly reduce the ΔG of *COOH → *(CO + OH) from 0.89 eV to 0.37 eV when compared to perfect Bi_2WO_6,[128] which is because that vacancies change the chemical state of the Bi_2WO_6 surface to enhancing the adsorption of CO_2 on Bi_2WO_6 surface. In addition, defects can also change the path of CO_2 reduction to realize the selectivity of reaction products. Yang *et al.* found that for perfect Bi_2MoO_6 surface, the reaction intermediate *CO + H_2O is more inclined to desorb on the surface to form CO.[129] On the surface of O-vacancy Bi_2MoO_6, *CO + H_2O is more inclined to hydrogenate to generate *CHO due to the lower ΔG (0.53 eV) than form CO (3.26 eV). Thus, the *CHO intermediate can undergo subsequent hydrogenation to form CH_4. The above studies indicate that the effect of oxygen vacancies on photocatalysis. In fact, some other defects, such as S vacancies in $CuIn_5S_8$ and $AgInP_2S_6$ or grain boundary, also play a great role in the photocatalytic process.[130,131] This is may related to the defect-tuning surface electronic structure of photocatalysts.

11.3.3.2 Doping

Besides defects, doping is also one of ways to tune the performance of photocatalysts.[132–137] Doping elements may provide additional electrons or holes to introduce defect trapped states inside photocatalysts, which changes the transition barrier of photogenerated carriers to enhance the light absorption. In addition, doping elements can provide reactive sites or change the electronic structure of photocatalysts to promote the photocatalytic reaction or to achieve selectivity of reaction products. For example, Cao *et al.* design Pd-doped/g-C_3N_4 (Pd/g-C_3N_4) and Pt-doped/g-C_3N_4 (Pt/g-C_3N_4) photocatalysts to reduce CO_2 to HCOOH ($\Delta G = 0.66$ eV) and CH_4 ($\Delta G = 1.16$ eV), respectively. This might be related to the different amount of charge transfer between Pt/g-C_3N_4 and Pd/g-C_3N_4. Our group reported the first metal-free B-doped g-C_3N_4 (B/g-C_3N_4) photocatalyst for NRR.[138] As shown in Figure 11.7, the B atom acts as an active center to achieve the charge transfer between the photocatalyst and N_2 molecules. The onset potential of the NRR is only 0.20 V, suggesting that B/g-C_3N_4 has broad application prospects in the field of photocatalysis to generate NH_3. Further, our group used DFT to design a photocatalyst with dual-metal sites and found that FeMo/g-C_3N_4 has low limit potential of NRR (−0.23 V) and high potential of HER (−0.79 V), thereby promoting NRR to inhibit HER and achieving high efficiency NH_3 yield.[139]

11.3.3.3 Heterostructures

Heterostructures that can combine advantages of different materials often exhibit higher carrier mobility, proper band edge positions, and more reactive sites, resulting in higher photocatalytic efficiency. For instance, Li *et al.* found that the reaction barrier ($CO_2 + {}^*H + 2e^- \rightarrow HCOO^-$) of (001) TiO_2// (100) GaP heterostructure is significantly lower of that of pure (001) TiO_2 facet (reduced from 1.93 eV to 0.82 eV), which is caused by the interface effect between TiO_2 and GaP.[140] In addition to three dimensional (3D) heterostructure photocatalysts, 2D material heterostructures have received extensive attention in the field of photocatalysis due to their large surface area. Lang *et al.* used DFT calculations to design blue phosphorus/PN photocatalyst for photocatalytic water splitting.[141] Liu *et al.* designed 3D/2D photocatalysts $CuIn_5S_8$/g-C_3N_4 for CO_2 reduction to CH_4,[130] which has a lower reaction barrier in the process of $CO_2 + {}^*H + 2e^- \rightarrow COOH^-$ (1.34 eV for C_3N_4 *vs.* 0.29 eV for $CuIn_5S_8$/g-C_3N_4).

11.4 Conclusion

This chapter provides a theory analysis of three steps in the photocatalytic process: (1) light absorption; (2) photogenerated carrier separation; and (3) recombination and surface redox reaction. Specifically, we discuss the influence of energy band on photocatalysis, how NAMD evaluate photogenerated carriers, and the calculation of reaction barrier during redox reaction. However, there are still some studies worthy of the attention of theoretical researchers.

(1) Most of current theoretical studies on the design of photocatalysts only consider PBE or HSE functional. The effect of the GW approach on the energy band needs to be elaborated more clearly.
(2) Regarding the issue of carrier transfer and recombination in photocatalysts, computational methods for exciton effect, spin–orbit coupling effect and many-body effect, need to be further developed.

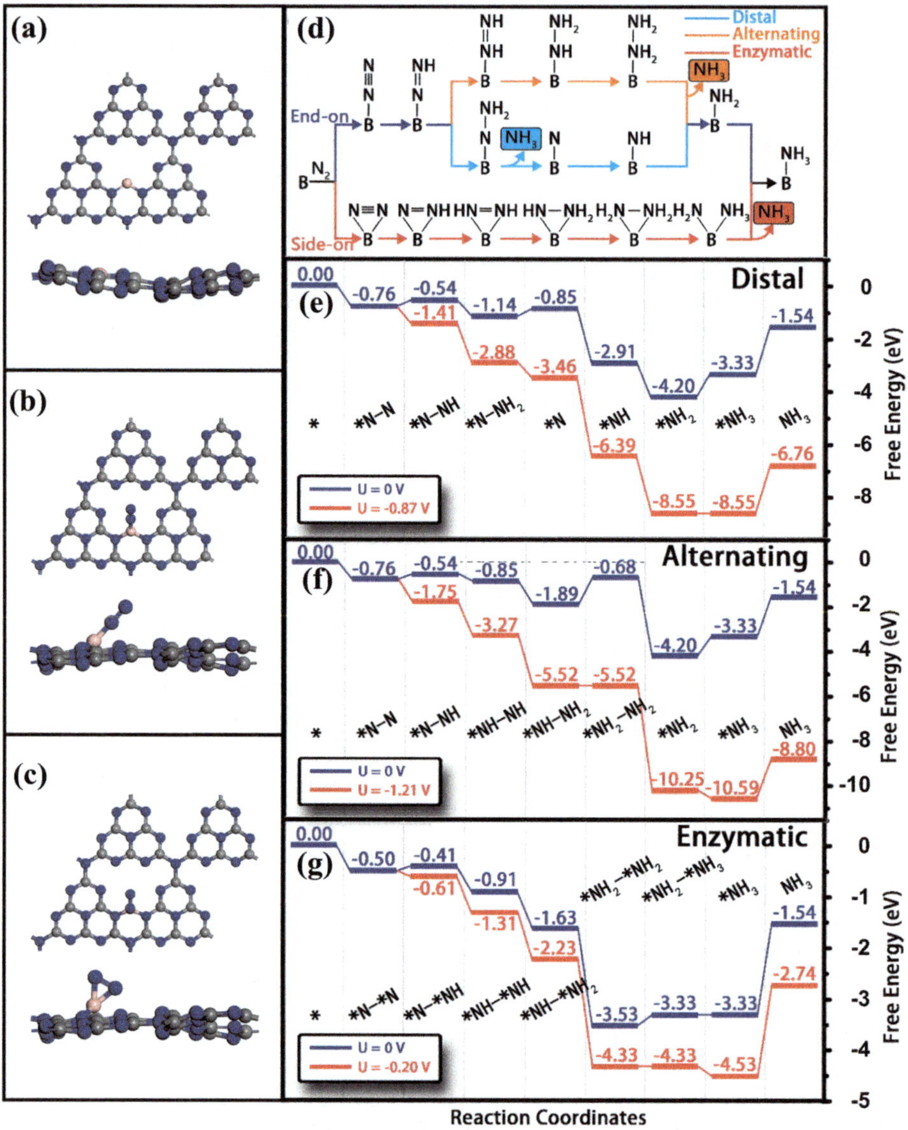

Figure 11.7 Atomic structure of B-doped C_3N_4 (a), N_2 adsorption end-on (b) and side-on (c) in B-doped C_3N_4. (d) Schemes of distal, alternating, and enzymatic mechanisms for NRR.[138] Free energy diagram for NRR of B-doped C_3N_4 through distal (e), alternating (f), and enzymatic (g) mechanism. Reproduced from ref. 138 with permission from American Chemical Society, Copyright 2018.

(3) The question of electron transfer from photocatalysts to reactants is still an open question.
(4) The solvation effect needs to be considered during the simulation of redox reaction.

Conflict of Interest

There are no conflicts to declare.

Acknowledgements

This work is supported by the National Key R&D Program of China (Grant No. 2022YFB3807200), National Natural Science Foundation of China (Grant No. 21973011 and 22033002). The authors acknowledge the computational resources from the Big Data Center of Southeast University and National Supercomputing Center of Tianjin.

References

1. E. Zhao, K. Du, P. F. Yin, J. Ran, J. Mao, T. Ling and S. Z. Qiao, *Adv. Sci.*, 2022, **9**, e2104363.
2. L. Lannelongue, J. Grealey and M. Inouye, *Adv. Sci.*, 2021, **8**, 2100707.
3. W. Wang, H. Zhang, S. Zhang, Y. Liu, G. Wang, C. Sun and H. Zhao, *Angew. Chem., Int. Ed.*, 2019, **58**, 16644–16650.
4. J. Shi and L. Guo, *Prog. Nat. Sci.: Mater. Int.*, 2012, **22**, 592–615.
5. A. V. Akimov, R. Asahi, R. Jinnouchi and O. V. Prezhdo, *J. Am. Chem. Soc.*, 2015, **137**, 11517–11525.
6. K. Maeda and K. Domen, *J. Phys. Chem. Lett.*, 2010, **1**, 2655–2661.
7. A. Kudo and Y. Miseki, *Chem. Soc. Rev.*, 2009, **38**, 253–278.
8. S. Chen, T. Takata and K. Domen, *Nat. Rev. Mater.*, 2017, **2**, 17050.
9. K. Takanabe, *ACS Catal.*, 2017, **7**, 8006–8022.
10. T. Su, Q. Shao, Z. Qin, Z. Guo and Z. Wu, *ACS Catal.*, 2018, **8**, 2253–2276.
11. W. Wang, H. Zhou, Y. Liu, S. Zhang, Y. Zhang, G. Wang, H. Zhang and H. Zhao, *Small*, 2020, **16**, e1906880.
12. G. Swain, S. Sultana and K. Parida, *ACS Sustainable Chem. Eng. J.*, 2020, **8**, 4848–4862.
13. G. Zhang, Y. Meng, B. Xie, Z. Ni, H. Lu and S. Xia, *Appl. Catal., B*, 2021, **296**, 120379.
14. Z. Wang, X. Hu, Z. Liu, G. Zou, G. Wang and K. Zhang, *ACS Catal.*, 2019, **9**, 10260–10278.
15. J. Luo, X. Bai, Q. Li, X. Yu, C. Li, Z. Wang, W. Wu, Y. Liang, Z. Zhao and H. Liu, *Nano Energy*, 2019, **66**, 104187.
16. A. Li, T. Wang, X. Chang, Z. J. Zhao, C. Li, Z. Huang, P. Yang, G. Zhou and J. Gong, *Chem. Sci.*, 2018, **9**, 5334–5340.
17. C. Du, X. Wang, W. Chen, S. Feng, J. Wen and Y. A. Wu, *Mater. Today Adv.*, 2020, **6**, 100071.
18. Y. Ma, Z. Wang, X. Xu and J. Wang, *Chin. J. Catal.*, 2017, **38**, 1956–1969.
19. Q. Liu, H. Cheng, T. Chen, T. W. B. Lo, J. Ma, A. Ling and F. Wang, *Appl. Catal., B*, 2021, **295**, 120289.
20. A. Li, Q. Cao, G. Zhou, B. Schmidt, W. Zhu, X. Yuan, H. Huo, J. Gong and M. Antonietti, *Angew. Chem., Int. Ed.*, 2019, **58**, 14549–14555.

21. A. L. Luna, D. Dragoe, K. Wang, P. Beaunier, E. Kowalska, B. Ohtani, D. Bahena Uribe, M. A. Valenzuela, H. Remita and C. Colbeau-Justin, *J. Phys. Chem. C*, 2017, **121**, 14302–14311.
22. Y. Song, X. Xin, S. Guo, Y. Zhang, L. Yang, B. Wang and X. Li, *Chem. Eng. J.*, 2020, **384**, 123337.
23. K. Maeda, A. Xiong, T. Yoshinaga, T. Ikeda, N. Sakamoto, T. Hisatomi, M. Takashima, D. Lu, M. Kanehara, T. Setoyama, T. Teranishi and K. Domen, *Angew. Chem., Int. Ed.*, 2010, **49**, 4096–4099.
24. R. Marschall, *Adv. Funct. Mater.*, 2014, **24**, 2421–2440.
25. S. Bai, N. Zhang, C. Gao and Y. Xiong, *Nano Energy*, 2018, **53**, 296–336.
26. Q. Wang and K. Domen, *Chem. Rev.*, 2020, **120**, 919–985.
27. N. Feng, Q. Wang, A. Zheng, Z. Zhang, J. Fan, S. B. Liu, J. P. Amoureux and F. Deng, *J. Am. Chem. Soc.*, 2013, **135**, 1607–1616.
28. G. Wang, H. Wang, Y. Ling, Y. Tang, X. Yang, R. C. Fitzmorris, C. Wang, J. Z. Zhang and Y. Li, *Nano Lett.*, 2011, **11**, 3026–3033.
29. S. Hu, L. Ma, J. You, F. Li, Z. Fan, G. Lu, D. Liu and J. Gui, *Appl. Surf. Sci.*, 2014, **311**, 164–171.
30. T. Tachikawa, S. Yamashita and T. Majima, *J. Am. Chem. Soc.*, 2011, **133**, 7197–7204.
31. T. Tachikawa, T. Ochi and Y. Kobori, *ACS Catal.*, 2016, **6**, 2250–2256.
32. J. Zhang, G. Huang, J. Zeng, X. Jiang, Y. Shi, S. Lin, X. Chen, H. Wang, Z. Kong, J. Xi and Z. Ji, *J. Alloys Compd.*, 2019, **775**, 726–735.
33. D. Chen, F. Huang, G. Ren, D. Li, M. Zheng, Y. Wang and Z. Lin, *Nanoscale*, 2010, **2**, 2062–2064.
34. A. J. Cohen, P. Mori-Sanchez and W. Yang, *Chem. Rev.*, 2012, **112**, 289–320.
35. M. Levy, *Phys. Rev. A*, 1982, **26**, 1200–1208.
36. J. P. Perdew, K. Burke and M. Ernzerhof, *Phys. Rev. Lett.*, 1996, **77**, 3865–3868.
37. K. Kirkpatrick, C. D. Gelat and M. P. Vecchi, *Science*, 1983, **220**, 4598.
38. S. Azam, T. V. Vu, D. H. Mirza, M. Irfan and S. Goumri-Said, *J. Solid State Chem.*, 2020, **290**, 121522.
39. V. Ivády, R. Armiento, K. Szász, E. Janzén, A. Gali and I. A. Abrikosov, *Phys. Rev. B*, 2014, **90**, 035146.
40. B. Himmetoglu, A. Floris, S. de Gironcoli and M. Cococcioni, *Int. J. Quantum Chem.*, 2014, **114**, 14–49.
41. J. Paier, M. Marsman and G. Kresse, *J. Chem. Phys.*, 2007, **127**, 024103.
42. C. Adamo and V. Barone, *J. Chem. Phys.*, 1999, **110**, 6158–6170.
43. P. Deák, B. Aradi, T. Frauenheim, E. Janzén and A. Gali, *Phys. Rev. B*, 2010, **81**, 153203.
44. X. Li, J. C. Tully, H. B. Schlegel and M. J. Frisch, *J. Chem. Phys.*, 2005, **123**, 084106.
45. J. R. Schmidt, P. V. Parandekar and J. C. Tully, *J. Chem. Phys.*, 2008, **129**, 044104.
46. X. Li, Z. Li and J. Yang, *Phys. Rev. Lett.*, 2014, **112**, 018301.
47. W. Yin, B. Wen, Q. Ge, D. Zou, Y. Xu, M. Liu, X. Wei, M. Chen and X. Fan, *Prog. Nat. Sci.: Mater. Int.*, 2019, **29**, 335–340.

48. C. Liang, H.-Y. Niu, H. Guo, C.-G. Niu, D.-W. Huang, Y.-Y. Yang, H.-Y. Liu, B.-B. Shao and H.-P. Feng, *Chem. Eng. J.*, 2020, **396**, 125395.
49. Y. Zhang, W. Qiu, Y. Ma, Y. Luo, Z. Tian, G. Cui, F. Xie, L. Chen, T. Li and X. Sun, *ACS Catal.*, 2018, **8**, 8540–8544.
50. S. Nitopi, E. Bertheussen, S. B. Scott, X. Liu, A. K. Engstfeld, S. Horch, B. Seger, I. E. L. Stephens, K. Chan, C. Hahn, J. K. Norskov, T. F. Jaramillo and I. Chorkendorff, *Chem. Rev.*, 2019, **119**, 7610–7672.
51. Ž. Kovačič, B. Likozar and M. Huš, *ACS Catal.*, 2020, **10**, 14984–15007.
52. X. Zhang, A. Chen, L. Chen and Z. Zhou, *Adv. Energy Mater.*, 2021, 2003841.
53. Z. Li, N. H. Attanayake, J. L. Blackburn and E. M. Miller, *Energy Environ. Sci.*, 2021, **14**, 6242–6286.
54. S. A. Rawool, K. K. Yadav and V. Polshettiwar, *Chem. Sci.*, 2021, **12**, 4267–4299.
55. Z. Sun, N. Talreja, H. Tao, J. Texter, M. Muhler, J. Strunk and J. Chen, *Angew. Chem., Int. Ed.*, 2018, **57**, 7610–7627.
56. C. F. Fu, R. Zhang, Q. Luo, X. Li and J. Yang, *J. Comput. Chem.*, 2019, **40**, 980–987.
57. Y. Lee, Y. Hwang and Y. C. Chung, *ACS Appl. Mater. Interfaces*, 2015, **7**, 7163–7169.
58. B. Wang, H. Yuan, J. Chang, X. Chen and H. Chen, *Appl. Surf. Sci.*, 2019, **485**, 375–380.
59. J. Low, J. Yu, M. Jaroniec, S. Wageh and A. A. Al-Ghamdi, *Adv. Mater.*, 2017, **29**, 1601694.
60. S. Shen and S. S. Mao, *Nanophotonics*, 2012, **1**, 31–50.
61. Y. Liu, P. Lv, W. Zhou and J. Hong, *J. Phys. Chem. C*, 2020, **124**, 9696–9702.
62. J. Heyd and G. E. Scuseria, *J. Chem. Phys.*, 2003, **118**, 8207–8215.
63. P. Deák, B. Aradi and T. Frauenheim, *Phys. Rev. B*, 2011, **83**, 155207.
64. E. J. Baerends and O. V. Gritsenko, *J. Phys. Chem. A*, 1997, **101**, 5384–5403.
65. F. M. BICKELHAUPT, *J. Comput. Chem.*, 1999, **20**, 114–128.
66. X. Li, J. Yu, J. Low, Y. Fang, J. Xiao and X. Chen, *J. Mater. Chem. A*, 2015, **3**, 2485–2534.
67. X. Chen, S. Shen, L. Guo and S. S. Mao, *Chem. Rev.*, 2010, **110**, 6503–6570.
68. N. Queyriaux, E. Giannoudis, C. D. Windle, S. Roy, J. Pécaut, A. G. Coutsolelos, V. Artero and M. Chavarot-Kerlidou, *Sustainable Energy Fuels*, 2018, **2**, 553–557.
69. K. Ren, W. Tang, M. Sun, Y. Cai, Y. Cheng and G. Zhang, *Nanoscale*, 2020, **12**, 17281–17289.
70. I. C. Man, H. Y. Su, F. Calle-Vallejo, H. A. Hansen, J. I. Martínez, N. G. Inoglu, J. Kitchin, T. F. Jaramillo, J. K. Nørskov and J. Rossmeisl, *ChemCatChem*, 2011, **3**, 1159–1165.
71. Z. H. Lu, D. J. Lockwood and J.-M. Baribeau, *Nature*, 1995, **378**, 258–260.
72. S. J. A. Moniz, S. A. Shevlin, D. J. Martin, Z.-X. Guo and J. Tang, *Energy Environ. Sci.*, 2015, **8**, 731–759.
73. S. Bai, J. Jiang, Q. Zhang and Y. Xiong, *Chem. Soc. Rev.*, 2015, **44**, 2893–2939.
74. J. Liu, B. Cheng and J. Yu, *Phys. Chem. Chem. Phys.*, 2016, **18**, 31175–31183.

75. Y.-L. Wang, Y. Tian, Z.-L. Lang, W. Guan and L.-K. Yan, *J. Mater. Chem. A*, 2018, **6**, 21056–21063.
76. T. Di, B. Zhu, B. Cheng, J. Yu and J. Xu, *J. Catal.*, 2017, **352**, 532–541.
77. J. Hu, D. Chen, Z. Mo, N. Li, Q. Xu, H. Li, J. He, H. Xu and J. Lu, *Angew. Chem., Int. Ed.*, 2019, **58**, 2073–2077.
78. F. Liu, R. Shi, Z. Wang, Y. Weng, C. M. Che and Y. Chen, *Angew. Chem., Int. Ed.*, 2019, **58**, 11791–11795.
79. W. Yu, J. Chen, T. Shang, L. Chen, L. Gu and T. Peng, *Appl. Catal., B*, 2017, **219**, 693–704.
80. Q. Yuan, D. Liu, N. Zhang, W. Ye, H. Ju, L. Shi, R. Long, J. Zhu and Y. Xiong, *Angew. Chem., Int. Ed.*, 2017, **56**, 4206–4210.
81. C. Cheng, W.-H. Fang, R. Long and O. V. Prezhdo, *JACS Au*, 2021, **1**, 550–559.
82. Y. Wei and R. Long, *J. Phys. Chem. Lett.*, 2018, **9**, 3856–3862.
83. Y. Wei, W. H. Fang and R. Long, *J. Phys. Chem. Lett.*, 2020, **11**, 478–484.
84. Z. Ji, H. Hong, J. Zhang, Q. Zhang, W. Huang, T. Cao, R. Qiao, C. Liu, J. Liang, C. Jin, L. Jiao, K. Shi, S. Meng and K. Liu, *ACS Nano*, 2017, **11**, 12020–12026.
85. S. Pal, D. Casanova and O. V. Prezhdo, *Nano Lett.*, 2018, **18**, 58–63.
86. Z. Zhou, S. Yuan and J. Wang, *Front. Phys.*, 2021, **16**, 43203.
87. Q. Xu, L. Zhang, J. Yu, S. Wageh, A. A. Al-Ghamdi and M. Jaroniec, *Mater. Today*, 2018, **21**, 1042–1063.
88. P. Zhou, J. Yu and M. Jaroniec, *Adv. Mater.*, 2014, **26**, 4920–4935.
89. J. Low, C. Jiang, B. Cheng, S. Wageh, A. A. Al-Ghamdi and J. Yu, *Small Methods*, 2017, **1**, 1700080.
90. Z. Zhou, X. Niu, Y. Zhang and J. Wang, *J. Mater. Chem. A*, 2019, **7**, 21835–21842.
91. C. F. Fu, X. Li and J. Yang, *Chem. Sci.*, 2021, **12**, 2863–2869.
92. X. Niu, X. Bai, Z. Zhou and J. Wang, *ACS Catal.*, 2020, **10**, 1976–1983.
93. Y. Gao, C. Fu, W. Hu and J. Yang, *J. Phys. Chem. Lett.*, 2022, **13**, 1–11.
94. R. Zhang, L. Zhang, Q. Zheng, P. Gao, J. Zhao and J. Yang, *J. Phys. Chem. Lett.*, 2018, **9**, 5419–5424.
95. J. Choi, M. Florian, A. Steinhoff, D. Erben, K. Tran, D. S. Kim, L. Sun, J. Quan, R. Claassen, S. Majumder, J. A. Hollingsworth, T. Taniguchi, K. Watanabe, K. Ueno, A. Singh, G. Moody, F. Jahnke and X. Li, *Phys. Rev. Lett.*, 2021, **126**, 047401.
96. H. Guo, X. Zhang and G. Lu, *Proc. Natl. Acad. Sci. U. S. A.*, 2021, **118**, e2105468118.
97. H. Niu, X. Wang, C. Shao, Y. Liu, Z. Zhang and Y. Guo, *J. Mater. Chem. A*, 2020, **8**, 6555–6563.
98. F. He, H. Li, Y. Ding, K. Li, Y. Wang and Z. Wu, *Carbon*, 2018, **130**, 636–644.
99. X. Niu, Y. Yi, X. Bai, J. Zhang, Z. Zhou, L. Chu, J. Yang and X. Li, *Nanoscale*, 2019, **11**, 4101–4107.
100. X. Zhang, T. Wu, C. Yu and R. Lu, *Adv. Mater.*, 2021, **33**, e2104695.

101. Z. Xue, X. Zhang, J. Qin and R. Liu, *Appl. Surf. Sci.*, 2020, **510**, 145489.
102. Z. Fu, C. Ling and J. Wang, *J. Mater. Chem. A*, 2020, **8**, 7801–7807.
103. Y. Fan, X. Song, H. Ai, W. Li and M. Zhao, *ACS Appl. Mater. Interfaces*, 2021, **13**, 34486–34494.
104. J. Hong, Z. Hu, M. Probert, K. Li, D. Lv, X. Yang, L. Gu, N. Mao, Q. Feng, L. Xie, J. Zhang, D. Wu, Z. Zhang, C. Jin, W. Ji, X. Zhang, J. Yuan and Z. Zhang, *Nat. Commun.*, 2015, **6**, 6293.
105. Y. C. Lin, B. Jariwala, B. M. Bersch, K. Xu, Y. Nie, B. Wang, S. M. Eichfeld, X. Zhang, T. H. Choudhury, Y. Pan, R. Addou, C. M. Smyth, J. Li, K. Zhang, M. A. Haque, S. Folsch, R. M. Feenstra, R. M. Wallace, K. Cho, S. K. Fullerton-Shirey, J. M. Redwing and J. A. Robinson, *ACS Nano*, 2018, **12**, 965–975.
106. W. H. Lin, W. S. Tseng, C. M. Went, M. L. Teague, G. R. Rossman, H. A. Atwater and N. C. Yeh, *ACS Nano*, 2020, **14**, 1350–1359.
107. L. Li, R. Long, T. Bertolini and O. V. Prezhdo, *Nano Lett.*, 2017, **17**, 7962–7967.
108. W. Pei, S. Zhou, J. Zhao, Y. Du and S. X. Dou, *J. Mater. Chem. A*, 2020, **8**, 20570–20580.
109. H. Guo, W. Chu, O. V. Prezhdo, Q. Zheng and J. Zhao, *J. Phys. Chem. Lett.*, 2021, **12**, 3960–3967.
110. H. Guo, W. Chu, Q. Zheng and J. Zhao, *J. Phys. Chem. Lett.*, 2020, **11**, 4662–4667.
111. L. Zhang, W. Chu, Q. Zheng, A. V. Benderskii, O. V. Prezhdo and J. Zhao, *J. Phys. Chem. Lett.*, 2019, **10**, 6151–6158.
112. M. Xue, D. Jiang, H. Du, Z. Li, H. Bi and Y. Yuan, *Catal. Sci. Technol.*, 2021, **11**, 5663.
113. Y. Sui, S. Liu, T. Li, Q. Liu, T. Jiang, Y. Guo and J.-L. Luo, *J. Catal.*, 2017, **353**, 250–255.
114. K. Li, Y.-Z. Lin, K. Wang, Y. Wang, Y. Zhang, Y. Zhang and F.-T. Liu, *Appl. Catal., B*, 2020, **268**, 118402.
115. F. Wang, Y. Wang, Y. Feng, Y. Zeng, Z. Xie, Q. Zhang, Y. Su, P. Chen, Y. Liu, K. Yao, W. Lv and G. Liu, *Appl. Catal., B*, 2018, **221**, 510–520.
116. K. Fujiwara and S. E. Pratsinis, *Appl. Catal., B*, 2018, **226**, 127–134.
117. H. Cheng, B. Huang and Y. Dai, *Nanoscale*, 2014, **6**, 2009–2026.
118. B. Pare, B. Sarwan and S. B. Jonnalagadda, *Appl. Surf. Sci.*, 2011, **258**, 247–253.
119. J. Tian, Z. Zhou, S. Zhang, Z. Li, L. Shi, Q. Li and J. Wang, *J. Mater. Chem. A*, 2021, **9**, 11753–11761.
120. Y. Zhao, S. Zhou, J. Zhao, Y. Du and S. X. Dou, *J. Phys. Chem. Lett.*, 2020, **11**, 9304–9312.
121. L. Xie, X. Li, X. Wang, W. Ge, J. Zhang, J. Jiang and G. Zhang, *J. Phys. Chem. C*, 2018, **123**, 1846–1851.
122. Y. Fan, J. Yu, X. Song, Y. Liu, S. Qi, W. Li and M. Zhao, *J. Mater. Chem. A*, 2021, **9**, 26077–26085.
123. X. Wang, X. Jiang, E. Sharman, L. Yang, X. Li, G. Zhang, J. Zhao, Y. Luo and J. Jiang, *J. Mater. Chem. A*, 2019, **7**, 6143–6148.

124. N. Liu, S. Zhou and J. Zhao, *Mater. Today Phys.*, 2021, **16**, 100312.
125. J. Zheng, W. Sheng, Z. Zhuang, B. Xu and Y. Yan, *Sci. Adv.*, 2016, **2**, e1501602.
126. R. Wang, K.-Q. Lu, Z.-R. Tang and Y.-J. Xu, *J. Mater. Chem. A*, 2017, **5**, 3717–3734.
127. Q. Zhang, P. Yang, H. Zhang, J. Zhao, H. Shi, Y. Huang and H. Yang, *Appl. Catal., B*, 2022, **300**, 120729.
128. T. Liu, H. Li, J. Gao, S. Ding, X. Liu, H. Jia and J. Xue, *Appl. Surf. Sci.*, 2022, **579**, 152135.
129. X. Yang, S. Wang, N. Yang, W. Zhou, P. Wang, K. Jiang, S. Li, H. Song, X. Ding, H. Chen and J. Ye, *Appl. Catal., B*, 2019, **259**, 118088.
130. S. Liu, L. Chen, T. Liu, S. Cai, X. Zou, J. Jiang, Z. Mei, Z. Gao and H. Guo, *Chem. Eng. J.*, 2021, **424**, 130325.
131. W. Gao, S. Li, H. He, X. Li, Z. Cheng, Y. Yang, J. Wang, Q. Shen, X. Wang, Y. Xiong, Y. Zhou and Z. Zou, *Nat. Commun.*, 2021, **12**, 4747.
132. L. Piccolo, P. Afanasiev, F. Morfin, T. Len, C. Dessal, J. L. Rousset, M. Aouine, F. Bourgain, A. Aguilar-Tapia, O. Proux, Y. Chen, L. Soler and J. Llorca, *ACS Catal.*, 2020, **10**, 12696–12705.
133. Y. Li, S. Wang, W. Chang, L. Zhang, Z. Wu, R. Jin and Y. Xing, *Appl. Catal., B*, 2020, **274**, 119116.
134. T. Boningari, S. N. R. Inturi, M. Suidan and P. G. Smirniotis, *Chem. Eng. J.*, 2018, **350**, 324–334.
135. H. Zhao, Z. Jiang, K. Xiao, H. Sun, H. S. Chan, T. H. Tsang, S. Yang and P. K. Wong, *Appl. Catal., B*, 2021, **280**, 119456.
136. H. Li, Z.-W. Wang, S. Zhang, G.-L. Li, W. Ju, T. Li and L. Li, *Phys. E*, 2021, **131**, 114730.
137. G. Gao, Y. Jiao, E. R. Waclawik and A. Du, *J. Am. Chem. Soc.*, 2016, **138**, 6292–6297.
138. C. Ling, X. Niu, Q. Li, A. Du and J. Wang, *J. Am. Chem. Soc.*, 2018, **140**, 14161–14168.
139. S. Wang, L. Shi, X. Bai, Q. Li, C. Ling and J. Wang, *ACS Cent. Sci.*, 2020, **6**, 1762–1771.
140. L.-F. Li, Y.-F. Li and Z.-P. Liu, *ACS Catal.*, 2019, **9**, 5668–5678.
141. J. Lang and Y. H. Hu, *Phys. Chem. Chem. Phys.*, 2020, **22**, 9250–9256.

CHAPTER 12

Scaling Up Photocatalysts and New Devices for Solar Water Splitting and CO_2 Reduction

TSUYOSHI TAKATA[a], KAZUNARI DOMEN[a,b], SAYURI OKUNAKA*[c,d,e], HIROMASA TOKUDOME*[c,d], NAOHIKO KATO[f], TAKESHI MORIKAWA[f] AND YASUHIKO TAKEDA[f]

[a]Research Initiative for Supra-Materials, Interdisciplinary Cluster for Cutting Edge Research, Shinshu University, 4-17-1 Wakasato, Nagano, 380-8553, Japan; [b]Office of University Professors, The University of Tokyo, 7-3-1 Hongo, Bunkyo-ku, Tokyo 113-8656, Japan; [c]Research Institute, TOTO Ltd, 2-8-1 Honson, Chigasaki, Kanagawa 253-8577, Japan; [d]Japan Technological Research Association of Artificial Photosynthetic Chemical Process (ARP-Chem), 2-11-9 Iwamotocho, Chiyoda-ku, Tokyo 101-0032, Japan; [e]Global Zero Emission Research Center (GZR), National Institute of Advanced Industrial Science and Technology (AIST), 16-1 Onogawa, Tsukuba, Ibaraki 305-8559, Japan; [f]Toyota Central Research and Development Laboratories Inc., 41-1 Yokomichi, Nagakute, Aichi, 480-1192 Japan
*E-mail: sayuri.okunaka@aist.go.jp, hiromasa.tokudome@jp.toto.com

12.1 Introduction

Recent global warming forces us to challenge the effective reduction of CO_2 emissions for constructing a sustainable society. Renewable fuel production technologies that do not rely on fossil fuels have received considerable

Sustainable Energy No. 1
Recent Developments in Functional Materials for Artificial Photosynthesis
Edited by Srabanti Ghosh and Qian Wang

Published by the Royal Society of Chemistry, www.rsc.org

attention in recent years as a means of resolving both energy and environmental problems. Among them, H_2 production through water splitting and reduction of CO_2 into useful compounds are highly expected due to their large-scale production availability. H_2 can be stored and converted to energy while emitting only water, allowing for CO_2 emissions to be reduced by actively introducing it as an energy carrier. Converting CO_2 to high-value-added chemical fuels has the potential to significantly reduce atmospheric CO_2 and achieve a carbon-neutral cycle. One of the most viable solutions is to use abundant sunlight to drive these chemical conversions. Artificial photosynthesis using semiconductor materials is considered to be one of the most promising and attractive approaches to convert solar energy into chemical fuels *via* water splitting and CO_2 reduction. This is because photocatalytic or photoelectrochemical reactions have advantages of convenience for the filtration or recovery process, and no need for harsh reaction conditions (*e.g.*, high temperature, high O_2 pressure, the use of strong acid or base). To date, various semiconductor photocatalysts and photoelectrodes that can work efficiently under solar-light irradiation have been developed.

The development of a scalable artificial photosynthesis system is an important issue in the practical application because of the low flux density of solar energy when compared to other energy forms. Recently, solar-light-driven photocatalysts and photoelectrodes have been applied to scalable panel devices. Instead of conventional powder suspension systems studied in a laboratory, those systems have some merits of no need to keep the photocatalyst particles suspended during the reaction, or separate particles from suspension for recycling. For overall water splitting, aluminum-doped strontium titanate ($SrTiO_3$:Al) has been reported as a material that can split water with high efficiency (an apparent quantum efficiency of almost 100%) under UV light illumination, and demonstrated large-scale H_2 production by fabricating the film onto a glass substrate.[1,2] Additionally, visible-light-driven Z-schematic photocatalyst sheets, wherein two semiconductor photocatalyst particles (H_2 evolution photocatalysts and O_2 evolution photocatalysts) are fixed onto a glass substrate, have been reported. Photocatalyst sheets can be prepared using a physical vapor deposition process under vacuum (*e.g.*, sputtering or vacuum evaporation), and those composed of La- and Rh-co-doped $SrTiO_3$ and $BiVO_4$:Mo embedded in a Au, C or ITO layer enables overall water splitting with the highest solar-to-hydrogen conversion efficiency (STH) of ~1.1%.[3–5] Photocatalyst sheets can also be prepared by a simple screen printing process. These printable photocatalyst sheets have the advantage of low cost because there is no need to use vacuum processes during the preparation. Alternatively, the separation of two functions of the photocatalysts has significantly improved the solar-to-chemical conversion efficiency, for example, the photogeneration of electron–hole pairs and reduction/oxidation reactions occur on solar cells and electrochemical reactors, respectively. This would be a particularly viable method for highly efficient CO_2 conversion, owing to the additional difficulties associated with the higher activation energy. However, scale-up of the reactors required improvements of the catalyst materials, electrolyte flow channels, *etc.*

In this chapter, challenges for the practical use of solar H_2 production *via* photocatalytic water splitting; the development of (i) 100 m^2 sized panel reactor using $SrTiO_3$:Al photocatalyst, and (ii) inexpensive visible-light-driven printable photocatalyst sheets, composed of $SrTiO_3$:Rh and $BiVO_4$ photocatalyst are described. In addition, the development of a 1 m^2 sized device consisting of crystalline silicon solar cells and an electrochemical reactor that is responsive to visible light for CO_2 reduction reaction is presented.

12.2 The Development of an Al-doped $SrTiO_3$ Photocatalyst for Overall Water Splitting and Its Application to Large-scale Solar Hydrogen Production

12.2.1 Photocatalytic Overall Water Splitting with a Quantum Efficiency of Almost Unity

Overall water splitting with a quantum efficiency approaching 100% was achieved in a recent study.[1] Although this high level of efficiency was the result of utilizing a wide-bandgap oxide requiring ultraviolet light excitation, it represents an important milestone in the development of active photocatalysts. This prior research employed a $SrTiO_3$-based photocatalyst having a band potential capable of splitting water that has been examined as a water-splitting photocatalyst from the very beginning of work in this field.[6–9] The bandgap of $SrTiO_3$ is 3.2 eV and it thus absorbs wavelengths shorter than approximately 390 nm. Even though $SrTiO_3$ functions only in response to ultraviolet light, it has the potential to split water under sunlight because the solar spectrum contains some ultraviolet radiation. The high quantum efficiency obtained from this catalyst resulted from strategic preparation of $SrTiO_3$ and the use of an optimal combination of co-catalysts loaded using carefully selected methods. Specifically, a flux technique was used to obtain single-crystalline $SrTiO_3$ particles with defined morphologies[10,11] in addition to defect engineering *via* flux-mediated doping[11] and the photodeposition of hydrogen evolution reaction (HER) and oxygen evolution reaction (OER) co-catalysts on selected facets.[1] All of these factors effectively enhanced the water-splitting activity of the $SrTiO_3$.

Flux methods are typically employed for the growth of single-crystal ingots and large crystals can be obtained *via* repeated dissolution and recrystallization processes. During these processes, the ratio of nucleation and growth rates can be controlled by adjusting the synthetic conditions, including the proportion of the target material to flux (that is, the solute–solvent ratio) and the temperature program applied during heating. By optimizing these factors, fine and highly crystalline semiconductor particles can be obtained that are suitable for use as powder photocatalysts. Well-crystallized $SrTiO_3$

particles can be synthesized by heating commercially available $SrTiO_3$ in molten $SrCl_2$ to produce cuboid particles likely to have exposed [100] facets based on the perovskite structure of the material (see Figure 12.1(a)). Interestingly, contamination from the crucible used to contain the flux has also been shown to effectively increase photocatalytic activity. Specifically, Al leached from the crucible is incorporated into the product, leading to the unintentional Al-doping of the $SrTiO_3$.[11] In this case, Al^{3+} ions preferentially occupy Ti^{4+} sites rather than Sr^{2+} sites because of the similar ionic radii of the former two elements. The doping of $SrTiO_3$ with lower valence cations is known to suppress its n-type semiconductor character and to remarkably enhance its water splitting activity.[12]

Because this accidental Al-doping was found to effectively increase water splitting activity, the effect of adding Al_2O_3 powder to $SrTiO_3$ was examined. Intentional addition of Al_2O_3 powder into the starting mixture of $SrTiO_3$ and $SrCl_2$ prior to the flux treatment further enhanced photocatalytic activity.[10] The addition of Al_2O_3 resulted in the formation of smaller crystalline particles compared with the product obtained without Al_2O_3, as is evident from

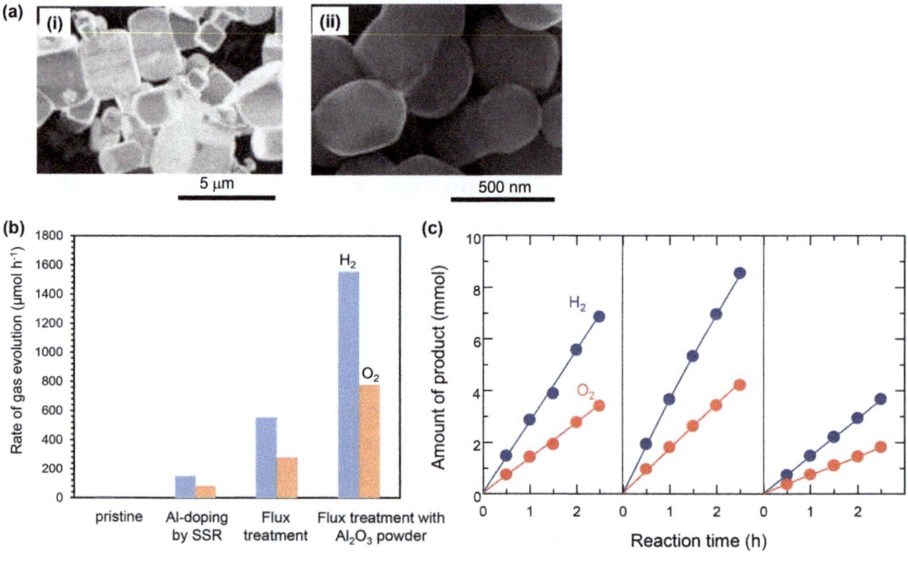

Figure 12.1 (a) SEM images of $SrTiO_3$ treated with $SrCl_2$: (i) with and (ii) without the addition of Al_2O_3 powder. (b) Photocatalytic water splitting activities of $SrTiO_3$ after various treatments. Reaction conditions: catalyst 0.1 g, Rh–Cr oxide co-catalyst (0.1 + 0.1 wt% loading), 300 W Xe lamp (full arc). (c) H_2 and O_2 evolutions over time on $SrTiO_3$:Al loaded with various co-catalysts during photoirradiation. Left: loaded with Rh (0.1 wt%)/Cr_2O_3 (0.05 wt%) by two-step photodeposition. Middle: loaded with Rh (0.1 wt%)/Cr_2O_3 (0.05 wt%)/CoOOH (0.05 wt%) by three-step photodeposition. Right: loaded with Rh–Cr oxide (0.1 + 0.1 wt%) by co-impregnation. Reproduced from ref. 1 with permission from Springer Nature, Copyright 2020.

Figure 12.1(b). In addition, the particles obtained with Al_2O_3 were somewhat rounded from the cuboid morphology that was otherwise observed. It should be noted that the amount of Al incorporated into the product was almost unchanged, regardless of the addition of Al_2O_3 powder and the added Al_2O_3 in this material appeared to prevent excessive crystal growth. The sample with added Al_2O_3 also showed the best water splitting activity among various samples (Figure 12.1(c)). The photocatalytic activity of this material was examined by loading the Rh–Cr oxide catalyst *via* a conventional impregnation method[13] and the water splitting performance was further increased as described below.

The loading of a co-catalyst onto the Al-doped $SrTiO_3$ (denoted herein as $SrTiO_3$:Al) *via* photodeposition resulted in enhanced water splitting, as shown in Figure 12.1(d). The photodeposition of Rh using $RhCl_3$ as a precursor followed by Cr_2O_3 photodeposition from K_2CrO_4 formed an HER co-catalyst having a Rh-core/Cr_2O_3-shell structure.[14] The loading of the co-catalyst by photodeposition in this manner was determined to be more effective than using an impregnation method, resulting in an approximate doubling of the photocatalytic activity. A Cr_2O_3 layer is known to selectively inhibit the access of O_2 to the Rh interfaces serving as active sites for the HER and oxygen reduction reaction (ORR), such that water splitting proceeds efficiently without competing with the ORR.[15] In addition, the photodeposition of Co on the Rh/Cr_2O_3-photodeposited sample using a $Co(NO_3)_2$ precursor provided an additional (albeit moderate) increase in the photocatalytic activity because these Co species promoted the OER.

Figure 12.2 presents the action spectrum for water splitting on the optimized photodeposited sample. Here, the external quantum efficiency is defined as the number of converted photons ratioed to the number of incident photons. This definition is used because the exact number of absorbed photons can be difficult to determine exactly as a result of scattering. The external quantum efficiency was found to increase at lower wavelengths, in general agreement with the light absorption characteristics of this material. In the region of 365–350 nm, over which the maximum absorption was observed, the external quantum efficiency was approximately 96%. Theoretically, the internal quantum efficiency would have been even higher (close to 100%), although the exact value could not be determined. The low-efficiency values in the wavelength range associated with low light absorption can be at least partially attributed to this decreased absorption. The external efficiencies near the absorption edge wavelengths are inevitably low for such materials unless the onset of absorption is steep. The STH determined for this catalyst under simulated sunlight was 0.65%.

The structural and functional characteristics of this material that resulted in its high quantum efficiency were examined. The scanning electron microscopy (SEM) images in Figure 12.3(a)–(d) demonstrate that the Rh co-catalyst particles were not homogeneously dispersed but rather were situated in specific areas. After the subsequent photodeposition of Cr_2O_3, the locations of these co-catalyst particles were unchanged, indicating that Cr was deposited

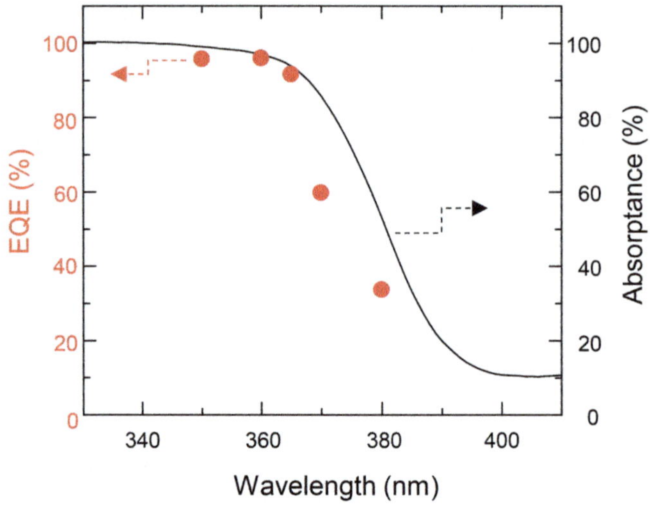

Figure 12.2 The UV-vis diffuse reflectance spectrum of bare $SrTiO_3$:Al (black solid line) and the wavelength dependence of external quantum efficiency (EQE) during water splitting on Rh (0.1 wt%)/Cr_2O_3 (0.05 wt%)/CoOOH (0.05 wt%)-loaded $SrTiO_3$:Al (red symbols). Reproduced from ref. 1 with permission from Springer Nature, Copyright 2020.

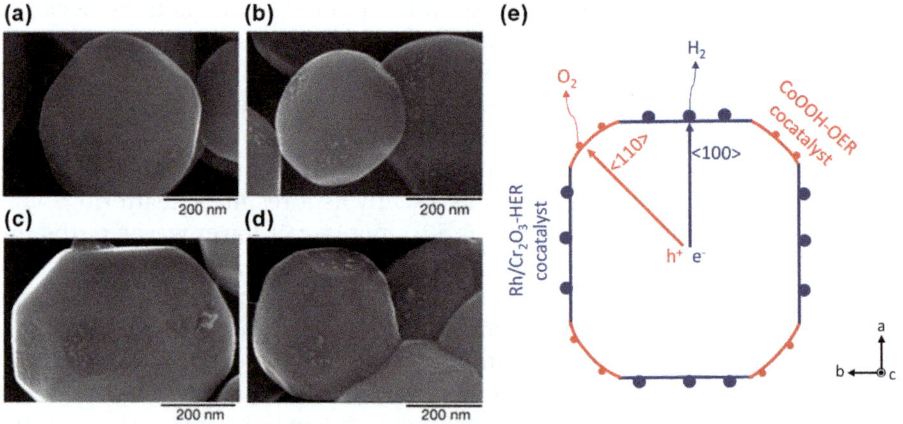

Figure 12.3 SEM images of $SrTiO_3$:Al loaded with various co-catalysts. (a) Unloaded and loaded with (b) Rh (0.1 wt%), (c) Rh (0.1 wt%)/Cr (0.05 wt%) and (d) Rh (0.1 wt%)/Cr_2O_3 (0.05 wt%)/CoOOH (0.05 wt%). (e) Schematic showing the overall *water* splitting reaction mechanism on the optimized $SrTiO_3$:Al photocatalyst. Reproduced from ref. 1 with permission from Springer Nature, Copyright 2020.

on the Rh particles. In fact, the Rh/Cr_2O_3 HER co-catalyst was found to be deposited preferentially on [100] facets. During the photodeposition, Rh^{3+} ions were reduced to Rh^0 metal by photoexcited electrons on the surfaces of the semiconductor particles. Similarly, CrO_4^{2-} ions containing Cr^{6+} were reduced to Cr_2O_3 containing Cr^{3+} by photoexcited electrons. These results confirm that the Rh/Cr co-catalyst was located on facets capable of attracting electrons and that the Rh/Cr_2O_3 functioned as an HER co-catalyst. In contrast, the Co co-catalyst was selectively deposited on different facets. The surface regions on which the Co co-catalyst was situated could not be clearly indexed because of the curved surface exposure, but likely comprised high-index facets in the range of [110]–[111]. During the deposition of this co-catalyst, the original Co^{2+} ions were not reduced but rather were oxidized by holes to Co^{3+} and then precipitated as CoOOH.[16] Thus, the CoOOH co-catalyst would have been preferentially deposited on hole-attracting facets and so would have functioned as an OER co-catalyst.

Figure 12.3(e) shows the reaction mechanism for overall water splitting on this photocatalyst. As noted, the co-catalysts were deposited on specific facets and so it is evident that photogenerated electrons and holes migrated anisotropically in this material such that they were spatially separated in each semiconductor particle.[1,17] Moreover, the HER and OER co-catalysts were located on those facets to which electrons and holes preferentially migrated, respectively. This structural feature helped to suppress the accumulation of photogenerated carriers in the semiconductor particles, meaning that the recombination of photogenerated charges was almost completely inhibited and the quantum efficiency was increased. Because photocatalytic water splitting involves reversible, multi-electron transfer steps, there are multiple opportunities for back electron transfer during this process. On a thermodynamic basis, this back electron transfer is preferred and so should be dominant, which inevitably causes a loss of quantum efficiency. Thus, whether a photocatalyst can enable water splitting with 100% quantum efficiency has remained an open question. The research with $SrTiO_3$:Al discussed herein has demonstrated that 100% efficiency is indeed possible. In addition, the functional and structural features of this photocatalyst enabling this exceptional quantum efficiency have been elucidated. As such, this work represents an important milestone in the development of active photocatalysts for overall water splitting. It is also expected that the knowledge gained from this research will be applicable to the study of narrow-bandgap photocatalysts allowing more efficient utilization of visible light.

12.3 Photocatalytic Solar Hydrogen Production from Water on a 100 m^2 Scale

In addition to the development of efficient photocatalysts in laboratory experiments, actual sunlight-driven photocatalytic water splitting in an outdoor environment has recently been explored.[2,10] The construction and operation of prototype photocatalytic solar hydrogen production stations are vital to examining issues related to eventual industrialization and to addressing these challenges. To date, these have not been attempted primarily because photocatalysts capable of steadily producing hydrogen and oxygen with the required efficiency have not been developed. However, the aforementioned state-of-the-art $SrTiO_3$:Al photocatalyst provides an STH as high as 0.65% under simulated sunlight. Although this efficiency is still insufficient to satisfy the requirements for commercialization, it is high enough to allow the fabrication of a prototype solar hydrogen production system.

In a conventional photocatalytic laboratory system, the photocatalyst powder is kept suspended by stirring and the gas phase of the reaction system is kept at a lower pressure than ambient to mitigate the effects of mass transfer limitations. In contrast, a large-scale solar hydrogen plant cannot realistically provide the mechanical power for stirring and pressure reduction. As such, it is vital to develop fixed powder processes, although the practical design and operation of such reactors have only recently been reported.[10] As noted, the reverse reaction to form H_2O from the H_2 and O_2 evolved by water splitting can be successfully inhibited by the addition of a Cr-based co-catalyst. In such cases, the rate of the water-splitting reaction will be almost independent of the partial pressures of H_2 and O_2 in the gas phase of the reactor, which could eliminate the need for reduced pressure. Despite this, the coevolution of gaseous H_2 and O_2 still represents an explosion hazard and so downstream gas separation is indispensable. It is also important to evaluate the potential outcome of an explosion in such systems and to establish means of minimizing any hazards. On the basis of the above, a prototype photocatalytic solar hydrogen production facility involving sequential gas generation, transport and separation processes was constructed and subsequently operated.[2]

This prototype photocatalytic reactor was based on a panel-type design using photocatalyst sheets in which photocatalyst powder was fixed on frosted glass substrates. These units were fabricated by spraying glass sheets with a slurry of the $SrTiO_3$:Al photocatalyst mixed with chained colloidal silica and $CaCl_2$, followed by drying and pre-heating. In this manner, a Rh, Cr, Co-ternary co-catalyst could be loaded *via* a one-pot impregnation method that provided superior stability and easier large-scale preparation, despite lowering the quantum efficiency of the unit compared with stepwise photodeposition. The silica and $CaCl_2$ were added as binders to assist in embedding the photocatalyst particles. The porous silica particles also provided a hydrophilic layer that allowed the ingress of water and egress of gaseous products from the photocatalyst layer. Consequently, an external energy input was not required because stirring and air evacuation were not required. In preliminary trials, the durability of these photocatalyst sheets was examined under simulated sunlight in a laboratory setting. A maximum STH of 0.51% was obtained followed by a gradual decrease to a value of 0.4% that was maintained for longer than 1600 h (Figure 12.4).

Figure 12.5(a) and (b) present a photographic image and diagram of the panel reactor used in the field tests. The photocatalyst sheets used in these trials had a light-receiving area of 625 cm^2 (25 cm × 25 cm). The upper surface of the panel reactor comprised an ultraviolet-transparent glass window and the gap between the inner surface of the window and the upper surface of the photocatalyst sheet was adjusted to 0.1 mm. This structure minimized both

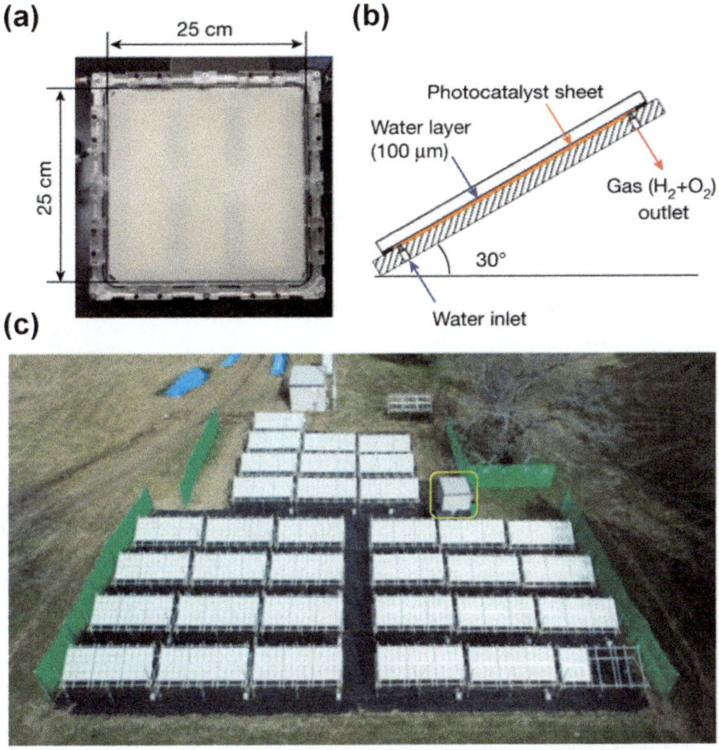

Figure 12.4 (a) A photographic image of a panel reactor unit (625 cm^2). (b) The structure of the panel reactor unit viewed from the side. (c) An overhead view of the 100 m^2 solar hydrogen production system consisting of 1600 panel reactor units and a hut housing a gas separation facility (indicated by the yellow box). Reproduced from ref. 2 with permission from Springer Nature, Copyright 2021.

the water load of the unit and the accumulation of the flammable mixture of H_2 and O_2 that was produced. Upon exposure to light, bubbles of this gaseous mixture 0.1–0.6 mm in size were generated on the sheet surfaces and many small bubbles coalesced to form larger bubbles. This bubble growth reduced the gas–liquid interface area and the resulting scattering of incident light. The narrow depth of the water layer in this reactor design also promoted the release of the gas bubbles.

A total of 1600 such panel reactor units with a total light accepting area of 100 m^2 were used to form an array, as shown in Figure 12.5(c), to investigate the feasibility of large-scale solar hydrogen production by photocatalytic water splitting. This solar hydrogen production system was built at the Kakioka Research Facilities, a branch of the University of Tokyo in Ishioka City, Ibaraki prefecture, Japan (140.32° E. 36.38° N). Each panel was positioned to face south and tilted upwards at an angle of 30° and all the reactor

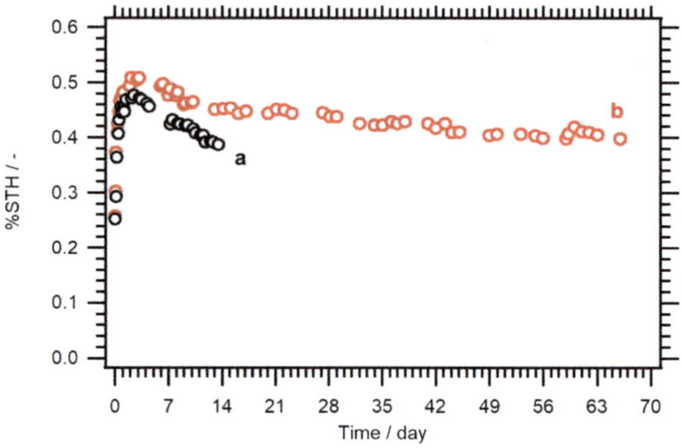

Figure 12.5 The STH values of small photocatalyst sheets (25 cm^2) prepared using (a) manual spraying onto flat, clear glass and (b) a program-controlled sprayer with frosted glass, during the overall water splitting reaction under continuous irradiation with simulated standard sunlight. Reproduced from ref. 2 with permission from Springer Nature, Copyright 2021.

units were connected together by plastic tubing to deliver the moist oxyhydrogen to the separator. A hydrogen-permeable filter element made of commercial polyimide hollow fibers was used as a membrane separator.[18] A dual gas storage tank unit used to control the gas feed rate, and quantification tanks for analysis of the filtrate and residual gases, were situated upstream and downstream of the membrane unit, respectively. A diaphragm pump was used for filtrate suction.

During the field tests, a maximum STH of 0.76% was obtained on a summer day with limited cloud cover (11:00–11:30 am, August 29, 2020, 0.88 kW m^{-1}). This STH value was higher than that obtained in the preliminary laboratory test using simulated sunlight derived from an artificial light source, because the natural sunlight at that point in time contained more intense UV radiation than the simulated sunlight. Solar irradiance and photocatalytic activity data were continuously acquired from September 22 to December 21, 2020 and the proportion of ultraviolet light in the natural sunlight was found to decrease from 5% to 2% during this time span. As a result, the STH also decreased, from 0.5% to 0.2%. It is evident from these results that the performance of the catalyst itself was essentially constant throughout the trial. When the solar water splitting and gas separator units were linked, filtrate gas with a hydrogen content of approximately 94% was obtained (Figure 12.6), which was close to the range out of the explosion limit. Although moist oxyhydrogen was present in the headspace of each panel and in the connecting gas delivery tubes, no unintentional ignitions or explosions occurred during the test period. In addition, an intentional ignition did not generate any noticeable damage because of the small volume of gas in the reaction

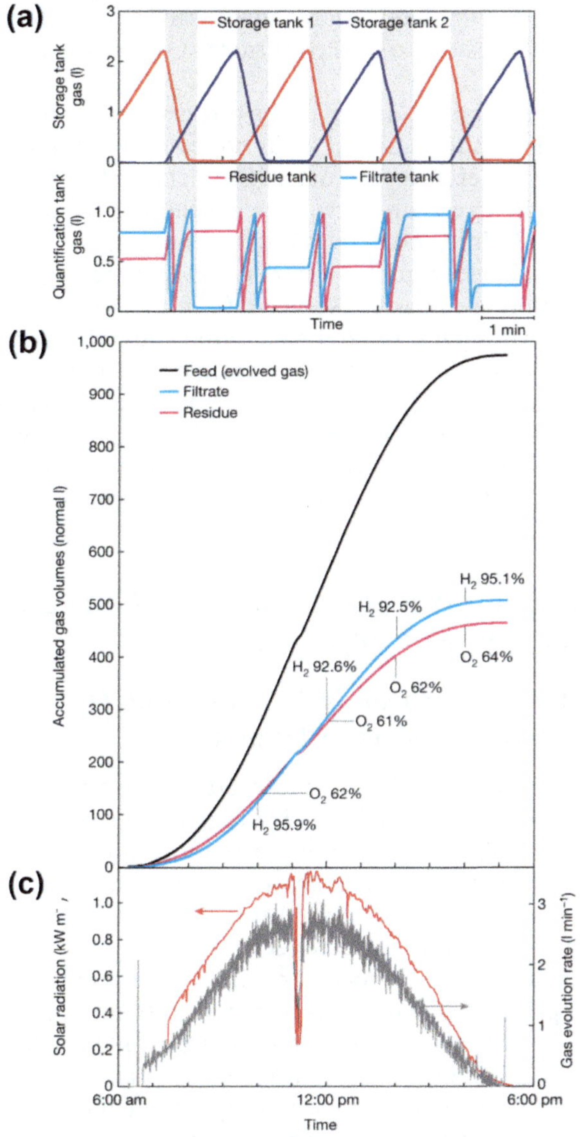

Figure 12.6 (a) Time courses of the gas volumes contained in the storage and quantification tanks. The periods during which the filter cartridge was operational are shaded. (b) The accumulated amounts of moist oxyhydrogen feed, filtrate and residue gases. The proportions of hydrogen and oxygen in the filtrate and residue gases (excluding water vapor), respectively, are noted along the curves. (c) Variations in solar illumination intensity (red) and gas evolution from the water splitting panel reactor (grey) over a 24 h cycle. Reproduced from ref. 2 with permission from Springer Nature, Copyright 2021.

system. Although a complete hazard analysis of this system has not yet been performed, the results to date confirm that this process presents a very low risk.

It should be noted that the design of the prototype solar hydrogen production facility described above did not focus on reducing operational or capital construction costs. Therefore, the use of inexpensive but durable materials to fabricate the panel reactors as well as developing means of prolonging the photocatalyst lifetime and of recycling/regenerating this material, reducing the power inputs necessary for water and gas processing, and optimizing safety and operation are possible subjects for future research. Even so, the scale-up of solar hydrogen production *via* photocatalytic overall water splitting to a size of 100 m^2 was shown to be feasible. Theoretically, further scaling would be possible using a similar system without any loss of efficiency. In conjunction with the development of new photocatalysts with STH values of 5–10%, the scale-up of such facilities is another important step toward the eventual industrialization of solar hydrogen production.

12.4 Printable Photocatalyst Sheet Composed of $SrTiO_3$:Rh and $BiVO_4$:Mo

12.4.1 Printable Photocatalyst Sheet Using Direct Interparticle Electron Transfer

Among various visible-light-driven photocatalysts, we used $SrTiO_3$:Rh and $BiVO_4$(:Mo) photocatalyst particles as hydrogen and oxygen evolution photocatalysts, respectively, for fabricating a photocatalyst sheet, because the combination of these two photocatalysts has reported being able to proceed Z-schematic water-splitting reaction through interparticle electron transfer.[19] We prepared $SrTiO_3$:Rh and $BiVO_4$ photocatalyst nanoparticles *via* a newly-developed water-based process.[20,21] Both the photocatalyst nanoparticles have small primary particle sizes below *ca.* 100 nm with high crystallinity after calcination at an appropriate temperature (500–1100 °C).

Printed $SrTiO_3$:Rh/$BiVO_4$ photocatalyst sheets can be fabricated as shown in Figure 12.7(a).[22,23] Ru nanoparticles for hydrogen generation co-catalysts were pre-loaded on $SrTiO_3$:Rh particles using photodeposition. The $SrTiO_3$:Rh and $BiVO_4$ particles were mixed with viscous organic solution (*e.g.*, terpineol–butyl carbitol–acrylic resin = 2 : 2 : 1) as a vehicle, yielding the paste of $SrTiO_3$:Rh and $BiVO_4$ particles with an appropriate viscosity for printing. The mixed films were prepared by screen-printing using the paste, followed by calcination in air. In the case of controlling the thickness, the screen-printing procedure was repeated (1–10 times), or another setup of screen-printing with a thicker screen was used before calcination.

The obtained $SrTiO_3$:Rh/$BiVO_4$ sheet was found to have a densely packed porous structure and possess a homogeneous thickness of *ca.* 10 μm with a flat surface (Figure 12.7(b)). In addition, both the $SrTiO_3$:Rh and $BiVO_4$

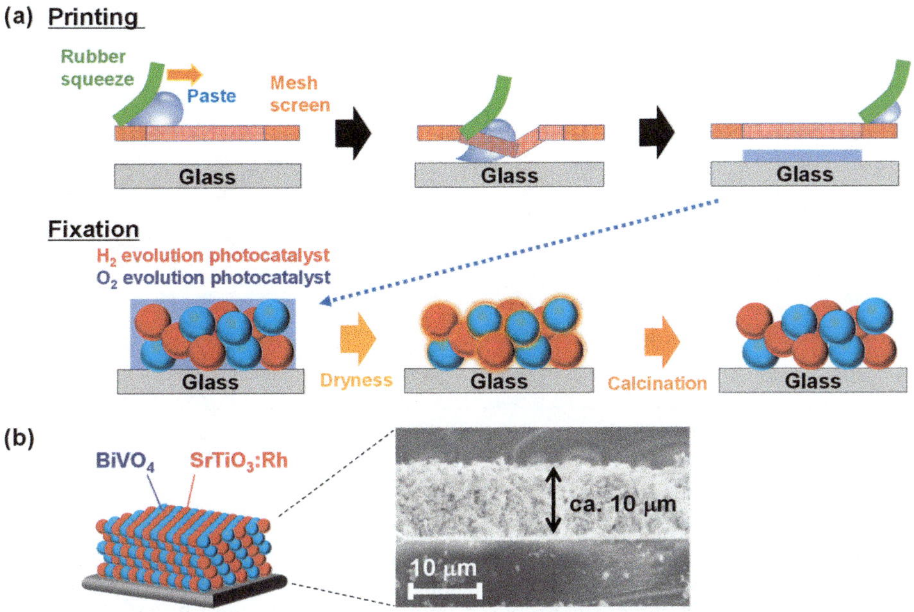

Figure 12.7 (a) Preparation processes of printed photocatalyst sheet, and (b) cross-sectional SEM image of nanopaticulate $SrTiO_3$:Rh/$BiVO_4$ sheet, which was prepared by calcination at 300 °C.

particles were not changed in size compared to the original ones even after calcination at temperatures up to 500 °C. We evaluated the water splitting activity on the obtained $SrTiO_3$:Rh/$BiVO_4$ sheets under visible light. A Xe lamp equipped with a long-pass filter (L-42) is used as a light source. When we examined the water splitting reaction, the $SrTiO_3$:Rh/$BiVO_4$ sheet was horizontally fixed at the bottom of a reaction vessel in 100 mL of pure water or a pH-adjusted aqueous solution. The amounts of H_2 and O_2 gases evolved *via* photocatalytic water splitting are measured using a gas chromatograph directly connected to a closed circulation system. Under visible light irradiation, the $SrTiO_3$:Rh/$BiVO_4$ sheets generated H_2 and O_2 simultaneously from pure water, while the sheets consisting of their single photocatalyst counterparts could not generate any gasses (Figure 12.8). The water splitting activities on $SrTiO_3$:Rh/$BiVO_4$ sheets were dependent on the calcination temperature, the sheet calcined at 300 °C showed the best performance. From the viewpoint of avoiding passivation formation between the two different photocatalysts and co-catalyst deactivation, it is desirable that the printed sheets should be calcined at a low temperature. Indeed, the rate of gas evolution on the $SrTiO_3$:Rh/$BiVO_4$ sheets significantly decreased with increasing calcination temperature up to 500 °C. However, the sheet calcined at <250 °C showed low activity for water splitting, because of forming the residual carbons derived from incomplete burning of the organic vehicle. Thus, the calcination temperature is an important factor for preparing an efficient

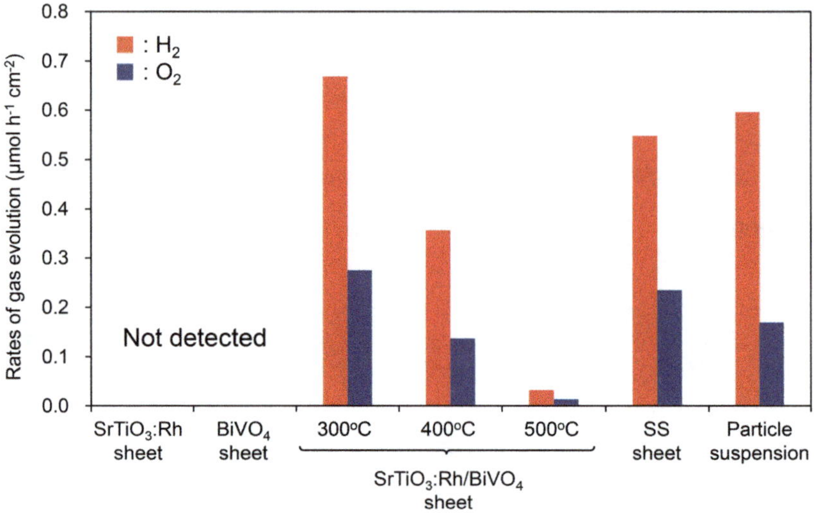

Figure 12.8 (a) Water splitting activity over $SrTiO_3$:Rh, $BiVO_4$, and $SrTiO_3$:Rh/$BiVO_4$ sheet and these particle suspension under visible-light irradiation. The SS-sheet was prepared by using $SrTiO_3$:Rh and $BiVO_4$ photocatalyst obtained *via* solid-state reaction. Sheet size, 5 × 5 cm; thickness, 10 μm; calcination temperature, 300 °C; reactant solution, 100 mL of pure water at 25 °C; reactant cell, top-irradiated separable cell; light source, 300 W Xe lamp with cut-off filters ($\lambda > 410$ nm).

photocatalyst sheet. The optimized photocatalyst sheet using the nanoparticulate $SrTiO_3$:Rh and $BiVO_4$ photocatalysts exhibited higher water splitting activity under visible-light irradiation compared to that using the large particles (>300 nm) prepared *via* the conventional solid-state reaction. Figure 12.8 compared the water splitting activities of $SrTiO_3$:Rh/$BiVO_4$ sheet and particle suspensions. The water splitting activity on the $SrTiO_3$:Rh/$BiVO_4$ sheet has been found to be almost the same as that on the corresponding suspended particles. It is considered that the porous $SrTiO_3$:Rh/$BiVO_4$ sheet could be maintained the original activity of photocatalyst particles for H_2/O_2 evolution under visible light even after the immobilization on a substrate.

12.4.2 Printable Photocatalyst Sheet with Colloidal Au Mediator

To date, when the $SrTiO_3$(:La), Rh and $BiVO_4$(:Mo) photocatalysts are embedded in an appropriate conductive layer such as gold (Au) or carbon (C), prepared *via* a physical vapor deposition process under vacuum, highly efficient water splitting reaction is found to proceed *via* a Z-scheme mechanism based on two-step excitation.[3,4] Introducing a conductive material having both appropriate working function and conductivity, works as a solid-state mediator for efficient electron transfer between the two photocatalysts, resulting in the promotion of the water splitting activity. Efficient water splitting can be also realized by using printed photocatalyst sheets consisting of $SrTiO_3$:Rh, $BiVO_4$(:Mo), and conductive nanocolloid such as Au, which were prepared *via* the similar screen-printing method as above using a paste that contained the respective nanocolloid.[3,5] We first investigated the printed photocatalyst sheet using $SrTiO_3$:Rh, $BiVO_4$:Mo, and Au nanocolloid with the size of <20 nm as a colloidal mediator (denoted as $SrTiO_3$:Rh/Au/$BiVO_4$:Mo).

SEM images of printed $SrTiO_3$:Rh/Au/$BiVO_4$:Mo sheet after calcination at 300 °C are displayed in Figure 12.9(a) as representative examples. It should be noted that the Au nanoparticles dispersed on the surface of both the $SrTiO_3$:Rh and $BiVO_4$:Mo particles to bridge between those particles effectively. It is evident that two photocatalyst particles and Au nanocolloid are

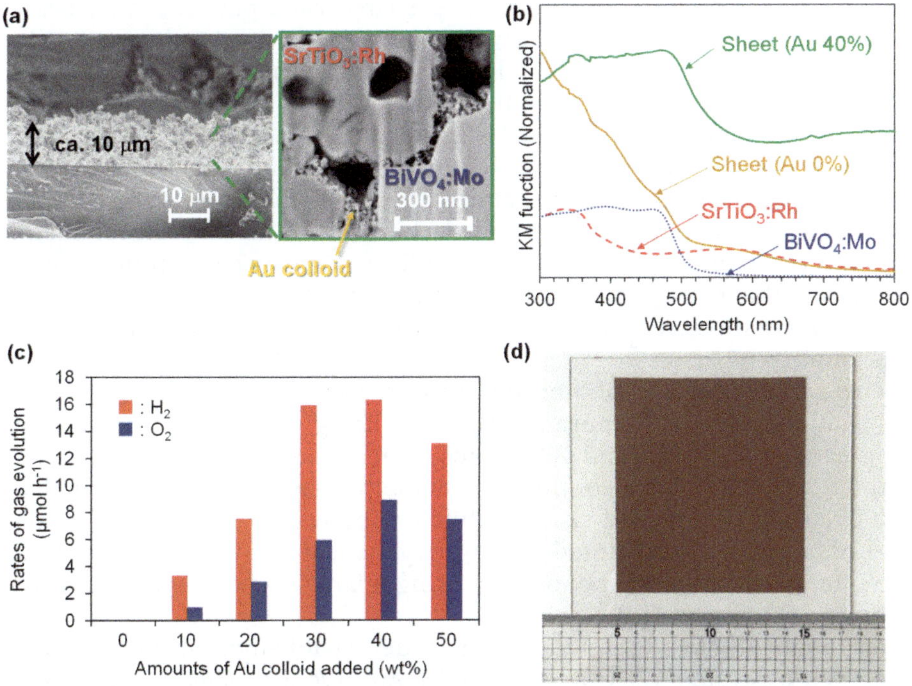

Figure 12.9 (a) SEM image of printed $SrTiO_3$:Rh/Au/$BiVO_4$ photocatalyst sheet. (b) UV-vis spectra of $SrTiO_3$:Rh/$BiVO_4$ photocatalyst sheet with/without containing Au colloid (40 wt%) and the raw materials ($SrTiO_3$:Rh and $BiVO_4$ particles), which were used to prepare the sheet. (c) The gas evolution rates under the visible-light irradiation over the printed $SrTiO_3$:Rh/Au/$BiVO_4$ photocatalyst sheet prepared by adding different amount of Au colloid (Au colloid: 0–50 wt%). Sheet size, 2.5 × 2.5 cm; thickness, 10 μm; reactant solution, 40 mL of pure water at 25 °C; reactant cell, top-irradiated separable cell; light source, 300 W Xe lamp with cut-off filters (λ > 410 nm). (d) Picture of printed $SrTiO_3$:Rh/Au/$BiVO_4$ photocatalyst sheet with large area (10 × 10 cm).

directly contacted with each other. As a result, excited electrons in $BiVO_4$:Mo can be transferred efficiently to recombine with excited holes in $SrTiO_3$:Rh particles through Au colloids. In the diffuse reflection spectrum of the $SrTiO_3$:Rh/Au/$BiVO_4$:Mo sheet shown in Figure 12.9(b), the absorption derived from Au colloid appeared in addition to the spectrum of the sheet without Au colloid, which roughly agreed with the sum of each diffuse reflection spectrum of $SrTiO_3$:Rh and $BiVO_4$:Mo. Since the broad absorption attributed to the localized surface plasmon resonance of the Au nanocolloid appears in the visible-light region, it is considered that the absorption of the photocatalyst can be slightly inhibited. However, both photocatalysts can absorb most of incident visible light because Au nanocolloid have a sufficiently small size, allowing efficient water splitting reactions.

The printed $SrTiO_3$:Rh/Au/$BiVO_4$:Mo sheets produced H_2 and O_2 from pure water through the Z-scheme mechanism based on two-step excitation under visible-light irradiation, like the sheet prepared *via* a physical vapor deposition process. In addition, their water splitting activity was influenced by the calcination temperature and the amount of Au nanocolloid added. For example, the water splitting activities of the printed $SrTiO_3$:Rh/Au/$BiVO_4$:Mo sheet with different amounts of Au nanocolloid mediator (0–50 wt%) are shown in Figure 12.9(c). The water splitting activity of the printed sheet was enhanced with increasing added amounts of Au nanocolloid up to 40 wt%, indicating that Au nanoparticles effectively served as an electron conductor between two photocatalysts. The water splitting activity of the printed sheet was also enhanced by loading co-catalyst, the STH of the printed $SrTiO_3$:Rh/Au/$BiVO_4$:Mo sheet loaded with Ru co-catalysts using photodeposition was found to be *ca.* 0.1%.

The present printed photocatalyst sheet can be extended to scalable applications. This is because the water splitting reaction can proceed as long as the $SrTiO_3$:Rh and $BiVO_4$:Mo particles are bridged in proximity, without the need of the continuous conductive layers as an external electric circuit in the printed sheet, unlike in photoelectrochemical systems. Indeed, a 10 × 10 cm printed photocatalyst sheet was easily prepared *via* the screen-printing method, only by using a large-size of screen and glass substrate (Figure 12.9(d)).

Although efficient water splitting under visible-light irradiation has been demonstrated by printed $SrTiO_3$:Rh/Au/$BiVO_4$:Mo sheets, the activity decreased under ambient pressure due to reverse reactions. It is caused by the ORR and the water formation from H_2 and O_2 that occurred on the sheet during the reaction and hindered their practical application. Therefore, we tried to realize efficient water splitting at the ambient pressure using an inexpensive and extensible process by coating amorphous oxide onto the printed sheet.[24] Amorphous oxyhydroxide covering the whole surface of the photocatalyst particles have known to be able to suppress reverse reactions involving O_2 effectively thanks to its permselectivity.[25–27] Indeed, forming the zirconium oxide layer, which was prepared by facile drop-casting of 2-propanol diluted zirconium alkoxide solution and followed by dryness, effectively suppresses the reverse reactions without blocking the active sites. It allowed a printed photocatalyst sheet to split water efficiently under visible light under near ambient pressure as shown in Figure 12.10. The water splitting activity of printed photocatalyst sheets depended on loaded amounts of zirconium oxide and solvents of zirconium(IV) butoxide diluted by terpineol used as a precursor. As a result, the photocatalytic performance of the optimized photocatalyst sheet at 90 kPa retained 84% of that at 10 kPa, whereas the pristine photocatalyst sheet maintained only half the activity. The concept of a facile modification onto the sheets opens up the prospect for large-scale application of photocatalyst sheets using other visible-light-driven photocatalysts.

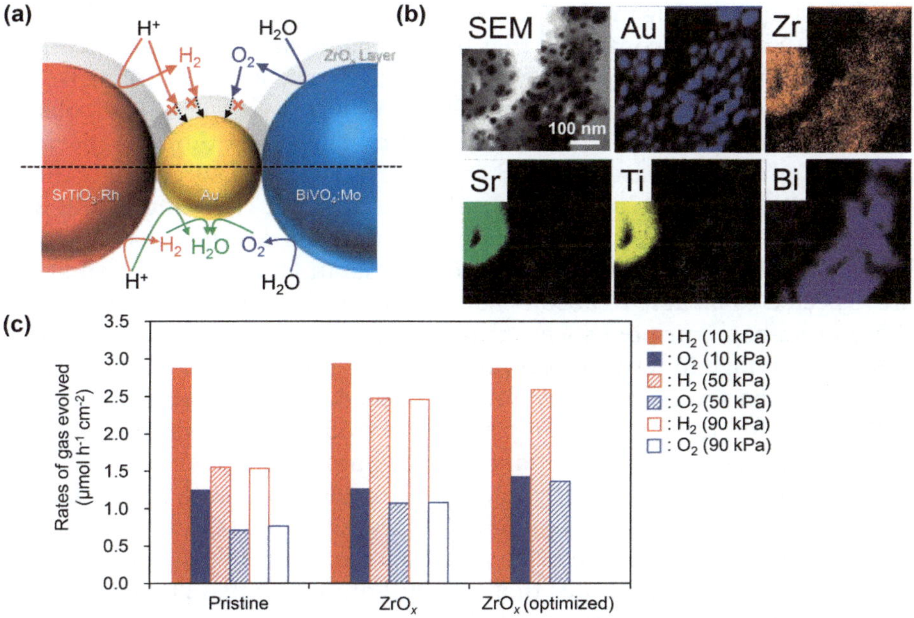

Figure 12.10 (a) Conceptualization of ZrO_x modification over the printed $SrTiO_3$:Rh/Au/$BiVO_4$ photocatalyst sheet. (b) SEM-EDX images of ZrO_x coated $SrTiO_3$:Rh/Au/$BiVO_4$ photocatalyst sheet. (c) The gas evolution rates on pristine $SrTiO_3$:Rh/Au/$BiVO_4$:Mo sheets and those coated with ZrO_x at background pressures of 10, 50 and 90 kPa. The optimized ZrO_x coated $SrTiO_3$:Rh/Au/$BiVO_4$ photocatalyst sheet was prepared by using the solution, which zirconium(IV) butoxide diluted in terpineol, as precursor and followed by dryness at r.t. Sheet size, 2.5 × 2.5 cm; thickness, 10 μm; reactant solution, 40 mL of pure water at 25 °C; reactant cell, top-irradiated separable cell; light source, 300 W Xe lamp with cut-off filters ($\lambda > 410$ nm). Reproduced from ref. 24 with permission from John Wiley & Sons, Copyright © 2020 Wiley-VCH GmbH.

12.4.3 Printable Photocatalyst Sheets with Colloidal Transparent Conductive Mediators

As mentioned previously, printed $SrTiO_3$:Rh/Au/$BiVO_4$:Mo photocatalyst sheets for efficient Z-scheme water splitting can be successfully prepared *via* simple screen-printing, and can be applied to large areas easily. However, incorporation of such metal colloids as Au leads to intercepting incident visible-light and accelerating reverse reactions. Therefore, it is desired to replace the Au colloid with other transparent and electrochemically inert conductive colloids. Recently, printed photocatalyst sheets introducing indium tin oxide (ITO) colloid as a transparent conductive material having appropriate work function (*ca.* 5.0 eV) close to Au colloid (*ca.* 5.1 eV),[5] which is sufficient to moderate the Schottky barrier of the $SrTiO_3$:Rh photocatalyst at the interface with the conductor, have been developed to achieve efficient water splitting. These $SrTiO_3$:Rh/ITO/$BiVO_4$:Mo photocatalyst sheets can be also prepared *via* a similar screen-printing method using nanocolloidal ITO particles with a size of about 10 nm. It is confirmed that the ITO nanoparticles are not changed in size and bridged between the two photocatalysts by SEM and transmission electron microscopy-energy dispersive X-ray analysis (TEM-EDX) observations. Different from the Au-based printed sheet, the diffuse reflection spectrum of the ITO-based printed sheet agreed with the sum of only those of $SrTiO_3$:Rh and $BiVO_4$:Mo photocatalyst because the ITO nanocolloid was optically transparent in the visible light region (400–800 nm), as shown in Figure 12.11(a). Therefore, both photocatalysts in ITO-based printed sheet are expected to be able to absorb visible light more effectively compared to those in Au-based printed sheet. Also, the use of ITO nanocolloid mediator has an advantage in terms of suppressing the reverse reaction. The characteristics of the reverse reaction were confirmed by the investigation of the ORR activity on the nanocolloid-printed electrodes, which was prepared by coating the paste of Au or ITO nanocolloids onto a fluorine-doped tin oxide (FTO) substrate and followed by calcination at 300 °C. As seen in the current density of ORR on the printed Au or ITO sheets shown in Figure 12.11(b), the onset potential on the printed ITO sheet was more negative than that on the Au sheet. This result indicates that the ORR on the printed ITO-based photocatalyst sheet can be suppressed efficiently, expecting to proceed with the water splitting reaction under even ambient pressure.

Similar to the printed $SrTiO_3$:Rh/Au/$BiVO_4$:Mo sheet, the printed ITO-based sheets produced H_2 and O_2 in a stoichiometric ratio from pure water. As shown in Figure 12.11(c), the water splitting activity on the printed ITO-based sheets showed much higher compared to the sheet without ITO nanocolloid. The optimized ITO-based sheet was obtained by adding 20 wt% of ITO nanocolloid and followed by calcination at 300 °C under air. Unlike Au colloid, the conductivity of doped conductive oxides with many oxygen vacancies such as ITO colloid tends to decrease with increasing the calcination temperature in air, because of the reduction of free electrons in the ITO colloid by thermal oxidation. So, the calcination condition is a key

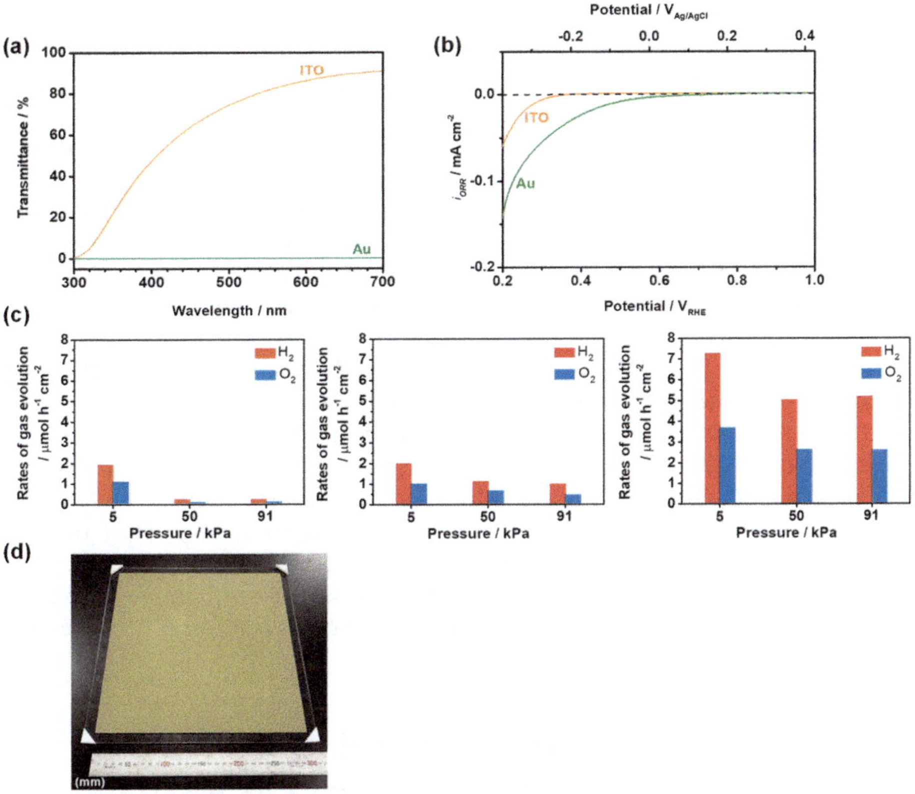

Figure 12.11 (a) UV-vis spectra of $SrTiO_3$:Rh/ITO/$BiVO_4$ photocatalyst sheet and the raw materials (ITO nanocolloid, $SrTiO_3$:Rh and $BiVO_4$ particles), which were used to prepare the sheet. (b) The O_2 reduction currents (i_{ORR}) of an Au sheet and ITO sheet. The electrochemical measurements were performed in a 0.1 M aqueous Na_2SO_4 electrolyte (pH 6.4) saturated with either Ar or O_2 at 25 °C. The potential was swept cathodically at a rate of 10 mV s^{-1}. (c) Water splitting activities under visible-light irradiation over the $SrTiO_3$:Rh/Au/$BiVO_4$, $SrTiO_3$:Rh/ITO/$BiVO_4$, and Ru,CrO_x loaded $SrTiO_3$:Rh/ITO/$BiVO_4$ sheet under various background pressures. (d) Picture of printed $SrTiO_3$:Rh/ITO/$BiVO_4$ photocatalyst sheet with large area (30 × 30 cm). Reproduced from ref. 5 with permission from Elsevier, Copyright 2020.

factor in the preparation of efficient ITO-based sheets for maintaining the conductivity of the ITO, as well as satisfying the complete decomposition of the organic compounds contained in the paste. This is because that insufficient decomposition leads to reduction in the conductivity of the ITO colloid by the carbon residues. Controlling the thickness of the printed photocatalyst sheet is another factor affecting the watersplitting activity; it is important to prepare the structures for efficient light absorption. Actually, the water splitting activity on the ITO-based sheet increased with increasing its thickness up to 10 μm, indicating that the two photocatalysts in the sheet utilize the incident light effectively. The photocatalytic water splitting activity on the printed $SrTiO_3$:Rh/ITO/$BiVO_4$:Mo sheet was also improved by loading Rh, CrO_x as appropriate co-catalysts, reaching the STH of the sheet to *ca.* 0.4%. It is also noted that the printed $SrTiO_3$:Rh/ITO/$BiVO_4$:Mo sheet can produce H_2 and O_2 efficiently even at an elevated background pressure of 91 kPa; 49% of base performance (at 5 kPa) on the $SrTiO_3$:Rh/ITO/$BiVO_4$:Mo sheet was retained at 91 kPa, whereas the Au-based sheet was retained only 12% of its performance (Figure 12.11(c)). The result indicates that the higher activity on the ITO-based sheet compared to the Au-based sheet is attributed to the lower ORR activity over the ITO nanocolloid mediator.

This printed ITO-based sheet is expected as one of the efficient Z-scheme photocatalyst sheets for large-scale applications driven under ambient pressure. Indeed, as shown in Figure 12.11(d), we have prepared the printed ITO-based sheet having a large area (30 × 30 cm) *via* similar screen-printing. These printable photocatalyst sheets using transparent conductive colloidal mediator will be expected to construct a practical and inexpensive route to solar fuel production on a large scale, and we are now examining the large-scale solar water splitting over these visible-light-driven printed photocatalyst sheets, using the panel reactor mentioned in Section 12.2.

12.5 Large-sized Electrochemical Reactors Integrated with Photovoltaic Cells for CO_2 Reduction with High Solar-to-formate Conversion Efficiencies

12.5.1 Background

Natural photosynthesis produces organic compounds such as sugar from CO_2 and H_2O using photon energy of the Sun, in which CO_2 reduction reaction (CO_2RR) is performed using electrons and hydrogen ions (protons) extracted from water molecules. CO_2RR is an uphill reaction, and therefore photosynthesis is a technology to store solar energy in chemical bonds within molecules. However, it is not necessary to synthesize sugars like photosynthesis in plants for the future practical implementation of artificial photosynthesis in terms of social acceptability. CO_2RR is much more difficult than hydrogen production by proton reduction, because CO_2 is the final form of combustion and oxidation of various organic substances and is hence extremely stable. Therefore, CO_2RR requires intrinsically higher energy and thus faces issues including excessively high reaction overpotentials over conventional catalysts.[28,29] Nevertheless, one-step direct synthesis of carbon-based compounds from CO_2 and water is required from viewpoints of future low cost and low lifecycle CO_2 emission in a system.

We are aiming to construct a practical system of solar-driven CO_2RR with a high solar-to-chemical conversion efficiency (STC) of over 10% that would be requisite for practical realization[30] and a reactor size scaled up to 1 m^2. One of our target products is formic acid. It is used as a feedstock for synthesis of various chemicals and as a hydrogen carrier for fuel cells.[31–33] Solar-driven CO_2RR devices are categorized into three types: (1) electrochemical reactors (EC) powered by photovoltaic (PV) cells (EC-PV); (2) photoelectrochemical (PEC) reactors; and (3) photocatalytic (PC) reactors. These three types, which are driven under visible light irradiation, have shown remarkable progress in recent years.[29] Among them, we are focusing on the PEC and EC-PV systems because these two types currently yield higher STC. The primal key toward our goal is new catalysts that promote CO_2RR at a low overpotential comparable to that of hydrogen production. We have worked on the development of the catalysts that meet the requisites for scale-up described in the next subsection, demonstrated a 1 cm^2 sized PEC device that functions free from external bias, and then fabricated 0.25–1 cm^2 sized monolithic PEC devices to achieve STC of 4.3–4.6%,[34,35] as summarized in Figure 12.12 and Table 12.1.[28,29,34,35] Based on this material technology, we tackled the problems unique to scale-up of the systems.[36,37]

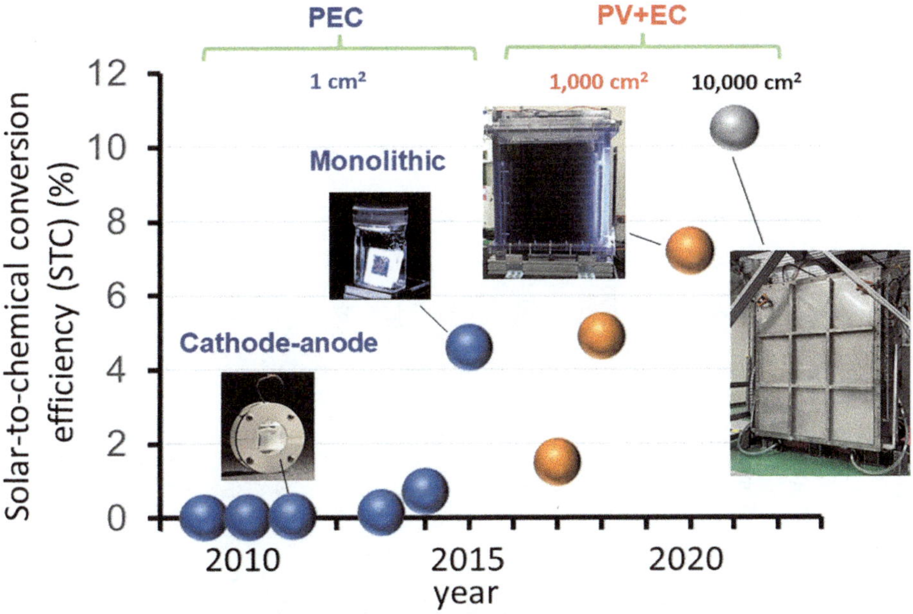

Figure 12.12 Historical progress of solar-driven CO_2 reduction devices.

Table 12.1 Comparison of performance of PV+EC and PEC devices for solar-driven CO_2 reduction.

Reference	37	36	34	41	43	38	39	40	42
Irradiation area/cm^2	8760	973	0.25	0.03	14 700	0.563	0.31	0.946	13.68
Planar area of a single electrode/cm^2	9600	1024	0.25	0.04	109	4.5	0.31	1.0	19.27
Main product	Formate	Formate	Formate	Formate	Formate	CO	CO	CO	CO
Solar to chemical efficiency STC (%)	10.5	7.2	4.6	10	1.86	13.4	19.1	20.4	8.05
Production rate/ mmol h^{-1}	1167	93.5	15.3×10^{-3}	4.5×10^{-3}	317	0.0766	0.0819	0.270	1.06
Production rate per irradiation area/ mmol h^{-1} cm^{-2}	0.133	0.096	0.061	0.15	0.022	0.136	0.264	0.285	0.077
Operating current density (per irradiation area)/mA cm^{-2}	7.42	6.44	3.3	8.5	0.68	11.6	14.4	15.4	6.5
Cell operating voltage V_{op}/V	1.65–1.69	1.85	1.79	2.04	3.80	2.38	2.23	2.12	2.75
Faradaic efficiency FE of main product (%)	96	80	94	94	67	86	99	100	93
PV cell	c-Si	c-Si	SiGe-jn	GaAs/InGaP/ TiO_2	c-Si	Ge/GaInAs/ GaInP	Ge/GaInAs/ GaInP	Ge/GaInAs/ GaInP	c-Si
Anode material	Ti/IrO_x	FTO/Ag/ IrO_x	ITO/IrO_x	Ni	Ti/IrO_x	SnO_2/CuO	Ni foil	Ni foam/NiFe	IrO_x
Cathode material	Ti/CS/MWCNT/RuCP		CC/RuCP	Ti mesh/C/ Pd	SUS/In/Cu mesh/In	SnO_2/CuO	GDE/Ag	PP/Ag	Ag
Anolyte	0.4 M KPi		0.1 M KPi	1 M KOH	1 M H_2SO_4	0.25 M CsOH	1 M KOH	1 M KOH	1 M KOH
Catholyte	(single compartment)		(single compartment)	2.8 M $KHCO_3$	1 M $KHCO_3$	0.1 M $CsHCO_3$	1 M KOH	1 M KOH	0.1–1 M $KHCO_3$
Ion-exchange membrane	Not used (Nanoporous film)	Not used	Not used	Bipolar membrane	Nafion	Bipolar membrane	Anion-exchange membrane	Anion-exchange membrane	Bipolar membrane

12.5.2 Design of Scaled-up EC-PV Systems

12.5.2.1 Requisites for Scale-up and Practical Uses

For constructing a practical scaled-up EC-PV system, low cost, ease of handling, high durability, *etc.* are essential in addition to high STC. To date, small-seized EC reactors connected with Ge/GaInAs/GaInP triple-junction PV cells[38–40] and PEC reactors composed of GaAs/InGaP/TiO_2/Ni photoanodes[41] demonstrated high STC of over 10% for solar-driven CO_2RR, as summarized in Table 12.1. However, the use of III–V group compound semiconductors for these reactors raises the material cost. Promising alternatives are single-crystalline silicon (c-Si) PV cells because of their high cost-performance, *i.e.*, low price per watt.[42,43] Another issue involved in these previous ones is the use of reactors of two-compartment types with high-cost ion-exchange membranes; the complicated structures of the reactors also drive up the cost. In addition, strong-acid and strong-alkaline electrolytes require careful handling. Thus, single-compartment and membrane-less EC reactors using electrolytes of near-neutral pH are suitable for scale-up and practical use. However, this type of reactor often causes the crossover reaction, and therefore countermeasures are needed.

Further, scaling-up of the reactors often lowers STC. A previously reported STC of a large-scale PV-EC system producing formate was 1.86% using 1.47 m^2 c-Si PV cells and three series-connected 109 cm^2 EC reactors of a two-compartment type.[43] Because the total area of the electrode catalysts was much smaller than the PV area, the current density was large, and as a result, the operating voltage V_{op} (3.7 V) was extremely higher than the thermodynamic threshold V_0 (1.43 V) of CO_2RR to formate. Therefore, the critical factor for achieving a high STC is a decrease in V_{op} because STC is proportional to V_{op}/V_0. For this purpose, lower electric resistance of the electrode catalysts is required for scale-up. In addition, sufficient supply of CO_2 and suppressing crossover reactions are more serious problems for larger reactors.[36,37]

12.5.2.2 Catalyst Materials for CO_2RR and Water Oxidation

Metal and metal-oxide catalysts have been developed for CO_2RR and adopted for scale-up, as summarized in Table 12.1. The priority for high STC is to reduce V_{op}, as described above. To this end, we have developed a Ru-complex polymer (RuCP; [Ru{4,4′-di(1-*H*-1-pyrrolypropyl carbonate)-2,2′-bipyridine}$(CO)_2Cl_2$]) as the cathode molecular catalyst, and IrO_x nanocolloids as the anode catalyst.[34,35] The RuCP drives CO_2RR to form formate with an extremely low overpotential (0.9 mA cm^{-2} at 0.02 V) in a potassium phosphate buffer aqueous electrolyte at a near-neutral pH (pH 6.3 at 0.4 M with saturated CO_2), owing to the two following mechanisms. The redox potential for CO_2RR is precisely tuned by structural control of ligands of Ru complexes. Presence of K^+ cations and conductive carbon nanotubes lowers the activation energy for CO_2RR at Ru active sites.[35,44] High reaction selectivity, *i.e.*, less production of

H_2 or CO is also an advantage of the RuCP loaded on carbon fiber supports. This improves the Faradaic efficiency (FE) of formate production and consequently STC. Another important point is the high catalytic activity of the IrO_x nanocolloids for water oxidation over a wide pH range,[45] which allows us to adopt a single-compartment reactor. Suppression of the crossover reactions in a wide V_{op} range is also a great point for the combination of the RuCP and IrO_x.[34] These also lead to high FE and STC for formate production *via* CO_2RR.

12.5.2.3 Configuration of Anode/Cathode Electrodes

We have succeeded in achieving an STC of 4.3–4.6% using the 0.25–1 cm^2 sized monolithic tablet-shaped PEC devices referred to as Type A in Figure 12.13(a).[34,35] This structure is free from the electric resistance problem regardless of the device size. However, the protons transfer from the anode to the cathode over a longer distance in the electrolyte, leading to a notable increase in the diffusion resistance. The detrimental impact of this larger resistance was experimentally confirmed using a pair of facing anode and cathode; the current density at a given applied voltage was halved when the anode/cathode distance increased to several tens of cm.[36] To eliminate this problem, we adopted an EC device structure of Type B illustrated in Figure 12.13(b). Anodes and cathodes are facing each other in a housing, and thus the proton transfer distance is always as short as 1 cm independently of the anode/cathode size. Instead, Type B suffers from the problem of large electric resistance, in addition to the inhomogeneous electrolyte flow in the narrow gaps, necessitating further development to address these issues.

Another factor in achieving a high STC is matching the EC activity to the PV property, which is especially important when the anode/cathode electrodes are directly connected to the PV cells as in Type B. The size of the electrode catalysts must be large enough so that the photocurrent from the PV cells is perfectly consumed for the target reaction. However, a larger electrode makes both the electric resistance problem and CO_2 supply problem more serious. Therefore, we employed electrically parallel-connected stacking of the electrode catalysts. It should be noted that the single-compartment and membrane-less reactor enabled this configuration.

12.5.3 Development of Scaled-up EC-PV Systems

12.5.3.1 Large-sized Anode/Cathode Electrode Catalysts

The substrates used for large-sized electrode catalysts must be low-resistive, in addition to chemically stable. Glass plates are covered with FTO layers, which are often used for dye-sensitized solar cells (DSCs),[46] have excellent chemical stability and hence they are promising candidates. The FTO glasses offer an additional advantage in that high optical transparency allows us to observe the O_2 production under solar irradiation; see Figure 12.13(c).

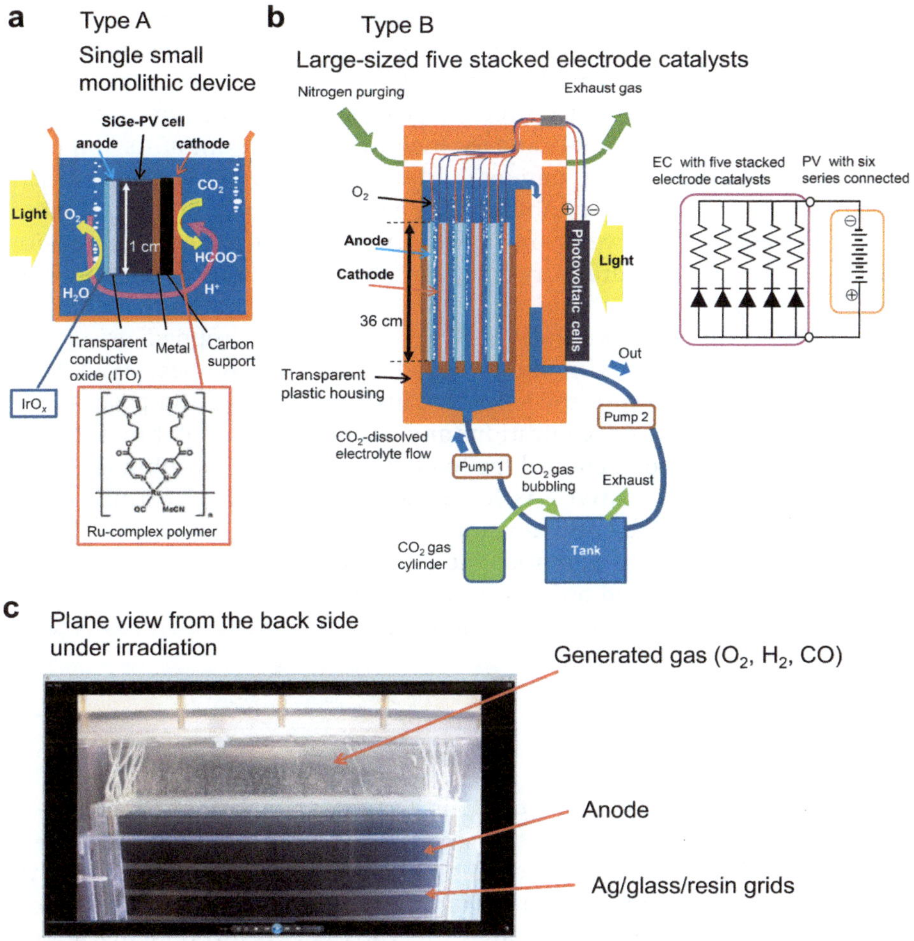

Figure 12.13 Two structures of reactors for solar-driven CO_2 reduction. (a) Single monolithic PEC device of 1 cm^2 in size (Type A). (b) EC reactor of 1000 cm^2 in size using stacked electrode catalysts (Type B). (c) Photographs of back-side view of the Type B reactor under solar irradiation (AM.1.5G, 1 sun). Reproduced from ref. 36 with permission from Elsevier, Copyright 2021.

However, since the sheet resistance (10 Ω $square^{-1}$) of the FTO is not sufficiently low when the electrodes are as large as 1000 cm^2, Ag-grids on the FTO are required. We designed the Ag-grid patterns in analogy with DSC modules,[46] and fabricated them by screen printing. Double protecting layers composed of low-melting glass and epoxy resin were applied to the Ag grids against corrosion by the electrolyte. The effect was confirmed by durability tests using model samples. Ti plates are another type of the candidates because of their lower electric resistance (specific resistance of 5 ×

10^{-7} Ωm) and high chemical stability. The synthesis and loading processes of the IrO_x and RuCP catalysts were developed based on the previously reported ones used for small-sized electrodes.[34,35] In particular, mixing the raw materials, heating to promote synthetic reactions, and drying to evaporate solvents were fine-tuned for meeting the large amounts and large areas. The synthesized IrO_x nanocolloid was dropped onto the substrates and dried. By adopting a suitable drying condition, the amount of loaded IrO_x increased and the water oxidation activity improved with increasing the repeat count of the drop-dry processes, since the IrO_x covered the substrate perfectly.[36]

An acetonitrile-ethanol solution of the RuCP was dropped on carbon sheets (CSs) decorated with multi-walled carbon nanotubes (MWCNTs), followed by evaporation of the solvent. This drop-evaporation process was repeated 10 times. The concentrations and amounts of the solution for a single drop were adjusted to control the loaded amount and morphology of the RuCP. The resultant CS/MWCNTs/RuCPs were fixed on the substrates using a graphite adhesive. The optimal amount of the RuCP increased the reaction current density at a given potential and FE for formate production.[36,37] However, excess RuCP formed a dense layer on the surface of the CS/MWCNT, which obstructed the protons, CO_2, and/or formate transport, and as a result, deteriorated the performance.

12.5.3.2 Flow Channels for Electrolytes

Homogeneous flow of the electrolyte is extremely important for sufficient supply of CO_2 and rapid extraction of formate, and a unique problem to scale up. As illustrated in Figure 12.13(b), the CO_2-dissolved electrolyte from the tank is injected into the reactor from the entrance on the bottom face of the housing, passes through the gaps between the facing anode/cathode electrodes dissolving the produced formate, and overflows to the exit on the side face connected to the tank. O_2 gas produced on the anodes and by-product gases (H_2 and CO) from the cathodes flow up with the electrolyte and then they are separated in the upper room; see Figure 12.13(c). The overflow structure and gas–liquid separation stabilize and homogenize the electrolyte flow.

Another key point is the flow rate. It has been reported that a high flow rate suppresses the crossover reaction in membrane-less EC reactors.[47] Therefore, the flow rate was set to be 4×10^{-4} L min^{-1} cm^{-2} that is significantly larger than 6×10^{-6} L min^{-1} cm^{-2} corresponding to the CO_2 consumption rate at STC = 10% and FE = 100%. We adopted an orifice plate at the bottom of the housing and guiding plates that partition the gaps between the facing anode/cathode electrodes. The former dispenses the electrolyte to all the gaps equally, and the latter delivers throughout the gaps. The optimized orifice pattern designed by fluid simulation realized a homogenous flow distribution, as depicted in Figure 12.14.[36]

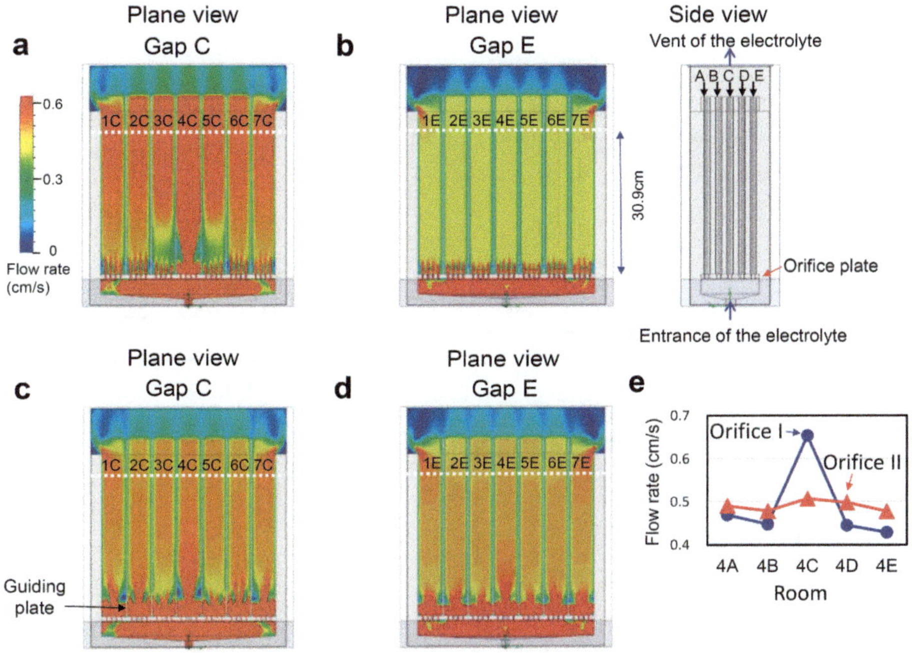

Figure 12.14 Distributions of the Flow rates in the 1000 cm^2 sized reactors of Type B using stacked electrodes. The five gaps between the facing anode/cathode electrodes are labelled as A, B, ..., and E. The seven channels partitioned by the guiding plates in each gap are numbered as 1, 2, ..., and 7. (a) and (b) Larger orifices (Orifice I). (c) and (d) Optimized orifices (Orifice II). (e) Flow rates at a distance of 30.9 cm from the orifice plates. Reproduced from ref. 36 with permission from Elsevier, Copyright 2021.

12.5.4 Performance of Scaled-up EC-PV Systems

We first constructed a combination of an EC reactor consisting of five-stacked electrode catalysts and six-series-connected PV cells of approximately 1000 cm^2 in size, and evaluated the solar-driven CO_2RR performance under AM1.5G 1 sun irradiation using a solar simulator.[36] The concentration of the produced formate in the electrolyte was measured with an ion-chromatograph, and the STC for formate production was calculated according to the standard protocol. The composition of the exhaust gas including CO, H_2, O_2, N_2, and CO_2 was determined using a gas chromatograph.

The operating condition (V_{op} = 2.23 V) determined by the cross point of the linear-sweep voltammetry curve of the EC reactor and current–voltage curve of the PV cells was close to the maximal power point of the PV cells, as clearly shown in Figure 12.15, when the FTO/Ag grid cathodes were used. This indicates that the EC and PV systems are well matched, owing to the suitable area ratio of the electrode catalysts to the PV cells. The operation was stable

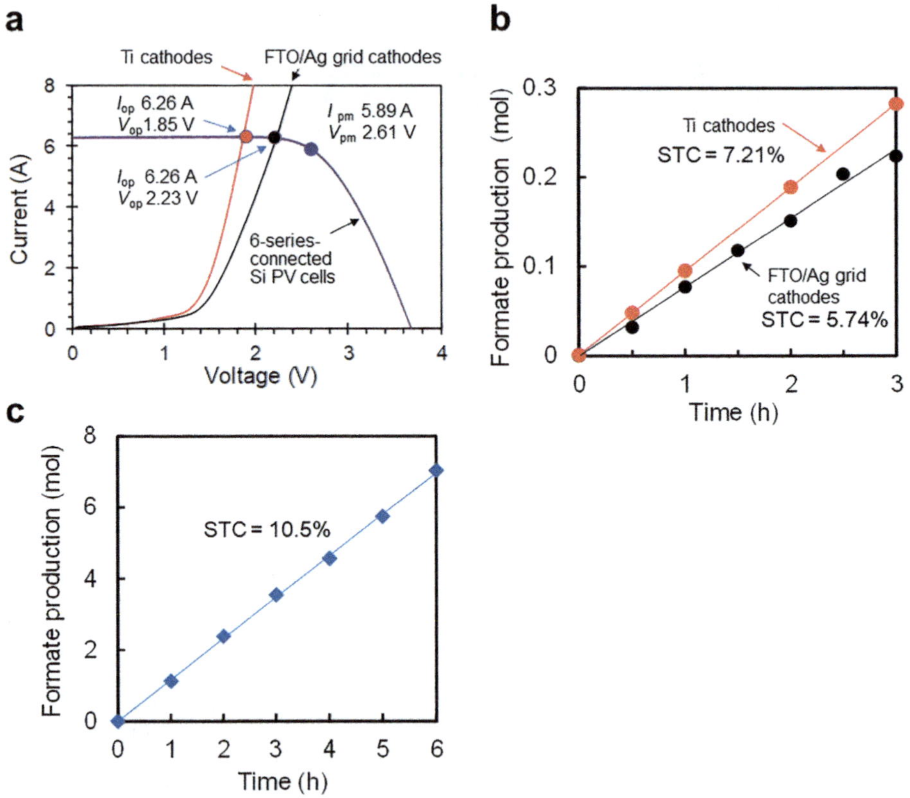

Figure 12.15 CO_2 reduction performance of the large-sized EC reactors powered by c-Si PV cells. (a) Linear-sweep voltammetry curves of the 1000 cm^2 sized Type B reactors, and current–voltage curve of the six series-connected c-Si-PV cells under solar irradiation (AM.1.5G, 1 sun). (b) Formate production of the 1000 cm^2 sized reactors under the solar irradiation. Reproduced from ref. 36 with permission from Elsevier, Copyright 2021. (c) Formate production of the 1 m^2 sized reactors under white LED irradiation equivalent to 0.95 sun.

for 3 h, resulting in FE = 64% and STC = 5.74%. Replacement with the Ti cathodes lowered V_{op} to 1.85 V with almost the same operating current, and consequently improved the performance; FE = 80% and STC = 7.21%. The lower potential near the terminals of the Ti cathodes suppressed H_2 and CO production, resulting in a higher FE.

We further scaled up the EC-PV system to 1 m^2 in size, in which eight-stacked electrode catalysts were mounted. Ti plates were used as both the anode and cathode substrates for lower resistivity.[37] Another main improvement was insertion of nanoporous films between the facing anode/cathode electrodes that prevented O_2 bubbles produced on the anodes from reaching the cathodes, suppressing the crossover reaction of O_2 reduction. These further lowered V_{op} to 1.65–1.69 V and increased FE to 96% with high stability during 6 h operation under white LED irradiation equivalent to 0.95 sun, as shown in Figure 12.15 and Table 12.1. Thus, we have realized our target of a 1 m^2 sized EC-PV system for solar-driven CO_2RR with an STC of 10.5% and a resultant formate production rate of as high as 1.2 mol h^{-1}.

12.5.5 Future Issues and Outlook

Two future issues remain for economically viable CO_2RR. One is an improvement in durability. The operating current of the EC reactors gradually lowered after the initial several hour operation shown in Figure 12.15, which would be due to desorption of the catalysts. Therefore, immobilization of the catalysts should be developed. The other is the reduction of material cost for the catalysts, because the present catalysts are composed of rare elements of Ir and Ru. β-FeOOH(Cl)[48] and β-FeOOH(Cl):Ni[49] are promising candidates as the low-cost anode catalyst, and Mn complexes as the cathode catalyst.[44,50] These Fe-anode and Mn-cathode also work efficiently in a single aqueous electrolyte, and thus could be applicable to the large-sized reactors of our concept.[51]

Higher STC is always required. The PV performance of c-Si PV cells lowers notably under irradiation owing to temperature rise. Replacement with other types of PV cells that are more tolerant of high temperature, including heterojunction cells with intrinsic thin-layer (HIT) and Cu(In,Ga)Se_2 (CIGS) cells,[52] can improve STC and the annual formate production.

12.6 Summary and Future Prospects

This chapter has summarized the recent developments of highly active and scalable artificial photosynthesis systems. It is required that photocatalytic solar-to-chemical conversion system must be both efficient and scalable in future applications. In overall water splitting using a single photocatalyst of $SrTiO_3$:Al, loading of Rh/Cr_2O_3 and CoOOH as efficient HER and OER co-catalyst, respectively, onto the specific facets enabled significantly improve water splitting activity, and the internal quantum efficiency was almost 100% at UV-light region (350–365 nm). In addition, the STH of obtained $SrTiO_3$:Al photocatalyst was 0.65%. Solar hydrogen production *via* photocatalytic overall water splitting over a photocatalyst panel using the $SrTiO_3$:Al photocatalyst scaled up to a size of 100 m^2 was demonstrated without decreasing the original catalyst performance. As for the overall water splitting using two kinds of photocatalysts, Z-schematic photocatalyst sheets, specifically composed of $SrTiO_3$:Rh and $BiVO_4$(:Mo) with or without conductive nanocolloids, were developed for efficient pure water splitting under visible-light irradiation. The Z-schematic photocatalyst sheets can be prepared by simple screen-printing, and easily scaled up to about 0.1 m^2. Theoretically, it can be further scaled up without efficiency loss as well as the above photocatalyst panel. Although the present STH value of the printed photocatalyst sheet is achieved to 0.4%, it is necessary to reach 5–10% for realizing economically viable solar hydrogen production. Therefore, the development of photocatalysts that can efficiently utilize visible light is needed to achieve the targeted STH.

Along with the development of photocatalyst materials and reactors for water splitting, those for CO_2 reduction have been successfully developed.[29] A novel molecular catalyst, RuCP, was developed and applied to small-scale devices, culminating in the demonstration of a 1 m^2 scale EC-PV system with an STC of 10.5% and a large formate production rate of 1.2 mol h^{-1}. The high STC is due to the very low V_{op} of 1.65 V and high FE of 96% at an operating current of 65 A. The present V_{op} is extremely lower than those reported in previous EC reactors for liquid organic products from CO_2. These great features were realized by the low-resistive electrode substrates and sufficient CO_2 supply, *etc.* that are more important for larger reactors, in addition to the high catalytic activity of the RuCP. The practically efficient large-sized EC-PV system demonstrated here opens the door to solar-driven CO_2 recycling plants.

Acknowledgements

A part of this work was financially supported by the Artificial Photosynthesis Project of the New Energy and Industrial Technology Development Organization (NEDO).

References

1. T. Takata, J. Jiang, Y. Sakata, M. Nakabayashi, N. Shibata, V. Nandal, K. Seki, T. Hisatomi and K. Domen, *Nature*, 2020, **581**, 411.
2. H. Nishiyama, T. Yamada, M. Nakabayashi, Y. Maehara, M. Yamaguchi, Y. Kuromiya, Y. Nagatsuma, H. Tokudome, S. Akiyama, T. Watanabe, R. Narushima, S. Okunaka, N. Shibata, T. Takata, T. Hisatomi and K. Domen, *Nature*, 2021, **598**, 304.
3. Q. Wang, H. Hisatomi, Q. Jia, H. Tokudome, M. Zhong, C. Wang, Z. Pan, T. Takata, M. Nakabayashi, N. Shibata, L. Yanbo, l-D. Sharp, A. Kudo, T. Yamada and K. Domen, *Nat. Mater.*, 2016, **15**, 611.
4. Q. Wang, T. Hisatomi, Y. Suzuki, Z. Pan, J. Seo, M. Katayama, T. Minegishi, H. Nishiyama, T. Takata, K. Seki, A. Kudo, T. Yamada and K. Domen, *J. Am. Chem. Soc.*, 2017, **139**, 1675.
5. Q. Wang, S. Okunaka, H. Tokudome, T. Hisatomi, M. Nakabayashi, N. Shibata, T. Yamada and K. Domen, *Joule*, 2018, **2**, 2667.
6. M. S. Wrighton, A. B. Ellis, P. T. Wolczanski, D. L. Morse, H. B. Abrahamson and D. S. Ginely, *J. Am. Chem. Soc.*, 1976, **98**, 2774.
7. F. T. Wanger and G. A. Somorojai, *J. Am. Chem. Soc.*, 1980, **102**, 5494.
8. K. Domen, S. Naito, M. Soma, T. Ohnishi and K. Tamaru, *J. Chem. Soc., Chem. Commun.*, 1980, 543.

9. J. M. Lehn, J. P. Sauvage, R. Ziessel and L. Hilaire, *Isr. J. Chem.*, 1982, **22**, 168.
10. Y. Goto, T. Hisatomi, Q. Wang, T. Higashi, K. Ishikiriyama, T. Maeda, Y. Sakata, S. Okunaka, H. Tokudome, M. Katayama, S. Akiyama, H. Nishiyama, Y. Inoue, T. Takewaki, T. Setoyama, T. Minegishi, T. Takata, T. Yamada and K. Domen, *Joule*, 2018, **2**, 1.
11. Y. Ham, T. Hisatomi, Y. Goto, Y. Moriya, Y. Sakata, A. Yamakata, J. Kubota and K. Domen, *J. Mater. Chem. A*, 2016, **4**, 3027.
12. T. Takata and K. Domen, *J. Phys. Chem. C*, 2009, **113**, 19386.
13. K. Maeda, K. Teramura, H. Masuda, T. Takata, N. Saito, Y. Inoue and K. Domen, *J. Phys. Chem. B*, 2006, **110**, 13107.
14. K. Maeda, K. Teramura, D. Lu, N. Saito, Y. Inoue and K. Domen, *Angew. Chem., Int. Ed.*, 2006, **45**, 7806.
15. M. Yoshida, K. Takanabe, A. Ishikawa, J. Kubota, Y. Sakata, Y. Ikezawa and K. Domen, *J. Phys. Chem. C*, 2009, **113**, 10151.
16. H. Lyu, T. Hisatomi, Y. Goto, M. Yoshida, T. Higashi, M. Katayama, T. Takata, T. Minegishi, H. Nishiyama, T. Yamada, Y. Sakata, K. Asakura and K. Domen, *Chem. Sci.*, 2019, **10**, 3196.
17. L. Mu, Y. Zhao, A. Li, S. Wang, Z. Wang, J. Yang, Y. Wang, T. Liu, R. Chen, J. Zhu, F. Fang, R. Li and C. Li, *Energy Environ. Sci.*, 2016, **9**, 2463.
18. N. Tanihara, S. Nakanishi and T. Yoshinaga, *J. Jpn. Pet. Inst.*, 2016, **59**, 276.
19. Y. Sasaki, H. Nemoto, K. Saito and A. Kudo, *J. Phys. Chem. C*, 2009, **113**, 17536.
20. S. Okunaka, H. Tokudome and R. Abe, *J. Mater. Chem. A*, 2015, **3**, 14794.
21. S. Okunaka, H. Tokudome, Y. Hitomi and R. Abe, *J. Mater. Chem. A*, 2016, **4**, 3926.
22. S. Okunaka, H. Tokudome and R. Abe, *Catal. Sci. Technol.*, 2016, **6**, 254.
23. S. Okunaka, H. Tokudome and R. Abe, *Chem. Lett.*, 2016, **57**, 57.
24. S. Okunaka, H. Kameshige, T. Ikeda, H. Tokudome, T. Hisatomi, T. Yamada and K. Domen, *ChemSusChem*, 2020, **13**, 4906.
25. T. Takata, C. Pan, M. Nakabayashi, N. Shibata and K. Domen, *J. Am. Chem. Soc.*, 2015, **137**, 9627.
26. M. Yoshida, K. Maeda, D. Lu, J. Kubota and K. Domen, *J. Phys. Chem. C*, 2013, **117**, 14000.
27. A. Garcia-Esparza, T. Shinagawa, S. Ould-Chikh, M. Qureshi, X. Peng, N. Wei, D. H. Anjum, A. Clo, T. C. Weng, D. Nordlund, D. Sokaras, J. Kubota, K. Domen and K. Takanabe, *Angew. Chem., Int. Ed.*, 2017, **56**, 5780.
28. S. Sato, T. Arai, T. Morikawa, K. Uemura, T. M. Suzuki, H. Tanaka and T. Kajino, *J. Am. Chem. Soc.*, 2011, **133**, 15240.
29. T. Morikawa, S. Sato, K. Sekizawa, T. M. Suzuki and T. Arai, *Acc. Chem. Res.*, 2022, **55**, 933.
30. J. B. Greenblatt, D. J. Miller, J. W. Ager, F. A. Houle and I. D. Sharp, *Joule*, 2018, **2**, 381.
31. O. S. Bushuyev, P. De Luna, C. T. Dinh, L. Tao, G. Saur, J. van de Lagemaat, S. O. Kelley and E. H. Sargent, *Joule*, 2018, **2**, 825.

32. J. Hull, Y. Himeda, W. H. Wang, B. Hashiguchi, R. Periana, D. J. Szaida, J. T. Muckerman and E. Fujita, *Nat. Chem.*, 2012, **4**, 383.
33. M. Jouny, W. Luc and F. Jiao, *Ind. Eng. Chem. Res.*, 2018, **57**, 2165.
34. T. Arai, S. Sato and T. Morikawa, *Energy Environ. Sci.*, 2015, **8**, 1998.
35. S. Sato, T. Arai and T. Morikawa, *Nanotechnology*, 2018, **29**, 034001.
36. N. Kato, S. Mizuno, M. Shiozawa, N. Nojiri, Y. Kawai, K. Fukumoto, T. Morikawa and Y. Takeda, *Joule*, 2021, **5**, 687.
37. N. Kato, Y. Takeda, Y. Kawai, N. Nojiri, M. Shiozawa, S. Mizuno, K. Yamanaka, T. Morikawa and T. Hamaguchi, *ACS Sustainable Chem. Eng.*, 2021, **9**, 16031.
38. M. Schreier, F. Herouguel, L. Steier, S. Ahmad, J. S. Luterbacher, M. T. Mayer, J. Luo and M. Grätzel, *Nat. Energy*, 2017, **2**, 17087.
39. W.-H. Cheng, M. H. Richiter, I. Sullivan, D. M. Larson, C. Xiang, B. S. Brunschwig and H. A. Atwater, *ACS Energy Lett.*, 2020, **5**, 470.
40. Y. Xiao, Y. Qian, A. Chen, T. Qin, F. Zhang, H. Tang, Z. Qiu and B.-L. Lin, *J. Mater. Chem. A*, 2020, **8**, 18310.
41. X. Zhou, R. Liu, K. Sun, Y. Chen, E. Veriage, S. A. Francis, N. S. Lewis and C. Xiang, *ACS Energy Lett.*, 2016, **1**, 764.
42. S. Y. Chae, S. Y. Lee, S. G. Han, H. Kim, J. Ko, S. Park, O. Joo, D. Kim, Y. Kang and U. Lee, *Sustainable Energy Fuels*, 2020, **4**, 199.
43. J. L. White, J. T. Herb, J. J. Kaczur, P. W. Majsztrik and A. B. Bocarsly, *J. CO_2 Util.*, 2014, **7**, 1.
44. S. Sato, K. Saita, K. Sekizawa, S. Maeda and T. Morikawa, *ACS Catal.*, 2018, **8**, 4452.
45. Y. Zhao, E. A. Hernadez-Pagon, N. N. Vargas-Barbosa, J. L. Dysart and T. E. Mallouk, *J. Phys. Chem. Lett.*, 2011, **2**, 402.
46. N. Kato, Y. Takeda, K. Higuchi, A. Takeichi, E. Sudo, H. Tanaka, T. Motohiro, T. Sano and T. Toyoda, *Sol. Energy Mater. Sol. Cells*, 2009, **93**, 893.
47. X. Pang, J. T. Davis, A. D. Harvey III and D. V. Esposito, *Energy Environ. Sci.*, 2020, **13**, 3663.
48. T. M. Suzuki, T. Nonaka, A. Suda, N. Suzuki, Y. Matsuoka, T. Arai, S. Sato and T. Morikawa, *Sustainable Energy Fuels*, 2017, **1**, 636.
49. T. M. Suzuki, T. Nonaka, K. Kitazumi, N. Takahashi, S. Kosaka, Y. Matsuoka, K. Sekizawa, A. Suda and T. Morikawa, *Bull. Chem. Soc. Jpn.*, 2018, **91**, 778.
50. B. Reuilard, K. H. Ly, T. E. Rosser, M. F. Kuehnel, I. Zebger and E. Reisner, *J. Am. Chem. Soc.*, 2017, **139**, 14425.
51. T. Arai, S. Sato, K. Sekizawa, T. M. Suzuki and T. Morikawa, *Chem. Commun.*, 2019, **55**, 237.
52. B. R. Paudyal and A. G. Imenes, *Sol. Energy*, 2021, **224**, 425.

Subject Index